FOUNDATIONS OF NATURAL HISTORY

Series Editors

Paula M. Mabee, San Diego State University
Kirk Fitzhugh, Los Angeles County Museum of Natural History

Foundations of Natural History is a series from the Johns Hopkins University Press for the republication of classic scientific writings that are of enduring importance for the study of origins, properties, and relationships in the natural world.

Published in the Series

Materials for the Study of Variation: Treated with Especial Regard to the Discontinuity in the Origin of Species by William Bateson,
with a new introduction by Peter J. Bowler and an essay by Garry Webster

Nicholas Copernicus: On the Revolutions translated by Edward Rosen

Nicholas Copernicus: Minor Works translated by Edward Rosen

NICHOLAS COPERNICUS COMPLETE WORKS

On the Revolutions

Minor Works

NICHOLAS COPERNICUS MINOR WORKS

Translation and Commentary by Edward Rosen
with the assistance of Erna Hilfstein

THE JOHNS HOPKINS UNIVERSITY PRESS
Baltimore and London

© 1985 Edward Rosen
All rights reserved
Printed in the United States on acid free paper

The original edition was published in 1985 as Volume III of *Nicholas Copernicus: Complete Works* by Polish Scientific Publishers, Warsaw and Cracow. It was edited by Paweł Czartoryski and sponsored by the Committee for the History of Science and by the Institute for the History of Science, Education, and Technology of the Polish Academy of Sciences. Volumes I and IV, copyright by PWN Polish Scientific Publishers.

Johns Hopkins Paperbacks edition, 1992

The Johns Hopkins University Press
701 West 40th Street
Baltimore, Maryland 21211-2190
The Johns Hopkins Press Ltd., London

Library of Congress Cataloging-in-Publication Data

Copernicus, Nicolaus, 1473–1543.
 [Selections. English. 1992]
 Minor works / Nicholas Copernicus ; translation and commentary by Edward Rosen, with the assistance of Erna Hilfstein.
    p.   cm.—(Foundations of natural history)
 Translated from the Latin.
 Includes bibliographical references and indexes.
 ISBN 0-8018-4516-5
 1. Astronomy—Early works to 1800. I. Rosen, Edward, 1906–
II. Hilfstein, Erna. III. Title. IV. Series.
QB41.C76213   1992
520–dc20   92-16612

CONTENTS

GENERAL INTRODUCTION *by Pawel Czartoryski* IX

EDITOR'S NOTE ON VOLUME III XI

INTRODUCTION *by Edward Rosen* XIII

ABBREVIATIONS OF WORKS FREQUENTLY CITED XV

COPERNICUS' TRANSLATION OF THEOPHYLACTUS SIMOCATTA

Introduction 3

   Copernicus' Study of Greek 3

   Copernicus and Theophylactus 5

   Copernicus' Knowledge of Greek 7

   Copernicus as a Translator 13

   The First Edition of Copernicus' Translation of Theophylactus' Letters 19

   The Later Editions of Copernicus' Translation of Theophylactus' Letters 24

Theophylactus' *Letters*

   Laurentius Corvinus' *Poem* 27

   Copernicus' Dedication 29

   Theophylactus' *Letters*, as translated by Copernicus 30

   Notes 51

COPERNICUS' MINOR ASTRONOMICAL WRITINGS

Copernicus' *Commentariolus*

   Introduction 75

   Commentariolus 81

   Notes 91

Copernicus' *Letter against Werner*

   Introduction 127

      Johann Werner 127

      Wapowski and Copernicus' *Letter against Werner* 131

      Manuscripts of the *Letter against Werner* 134

   Letter against Werner 145

   Notes 151

COPERNICUS AND THE MONEY QUESTION

Introduction 169

Copernicus' Writings about Money (Parallel texts of *Meditata*, *Denkschrift* with Appendix, and *Essay*) 176

CONTENTS

Copernicus' Comparison of Silver with Gold 194
Copernicus' Conclusion regarding the Restoration of the Coinage 195
Proclamation of 6 November 1418 197
The *Letter to Decius* 202
Notes 208

COPERNICUS' ADMINISTRATIVE DOCUMENTS

Copernicus' Scholastry 219
Copernicus' Leases of Abandoned Farmsteads 224
Copernicus' Approval of Four Financial Transactions 253
Copernicus' Inventory of 1520 258
Copernicus' Bread Tariff 281
Copernicus as a Guardian of his Chapter's Counting Table 285
Copernicus as a Physician 301

LETTERS WRITTEN BY COPERNICUS

	Introduction		307
I	To the Varmia Chapter	22 October 1518	309
II	To Bishop Ferber	29 February 1524	312
IV	To Canon Felix Reich	28 April 1527	315
V	To Bishop Ferber	27 July 1531	319
VI	To Bishop Dantiscus	11 April 1533	321
VII		8 June 1536	326
VIII		9 August 1537	327
IX		25 April 1538	331
X		2 December 1538	332
XI		11 January 1539	334
XII		3 March 1539	339
XIII		11 March 1539	342
XIV	To Duke Albert of Prussia	15 June 1541	343
XV		21 June 1541	348
XVI	To Bishop Dantiscus	27 June 1541	349
XVII		28 September 1541	352

INDEX TO THEOPHYLACTUS 355
INDEX OF PERSONS 359
INDEX OF PLACES 366
INDEX OF SUBJECTS 369
POLISH PLACES IN COPERNICUS' LIFE AND WORK 371

GENERAL INTRODUCTION

The English version of Nicholas Copernicus' Complete Works *is brought to a close by the present book. This project was undertaken by the Polish Academy of Sciences on the occasion of the celebration in 1973 of the 500th anniversary of the astronomer's birth. Volume I offered a facsimile of the* Revolutions *written by Copernicus' own hand. Volume II presented an English translation of the* Revolutions, *together with a commentary. Volume III now provides an English translation of, and commentary on, everything else written by Copernicus. As a result the English-speaking reader has available to him the entire output of Copernicus as an author, many of his writings being translated into English here for the first time.*

The research required to establish all the authentic underlying texts was carried on for many years by those responsible for the English version in wholehearted cooperation with those responsible for the Polish version, of which Volume III is now in press. The Latin version of Volume III will soon be ready. Facsimiles of all the underlying documents, many of them written by Copernicus himself, will be published in Volume IV, which was not foreseen in the preliminary plan of the Complete Works *(see Introduction to Volume I). Volume IV, which is now also in press, will form part of the* Complete Works, *whether in Latin, Polish or English. A list of all five language versions will be found at the end of this volume.*

The idea of a complete edition of Copernicus' works was first proposed during the XIth International Congress of the History of Science held in Warsaw and Cracow in 1965. Since that time many institutions and individuals have contributed to the realization of this edition, sponsored by the Polish Academy of Sciences and published by the Polish Scientific Publishers. Distribution in the English-speaking world was undertaken by the Macmillan Company in London, which assigned distribution in the United States to the Johnson Reprint Company for Volume I, and for Volume II to the Johns Hopkins University Press.

Copernican scholarship will always keep in grateful memory the late Professor Jerzy Bukowski, who initiated cordial international cooperation as President of the Comité Nicolas Copernic *of the History of Science Division of the International Union for the History and Philosophy of Science in 1963–1977. Special acknowledgments are due to Professor Bogdan Suchodolski, former Director of the Polish Academy of Sciences' Institute for the History of Science and Technology; to Professor Józef Miąso, the present Director of the same Institute; to Professor Jerzy Zathey, author of the Introduction to Volume I, and to Professor Jerzy Dobrzycki, editor of Volume II; to Marian Dobrowolski, of the former Front of National Unity, for helping to arrange official relations; to Władysław Negrey, Director of the Cracow Branch of the Polish Scientific Publishers, for expert supervision of Volumes I and II; to Anna Kotulecka of Cracow, for her splendid editing of Volumes I and II; to Maria*

M. Berger of Warsaw, for her editing of Volume III, and to Maria Bujnowska of Warsaw for its expert supervision; to Zygmunt Gebethner of the Polish Scientific Publishers in Warsaw and Malcolm Stewart of the Macmillan Publishers in London, for smoothly organizing the international distribution; to the National Press in Cracow and its Director Zbigniew Janus, for their skill in printing the successive volumes; to Jan Durociński for supervising the production of the facsimile in Volume I; to Małgorzata Golińska-Gierych of the Polish Academy of Sciences' Center for Copernican Studies for assembling the plates in Volume IV and helping to prepare the indexes in the present volume together with Marek Troszyński, who also prepared the maps; and to Dr. Grażyna Rosińska and Dr. Anna Słomczyńska of the same Center for current cooperation. Mention should also be made of former workers of the Center for Copernican Studies, Dr. Andrzej Kempfi, Jadwiga Kozicka-Gerould, Dr. Małgorzata Malewicz, Aleksandra Szpilewicz, and Dr. Zofia Wardęska, who were very helpful at various stages of the project. Finally thanks are due to the Astronomical Observatory of the Nicholas Copernicus University at Toruń for supplying splendid photographs of the sky used for the endpapers in all the volumes.

Professors Marian Biskup and Jerzy Dobrzycki, Bishop Jan Obłąk, Małgorzata Golińska-Gierych, and Dr. Jerzy Drewnowski, the authors of the Polish and Latin versions of Volume III, generously submitted their own typescripts to the authors of the English version, Professor Edward Rosen and Dr. Erna Hilfstein, who are to be thanked for their indefatigable enthusiasm and unfailing cooperation during many years of common effort.

Professor Rosen passed away on March 28 1985. Nicholas Copernicus' Complete Works *remain an everlasting monument of his creative work.*

Paweł Czartoryski
Chief, Center for Copernican Studies
Polish Academy of Sciences

EDITOR'S NOTE ON VOLUME III

The research for the English version of Volume III was done by Professor Edward Rosen and Dr. Erna Hilfstein, and for the Polish version by several Polish scholars. The problems confronting both teams were not exactly the same, and therefore their results are not interchangeable. Accordingly, I shall briefly indicate wherein the Polish version differs from the English.

In his analysis of Copernicus' Latin translation of Theophylactus Simocatta's Greek *Letters*, Rosen used the translator's marginal notes in his copy of a Greek-Latin dictionary. The results so attained were incorporated in the Polish version prepared by Paweł Czartoryski. On the other hand, the Polish version examined the variant readings in Greek manuscripts of Theophylactus, as presented in Giuseppe Zanetto's recent study. Moreover, the Introduction to the Polish version lists pre-Copernican translations of Greek epistolographers into Latin, and also surveys the teaching of epistolography at the University of Cracow in Copernicus' time, in an effort to depict the cultural background of his translation of Theophylactus' *Letters*.

To his Polish translation of Copernicus' *Commentariolus*, Professor Jerzy Dobrzycki added a translation of Copernicus' notes in his copy of the *Alfonsine Tables*. Professor Dobrzycki clearly expounds the basic ideas set forth in the *Commentariolus* and also in Copernicus' *Letter against Werner*, which was translated into Polish by Dr. Jerzy Drewnowski. Rosen analyzes the manuscript tradition of both these texts, thereby preparing the way for definitive Latin editions, no previous edition having been based on all the available manuscripts.

Monetary problems were discussed by Copernicus in a number of texts. Their mutual relations are analyzed here on the basis of all the available manuscripts and printed sources. But they are presented differently in the English and Polish versions. The English version offers one principal text, with the variants assigned to the several redactions. On the other hand, the Polish version, prepared by Professors Marian Biskup and Paweł Czartoryski, prints each redaction separately, indicating their mutual relations. The Polish version includes "In libro actorum antiquo," an anonymous note first published in Erich Sommerfeld's recent book. Copernicus' letter to Felix Reich is grouped with his monetary writings in the Polish version, whereas it appears in the Correspondence section of the English version. This letter, written hurriedly, lacks a precise date. It is assigned to 19 April 1528 by Professor Biskup, and to 28 April 1527 by Rosen. A similar difference of opinion concerns Copernicus' undated *Essay*, which he wrote between April 1525 and April 1526, according to Rosen, and in 1528, before 29 March, according to Professor Biskup. The Gdańsk copy of Copernicus' 1517 memorandum according to Professor Biskup is in Reich's handwriting, an ascription rejected by Rosen.

Professor Biskup's edition of the *Leases of Abandoned Farms* in 1970 (second edition 1983) advanced the investigation of that important administrative document, which he treats in the Polish version. In 1972 Bishop Obłąk's discussion of the *Inventory of 1520* was the first to recognize that Copernicus' hand wrote that list, which is edited by Bishop Obłąk in the Polish version. By scrutinizing the holdings of the Olsztyn Diocesan Archives, Dr. Hilfstein improved the identification of the documents inventoried by the astronomer.

In addition to the medical prescription ascribed to Arnold of Villanova, prescriptions found in Montagnana's *Consilia* are published in the Polish version by Małgorzata Golińska-Gierych, while in the English version stress is laid on Copernicus' activities as a physician.

In addition to the seventeen letters of Copernicus presented in the English version, the Polish version prints ten letters written for the Varmia Chapter and one for a military commander in Olsztyn. On the basis of the handwriting Dr. Drewnowski assigns these eleven letters to Copernicus.

In conclusion, by correcting many errors deeply rooted in the literature, the English version makes a notable contribution to the advancement of Copernican scholarship.

Pawel Czartoryski

INTRODUCTION

Copernicus is universally ranked with those minds who in the course of the centuries have radically changed the thinking of mankind. His principal contribution altered our ancestors' conception of the cosmos, concerning which his ideas were set forth not only in his *Revolutions* (Volume II in this edition), but also in his *Commentariolus* and *Letter against Werner*, the minor works that form the astronomical section of the present volume.

Like ourselves, Copernicus lived at a time of grave financial and social upheaval, from which he did not hold aloof. His writings on monetary theory and policy are justly prized by economic historians. As a faithful canon of the Varmia diocese, he was charged with the responsibility of resettling farms abandoned on account of invasion, war, and harsh conditions. His official efforts on behalf of the Varmia Chapter are recorded in the *Leases of Abandoned Farmsteads*. Related administrative documents include his *Inventory* of the valuable papers preserved in the Chapter's treasure chest.

What remains of Copernicus' correspondence is one-sided. The letters he received from others have not survived. Still extant, however, either as originals or as copies, are some of the communications he wrote to the Chapter, to fellow-canons, to bishops of Varmia, and to the duke of Prussia.

Affected by the contagious influence of the powerful intellectual movement sweeping across Europe in his student days, Copernicus became an ardent humanist. Learning the language of ancient Greece, he published his own translation of a Greek letter-writer into Latin. His little book marked the beginning of a specialty in the history of Polish printing, being the first translation made in Poland of a Greek literary work.

These literary, epistolary, administrative, monetary, and scientific documents, all that is left of Copernicus' intellectual activities apart from the *Revolutions*, have been assembled here in Volume III. Together with Volume II, it constitutes what is known today about *Nicholas Copernicus' Complete Works*, the first such collection.

Special thanks are due to Professor Eugeniusz Rybka for stimulating Dr. Jan Pirożyński to recover Leonard Niedźwiedzki's copy of Antoni Makowski's copy of the (lost) Strasbourg copy of Simon Hájek's (lost) copy of Copernicus' *Letter against Werner*. Professor Vera Lachmann kindly prepared an analytical table of Copernicus' Latin translation of Theophylactus Simocatta's eighty-five Greek letters. In the production of this volume, Erna Hilfstein shared in all the editorial activities. Tadeusz and Wanda Pochopień rendered invaluable assistance in doing essential research in various Polish libraries.

Edward Rosen
Distinguished Professor Emeritus
of the History of Science
City University of New York

ABBREVIATIONS OF WORKS FREQUENTLY CITED

Acten	*Acten der Ständetage Preussens unter der Herrschaft des Deutschen Ordens* (Leipzig, 1874–1886; reprint, Aalen, 1974)
AcTom	*Acta Tomiciana* (Poznań/Wrocław, 1852–)
ASPK	*Akta Stanów Prus Królewskich* (Towarzystwo naukowe w Toruniu, Fontes; Toruń/Poznań/Warsaw, 1955–)
Baranowski,	Jan, ed., Copernicus, *De revolutionibus orbium coelestium* (Warsaw, 1854)
CDP	*Codex diplomaticus prussicus*, ed. Johannes Voigt (Koenigsberg, 1836–1861)
CDW	*Codex diplomaticus warmiensis*, I = MHW, I (Mainz, 1860); II = MHW, II (Mainz, 1864); III = MHW, V (Braunsberg/Leipzig, 1874)
Dmochowski,	Jan, *Mikołaja Kopernika rozprawy o monecie i inne pisma ekonomiczne* (Warsaw, [1924])
DSB	*Dictionary of Scientific Biography*, 16 vols. (New York, 1970–1980)
Epitome	Peurbach, George and Regiomontanus, Johannes, *Epytoma Joannis de Monte regio in almagestum Ptolomei* (Venice, 1496)
Eubel,	Conrad, *Hierarchia catholica medii aevi*, 2nd ed. (Münster, 1913–1923; reprint, Padua, 1960)
GG	Galilei, Galileo, *Opere*, national edition (Florence, 1890–1909; reprinted, 1929–1939, 1968)
GV	Valla, Giorgio, *De expetendis et fugiendis rebus* (Venice, 1501)
GW	Kepler, Johannes, *Gesammelte Werke* (Munich, 1937–)
Hipler,	Franz, *Spicilegium Copernicanum* (Braunsberg, 1873)
JHA	*Journal for the History of Astronomy*
KHNT	*Kwartalnik Historii Nauki i Techniki*
KMW	*Komunikaty Mazursko-Warmińskie*
Koestler,	Arthur, *The Sleepwalkers* (London, 1959; Pelican reprint, 1968)
MCV	*Mitteilungen des Coppernicus-Vereins für Wissenschaft und Kunst zu Thorn*
MHW	*Monumenta historiae warmiensis*
MK	Birkenmajer, Ludwik Antoni, *Mikołaj Kopernik* (Cracow, 1900)
NCCW	*Nicholas Copernicus Complete Works*, I (London/Warsaw, 1972); II (London/Baltimore, 1978)
NCOO	*Nicolai Copernici opera omnia*, I (Warsaw, 1973); II (Warsaw, 1975)
NM	*Nowe materiały do działalności publicznej Mikołaja Kopernika z lat 1512–1537*, ed. Marian Biskup (Warsaw, 1971; Studia i materiały z dziejów nauki polskiej, Series C, 15)
P	Prowe, Leopold, *Nicolaus Coppernicus* (Berlin, 1883–1884; reprint, Osnabrück, 1967) PI^1: vol. I, part I; PI^2: vol. I, part II; PII: vol. II
Polkowski,	Ignacy, *Kopernikijana*, 3 vols. (Gniezno, 1873–1875)
PS	Ptolemy, *Syntaxis* (miscalled "Almagest")
PS 1515	the first printed translation into Latin (Venice, 1515)
Revolutions	Copernicus, Nicholas, *De revolutionibus orbium caelestium*
Sikorski,	Jerzy, *Mikołaj Kopernik na Warmii* (Olsztyn, 1968; Nicholas Copernicus in Varmia)
SRW	*Scriptores rerum warmiensium*, I = MHW, III^2 (Braunsberg, 1866); II = MHW, VIII (Braunsberg, 1889)
StC	*Studia Copernicana* (Wrocław, 1970–)
Summaria	*Matricularum regni Poloniae summaria*, ed. Teodor Wierzbowski (Warsaw, 1905–1919)
TB	*Tychonis Brahe dani opera omnia* (Copenhagen, 1913–1929)
TCT	Rosen, Edward, *Three Copernican Treatises*, 3rd ed. (New York, 1971)
Wasiutyński,	Jeremi, *Kopernik, twórca nowego nieba* (Warsaw, 1938)
Z	Zinner, Ernst, *Entstehung und Ausbreitung der coppernicanischen Lehre* (Erlangen, 1943)
ZGAE	*Zeitschrift für die Geschichte und Altertumskunde Ermlands*
Zimmermann,	Gerhard, *Das Breslauer Domkapitel im Zeitalter der Reformation und Gegenreformation* (Historisch-diplomatische Forschungen, II), Weimar, 1938

COPERNICUS' TRANSLATION
OF THEOPHYLACTUS SIMOCATTA

INTRODUCTION

I

COPERNICUS' STUDY OF GREEK

The first institution of higher learning attended by Copernicus was the University of Cracow. While he was a student there, no courses were offered in the Greek language. Some influential humanists were members of the faculty, but they had not yet succeeded in expanding the curriculum to include direct contact with the cultural achievement of ancient Hellas. Hence, during Copernicus' stay in Cracow, whatever he learned about the Greek writers of antiquity came to him through the medium of Latin translations and commentaries.

Copernicus, however, was no ordinary humanist. To be sure, in company with the other supporters of that vibrant movement quickening the intellectual pulse throughout Europe, he prized the epic and lyric poets, tragedians and comedians, orators and historians, philosophers and scientists. But, first and foremost, he was an astronomer.

In his day ancient Greek astronomy did not occupy the position it holds today, when it is widely regarded as a fascinating relic of the past. For the purpose of grappling successfully with our expanding galaxies and the red shift, the modern would-be astronomer does not have to master Aristotle's homocentric spheres and Ptolemy's equants. But in the age of Copernicus, nearly five hundred years ago, the situation was entirely different.

At that time, what the Greek astronomers had said a millennium or more earlier still presented the most advanced state of the science. Far from being obsolete, ancient Greek thinking remained in the forefront of current discussion and research. If Copernicus was to be active on that front, he would have to familiarize himself with the writings of the Greeks, in particular, their greatest astronomer, Ptolemy.

His masterpiece, the *Mathematical Syntaxis* (or "Almagest," as that consummation of ancient astronomy was long miscalled), remained in manuscript form throughout most of Copernicus' life. The Greek text of Ptolemy's *Syntaxis* was first printed, nearly a century after the invention of movable type, in 1538. In the spring of the following year Copernicus' only disciple brought him a copy of this first edition. By that time Copernicus had nearly finished his *Revolutions*. For this reason the printed Greek text of Ptolemy's *Syntaxis* had only a minimal effect on the composition of Copernicus' *Revolutions*, the work that superseded the *Syntaxis* after its reign of 1400 years.

Nevertheless, Ptolemy's thinking exerted a powerful influence on the development of Copernicus' ideas. In the absence of the printed Greek text, the means of transmission was the *Epitome* of Ptolemy by Peurbach and Regiomontanus. This immensely valuable collaborative work was first printed, more than thirty years after it was finished, in 1496. The place of publication was Venice. In the autumn of that year Copernicus arrived in nearby Bologna, in order to study at the local university. He was deeply affected by his reading of the *Epitome*. For besides summarizing Ptolemy, the *Epitome* also reported the findings of astronomers later than Ptolemy, in particular the Muslims. They had detected a fatal flaw in Ptolemy's lunar theory. This flaw, as explained in the *Epitome*, made a deep impression on Copernicus, who repeated the explanation in his earliest astronomical treatise, the *Commentariolus*, as well as in his mature masterpiece, the *Revolutions*.

While Copernicus was writing the *Commentariolus*, he had no access to any printed Latin translation of the *Syntaxis*. The typesetting of the earliest such publication was completed on 10 January 1515. But the *Commentariolus* was listed in a Cracow professor's inventory that was dated 1 May 1514. Hence, the recent suggestion that the 1515 translation was a possible source of information for the *Commentariolus* is chronologically preposterous.[1]

In the *Commentariolus*, which was quite brief, Copernicus announced his intention of producing a larger volume, the *Revolutions*.[2] He toiled at this long and complicated task nearly thirty years. During those three decades, not only was the Greek text of the *Syntaxis* printed in 1538, and the first Latin translation in 1515, but also a second, and much better, Latin translation in 1528. Long before the appearance of these three publications, however, Copernicus had made up his mind that he had to learn the Greek language.

When he was studying Greek, there was no printed Latin translation of Ptolemy nor of many another important Greek author. Even if a Latin translation was available, it might be so bad that the poor reader often had to guess what meaning lay concealed beneath its baffling convolutions. This was certainly true of the 1515 translation of Ptolemy's *Syntaxis* and, to a lesser degree, of Giorgio Valla's *De expetendis et fugiendis rebus* (Venice, 1501). This vast conglomeration of treatises on *What to Seek and Avoid* was consulted by Copernicus. But as a body of translations from various Greek authors, it was far from satisfactory. In the absence of translations of some authors, therefore, and in the presence of unsatisfactory translations of other authors, Copernicus decided that he must learn Greek.

An important ingredient in this decision was the steady outpouring of a magnificent series of editions of Greek authors by the highly learned and immensely successful Venetian publisher Aldo Manuzio (1449-1515). One such Aldine publication became Copernicus' gateway to the study of Greek. It was a collection of epistolographers which Manuzio dedicated to his former classmate Antonio Urceo (1446-1500), known affectionately by his nickname "Codro" (pauper). The dedication suggested the collection's usefulness to students of Greek:

Recently, most learned Codro, I gathered as many Greek letters as I could lay my hands on. I am publishing them as printed on my own press in two volumes. I have omitted the numerous letters of Basil, Gregory, and Libanius, which I am keeping at home to be printed at the earliest opportunity. The authors whose letters are presented here number about thirty-five, as may be seen in the volumes themselves.

I am sending these letters as a gift to you, the public professor of Latin and Greek literature at the very illustrious University of Bologna, to be shown to your students who may be inspired thereby with a greater zeal to master more elegant writing, and also to be a reminder to you of your friend Aldo and a token of my esteem for you. Farewell. Venice, 17 April 1499

As the date of the dedication shows, the Aldine *Epistolographers*[3] reached the market at a time when Copernicus' finances were at a low ebb. For in September 1499 a bank loan of a hundred ducats had to be arranged for him and his brother, who also was studying at Bologna. Hence Copernicus could not afford to buy his own copy of the expensive Aldine volumes when they were available. Later on, however, he translated into Latin two sections

---

[1] See *Commentariolus*, Introduction, n. 25.

[2] See *Commentariolus*, n. 35.

[3] *Epistolae diversorum philosophorum, oratorum, rhetorum* (Letters of Various Philosophers, Orators, Rhetoricians; Venice, 1499), with Theophylactus at sig. φ2^r–ψ5^v, and Lysis at Γ6^v–7^v.

of the letters. Did he copy the Greek text of these two sections from the printed edition as portions of particular interest to himself?

The *Epistolographers* provided Copernicus with one of the two crutches by whose help he hobbled toward his goal of learning Greek. The other crutch was a Greek-Latin dictionary in two parts. The Greek part was published in nearby Modena on 20 October 1499. The Latin part was added in Reggio Emilia, not much farther from Bologna, on 5 July 1500. Whether Copernicus was still enjoying the proceeds of the bank loan or had received additional funds, he bought his own copy of the dictionary. On its flyleaf he inscribed his name as owner, in Greek characters.[4] His copy still survives, in the library of the University of Uppsala, Sweden.

To what extent, if any, Copernicus benefited, formally or informally, from the instruction of Professor Urceo before the latter's death on 11 February 1500 is a question that still remains to be answered. In any case, Copernicus' knowledge of Greek was mainly self-taught. In his dictionary he recorded various forms of irregular verbs, that perpetual bane of beginning students of classical Greek. These entries are not particularly connected with his translations into Latin, which were undertaken later on.

On 31 May 1503 Copernicus received his doctoral degree, his principal reason for going to Italy. When he returned home thereafter, he found a copy of the *Epistolographers* in the library of the Cathedral Chapter to which he belonged as a canon. But instead of participating in the Chapter's activities in Frombork, he joined his uncle, Bishop Lucas Watzenrode of Varmia, in the episcopal palace in Lidzbark. His duties there were light enough to permit him to undertake the translation of a short Greek work into Latin. For Copernicus knew that a handy way of teaching oneself a foreign language was to translate a literary composition written originally in that language.

II

COPERNICUS AND THEOPHYLACTUS

Partly for the purpose of teaching himself Greek in the process, Copernicus decided to translate into Latin the *Letters* of Theophylactus, one of the writers in the Aldine *Epistolographers*. This letter-writing Theophylactus was later surnamed Simocatta in order to distinguish him from several other Greek authors also called Theophylactus. To this day nobody has been able to pinpoint the exact meaning of "Simocatta."

Copernicus had no information about Theophylactus Simocatta apart from his *Letters*. He was not aware that this Theophylactus was one of several hundred authors whose works had been summarized by Photius (c. 820–c. 891). In his *Bibliotheke* (Library, Chapter 65) Photius began his report on Theophylactus as follows:

I have read the *History* in eight books by Theophylactus, an ex-prefect and referendary. This Theophylactus is an Egyptian by birth. While his style has a certain charm, his overabundant use of metaphorical expressions and allegorical ideas ends up with

---

[4] For a photocopy of Copernicus' ownership signature, see NCCW IV, Plate XXXIV; in StC XVI, 356/20, 366/No. 3, φ should be deleted; Copernicus' misplaced accent over his surname indicates that when he bought the dictionary, he was not yet familiar with the rules of Greek accentuation.

a certain frigidity and immature lack of taste. In addition, his poorly timed introduction of sententious utterances is a mark of his needless and excessive straining for renown. In other respects, however, there is no reason to reproach him.

Photius then added an extensive epitome of Theophylactus' *History of Emperor Maurice's Reign* (582–602) without saying a word about Theophylactus' *Letters*.

Like Photius, Copernicus had some reservations about Theophylactus' literary style. But he gave the epistolographer high marks for the contents of his *Letters*. As Copernicus wrote in dedicating his translation of them,

> Theophylactus has disposed so much of value in all of them that they seem to be not letters but rather laws and rules for the conduct of human life.

Copernicus' translation therefore served a dual purpose. Not only did it improve his knowledge of Greek, but it also disseminated a code of conduct in the best tradition of Greek ethics. The ancient proverb "Nothing in excess" is interwoven, like an unseen thread, throughout the warp and woof of Theophylactus' eighty-five *Letters*.

They are arranged in triads. In each group of three, the first Letter is ethical in nature; the second is rustic; and the third is erotic. After the close of the twenty-eighth triad, the last Letter, No. 85, terminates the work with a grim visit to a cemetery, to "behold man's greatest joys as in the end they take on the lightness of dust."

The mood, however, is not somber throughout. As Copernicus noted in the Dedication, Theophylactus "interspersed the gay with the serious, and the playful with the austere." Life on the farm can be joyful and prosperous. More often it is harsh and unrewarding: barren soil, floods and foxes, nasty neighbors, thieving wayfarers. The ethical Letters advocate sweet reasonableness, self-control, family loyalty, sincerity, and an active pursuit of worldly goods in moderate amount. "Noble birth is of no use to people, for all of them value nothing more than wealth." By the same token, the ethical Letters denounce greed, gambling, arrogance, false modesty, garrulity, grudges, indolence, and the quest for hollow fame. True philosophers are revered, lawyers reviled, and painters praised.

For Copernicus the cathedral canon, who dedicated his translation to his uncle, the bishop of Varmia, the third class of Letters, the erotic, posed a problem. "The title of the love letters seems to portend licentiousness," Copernicus remarked in the Dedication. He looked to his medical training and practice, however, for a solution of this problem: how to retain the erotic Letters and yet avoid being charged with promoting pornography. "Just as physicians usually moderate the bitterness of drugs by sweetening them to make them more palatable to patients, so these love letters have in like manner been rectified, with the result that they ought to receive the label 'moral' no less" than the others.

Built into the structure of some Letters is a comparison of animal and human behavior, recalling the fables of Aesop. His habit of teaching the moral lesson explicitly and didactically, however, is avoided by Theophylactus, who prefers the more sophisticated device of letting the thoughtful reader draw his own conclusion. This is not always immediately obvious. For instance, a cricket breaks its silence by chirping loudly under the influence of the noonday sun. A man sunk in an evil life is similarly inspired to acclaim the philosopher whose counsel drew him back to the path of morality. A rich uncle who ignores his poor nephew's poverty is unfavorably contrasted with the mare which nurses a motherless foal. The elephant lets itself be trained, whereas the son of an educated man refuses to follow in his father's footsteps. The peacock takes pride in displaying its beauty, unlike the antisocial writer who declines to publish his own works. The wolf, when sated, stops hunting, but the drunkard keeps right on drinking. Such are the Letters by which Copernicus hoped to draw his readers into the incipient humanist movement in Poland.

III

COPERNICUS' KNOWLEDGE OF GREEK

While Theophylactus' *Letters* appealed to Copernicus as literature, he was not adequately prepared to do them justice philologically. Perhaps he would have been able to cope more effectively with a writer whose diction was plain and unadorned. But Theophylactus did not belong to that unpretentious school. As an Egyptian, he strove to surpass the native Greeks in the elegant manipulation of their own language. He strained every nerve to achieve these dazzling effects not only in his *History* and *Letters* but also in his *Natural Questions*, a dialog designed to be recited to a sophisticated gathering in Greece. In the First Introduction to his *Natural Questions*, Theophylactus presented himself as a novice trying to master the style of Athens, and expressed confidence that he would succeed "even if not treading barbarian soil." On "barbarian soil," that is, outside of Greece, critical standards were less demanding than in the homeland of Plato and Demosthenes. Theophylactus remarks that he has heard about swallows being taught to sing by their mothers. "If I have not acquired [this] art as my mother, nevertheless my origins are the same as yours," he tells his native Greek audience, "since eloquence is my fatherland, and what is Greek is mine."[5]

Theophylactus buttressed this claim over and over again in his *Letters*. With deliberate inconspicuousness he artfully introduced echoes of the favorite ancient Greek authors. For the most part he refrained from naming them, and only twice did he resort to actual quotations. Such conscious concealment fascinated his audience, who were thus gently reminded of familiar passages without undergoing the embarrassment of being told who said what. Not the pedantic professor, but the popular lecturer was the role at which the youthful Theophylactus aimed in his *Natural Questions* and *Letters* before being admitted to the ranks of the

---

[5] By claiming eloquence as his fatherland, and what was Greek as his own, Theophylactus implied that he was a Greek by culture, not by birth (*Questioni naturali*, p. 8/11–12). This passage was misunderstood to be an assertion by Theophylactus that he was born in Greece, an error committed by Cimedoncius and repeated by an editor of Theophylactus' *Natural Questions* (both reprinted in Boissonade, p. xxxi/6up–5up, p. xxxiv/2–3). Photius' plain statement that "Theophylactus is an Egyptian by birth" was mistakenly applied only to his ancestors by PI¹, 392/n. */last 5 lines, who clung to Cimedoncius' error.

Photius described our Theophylactus as an ex-prefect (ἀπὸ ἐπάρχων; apo eparchon). The same description is applied to the Theophylactus who presided over the witnessing of a legal document, according to an inscription found in Aphrodisias, a city in the district of Caria in south-western Asia Minor (now Turkey); Henri Grégoire, *Recueil des inscriptions grecques-chrétiennes d'Asie Mineure* (Paris, 1922; reprint, Amsterdam, 1968), p. 88, No. 247. The inscription calls Theophylactus ἐνδοξότατος (endoxotatos, most honorable), a designation borne by city prefects, and also θεῖος δικαστής (theios dikastes, imperial judge). The inscription, however, does not assign to Theophylactus the title ἀντιγραφεύς (antigrapheus), as Photius does. An analogous inscription, labeling an official ἐνδοξότατος and θεῖος δικαστής, also refers to him as a *referendarius* (Grégoire, p. 114, No. 324). A referendary transmitted imperial decisions to functionaries, and submitted subjects' petitions to the imperial court. Photius' ἀντιγραφεύς is evidently a translation of *referendarius* into Greek. He did not transliterate the Latin word in Greek characters as the inscription did.

imperial bureaucracy. His avoidance of citation by chapter and verse has given rise to the false impression[6] that "apart from Homer, he read the ancient authors only a little," and that "classical reminiscences are rare" in his writings. Actually he read very widely, and made frequent use of his familiarity, not only with Homer, but also with Achilles Tatius, Aelian, Aeschylus, Aesop, Aristophanes, Arrian, Diogenes Laertius, Euripides, Julian, Menander, Plato, Plotinus, Plutarch, and Porphyry.

This catalog, which does not profess to be complete, lay outside the range of Copernicus' competence. He even misunderstood the true nature of Theophylactus' *Letters*, which he described as "collected from various authors." Blissfully unaware of the fictitious letter as a widespread form of ancient literature,[7] Copernicus falsely supposed that Theophylactus had assembled his *Letters* from writings by others, whereas in fact he composed them himself. Copernicus did not question the authenticity even of Letter 22, which pretends to advise Pericles to follow the example of Alexander the Great, who was born nearly three-quarters of a century after Pericles' death.

Like his familiarity with the history of Greece, Copernicus' knowledge of its language was somewhat imperfect. Moreover, his endeavor to translate Theophylactus' *Letters* was hampered by the typographical peculiarities of his Greek text and dictionary. The *Epistolographers*, for example, printed some proper nouns with an initial small letter. On the other hand, every word in the Greek part of his dictionary began with a capital letter. Hence Copernicus could not always be certain whether he was looking at a proper noun or a common noun. Letter 31, for instance, began by describing the peacock as Τὸ Μηδικὸν ὄρνεον (To Medikon orneon, The Persian bird), and added that it had acquired τῶν Μήδων τὴν ὑπεροψίαν (ton Medon ten hyperopsian, the arrogance of the Persians). In the Aldine edition, however, μηδικόν and μήδων were printed with small initial letters instead of capitals. In his dictionary Copernicus could not find μηδικόν, but he did see μῆδος and μῆδος (medos; Persian, plan). In choosing between these two alternatives, he was affected by μήδομαι (medomai, *operor*, be active). As a result, his peacock became an active (*operosa*) bird, instead of a Persian bird; and what it acquired was the "splendor of the energetic birds" (*industriarum avium spectabilitatem*), instead of "the arrogance of the Persians."

In Letter 31 Copernicus mistook a proper noun for a common noun. In the preceding Letter, No. 30, he made the opposite error. There Medea, who killed her children, is described as φονικῆς (phonikes). This was equated in Copernicus' dictionary with *mortifer* (murderous). By a slip, however, he looked at φοινίκεος (phoinikeos), which was matched with *puniceus* (Punic). Then, thinking of Rome's Punic rivals in Africa, Copernicus made Medea an African (*africana*) instead of a murderous woman.

The pretended writer of Letter 14 says: "I entertained Trygias magnificently yesterday." Trygias, a personal name made up by Theophylactus, was not recognized as such by Copernicus in the absence of an initial capital letter. But in his dictionary he saw τρύγητος (trygetos, vintage) matched with *vindemia*. This was what he worked into the passage somehow instead of a personal name. Later on in Letter 14 "Trygias gaped." Since a vintage cannot gape, Copernicus took his dictionary's definition of τρυγίας as a common noun (trygias,

---

[6] Karl Krumbacher, *Geschichte der byzantinischen Litteratur* (New York, 1958; reprint of 2nd ed., Munich, 1897), p. 250.

[7] Wolfgang Speyer, *Die literarische Fälschung im heidnischen und christlichen Altertum* (Munich, 1971; *Handbuch der Altertumswissenschaft*, Abt. I, Teil 2), pp. 79–84, cites less innocent examples of this popular sport.

faex, dregs) and applied this term of contempt to the individual whose misconduct is under attack in Letter 14. In Copernicus' time, *faex* was written as *fex*. But in his handwriting, initial *f* may easily be mistaken for *s* (for example, NCCW, I, folio 1ʳ, line 13: *si*; line 18: *fulcitur*). Hence, in the 1509 edition of Copernicus' translation, *fex* was misprinted as *sex*. How his readers interpreted this weird result must be left to the imagination.

In Letter 14 Copernicus converted a man's name into two different common nouns. He did just the opposite in Letter 5. There γεράνους (geranous) were described as a farmer's bad neighbors. On the assumption that such neighbors had to be human beings, Copernicus simply transliterated *Geranos* with an initial capital letter, as if they were an otherwise unknown hostile tribe. In so doing, he ignored the feminine gender (τάς, tas) of γεράνους in his Greek text as well as his dictionary's matching of γέρανος with *grus* (crane).

Whereas Copernicus could have found γέρανος in his dictionary, it did not contain Λεωκόριον (Leokorion), an Athenian temple built in honor of the daughters of Leos who sacrificed themselves in order to avert a famine. Letter 12's concerts occur ἐπὶ τὸ λεωκόριον (epi to Leokorion, in front of this temple). Being unfamiliar with Λεωκόριον, Copernicus translated it as though it were λεωφόρον (leophoron, *via qua it populus*, highway). In placing the concerts on the highway, he may have been guided by Letter 68's summoning of all local farmers to the highway (ἐπὶ τὴν λεωφόρον, epi ten leophoron, *in publicum*).

In Letter 43 Copernicus' Greek text not only began an individual's name with a small letter, but also made matters worse by tacking on to the end of that name the first two letters of the following name: σωφρονίσκουσω (sophroniskouso). The rest of the second name (κράτει, kratei), being separated from its first two letters, looked like a form of a word signifying "authority." In addition, the first name suggested some garbled misprinting of σωφρονισκός (sophroniskos, discreet). Accordingly, doing the best he could with his mangled text, Copernicus translated: "This surely is proclaimed with a discreet command." What was actually intended by Theophylactus never dawned on Copernicus: "This was beautifully discussed by Socrates, son of Sophroniscus."

In Letter 4, limits (ὅρους, horous) were imposed on the sea's turbulence by ὁ δημιουργός (ho demiourgos). This word was matched in Copernicus' dictionary with *artifex, opifex, creator*. Evidently unfamiliar with δημιουργός in Plato's *Timaeus*, where it denotes the Creator of the universe, Copernicus selected *artifex* (workman). On this workman he bestowed a name, Horus (Horus the workman was put in charge of the sea's turbulence: *Horus artifex praepositus fuit marinae illuvioni*). In this baptismal process the personal name Horus ousted the "limits" (horous). They disappeared a second time in the same letter: where an irate man is urged to impose limits (horous) on his wrath, Copernicus' "Horus dictated a law" (*Horus ... legem dictavit*). Copernicus' "Horus passed a law" (*Horus ... legem tulit*) again in Letter 25, where a mourner is advised to put limits (horous) on his grief. Yet between Letters 4 and 25, in Letter 17 Copernicus matched ὅρον (horon) accurately with *terminum*. Similarly, in Letter 56 a dissident is threatened with banishment beyond the ὅρων (horon), which Copernicus quite properly equated with *finibus* (boundaries). Yet in his characteristic manner he did not return with this correct equation to Letters 4 and 25 for the purpose of banishing Horus. Nevertheless he did banish a misprint in Letter 4, where he recognized that the sign for the breathing over ὅρος should not be rough (horos) but smooth (ὄρος, oros, peak). For Letter 4 referred to the combined action of hand and tongue as virtue's "loftiest peak" (ὄρος ἀκρότατος, oros akrotatos). Letter 79 similarly mentioned Fortune's "loftiest peak." In both these instances, by quietly reversing the sign for the breathing, Copernicus arrived at the correct translation *mons altissimus*. He may have been led to this happy result by Letter 26, where Etna is called the Sicilian ὄρος (*montem*).

Copernicus' ability to distinguish between ὅρος (horos, boundary) and ὄρος (oros, peak) did not extend to ὁδός (hodos, way) and ὀδούς (odous, tooth). Letter 62's growing children change (that is, acquire their second set of) teeth (ὀδόντας, odontas). Instead, Copernicus has them "mimic walking" (*ad gressum imitantur*). Letter 67's maker of false promises supposes that the matter ends with the clacking of his teeth (ὀδόντων, odonton). In Copernicus, this noise is made by the promises "as they go away" (*abeuntium*).

Copernicus correctly matched γυναικεῖον φῦλον (gynaikeion phylon) and γυναικείῳ φύλῳ (gynaikeioi phyloi) with *muliebre genus* and *feminino sexui* (female sex) in Letters 15 and 72. Yet the male lament "Everything is enslaved to the female sex" (φύλῳ, phyloi) became "Everything is enslaved by love of women" (*Omnia muliebri amori subdita sunt*) as a result of Copernicus' confusion in Letter 60 of φύλῳ (phyloi) with φίλῳ (philoi, beloved).

In Letter 54 a jilted girl upbraids her former lover: "No longer do you go without sleep night after night; you have given up your morning serenades." These serenades in the morning (ἑωθινά, heothina) became customary (*consueta*) serenades in Copernicus, who was thinking of εἰωθότα (eiothota). Contributing to his confusion of these two adjectives was his dictionary's misprinting of ἑωθινός with the smooth breathing. Copernicus had matched εἰώθασιν (eiothasin) and εἰωθώς (eiothos) in Letters 9 and 22 with *consueverunt* and *soleat* (accustomed).

Short ŏ (omicron) and long ō (omega) were not always correctly distinguished. For instance, Copernicus confused πολέω (pŏleo, stroll) with πωλέω (pōleo, do business). Thus, in Letter 82 the sounds "swirling around" (περιπολοῦσιν, peripolousin) were misinterpreted as "bothersome" (*negotientur*).

Another example of confusion between omicron and omega occurs in the future tense of two verbs meaning "enjoy" (ὀνήσομαι, ŏnesomai) and "buy" (ὠνήσομαι, ōnesomai). Thus, a bachelor predicts: "I'll enjoy" (ὀνήσομαι) marriage. By mistaking ὀνήσομαι for ὠνήσομαι, Copernicus has this bachelor say: "I'll buy" (*emam*) my marriage with gifts. Here in Letter 75 ὀνήσομαι was followed immediately by πρίασθαι (priasthai, buy), which affected Copernicus' understanding of the preceding word. Moreover, in Letter 81 a businesslike suitor brusquely informs the tender object of his affections: "I am buying" (ὠνούμενος, onoumenos) your maidenhead with gold.

A third instance of the interchange of omicron and omega concerns a wayward son, who is chided: "You have become" (γέγονας, gegŏnas) the end of good breeding in your family (Letter 13). However, finding γεγώνω (gegōno) equated in his dictionary with "shout," Copernicus mistranslated: "you brag" (*iactas te*) about your noble birth, as though his text's γέγονας (gegŏnas) were γέγωνας (gegōnas). Contributing to his error was the statement near the end of Letter 13: "you proclaim yourself to be the son" (*te filium ... pronuntias*, παῖδα σαυτὸν ... κηρύττεις, paida sauton kerytteis).

Copernicus had no trouble matching ᾆσαι (aisai) and ᾄδεις (aideis) with *cantare* and *cantas* (to sing, you sing; Letters 1, 21). Yet where a girl is said to sing (ᾄδειν, aidein; Letter 36), Copernicus instead made her pleasing (*placere*), as though his text read ἁδεῖν (hadein). His dictionary translated ᾄδω (ado, lacking the iota subscript) by "sing," and ἅδω (hado) by "please." As a result of his confusion of ᾄδειν with ἁδεῖν, Copernicus was unaware of the young lady's musical accomplishments. Her admirer, however, was aware of them. His "eyes have become ears also," because he has never seen her. He loves the girl without her being "seen" (ὁρωμένης, horomenes). This was translated by Copernicus as though it were ἐρωμένης (eromenes, *amatam*, loved). In his version, therefore, instead of the male loving the girl without her being seen, her admirer says paradoxically: "I love the girl I don't love" (*non amatam amo*).

Letter 68's farmer, having finally caught a marauding fox, declares: "I'll carry" (ἄξω, axo) her to a meeting of the local farmers. Instead, Copernicus' farmer says: "I think it worthwhile" (*dignum puto*) to call a meeting, because Copernicus has evidently confused ἀξιῶ (axio) with ἄξω (axo).

A decrease in game animals (Letter 65) and the behavior of the octopus (Letter 73) are both ascribed to ἀθηρία (atheria, scarcity of food). This word was not to be found in Copernicus' dictionary, where he wrote ἀθηρία in the margin alongside ἀθήρ, translated as *acutus* (acute). Hence in Letter 65 he resorted to *sagacitas*, with human astuteness being responsible for the decrease in the game animals. In Letter 73, however, the situation was entirely different. There, instead of ἀθηρίας, Copernicus' Greek text printed ἀθυρίας. This combination of letters which do not form a word seemed to Copernicus similar to ἀθύρω (athyro, play). Hence, unlike Theophylactus' octopus suffering from scarcity of food, Copernicus' octopus behaved in a laughable manner.

Once Copernicus slipped into a misunderstanding through inattention to the ending of a participle. Letter 58 deems a father qualified to be a teacher because "he learns (μανθάνοντα, manthanonta) through experience the conditions of procreation and the agonies of love." The ending of μανθάνοντα requires it to be linked with a singular noun (πατέρα, patera, father). Yet Copernicus mistakenly associated it with "sons" (*pueri*) in the plural. A participle ἠχῶν (echon, sounding) was confused by Copernicus with ἤχων (echon), a form of the related noun (Letter 76). Hence the derision of a wee bit of fame "as a plaything more insubstantial than echoes" was altered by Copernicus to fame "sounding like a game" (*ludum sonans*). In Letter 14 he mistook εἱστίον (heistion, I entertained) for a participle (*epulatus*, having dined). As a result, the splendid feast arranged by the host became the splendid dinner gobbled up by the guest. The participle εἰδώς (eidos, knowing) in Letter 67 was translated by Copernicus as though it were an indicative form, which he turned into a question, *videsne* (don't you see?). A participle puzzled him so much that on the page (sig. ? 2ᵛ) which he numbered 155ᵛ, in his dictionary's unnumbered pages, in the margin he wrote μονωθεν (monōthen), without any accent. His dictionary did not record μονόω (monoo), the verb of which μονωθέν is a participial form. In Letter 7 τὸ μονωθέν denotes the foal "that has been forsaken" by its mother and is nursed by the other mares. Contributing to Copernicus' misunderstanding of this construction is the similarity between the participial suffix -θεν and the adverbial suffix -θεν. Although μονόθεν (monŏthen, singly) did not appear in his dictionary, he falsely supposed that it, rather than μονωθέν (monōthen), was intended in Letter 7. As a result, his mares nurse the forsaken foal *unanimiter* (with a single purpose).

When Copernicus saw ὥσπερ (hosper) in Letter 1, he could not find it in his dictionary and mismatched it with *utpote* (because). But he soon realized that it is a non-causal intensified form of ὥς (hos, thus). Accordingly, he translated it by *quasi* (as if; Letters 5, 7, 10, 19, 33); *tamquam* (as though; Letters 6, 16, 28, 52, 73); *ad modum* and *quemadmodum* (like; Letters 14, 29, 46, 62; 16, 52); *ut* (as; Letter 49); and *sicut* (just as; Letter 76). Copernicus' corrected renderings of ὥσπερ illustrate how he improved his knowledge of Greek as he translated Theophylactus' *Letters*.

Copernicus' dictionary could have helped him in three passages where he failed to make proper use of the information it provided. In the first place, a disgusted farmer in Letter 26 is leaving Attica because it produced "myrtle instead of πυρῶν" (pyron, equated in Copernicus' dictionary with *triticum*, wheat). Yet, instead of "wheat," he said "pears" (*piris*). Was he misled by the resemblance of *piris* to πυρῶν as pronounced? Elsewhere he matched *piri* correctly with ἀχράδες (achrades, pear-trees; Letter 62).

Secondly, his dictionary defined ὀργάδα (orgada) as *terra aliqua dedicata* ("some land sacred" to a god). Despite this precise definition, Copernicus mistakenly connected ὀργάδα with ὀργή (orge, wrath). As a result, his farmers donate their land as if to an "embittered" god (Letter 5).

Thirdly, his dictionary matched διφθέρα (diphthera), not with a Latin equivalent, but only with a Greek synonym. Had he glanced at the synonym, he would have found it defined as a "garment made of skins sewn together." But instead of following this little detour, he relied on reasoning by analogy. In Letter 11 a farmer throws down his διφθέραν and δίκελλαν (dikellan, hoe). This tool owes its name to its prongs, the prefix di- signifying "two," as in "dichotomy." Falsely supposing that διφθέραν begins with the same prefix, Copernicus translated it by *bipennem*, a two-edged ax. But there is no prefix in διφθέρα, which is related to δέφω (depho, soften). Far from being a tool, it is a leather coat.

As a self-conscious stylist, Theophylactus sometimes used words not found in Copernicus' dictionary. Thus, a careless farm lad allows the herd to scatter hither and thither (τῇδε κἀκεῖσε, teide kakeise). The dictionary matched τῇδε with *huc* (hither), and ἐκεῖσε with *illuc* (thither), but it did not record κἀκεῖσε (καί + ἐκεῖσε, kai + ekeise, and + thither). Unfamiliar with this contraction, Copernicus mistakenly connected κἀκεῖσε with κακός (kakos, evil). Hence, his herd was scattered "by this evil" behavior, instead of hither and thither (Letter 14).

The same letter refers to a thief (λωποδύτης, lopodytes), someone who slips into (δύω, dyo) the clothes (λῶπος, lopos) of someone else who is swimming. Although λωποδύτης itself was not included in Copernicus' dictionary, the related action (λωποδυσία, lopodysia) was matched with *furtum* (theft), and the related verb (λωποδυτέω, lopodyteo) was matched with *furor* (steal). Nevertheless, overlooking these ready-made connections, Copernicus imaginatively combined po (from lopos) and d (from duo) to form "pod," the root of the word for "foot." In this way he transformed the thief into a pedestrian (*pedester*).

Like many another student of classical Greek, Copernicus was sometimes puzzled by the irregular forms of its verbs. Thus, one farmer rebukes another: "You are not at all different (διενήνοχας, dienenochas) from the drones." The connection between διενήνοχας and διαφέρω (diaphero, differ) eluded Copernicus. Instead, he related this highly irregular form to διανέμω (dianemo, distribute). Hence his farmer complained: "You have not passed any drones along to me" (*nullum ... mihi fucum tribuisti*; Letter 23).

In Letter 61, winter "came upon" (ἀπήντα, apenta) the earth. Not recognizing ἀπήντα as a form of the verb ἀπαντάω (apantao), Copernicus thought of ἁπάντῃ (hapantei, everywhere). But that left him without a verb, which he proceeded to invent on the basis of ἀπηνής (apenes, rough). By combining these two steps, he emerged with "winter raged everywhere" (*hiems ubique saeviebat*).

A jilted girl accuses her former lover: "You cast aside" (παρώσω, paroso) your innumerable messages, agreements, and oaths. This form of παρωθέω (parotheo) was not recognized by Copernicus here in Letter 54. Yet shortly thereafter, in Letter 60, he correctly equated παρώσαντο (parosanto) with *dereliquerunt* (they cast aside). Nevertheless, he did not turn back to improve his mistranslation of παρώσω in Letter 54, where he had just used *dereliquisti* (you cast aside) to match a synonym of παρώσω. This verb he mistranslated in Letter 54 by *adhibuisti*, which means just the opposite of "cast aside." Perhaps Copernicus misunderstood παρώσω to be some form of παρίημι (pariemi, admit).

He may likewise have thought he saw a form of ἵημι (hiemi, let go) in Letter 14's ᾔτει (eitei, asked). For whereas a diner, after eating his fill, asked (ᾔτει) to take some of the food home with him, Copernicus' diner put aside (*neglexit*) taking it home.

Again, in the same Letter, Copernicus imagined that he saw an imperative form of ἀφίημι (aphiemi, *concedat*, let him yield). But ἄπιτω (apito, let him stay away) is an imperative form of ἄπειμι (apeimi). Because this imperative was misconstrued by Copernicus as a transitive verb, he made λοιπόν (loipon) its object: "let him yield the rest" of the land. But the imperative is intransitive: "let him stay away from my land" hereafter, λοιπόν being used as a temporal adverb here, not as a noun. If he does stay away, Letter 14 concludes, "enduring an evil from afar is preferable to nourishing a hidden enemy at home." This eminently sensible conclusion was completely missed by Copernicus on account of his confusion of ἀφίημι with ἄπειμι. His conclusion ("let him yield the rest of the land to me"; *concedat mihi residuum agrorum*) is utterly inappropriate, since in Letter 14 nobody is yielding any land to anybody else.

The defects in Copernicus' knowledge of Greek have been blamed on the philosophy of education advocated by Manuzio (PI[1], 258). But this successful publisher was also a first-rate scholar whose pedagogical method was exactly the opposite of that which has been misattributed to him. As the foreword to his *Introduction to Grammar* (*Institutionum grammaticarum libri quatuor*) he addressed an open letter (dated October 1507) to teachers of the humanities, advising them

> to require pupils to memorize nothing but what is in the most highly trained writers, not even the rules of grammar, apart from certain very compact summaries which can easily be committed to memory. I am in favor of having pupils memorize, but only for the purpose of reading with attention and precision as well as knowing perfectly how to inflect nouns and verbs.

Copernicus' difficulties with the conjugation of verbs should be imputed, not to Manuzio's pedagogical principles, but to his own bad luck in never having had any rigorous formal instruction in Greek.

IV

COPERNICUS AS A TRANSLATOR

Because of his imperfect knowledge of Greek, Copernicus misunderstood some passages in Theophylactus. On the other hand, he knew enough to correct some misprints in the *Epistolographers*. For these two reasons his Latin translation is not a mirror image of the Aldine Greek text. In addition, there is a third explanation of some divergences between his Latin version and his Greek text.

Copernicus' conception of his task as a translator did not jibe in all respects with the view generally held today. While he proposed to present to his readers a fairly faithful reproduction of the Greek original, he did not feel himself obligated to follow it in every last detail. The erotic Letters, for instance, were "rectified" (*castigatae*) by him, as he emphasized in the Dedication. But he did not point out that he deliberately departed from his Greek text from motives having nothing to do with eroticism.

Copernicus was sensitive to the difference between his own readers and Theophylactus' audience. This consisted of Greeks, intensely proud of their cultural heritage. Hence when the epistolographer casually mentioned "the son of Meles," he took it for granted that Homer would instantly leap into the minds of his readers. Copernicus, on the other hand, proceeded on the opposite assumption. His readers were not Greeks and knew no Greek. They were completely unaware that several obscure biographies of Homer named Meles as his father.[8] Copernicus himself may have been unfamiliar with this tradition. Hence he abandoned the principle that the translator must preserve perfect fidelity to the original at whatever cost. Rather than alienate his readers by intruding obscure allusions, he sought to win them over by confining his classical references to familiar themes. In this instance, he replaces "the son of Meles says" by the impersonal "they say" (*aiunt*, Letter 82).

In the very same sentence Theophylactus refers to Ulysses twice, first by name, and then as "the son of Laertes." Copernicus, however, presuming that his readers might imagine Laertes' son to be someone other than Ulysses, repeated his name rather than run the risk of being misunderstood. In the same spirit, where Letter 55 alluded to "some Ulysses" (τὶς Ὀδυσσεύς, tis Odysseus), Copernicus came to his readers' aid by specifying "that Homeric Ulysses" (*Ulysses ille Homericus*).

If Copernicus' readers had not heard about the son of Meles, they were surely even less familiar with the calendar of ancient Athens. In Letter 9, however, a girl reminds her lover of his promise to come back "on the ninth day before the end of Anthesterion." Copernicus' readers of course did not know that Anthesterion was the eighth Athenian month, nor were they in the habit of counting the days of the month backward from the last day. On the contrary, they counted the days forward from the first of the month, just as we do, and the names of their months in Latin were Roman, like ours, not Greek. Hence Copernicus had to discard "the ninth day before the end of Anthesterion," and he inadvisedly replaced it by the "first of November" (*prima novembris*).[9]

Besides eliminating recondite references which Copernicus felt might deter his readers from joining the incipient humanist movement in Poland, he made additional excisions from the *Letters* for stylistic reasons. Theophylactus loved fullness, even exuberance, as Photius had remarked. But Copernicus leaned in the opposite direction, toward spareness.

Letter 10, for instance, provides three illustrations of Nature's even-handed generosity: (1) the sun shines equally on human beings; (2) a supply of fire is instantly available to everybody. But, given these two compact and clear cases, Copernicus chose to omit Theophylactus' third demonstration of Nature's undiscriminating bountifulness: "and the waters of the torrential and ever-flowing rivers are readily accessible to all." If two examples will do, why use three?

---

[8] *Homeri opera*, Oxford: Clarendon Press, V (ed. T. W. Allen, reprinted 1965), 228:75; 231:151; 243:78; 247:1; 254:631.

[9] In his dictionary Copernicus found Anthesterion matched with November. On a flyleaf of his dictionary he wrote: "The Athenians begin their year with the summer solstice." Thus, the first day of the first Athenian month fell about the ninth day before the end of the corresponding Roman month. Then, by a curious inversion, the ninth day before the end of Anthesterion became for Copernicus the first of November. On the same flyleaf of his dictionary he also indicated that he had taken some statements about the months εκ των Θεωδορου Γαζα (ek ton Theodorou Gaza, without accents or breathing). The treatise *On the Months* by Theodore Gaza (1398–1478) appeared in a composite volume, the printing of which was completed in the Venetian workshop of Aldo Manuzio on 25 December 1495, with the material borrowed by Copernicus appearing on folio a 1ᵛ. The preceding treatise in this volume is Theodore Gaza's *Introductivae grammatices libri quatuor*. This introduction to Greek grammar in four books by Gaza was published here in the original Greek and was, as Manuzio himself admitted, at first sight without any appeal to anybody at Copernicus' modest level of competence as a student of the Greek language. Since he used the genitive plural ton (των) to specify the treatise from which he was borrowing, he consulted Gaza's treatise *On the Months*, but not Gaza's Greek grammar, which would have required a feminine singular, not a plural.

INTRODUCTION

Letter 26's farmer is leaving for more fertile soil, because Attica produced "myrtle instead of wheat, ivy instead of barley." Copernicus' Greek text continued: "and my threshing-floor was full of famine." This expressive utterance was omitted by Copernicus, who certainly had no difficulty understanding the Greek: the words are familiar, the syntax is simple. He did not even know that his Greek text itself had left out: "My flour-jar [is] empty," immediately in front of his own omission.[10] Surely the failure of edible crops adequately explained the farmer's decision to emigrate. The bareness of the threshing-floor, however paradoxically phrased, was an expendable detail, in Copernicus' judgment as a translator.

Letter 61 retells the familiar fable of the busy ant and the idle grasshopper: the latter enjoyed listening to its own music, while the ant followed the reapers, gleaned the threshing-floor, and buried its food. In Copernicus' Greek text, this ample contrast is followed by the explanation: "inasmuch as the ant has more foresight than the grasshopper." He eliminated this rather obvious and unnecessary explanation.

Letter 43's eunuch is called "The artificial little woman, the half-man διγενές (digenes), whose nature conforms completely to no" model. Copernicus did not translate διγενές (bisexual), because he deemed it superfluous.

In Letter 49, recalling Homer's *Iliad*, Book 24, Priam asks for his dead son's corpse. The slayer, called "the son of Thetis" and "the son of Peleus," respected Priam and also felt ashamed (ἠδέσθη, eidesthe; ᾐσχύνετο, eischyneto). In Copernicus' spare translation, "son of Peleus" disappears, and so does "felt ashamed."

Copernicus thought himself authorized not only to remove excrescences but also to alter content. For instance, Letter 85's visit to the cemetery is recommended for the purpose of "overcoming sadness" (λύπης κρατεῖν, lypes kratein). This mood is reversed, however, by Copernicus, who advises the visitor to go if "you wish to be overcome by sadness" (*tristitia vis teneri*).

Although Theophylactus lived in the Christian Byzantine Empire, he sprinkled an antique flavor over his *Letters* by invoking pagan divinities and religious conceptions in order to create an aura of authenticity. In Copernicus' nostrils, this was no mere aura, but the genuine breath of antiquity. This is why he left undisturbed the Gods (Letters 29, 67), Jupiter (Letters 22, 34), Pan (Letters 2, 29), the Pythia (Letter 33), Nymphs (Letter 29) and Cupids (Letters 36, 72, 78). On the other hand, Copernicus unintentionally obscured a Christian concept in Letter 25, where someone has recently died. His brother is advised to be resigned to the parting, while awaiting "the reunion" (αὖθις τὴν ἕνωσιν, authis ten henosin). Mistaking αὖθις (again) for αὐτίκα (autika, at once), Copernicus wrote *statim*: (*unionem statim expectans*). But if the union (*unionem*) is to take place at once (*statim*), there is no need to wait (*expectans*). Not only is Copernicus' mistranslation illogical, but it also negates Theophylactus' reference to an ultimate (not immediate) reunion with the dead.

By contrast, Copernicus deliberately removed two invocations of pagan Providence (Πρόνοια, Pronoia). Letter 7's rich uncle ignores his nephew's poverty, while lavishly entertaining strangers who flatter him. He is therefore warned that, unless he reforms, he will some day have as an implacable foe Theophylactus' pagan Providence, which Copernicus changed to Christian conscience (*paenitentiam*). Somewhat similarly, the misdeeds of Letter 73's penitent will be forgotten by Theophylactus' Providence (Πρόνοιαν), which Copernicus replaced by an impersonal forgetfulness (*oblivionem*).

When Copernicus noticed an error in his Greek text, he did not hesitate to correct it.

---

[10] Zanetto, p. 77/6–7.

An example of such editorial intervention on his part occurs in Letter 8. Its heading, like the heading of every other Letter, names two persons, the author and the addressee. The author, farmer Daphno, complains to his neighbor Myron, who is preventing the rain water from reaching Daphno's parched land. In a tone of exasperation, Daphno appeals to Myron: "For heaven's sake, ask the clouds whether they drop their water only for ..." Here the person named in Copernicus' Greek text was Daphno, an obvious mistake which he rectified by substituting Myron. Letters 73 and 82 contain similar blunders. Letter 73, according to its heading, is addressed to Archimedes, who is called Xanthippus in the body of the document. By the same token, Letter 82 is addressed to Alcibiades, who is called Antimachus later on. Less alert in these two instances than he had been in Letter 8, Copernicus retained these inconsistencies in Letters 73 and 82. Elsewhere, on the other hand, he solved a related problem neatly. The writer of Letter 38, Tettigo, warns Tettigo in Copernicus' Greek text. He sagely omitted the name of the person being warned.

According to Copernicus' Greek text, "Love's tears are delightful, for mingled with pleasure is madness" (ἀνοίας, anoias, Letter 84). As Copernicus' translation (*dementia*) shows, he did not realize that this weird pronouncement is merely the result of misprinting ἀνοίας for ἀνίας (anias, sadness). He was of course thoroughly familiar with the difference between these two words. Thus, he matched ἀνοίας correctly with *dementia* in Letter 79. He also correctly equated ἀνιᾷ (aniai) with *tristaris* (you are sad; Letter 16); ἀνιᾷ with *contristat* (saddens; Letter 76); ἀνίαν (anian) with *tristitiam* and *maestitiam* (sadness; Letters 28, 29); ἀνιώμενοι (aniomenoi) and ἀνιωμένων (aniomenon) with *contristati* and *contristatis* (saddened; Letter 21); and ἀνιᾷς (aniais, you sadden; Letter 37) with *contristas*. Yet he failed to see the necessity of replacing ἀνοίας by ἀνίας in Letter 84.

On the other hand, he felt no compunction about trying to improve his Greek text. Thus, Letter 1's cricket starts to sing "at the appearance of spring" (ἦρος, eros). The same expression recurs in Letter 71, where Copernicus matched it perfectly (*imminente vere*). But in Letter 1 he wrote *aurora lucente* (at the break of dawn), as though his Greek text read ἠοῦς (eous). He was tempted into substituting ἠοῦς for ἦρος because Letter 1's cricket sings louder "at the noon hour," when it is "intoxicated by the sun's rays." Hence, Copernicus reasoned, it begins to sing, less loudly, when the sun rises at dawn. But he was unfamiliar with the timing of the cricket's song in Theophylactus' *Natural Questions*. Copernicus was not even aware that the epistolographer had written this dialog, where he said:

> Some crickets are cold by nature. Hence they begin to sing after the [summer] solstice, and as the sun moves away [from the solstice], they emit their clamor, and at the noon hour they are more vocal (Chapter 13).

Since the *Natural Questions*' cold crickets begin to sing at the start of summer, presumably Letter 1's cricket begins "at the appearance of spring," not "at the break of dawn," as Copernicus' would-be emendation proposed.

Elsewhere his emendations were more successful. Letter 34's birds conduct a beauty contest. "The magpie, being concerned about its misshapenness, falsified Nature's handiwork by decking out its own 'beauty' with another's charm." Copernicus' Greek text read εὐπρέπειαν (euprepeian, beauty). But he translated as though this word were ἀπρέπειαν (aprepeian, ugliness), which he rendered by *indecentiam*. Exactly the opposite situation occurred in Letter 69. There old age, illness, and suffering are grouped as adversaries of *pulchritudini* (good looks) in Copernicus' translation. Yet his Greek text read ἀπρεπείας (aprepeias, ugliness), which he recognized as a misprint for εὐπρεπείας (euprepeias). In Letter 31, an author refuses to publish his writings. Hence a complaint is addressed to him on behalf of his would-be readers: You make us sad (*contristando nos*, in Copernicus' version). Yet his Greek text had the writer mistakenly making "you" sad (ὑμᾶς, hymas), which Copernicus translated as though it were ἡμᾶς (hemas, us). Letter 16's sailors in danger of shipwreck during a storm seek refuge in a harbor. They are compared to a debtor looking around on the streets and inspecting the doors in his desire θεραπεύειν (therapeuein) his creditors. For θεραπεύειν, Copernicus' dictionary offered seven equivalents, none of them suitable to this situation. Rejecting them all, therefore, he wrote *declinare ... iram* (avoid the wrath). He had no way of knowing that, instead of θεραπεύειν, some manuscripts read δραπετεύειν (drapeteuein, run away from), a verb used in Letters 33 and 79.[11]

These efforts to correct and improve his Greek text caused Copernicus to diverge from the Aldine edition. His other deviations from it resulted from his many misunderstandings as well as from his deliberate omissions for the purpose of presenting to his readers a more compact and less abstruse version. Taken together, these various explanations account for the numerous differences between Copernicus' translation of Theophylactus' *Letters* and the Aldine text of the *Letters*.

These differences were first explained by Hipler's statement (SpC, p. 73/22-24) that Copernicus' "translation of these letters, which quite often differs from the Aldine edition, seems to have been done partly in agreement with a manuscript other" than that on which the Aldine edition was based. Hipler, however, knew of no particular Theophylactus manuscript with which Copernicus agreed and the Aldine edition disagreed. In like manner P II, 45/10up–7up, simply assumed without discussion or proof that Copernicus had a Theophylactus manuscript which "must have agreed in many passages with the Aldine text," but differed from it where Copernicus did. P's unproved assumption was repeated half a century later[12].

In the absence of any Theophylactus manuscript possessing these Hipler — Prowe — Nissen characteristics, an alternative explanation of Copernicus' departures from the Aldine text was based on its use in the classroom. Urceo, according to his biographer, was the only professor at the University of Bologna who was qualified to teach Greek while Copernicus was a student there[13]. Hence at Bologna the editor of the Aldine *Epistolographers* was not Copernicus' Greek teacher. Had he been, he would not have introduced variant readings so soon after the publication of the *Epistolographers*. For, as we saw above (p. 4), Manuzio dedicated the *Epistolographers* on 17 April 1499. The next academic year began about six months later, and after so short a time the editor would not have had variant readings for his students. When this Italian biographer's chapter on Copernicus in Bologna was translated into German, at this point a translator's footnote[14] suggested that it would be quite natural for Urceo to depart from the Aldine edition (as though he were a modern philologist training advanced students). But Copernicus was quite far from being an advanced student of Greek during the last three months or so of the academic life of Urceo, who died on 11 February 1500[15]. Did Urceo teach the *Epistolographers* before his death? If so, did he go as far as Theophylactus, the thirtieth of about three dozen epistolographers? And if Urceo lectured on Theophylactus,

---

[11] *Ibid.*, p. 75/8up–7up.
[12] Theodor Nissen, "Die Briefe des Theophylaktos Simokattes und ihre lateinische Uebersetzung durch Nikolaus Coppernicus," *Byzantinisch-Neugriechische Jahrbücher*, 1937, 13:41/10up–7up.
[13] Carlo Malagola, *Della vita e delle opere di Antonio Urceo, detto Codro* (Bologna, 1878), p. 337/last 5 lines.
[14] *Mittheilungen des Coppernicus-Vereins für Wissenschaft und Kunst zu Thorn*, 1880, 2: 27/n. 1 (Maximilian Curtze).
[15] Malagola, p. 193/12up–8up.

did he discuss variant readings? Affirmative answers to all three of these questions would have to be documented if the classroom use of the *Epistolographers* were to be invoked as the reason for Copernicus' deviations from the Aldine edition.

These were explained in still a third way (in addition to a separate manuscript and classroom variants). For in a catalog of the printed books in the library of the University of Uppsala, Sweden, Ludwik Antoni Birkenmajer noticed Theophylactus' *Letters* listed separately from the Aldine *Epistolographers*[16]. When he asked for the separate Theophylactus, it could not be found. Nevertheless he gratuitously assumed that it had been printed separately from the *Epistolographers*, and thus differed from the Theophylactus section in the *Epistolographers*. By these putative differences Birkenmajer sought to account for Copernicus' divergences from the Aldine edition, and he rejected the Hipler — Prowe thesis of a separate manuscript[17]. But the separate listing of Theophylactus in the Uppsala catalog resulted from a bibliographical whim on the part of the Uppsala editor, Aurivillius, and not from the physical presence of a Theophylactus fascicle in Uppsala. That is why no Theophylactus fascicle was found in Uppsala when Birkenmajer asked for it. Hence this supposititious fascicle explained Copernicus' divergences from the Aldine *Epistolographers* no better than the supposititious Hipler — Prowe manuscript or the classroom variant readings.

Jerzy Kowalski felt that Birkenmajer's separate Theophylactus fascicle would be identical with the Theophylactus section in the *Epistolographers*. Kowalski's feeling was supported later on by the statement that the Aldine edition started each epistolographer at the top of a right-hand page, thereby facilitating the production of separate fascicles.[18] This description does not fit about a third of the Aldine epistolographers, who begin either on a left-hand page or on a right-hand page, but below the closing piece of the preceding epistolographer. Discarding Birkenmajer's separate fascicle as identical with the corresponding section of the Aldine *Epistolographers* and therefore incapable of accounting for Copernicus' divergences from the Aldine edition, Kowalski invoked cheap and inferior copies of Greek manuscripts in the hands of students in Copernicus' Bologna. But at the end of the fifteenth century in Bologna and other Italian university towns, Greek manuscripts were rare, expensive, and jealously guarded.

To Kowalski's misrepresentation of student reading matter in Copernicus' Bologna, Ivan Ivanovich Tolstoi, professor of philology at the University of Leningrad, added the unsupported notion that Urceo had access to a Theophylactus manuscript tradition earlier than the Aldine edition.[19]

According to the first comprehensive analysis of the manuscript tradition of Theophylactus' *Letters*, the Aldine edition was based on a manuscript (now lost), which was derived through an intermediary from a (now lost) archetype.[20] Seven extant manuscripts, however, constitute a related family. But Copernicus' deviations from the Aldine edition do not conform to any member of this family, or to any other of the numerous manuscripts subjected to scrutiny.

---

[16] *Catalogus librorum impressorum bibliothecae r. academiae Upsaliensis*, ed. Pehr Fabian Aurivillius (Uppsala, 1814), sectio prior, p. 910; sectio posterior, p. 132.

[17] MK, p. 123. When Birkenmajer's proposal was brought to the attention of Nissen, he shifted from Prowe's explanation requiring a separate manuscript to Birkenmajer's separate fascicle; "Zur Theophylakt-Uebersetzung des Coppernicus," *Byzantinisch-Neugriechische Jahrbücher*, 1938, *14*: 41/13up–9up.

[18] Jerzy Kowalski, "Kopernik jako filolog i pisarz łaciński," in *Mikołaj Kopernik* (Lvov/Warsaw, 1924), p. 136. Ryszard Gansiniec's introduction to *Teofilakt Symokatta, Listy* (Wrocław, 1953), p. XVII.

[19] *Nikolai Kopernik*, ed. Aleksander Aleksandrovich Mikhailov (Moscow/Leningrad, 1947), p. 76.

[20] Zanetto, p. 64/n. 1; p. 78/n. 20; p. 86.

This recent analysis of the filiation of about three dozen manuscripts of Theophylactus' *Letters* explains why the Hipler — Prowe — earlier Nissen thesis has never been substantiated by pointing to a particular manuscript accessible to Copernicus. Nor did he ever have a cheap and inferior student copy of Theophylactus' letters, nor a separate fascicular imprint of them. Neither did he hear variant readings expounded by Urceo in the classroom. But Copernicus did depart from the Aldine edition when he could not understand it as a result of its shortcomings or his own, and also when he deliberately altered it to suit his own taste and that of his readers. The only Greek text of Theophylactus' *Letters* ever seen by Copernicus was the Aldine. He realized that it was less than perfect. But he had no access to, indeed no knowledge of, any superior or different Greek text. It was on the Aldine *Epistolographers*' text, and on it alone, that Copernicus based his Latin translation of Theophylactus' *Letters*.

V

THE FIRST EDITION OF COPERNICUS' TRANSLATION OF THEOPHYLACTUS' *LETTERS*

After Copernicus returned from Italy in 1503 to take up his duties as a canon, he managed to find the time to translate Theophylactus' *Letters*. This would become the first independent translation of a Greek author to be printed in Poland if Copernicus could put it in the hands of a publisher. But there was as yet no printing press in Frombork, the seat of the Cathedral Chapter to which Copernicus belonged as a canon. Nor was there a press in Lidzbark, the residence of the bishops of Varmia, where Copernicus assisted his uncle, Bishop Lucas. In fact, the relatively recent invention had not yet established itself anywhere in the entire diocese of Varmia. Even Gdańsk, the busy Baltic seaport profiting from its membership in the Hanseatic League, in 1503 lost its printer, who moved to Wrocław; his successor published only a few works in German.

Toruń, Copernicus' birthplace, had no printer at all. But the municipal secretary was Lawrence Corvinus (c. 1465-1527), whom Copernicus had known when they were both in the University of Cracow. Seven and a half years before Copernicus, Corvinus entered the university in the summer semester of 1484. The official register lists him as *Laurentius Bartholomei de Novo foro*[21] (Lawrence, son of Bartholomew, of Neumarkt). This town is associated with Wrocław by a footnote in the register in order to prevent it from being confused with other towns bearing the same name (equivalent to Newmarket). Corvinus' Neumarkt, now called Środa Śląska, is situated about twenty miles northwest of Wrocław. His surname, it will be observed, is not recorded in the register. Later on, after achieving fame as a writer and poet, he followed the humanist fashion of the time by dropping his surname Rabe (Raven, in German) in favor of Corvinus, the closest Latin equivalent. At the university he received the bachelor of arts degree in September 1486, and the master's degree in the winter semester of 1488-1489.[22] His academic ability was recognized when the university appointed him as an instructor in the winter semester of 1489-1490.[23] He was not a member of the regular faculty,

---

[21] *Album studiosorum universitatis cracoviensis*, I (Cracow, 1887), 260, ed. Adam Chmiel.

[22] *Statuta nec non liber promotionum philosophorum ordinis in universitate studiorum Jagellonica ab anno 1402 ad annum 1849*, ed. Józef Muczkowski (Cracow, 1849), pp. 96, 103.

[23] *Liber diligentiarum facultatis artisticae universitatis cracoviensis*, ed. W. Wisłocki (Cracow, 1886), p. 11.

but was classified as *extraneus de facultate*, roughly equivalent to our "adjunct." He taught during the summer semester of 1490 and the winter semester of 1490–1491. Whether he conducted a class during the summer semester of 1491 is not known, because the relevant records are missing. In the winter semester of 1491–1492, when Copernicus entered the university,[24] Corvinus was scheduled to teach, but did not do so for a reason that is not specified. This may be connected somehow with the absence of his name from the list of instructors during the summer semester of 1492. In the winter semester of 1492–1493 he is back again on the list. In the summer semester of 1493 he was scheduled to discuss Aristotle *On the Soul*. But he often talked about something else (the official record notes), perhaps because at the same hour, 4 P.M., in another room another instructor was scheduled to talk about Aristotle *On the Soul*. Corvinus appears on the roster of instructors for the last time in the winter semester of 1493–1494.[25]

At the end of that semester he planned to leave Cracow, as he made clear in a petition which he presented to the university rector. Several years before, a student from Wrocław had been thrown into jail. Acting jointly with an associate, Corvinus had the student released by promising on 27 August 1491 to stand surety for him. Now, on 5 May 1494, Corvinus sought to compel his associate to pay half of the fine. Meeting resistance, on the following day Corvinus deposited an appropriate sum of money and named a suitable proxy to act for him in his absence.[26]

Returning to his native Silesia, in 1495 Corvinus became the municipal secretary of a town sixty miles southwest of Wrocław, and also taught school there.[27] Later on he moved to Wrocław, where in 1499 he was the principal of the local school. He also qualified as a notary public, licensed by the Holy Roman Emperor to practice that profession.[28] In this capacity he served the diocese of Wrocław and then, giving up his teaching career, in the summer of 1503 he became an undersecretary of the city of Wrocław.

Between Copernicus' matriculation in the autumn of 1491 and Corvinus' departure in the spring of 1494, there were three semesters when Corvinus was a teacher and Copernicus was a student. Did Copernicus ever sit in a class taught by Corvinus at the University of Cracow? This question has often been answered affirmatively, despite the absence of hard evidence. Moreover, when Corvinus had a ready-made opportunity to claim Copernicus as a former pupil, he refrained from doing so. Nevertheless, there can be little doubt that the two men came to know each other while they were both at the University of Cracow.

Unlike Corvinus, Copernicus did not stay at the University of Cracow long enough to earn a degree. Exactly when he left is still an unresolved question. In any case, by 22 February 1496 he was back in his uncle's palace in Lidzbark, witnessing a power of attorney. Toward the end of 1502 he was appointed scholaster of the Church of the Holy Cross in Wrocław.[29] This office was a sinecure, whose duties were performed by a vicar designated by Copernicus. He himself never set foot in Wrocław. Yet he may have played a part in Corvinus' move from

---

[24] *Album*, II (Cracow, 1892), 12, ed. Chmiel.

[25] *Liber diligentiarum*, pp. 13, 16, 19, 23, 24, 26.

[26] *Acta rectoralia almae universitatis studii cracoviensis*, I (Cracow, 1893), p. 321, no. 1464; pp. 371–372, no. 1665–1667, ed. W. Wisłocki.

[27] Gustav Bauch, "Laurentius Corvinus, der Breslauer Stadtschreiber und Humanist," *Zeitschrift des Vereins für Geschichte und Alterthum Schlesiens*, 1883, *17*:230–302; 1898, *32*:390–391; here, *32*:391, *17*:248.

[28] *Ibid.*, p. 261.

[29] See Copernicus' Scholastry, under Administrative Documents, below.

INTRODUCTION

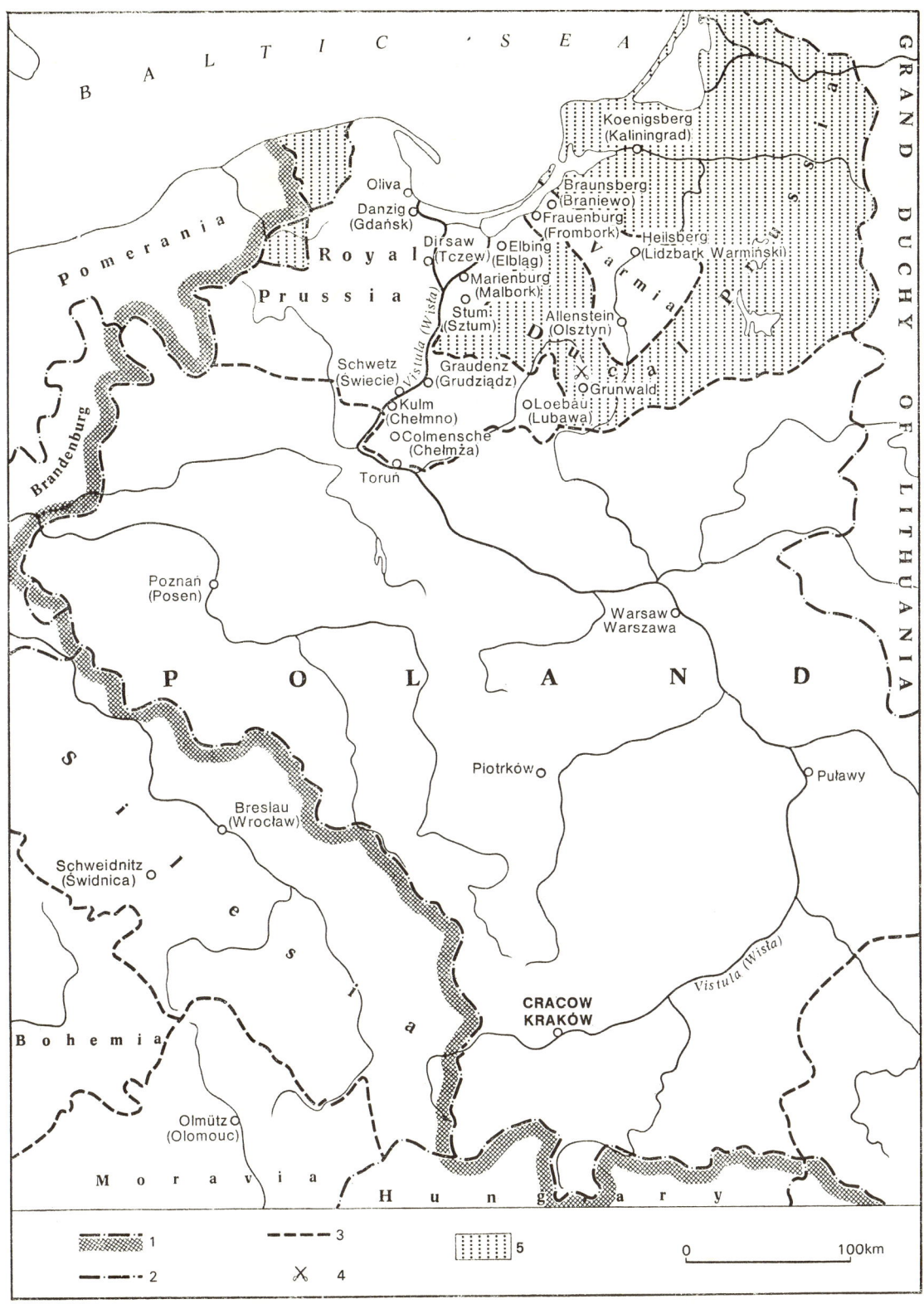

1. Polish Places in the Life and Work of Copernicus (ca. 1526)

1 — Boundaries of the Jagiellonian Commonwealth, 2 — Boundaries of States, 3 — Boundaries of Lands, 4 — Battlefield of Grunwald (1410), 5 — Fiefs

Wrocław to Toruń. On 18 June 1506 the City Council of Wrocław gave Corvinus an unqualified recommendation, armed with which he proceeded to his new post.[30]

Corvinus did not remain long in Toruń. His wife Anna, whom he dearly loved, became acutely homesick for Silesia, and longed to see her friends and relatives again. Hence, when a vacancy occurred in the higher ranks of the municipal secretariat in Wrocław, Corvinus decided to return to the Silesian capital, where he and his wife owned their own house. This sudden turn of events brought Copernicus in a great hurry to Toruń, carrying with him the manuscript of his translation of Theophylactus' *Letters*. He discussed with Corvinus the possibility of having the work published in Cracow by Johann Haller (c. 1467–1525).

It was in Haller's bookshop that Copernicus, while he was a student at the University of Cracow, bought his own copy of the second edition of the *Alfonsine Tables* (Venice, 1492), and had it bound by a local craftsman.[31] Like Corvinus and Copernicus, Haller too had been a student at the University of Cracow. *Johannes Johannis Haller de Werdea*[32] (Johann Haller Jr., son of Johann Haller Sr., of Donauwörth) was admitted in the winter semester of 1482. Like Copernicus, but unlike Corvinus, Haller left the university without obtaining a degree. Plunging into the world of trade, he bought and sold copper, tin, and wine. With the handsome profits derived from these transactions, he branched out into the book business, at first as a bookseller and then also as a publisher. Acquiring his own printing equipment, and exempted from taxes by a royal decree dated 25 February 1506 and renewed on 14 March 1507,[33] Haller soon became the most prolific publisher in Poland.

When Corvinus left Toruń, he took Copernicus' manuscript with him. The coach carrying Corvinus and his wife needed four days to go from Toruń to Wrocław. During that tedious trip he whiled away the time by reading the manuscript, which pleased him very much. By 19 June 1508 Corvinus was back at work in Wrocław.[34] In due course he arranged with Haller the publication of Copernicus' translation of Theophylactus' *Letters*, for which he himself wrote an introduction in the form of a long poem.

The entire production was carried out in Haller's own house, where he maintained his printing establishment. Had his employees read the proofs, the customary list of corrections would have been inserted at the end of the volume. Neither Corvinus in Wrocław nor Copernicus in Frombork received any sheets to be examined for printer's errors. That is why ludicrous mistakes were allowed to remain. For example, Letter 27 attributes to objects which are inanimate (*inanimatis*) emotional reactions felt by animate beings. But instead of *inanimatis*, *mammatis* (breasted) was printed. As an experienced author, two of whose works passed through twenty-five editions, and as a professional draftsman of official documents and diplomatic correspondence, Corvinus would never have permitted such an absurdity to appear in a book with which his name was connected. Copernicus would have been equally eager to purify the translation, his first publication, had he ever had the opportunity to scrutinize the proofs of the volume, which Haller issued in 1509 as his 63rd publication.[35]

---

[30] Bauch, p. 264.

[31] MK, between pages 26 and 27, prints two plates showing the front and back of the binding of Copernicus' copy of the *Alfonsine Tables*. See also Paweł Czartoryski, "The Library of Copernicus," in *Science and History*, StC XVI, 366, no. 2: *Alfonsine Tables*; no. 3: Greek-Latin dictionary.

[32] *Album*; I, 252. Its misprint "Werdra" was corrected by Józef Seruga, *Jan Haller wydawca i drukarz krakowski* (Cracow, 1933), p. 14.

[33] Joannes Ptaśnik, *Cracovia impressorum XV et XVI saeculorum* (Lvov, 1922; *Monumenta poloniae typographica*, I), Materiały źródłowe, p. 48, no. 108; p. 50, no. 112.

[34] Bauch, p. 266.

[35] *Polonia typographica saeculi sedecimi*, IV (Wrocław, 1962), Introduction, p. 47, misprints the epistolog-

INTRODUCTION

The explanation of the *inanimatis/mammatis* printer's error is to be found in Copernicus' faulty handwriting.[36] He wrote out his translation himself, without employing any professional assistant. The same holds true for his *Revolutions*, the original manuscript of which is still preserved. Every one of its 212 folios, recto and verso, is unmistakably in Copernicus' handwriting, as is evident in the photofacsimile published in NCCW, I. In like manner, he attended to his correspondence himself, his surviving letters being written with his own hand. We hear nothing about his having been helped throughout the seventy years of his long life by any secretary, scribe, copyist, or amanuensis. The outward circumstances of his career, therefore, suggest that what is true of the *Revolutions* and of his letters is likewise true of his Theophylactus manuscript: he wrote out the translation with his own hand.

But, by contrast, as the municipal secretary of the capital of Silesia, Corvinus did not lack the services of a professional copyist. His introductory poem was sent to Haller as an instantly legible manuscript. It was accordingly printed in Haller's house without any of the gross typographical errors that mar Copernicus' translation and in many cases are due to the peculiarities of his handwriting.

These printer's errors would have been diminished or eliminated, had Copernicus "himself supervised and taken care of the printing and proofreading." Supposedly he did so, according to Hipler,[37] who misunderstood Copernicus' report of his observation of the lunar eclipse of 2 June 1509[38]. Copernicus says that he observed this eclipse "on that same meridian of Cracow" (*sub eodem meridiano Cracoviensi*),[39] recalling his remark in *Revolutions*, IV, 7, that "Frombork, where I generally made my observations ... lies on the meridian of Cracow."[40] Hence the eclipse of 2 June 1509 was observed by Copernicus in Frombork on the Cracow meridian, but not in Cracow itself.[41] Since he was not in Cracow in 1509, he did not "himself supervise and take care of the printing and proofreading" of his Theophylactus translation. That is why it is marred by so many typographical errors.

Their presence convinced PI¹, 366/n.*, that Copernicus did not supervise the printing of his translation. Yet P (mistakenly) said that Copernicus handed his translation to Haller's print shop in Cracow,[42] despite Copernicus' previous delivery of it to Corvinus in Toruń (PI¹, 388/4–6). If Copernicus handed his translation to the Cracow print shop, why did he not supervise the printing? The reason, according to P, was that Copernicus was concerned with the eclipse of 2 June 1509. Even though P understood that Copernicus' *sub eodem meridiano Cracoviensi* can be referred to Frombork, P still preferred to place Copernicus in Cracow on 2 June 1509. For this preference P invoked Copernicus' personal feeling "that he would not fail to use the opportunity to observe the lunar eclipse with his scientific friends."[43]

---

rapher's name (Theophylact*es*) and also dates the release of the book in December without disclosing any reason for choosing that month in 1509.

[36] Additional misprints in the 1509 edition that are due to Copernicus' handwriting are discussed below in Letter 12, n. 2; Letter 13, n. 2; Letter 27, n. 4; Letter 34, n. 1; Letter 64, n. 3; Letter 68, n. 1; Letter 75, n. 5; Letter 78, n. 2; Letter 79, n. 3; Letter 81, n. 1; and Letter 84, n. 1.

[37] SpC, p. 73/21–29.

[38] SpC cites *Revolutions*, IV, 17, instead of IV, 13 (NCCW, II: 199/31–32).

[39] NCOO, II: 203/33.

[40] NCCW, II: 191/30–32.

[41] The observation of this eclipse by Copernicus was misplaced in Cracow by MK, p. 317/No. 10, and StC VIII, 52/No. 59. Had Copernicus made this observation in Cracow, he would have written *Cracoviae*, not *sub eodem meridiano Cracoviensi*.

[42] PI¹, 365/5up–4up, repeated by StC VIII, 51/No. 58/n. 1.

[43] PI¹, 366/n. **/last 3 lines.

But Copernicus himself said that "Frombork ... lies on the meridian of Cracow, as I learn from lunar ... eclipses observed simultaneously in both places."[44]

Corvinus' introductory poem throws a welcome light on the obscure question of Copernicus' intellectual development. In lines 27-30 the poet says about Copernicus:

> He discusses the swift course of the moon and the alternating movements of its brother as well as the stars together with the wandering planets — the Almighty's marvelous creation — and he knows how to seek out the hidden causes of phenomena by the aid of wonderful principles.

These wonderful principles (*miris ... principiis*) have often been identified with the main features of the Copernican astronomy. If this identification were correct, it would put the time when Copernicus discovered his new system not later than the first half of 1508.

In the Copernican system, however, the sun is stationary and has no alternating movements (*alternos ... meatus*). These belonged to the pre-Copernican sun, which was therefore regarded as the moon's brother (*fratris*). For example, Ambrose's *Hexameron* (IV, 7) labels the sun the moon's brother (*frater*). When Corvinus wrote that Copernicus' wonderful principles treated the sun as the moon's brother, the poet echoed Ambrose and did not announce the new astronomy. Precisely when that epoch-making discovery was made by Copernicus is a question that still awaits an answer. What Corvinus implies is that the Copernican astronomy was born not before the latter half of 1508 at the earliest.

VI

THE LATER EDITIONS OF COPERNICUS' TRANSLATION OF THEOPHYLACTUS' *LETTERS*

When Copernicus' translation of Theophylactus' *Letters* was first published in 1509, it attracted very little attention. More than a century later, when Simon Starowolski (1588-1656) published the enlarged version of his biography of Copernicus in 1627, he evidently had not heard of the translation. Had he ever seen a copy, he certainly would have refrained from saying that Copernicus had a "perfect knowledge of the Greek" language (StC XXI, 14). What was first noticed in the 1509 publication was Corvinus' poem, which was discussed by two works on Silesian writers.[45] Then a description of the rare items in the royal library in Dresden[46] reported Copernicus' translation of Theophylactus for the first time, and remarked on its absence from the then standard bibliography of Greek literature. From this Dresden report and a Polish bibliographer,[47] Copernicus' translation finally passed into an authoritative inventory of early printed books,[48] and thence into scholarly notice.

The only one of Theophylactus' three works to be printed during the century that saw the birth of that invention was his *Letters*, when they were included in the Aldine *Epistolog-*

---

[44] NCCW, II: 191/30-33.
[45] Martin Hanke, *De Silesiis indigenis eruditis* (Leipzig, 1707), pp. 204-205; Johann Jacob von Füldener, *Bio- et Bibliographia Silesiaca* (Lauban, 1731), p. 359.
[46] Johann Christian Götze, *Merckwürdigkeiten der k. Bibliothek zu Dresden*, II (Dresden, 1743), 6.
[47] Jan Daniel Janocki, *Janociana*, I (Warsaw/Leipzig, 1776), 45-46.
[48] Georg Wolfgang Panzer, *Annales typographici*, VI (Nuremberg, 1798), 452/No. 33.

raphers of 1499. Thereafter Theophylactus was ignored for nearly a century. Then Bonaventura de Smit (1538–1614), who latinized his surname as Vulcanius and was appointed professor of Greek at the University of Leiden, brought out the first edition of the Greek text of Theophylactus' *Natural Questions* (Leiden, 1596, 1597). This was accompanied by the second printing of the Greek text of Theophylactus' *Letters*.[49] Meanwhile his complete works, comprising his *History* in addition to the other two shorter pieces, were being prepared for publication in Heidelberg. When Vulcanius' Theophylactus became available in the bookshops, however, the Heidelberg publisher abandoned his overlapping project. But it was revived by Jan Gruter (1560–1627), professor of history at the University of Heidelberg, who did more than edit the complete works of Theophylactus in Greek (Heidelberg, 1599). For as the son-in-law of his colleague Jacob Kimmendonck Sr., Gruter had access to the literary remains of Jacob Kimmendonck Jr., a precocious youth who died of the plague in 1596 at the age of 17 years, 9 months, and 8 days. The younger Kimedoncius or Cimedoncius (as his surname was latinized) had left in manuscript his Latin translation of Theophylactus, which was printed with Gruter as editor (Heidelberg, 1598). This volume contained the second Latin translation of Theophylactus' *Letters*. Neither the translator Cimedoncius nor his editor Gruter had ever heard of the first Latin translation of Theophylactus' *Letters* by Copernicus.

The same may be said of the third translator, whose version was printed in *Epistolae graecanicae* (Geneva, 1606), pp. 397–416.[50] Even when Jean-François Boissonade (1774–1857)

---

[49] Besides editing the Greek text of Theophylactus' *Letters*, Vulcanius was also concerned with translations of them into Latin. He made (or acquired) translations of Letters 1–4, in addition to earlier translations of Letters 34 and 70; see Philip Christiaan Molhuysen, ed., *Bibliotheca universitatis Leidensis Codices manuscripti*, I, Codices Vulcaniani (Leiden, 1910), 11, 51. Letter 34, ascribed by Theophylactus to the Athenian political leader Themistocles, had been translated presumably for inclusion among the letters attributed to him. But it was pronounced spurious when published in the Leipzig 1710 edition of his letters (Preface, ¶ 12). As regards the "collection of twenty-one letters ostensibly written by Themistocles to his friends and associates ... their authenticity was first called into question ... in the early part of the seventeenth century" (Anthony J. Podlecki, *The Life of Themistocles* [Montreal/London, 1975], p. 129). Their spuriousness is no longer challenged. In Theophylactus' Letter 70, Plato writes to Axiochus, an interlocutor in a dialog attributed to Plato but classified as spurious as early as Diogenes Laertius (III, 62). Letter 70 was printed by Johann Konrad Orelli (1770–1826) in his edition of the letters attributed to Socrates and the Socratics, Pythagoras and the Pythagoreans (Leipzig, 1815, pp. 447–451), together with Letters 3, 22, 40, 52, 64, 82, 85. On the other hand, Letter 79, assigned by Theophylactus to the Attic orator Isocrates, was added in some manuscripts to his nine authentic letters, and printed in some editions of his speeches and letters; this supposed tenth letter "is rightly rejected by all modern editors"; see Leslie Francis Smith, *The Genuineness of the Ninth and Third Letters of Isocrates* (Lancaster, Pennsylvania, 1940), p. 43. Thus, Copernicus was not alone in failing to pierce the veil drawn by Theophylactus over his imaginative attributions.

Eilhard Lubin (1565–1621), a professor at the University of Rostock, published *Epistolae veterum graecorum* (Heidelberg, 1597, 1608), containing not only the text of ancient Greek letters but also his own translation of them into Latin. While preparing his edition, he translated into Latin the Greek letters of Theophylactus, whom he omitted, however, from the printed edition. Lubin's translation, dated 1595, survives in manuscript form in the Vatican Library; see *Codices manuscripti palatini graeci Bibliothecae Vaticanae* (Rome, 1885), p. 272/No. 419/fol. 180ᵛ–208.

[50] The Geneva editors claimed that they had received the translation as a work done mostly by a famous authority on Roman law. Four years after the appearance of their edition in 1606, however, their claim was denounced as a literary fraud:

> Under the name of Jacques Cujas [1522–1590], the most renowned of the students of jurisprudence in this age, ... certain works are published of which he never thought even in his dreams. Of this kind is that Latin translation of the Greek letters which certain persons ascribed to Cujas by lying and falsifying with unprecedented impudence. Under so illustrious a name they published a version by some obscure

published the first critical edition of the Greek text of Theophylactus' *Letters* (Paris, 1835), he reprinted Cimedoncius' translation without mentioning Copernicus', of which he too had never heard. When Anton Westermann (1806-1869) was preparing a new edition of the Greek epistolographers, he reworked the previous Latin translations. But finding himself too ill to complete the edition, Westermann turned over his material to his successor Rudolf Hercher (1821-1878). Hercher's *Epistolographi graeci* (Paris, 1873) printed Westermann's translation and critical apparatus, based on Boissonade's edition and supplementary notes; besides his two main manuscripts, Paris 2991 (which was copied in 1419 and lacks Letter 30) and Paris 3047 (which was copied in 1420 and lacks 36 of the 85 letters), a small number of additional manuscripts was consulted (Hercher, pp. lxxxii-lxxxv, 763-786). Like Boissonade's, the Westermann-Hercher edition made no mention of Copernicus' translation of Theophylactus' *Letters*.

Although this was overlooked by classical philologists, it was resuscitated by a Copernican scholar. Whereas the persons responsible for the first three editions of Copernicus' *Revolutions* (Nuremberg, 1543; Basel, 1566; Amsterdam, 1617) knew nothing about his translation of Theophylactus' *Letters*, Jan Baranowski (1800-1879), who brought out the fourth edition of the *Revolutions* (Warsaw, 1854), not only reprinted Copernicus' translation but also added his own Polish version (pp. 595-631). This was reprinted in its entirety by Ignacy Polkowski (1833-1888) in his *Kopernikijana*, I (Gniezno, 1873), 1-32. A part of Baranowski's Polish translation was reproduced in *M. Kopernik: Wybór pism* (Cracow, 1920), edited by L. A. Birkenmajer (pp. 120-124). Franz Hipler (1836-1898) reprinted Copernicus' Latin translation in his *Spicilegium Copernicanum* (Braunsberg, 1873), pp. 74-102. In his *Nicolaus Coppernicus* (Berlin, 1883-1884) Leopold Prowe (1821-1887) included a Greek text of Theophylactus' *Letters* facing Copernicus' translation (II, 52-127) as well as his own German version of ten Letters (I, 1, 395-397).[51] This biography of Copernicus by Prowe was reissued (Osnabrück, 1967) with its numerous errors left uncorrected. The Haller edition of 1509 was reproduced in photofacsimile (Wrocław, 1953), edited by Ryszard Gansiniec (1888-1958), with Polish translations of Theophylactus' Greek by Jan Parandowski (1895-1978) and of Corvinus' Latin by Ludwik Hieronim Morstin (1886-1966). A German translation of all of Theophylactus' *Letters* was included in Bernhard Kytzler (1929-), *Erotische Briefe der griechischen Antike* (Munich, 1967), pp. 227-270. What follows in the present volume should be viewed, not so much as the first English translation of Theophylactus' *Letters*, but rather as a translation of them as they were refracted through the prism of Copernicus' Latin version.

and barely educated German. They obtained it from François Pithou [1543-1621], who often showed it to me in his library, and said he had by some chance found it in Basel (Isaac Casaubon, Preface, sig. e 3r-v, in Joseph J. Scaliger, *Opuscula varia*, Paris, 1610).

[51] From Prowe's German version, the last three Letters were retranslated into English by Arthur Koestler, *The Sleepwalkers* (London, 1959; Pelican reprint, 1968, p. 140). Koestler described (p. 143) Theophylactus' *Letters* as full of Christian piety. This trait has not been noticed by anybody who has actually read Letters 1-82.

THEOPHYLACTUS SIMOCATTA, THE MORALIST
HIS ETHICAL, RUSTIC, AND LOVE LETTERS
IN A LATIN TRANSLATION

LAURENTIUS CORVINUS

Notary of the city of Wrocław, the capital [of Silesia]

A
POEM

Wherein he bids farewell to the Prussians, and
Describes how much pleasure the following letters of Theophylactus afforded him, and
How sweet it is for an exile from his native soil to return to his fatherland

O Prussia, land on which gaze the northern Bear with its shining stars and the bright Wain together with the Bear-keeper, land which possesses fertile soil, rivers and bays stocked with fish, hills covered by vines and rich in cattle, farewell! Where you are washed by the tide of the far northern sea, you collect precious amber, the tears of Helios' daughters. Not content with your own wealth, you gather greater riches from alien shores by land and sea.

Outstanding among Prussian cities, O Toruń, under an auspicious sign twice and thrice farewell! With unstinted generosity you cherished me while the golden sun twice traversed its curved path through the celestial signs. May your Council be happy and your people safe! May your every citizen lead a joyful life! When for your favors and the honor of your kindnesses worthy thanks cannot be returned by me, yet will you forever be glorified by me, as long as the River Vistula glides past your walls. Under a friendly star Atlas' grandson[1] fitly attends you, who as bountiful mother give birth to most splendid men.

Among them is conspicuous for his piety Lucas [Watzenrode], bishop and prelate, revered for his grave demeanor. Subject to him is Varmia, an extensive division of the land of Prussia that rightly rejoices under his rule. Linked to him, like faithful Achates to Aeneas, is the scholar who translates this work from Greek into Latin. He discusses the swift course of the moon and the alternating movements of its brother as well as the stars together with the wandering planets — the Almighty's marvelous creation — and he knows how to seek out the hidden causes of phenomena by the aid of wonderful principles.[2]

I make no mention of very many others since our driver has loosened the reins and does not check his steeds. He had been bidden to speed his swift wheels by Anna, my spouse, because of her deep affection for her exceedingly sweet fatherland. Hence, again and again, farewell, Toruń, and may your glory be ever enhanced by your deeds!

Thence through the pines of a fragrant grove, through deep woods, over valleys and many a ridge am I borne, and through towns subject to your mighty rule, O King Sigismund,³ am I transported across your broad realms. Traversing these spacious regions in three days, thus do I joyfully hasten toward the domiciles of my forefathers. However tedious the trip, still all by itself the perusal of the Hellene Simocatta notably shortened the journey, until I reach the boundaries of the territory of Poland. This is severed from our fields by a meandering stream [Prosna], a sandy stream, winding through rocks strewn with rushes, that adds to the somnolent sound the swish of its waters. Here we take shelter beneath the exceedingly scanty cover of a humble hut and rightly feed our weary horses. A generous beldam brings us in a basket of twigs morsels of cheese, not yet thoroughly dried. But since this product of the farm contained no fluid, I quench my parched thirst at a nearby torrent. Cupping her hands, Anna too lifts the silvery drops, watched as they wash the soil of Silesia, and exclaims: "Goblets of the winter's vintage did I fancy myself imbibing, so sweet to my gullet is this fluid."

When the fourth dawn breaks, we are received by Silesia with its gentle breeze and more propitious sky. Here my spouse, having reached the land she had so long yearned for, greets her native ground with the following discourse:

> Hail, O land of my salvation, desire, and joy, which nourished me from infancy with profound solicitude. Welcome me returning from the confines of the land of Prussia, and be gracious to your daughter with your customary kindness. As long as I was far away from you, naturally I clutched you to my mindful breast, the remembrance of you being sweet. Had Prussia furnished me with a sail-studded harbor, and a dark vessel brought me foreign treasure; had the Hermus lavished its auriferous waters on me or the Caucasian Mountains poured down on me the gold gathered from the driving rain by the realm of Colchis as I, who have never been to school, have often been told by my loving husband, you, dear parent, are far more agreeable to me, and nothing exists in the world which I esteem more highly than you.

Having uttered these words, with a vow of an offering she cheerfully asks the cherished local deities for their habitual aid. Thereupon I too spoke:

> O ye gods on high, most gentle divinities, under whose rule this land flourishes, and likewise you, O saintly Jadwiga,⁴ under whose sway in former times Silesia lay, and now the entire kingdom of heaven reclines, bring it to pass that the lord of almighty Olympus keeps us safe and restores us to our forefathers' hearth.

While I am speaking, the cloudy peaks of the mountain beyond loom into sight, as well as a dim tower atop the ridge. Little by little are revealed the lofty walls of Wrocław, rising toward the round moon. Thence with bounding step we approach the mighty city as the fading light withdrew into the western waters. And after our dear friends congratulate us on our return, we seek out the charming rooms of the house we left behind. Here, where the fish-laden Oława⁵ twists through seven turns and its gushing waters sweetly murmur, may a gracious god grant me leave to occupy this domicile securely for many a year with my cherished wife. Let some other exile seek his fortune amid the northern storms; travel to the Strait of Gibraltar, sailing past the shriveled traces of the burning [Tropic of] Cancer and beyond the parched places of the low-lying beach; press on toward the [southern constellation] Altar, famous for its brilliant stars; behold the hidden signs of the South Pole; toss about on a hitherto uncharted sea; bring us the wrinkled pepper from the other hemisphere; seize numerous acres in far-off lands; and heap up vast riches as a wealthy alien. Let me have quiet repose at home, a glowing hearth, and nourishment to still my hunger. Sweeter is it to possess a paltry plot in the fatherland than to cultivate soil abroad with a hundred oxen.

That the letters of sage Simocatta may circulate widely, they have been cast in the printer's mold. The first teaches morals; the second, rustic life; the third, love. Thus is the work continuously fabricated of intertwining branches whence, like variegated blossoms from a well-watered sprig, the reader may pluck a bouquet of virtues.

A LETTER OF NICHOLAS COPERNICUS TO THE RIGHT REVEREND LUCAS [WATZENRODE], BISHOP OF VARMIA

Most excellently, as it seems to me, O right reverend ruler and father of our country, did the moralist Theophylactus commingle ethical, rustic, and love letters, according to the principle that variety usually gives greater pleasure than anything else. Indeed, different minds delight in different things, since some enjoy seriousness, others gaiety, some austerity, others fantasy, with each one rejoicing in his own favorite. Theophylactus so interspersed the gay with the serious, and the playful with the austere, that every reader may pluck what pleases him most in these letters, like an assortment of flowers in a garden. Yet Theophylactus disposed so much of value in all of them that they seem to be not letters but rather laws and rules for the conduct of human life, clearly by reason of their compactness. He collected them from various authors as the briefest and most effective.

With regard to the ethical and rural letters, it may be, nobody at all will raise any questions. The title of the love letters, however, seems to portend licentiousness. Nevertheless, just as physicians usually moderate the bitterness of drugs by sweetening them to make them more palatable to patients, so these love letters have in like manner been rectified, with the result that they ought to receive the label "moral" no less [than the others]. In view of these considerations, deeming it improper that these letters were available only in Greek and were not more accessible in Latin, I have undertaken to translate them into Latin to the best of my ability.

To you, Right Reverend Bishop, do I dedicate this modest gift which, however, can by no means be compared to your generosity. For, everything of this sort which my meager talent attempts or produces may be properly considered yours, if that is true (as it surely is) which Ovid once said to Germanicus Caesar:

In rapport with your mien, my inspiration stands or falls.[6]

THEOPHYLACTUS SIMOCATTA, THE MORALIST
ETHICAL, RUSTIC, AND LOVE LETTERS
IN LATIN TRANSLATION

1. Ethical. From CRITIAS to PLOTINUS

The cricket is a musical being.[1] At the break of dawn[2] it starts to sing. But much louder[3] and more vociferous, according to its nature, is it heard at the noon hour, because[4] intoxicated by the sun's rays. As the songster chirps,[5] then, it turns the tree into a platform and the field into a theater, performing a concert for the wayfarers.

Accordingly, I too am impelled to celebrate your virtues, because they inspire and absolutely incite me to praise you. For once, while I lay perishing in a foul existence, by your letters you brought me back to a virtuous [life]. So may I, Critias, become Plotinus. Either disembodied,[6] he philosophizes on earth, or philosophy, become incarnate, lives among mankind in the guise of a human being.

2. Rustic. From DORCON to MOSCHON

The leader of my flock, a marvelous ram, died and my beasts are bereft of the head of their herd. I sustain a huge loss, and I presume that [the god] Pan is somewhat angry with me since I failed to honor[1] him with the first fruits of the beehives. Hence I am on my way to town to appease his wrath. And I shall tell the townsmen about his cruelty by saying: "For the sake of a honey-cake Pan destroyed the leader of my flock."

3. Love. From THEANO to EURYDICE

Your natural beauty has faded and your good looks[1] are approaching [the stage of] wrinkles. But you try to give the impression of truth when you deceive[2] your lovers with artificial cosmetics. Hearken to time, you hag, for in the autumn meadows are not the place for flowers. Be mindful of death too, for you have drawn close to it, and of necessity plan to practice discretion. For you do wrong to old age and youth alike. In promising the latter you lied, but having acquired[3] the former, you falsified it.

4. Ethical. From EVAGORAS to ANTIPATER

Horus[1] the workman was put in charge of the sea's turbulence, and he checked the watery tides by means of jetties. Some sand was also interposed between the land and the sea. As a result the sea was not allowed[2] to damage the land nearby. On the contrary, it turned back on itself its savage swell, which had threatened[3] a massive assault on the mainland.

Assuredly, O Antipater, Horus imposed a restraint upon your rage to prevent you from making your hand the accessory of your wrath. To unite the tongue with the hands in combined activity is indeed the loftiest peak of perfect virtue. But if you cannot achieve this state, assuage your temper with insults, provided you want to act like a barking dog. For thus does the angry sea show no further marks of its wrath than froth and commotion.

5. Rustic. From AEGIRUS to PLATANUS

We have cranes,[1] my friend, as bad neighbors. They wage war constantly around the farm. For they came to no agreement with our fathers, nor thereafter did they stop their campaign against us. Yet we have often honored them with the first fruits of our harvest. Moreover, we have even given a part of the property to these marauders, as if to an embittered[2] god. Yet the gifts were unacceptable to them, it so happened.[3] Hence we are all going away from here. For it is better for us to cultivate stones than to inhabit fields and hills with[4] nasty neighbors.

6. Love. From ERATO to TERPSITHEA

A valuable[1] portrait of you was painted by Callicrates, they say. But the picture seems to me[2] to represent, not Terpsithea, but Helen of Troy in Parrhasius' lifelike panel.[3] Therefore you have injured both art and nature by finding fault with the latter and falsifying the former. For you compelled Parrhasius' art to be deceptive and to blend[4] into the sketches of you what is no part of you at all, as though you were correcting the errors of Nature and showing its great ineptness. Yet I praise your painter[5] because he declined to depict your ugliness, and I admire the wisdom of Nature, which did not entrust a beautiful body to an exceedingly corrupt mind.

7. Ethical. From SOSIPATER to TERPANDER

Among mares there is a rule, and it seems to me quite wise. Indeed I praise their profound kindliness. But what is this rule? If they see that a foal lacks a teat and the mother is far away,[1] any one of them nurses the foal. For they do not forget their own species and, with a single purpose[2] and no ill will, they do their nursing as though having[3] to do with their own true descendant. They were granted this trait by Nature, for they were not coerced by a law of Solon [the famous Athenian lawgiver].

Now I shall apply this discourse to you. You scorn your brother's son as he roams from door to door, clad in most wretched rags. Your feelings are less sensible than the brutes'. You feed others' hounds, for that is what I would quite properly call the flatterers around you. For they appear to be completely loyal as long as they are stuffed full of your food, you wretch! Yet they constantly bark at you even while they are still belching out the booze they just drank. For, flatterers constitute a breed that is mindful of harm and most forgetful of favors. Therefore, O Terpander, take care of your nephew at last. If you do not, you will have your conscience[4] as your implacable foe, sharpening his sword with Nature's tears.

8. Rustic. From DAPHNO to MYRON

How long will you hollow out your field and drain off the rain water, you wretch? Or will you perhaps even contrive to have my sons grow faint with hunger on account of the drought? On the one hand your land is flooded, whereas mine is not even acquainted with the properties of water. For heaven's sake, ask the clouds whether they drop their water only for Myron.[1] A jealous man is a big nuisance.[2] But if he happens also to be one's neighbor, the misfortune is unyielding and death will scarcely lay it to rest.

9. Love. From EURIPA to DEXICRATES

You promised to visit me on 1 November[1] and you broke your word, Dexicrates. But my spirit grew faint[2] with love, and my heart flared up like a torch, and daily my tears poured forth, and every day I dreamed that you were arriving, and a knock at the door always seemed to me an intimation of your presence. But you, Dexicrates, share your love with some other girl, and you always find pleasure in new ones. For, the feelings of the fickle are wont to be depressed by satiety, which comes very quickly. Money,[3] passion, and love are untrustworthy. Some day you too will be affected. For, the misfortunes of those who have suffered injury often recoil upon those who inflicted the injury.

10. Ethical. From HERMAGORAS to SOSIPATER

You bemoan your poverty very disgracefully, I hear, and revile wealth because it is held inequitably by people, being easily acquired by some as their own possession, but unobtainable by others, as if Nature were ill-disposed to mankind in this respect. For if the sun shines equally on human beings and a supply of fire is instantly available to everybody,[1] why then, you ask, has Nature put gold so sparingly[2] at the disposition of people, and imparted[3] to those who dwell beneath the moon so contentious[4] a gift from which man's worst misfortunes arise?

As for me, I emit a loud heehaw at your mouthings. For what is praiseworthy in Nature, you have turned into a topic for censure, at which you have hooted with the owl.[5] For it supposes the cause of its blindness to be the all-encompassing brilliance of the sun.[6]

It is useful, O Sosipater, for mankind to be gnawed by a hunger for gold. For from this origin spring the arts of life, the development of cities, and the convenience of contracts. And if I must speak concisely, had gold not brought about man's mutual interdependence, the earth's habitations would have been shorn of all adornment. For neither would the sailor have embarked, nor the traveler undertaken a journey, nor the farmer acquired a plow ox, nor the scepter of royal power received respect, nor princes and potentates[7] obtained obedience, nor the commander led his army. But if you also want to learn the hidden truth, control over virtue and vice is exercised by gold. Through it is the soul's desire put to the test,[8] and it is the counterpart of the Celtic river,[9] for it is the truest indicator of fake virtue and roguery.

11. Rustic. From CALLISTACHUS to CYPARISSO

Simichidas set fire yesterday[1] to a species of non-fruitbearing and wild trees, for he passed this sentence on useless growths. But all-devouring fire by its nature launched an unrestrained assault and unexpectedly destroyed the estate of the neighboring farmer. Throwing down his two-edged ax[2] and hoe, he rushed to town to get a lawyer to help him, and summoned Simichidas to court.

I too shall play this game against you, O Cyparisso, unless you order your bees to stay away from my meadows. You will learn you do not have a just approach through intrusive forays.[3]

12. Love. From MELPOMENE to PRAXIMILLE

Chrysogone the flute-player gave her concerts on the public highway.[1] Maybe she thinks that she pleases my lovers too, and the little whore claims that I am greatly annoyed by this performance. But I don't attach much importance to this behavior. For Chrysogone's manners put my lovers, whoever they are, to the test.[2]

I beseech you, however, to be the completely honest bearer of my sincere reply, and to tell the lady from Sparta:

We owe you, O Chrysogone, the most profound thanks on this account. For your ugliness makes us look more beautiful, since even when the jackdaw does not show up, the raven is listed with the good-looking birds.[3]

13. Ethical. From ARISTO to NICIAS

The elephant, they say,[1] is the animal that is the most eager to learn, and quite an apt pupil of human teachings. For its astounding bodily mass is not in itself as famous as the elegance of its training, and these are the traits talked about by the sons of the Hindus.

I am amazed,[2] on the other hand, that in intellect Nicias is more irrational than brute beasts. For as the son of an educated man, you had access[3] to your father's knowledge. Yet you have wasted most of the leisure time of your life on gambling and sports, and you brag about your noble birth.[4] Therefore, if you want to be called Hermagoras' son, turn back at last to his way of life. For even[5] in old age it is good to have wisdom and reason, as Plato too believes.[6] But if you refuse to give up your old[7] vices and proclaim yourself to be the son of Hermagoras, I want you to know that you have become a defiler of your father's grave. For by your misconduct you dishonor his virtues.

14. Rustic. From MYRONIDES to DAMALUS

Your boy has ruined the whole herd, and always filling his pail with milk, proceeds on foot[1] to the plane trees, and bedding himself down, nonchalantly stretches out and embraces the soft life. Afterwards he produces his pipe, gives vent to a sweet[2] song as though inducing sleep, and violates the customs of the countryside. By this misbehavior[3] he has scattered the fodder,[4] and he is sluggish in haggling with a customer. The scoundrel even sold the fertilizer at a low price. Moreover, he is not ashamed to harm Myronides after dining[5] magnificently at the vintage[6] festival yesterday. For, my delicacies were dried figs and grasshoppers. Indeed that marvelous youth gobbled up most of the figs and I wonder how he swallowed the locusts.[7] The lout[8] also gaped, and after eating his fill, he put aside[9] taking home a certain[10] portion. Let him[11] yield the rest of the land[12] to me. For, enduring an evil from afar is preferable to nourishing a hidden enemy at home.

15. Love. From ATALANTA to CORINNA

In the gymnasium, O Corinna, I saw Augeas too. But that sight will not be described in words nor depicted by the hands of painters. For the young man was built strong and straight, with a solid[1] chest. His eyes [were like] a gazelle's. His face was not flushed with frenzy nor wan with softness, but manly and gentle at the same time. The color of his body was neither womanly white nor duskily dark. His hair was slightly wavy in a soft curl and resembled the azure sea in an hour of calmness, when it unfolds its quiet waves on the neighboring land before releasing[2] the savage fury of a storm. His cheeks were not very reddish, for this is effeminate, nor again did they exhibit an unbecoming sadness by their pallor. By contrast his nose was quite elegantly shaped and indicated the great skill of creative Nature. The oil with which he had been rubbed shone like the sun, and by the reflection of its splendor brightened the gymnasium, as though with dazzling rays. In my heart, O Corinna, I sighed and now I feel sharper pains, for the female sex is ashamed to reveal its erotic lust.

16. Ethical. From GORGIAS to ARISTIDES

After borrowing, you were happy. When you are asked to repay, you are sad. And when you encounter your creditors, you are paralyzed with fright as though thinking that you are falling into some dreadful horrors. You look about you at the crossroads and inspect the doors in your desire to avoid the wrath[1] of your creditors, just as people in danger of shipwreck during a big storm seek refuge in a harbor. Moreover, you add[2] misfortune to misfortune. For you pay back your debt to some by borrowing from others. This is the behavior of those who, out of fear of dying, destroy themselves. But[3] borrowing brings manifold troubles to people and is more dangerous than the spontaneous[4] regeneration of the legendary hydra. In accordance with all sound thinking,[5] beware of borrowing. For in that way you will be free to gaze at the sun's rays and quite calmly breathe the open air anywhere.

17. Rustic. From LOPHO to PEDIADES

May Leucippus drop dead! For near the top[1] of the hill he conducted the business[2] harmful to us. He summoned Sostratus and me to court.[3] But Leucippus' mind was completely corrupt,[4] and he wanted to see the courtroom full of gold, so strong a feeling of avarice gripped the wretch. This was understood also by Sostratus, who bought victory with gold and stuffed Leucippus' throat with gifts. The maiden, Justice, is ruined,[5] and gold purchases victory for people. Impartial judgment is dead, since bribes are valued more than what is right.

18. Love. From EROTYLUS to HYPSIPYLE

Even the palm trees are stirred by natural love. The male yearns for the female. The male is convulsed with passion as it embraces the female with its foliage. But if the female is far away,[1] they take the feminine parts,[2] put them in contact with the male, and by a certain trick revive its passion.[3]

Therefore, if you can't come to me soon, assuage my love with a portrait,[4] and let a painting or[5] a likeness furnish me with a view of your image. For even a suggestion is enough to fool those who love deeply.

19. Ethical. From DIOGENES to CHRYSES

You are the guardian of the wealth, I declare, not the owner of the money. For, this judgment of you was rendered by your character, since wicked souls should not share in anything good.[1] Therefore dig up the earth and guard the gold, you wretch. For, the wealth is believed[2] to be not yours but in your custody. In your thirst for riches,[3] you vie with Midas of Phrygia, you who are choked with gold as if by a cord.

20. Rustic. From CHLOAZO to MECO

I sent wild fruit, O Meco, to my sweetheart yesterday. But she pushed her loom aside, stood up at once from her weaving, and taking my presents threw them to the pigs. She also sent my messenger away as an unworthy[1] intermediary. But I lament, for I am attacked by a more dreadful love, striking on account of an impatient[2] girl. Both chance and love are blind; the latter hands out pain, and the former, pleasure, at random and fortuitously.

21. Love. From PERDICCAS to RHODOPE

You sing jarringly and make lovers sad, not happy. For you introduce tragedy, not some song pleasing to your listeners, and the lovers weep from sorrow. For your melodies scold them for their jollity, and you have given voice to no charming tune. Therefore, I beg you, spare us in our sadness. For to your audience you seem to be not a musician but a mourner. And we will all stop our ears with wax.[1] For we will listen to the Sirens rather than to the weeping Muses.

22. Ethical. From ANTISTHENES to PERICLES

Alexander [the Great], son of Philip [king of Macedonia], was not at all blinded[1] by his successes. On the contrary, he wisely recognized the haughtiness of Chance, which usually[2] entices the imprudent with great honors. For this reason, when in the ups and downs of war he saw Darius [the Persian ruler] falling,[3] Alexander covered his foe with his cloak, displaying the excellence of his character and of his fortunes at the same time. Thereupon his subjects[4] reproached Alexander, and the king's mercifulness was a fault. Hence Alexander, like a philosopher, feared the[5] unreliability of Chance. Accordingly, when many hailed[6] him for his victory, he declared: "Some adversity too, O Jupiter, is mixed[7] with things when they are at their best." Thus was Alexander very intelligently circumspect with regard to the vagaries[8] of Fortune at its height.

If, therefore, you have not learned the nature of fickle Fortune, you will soon see experience as your teacher.[9] But if you persist in your blindness, you will call down on yourself fiercer retributions by undergoing the punishments of fines[10] and a trial.

23. Rustic. From ASTACHYON to MILO

Clean the hemlocks off your land, for you[1] have harmed my bees. Don't make trouble for a working farmer, please! To me at any rate you have not passed along any drones.[2] Why do you annoy your neighbor so unreasonably, you scoundrel?[3] Unless you stop this mischief, I shall write up your misbehavior on my door, and show the damage to our neighbors so that they may all avoid you like some accursed evil.

24. Love. From TELESILLA to LAIS

Miners searching for veins of gold, and diggers of wells[1] trying to see pockets of water and the secrets of the earth, do not ply their trade as hard as I did in roaming through the entire city to catch sight, if possible, of Agesilaus anywhere. For I heard that disreputable Leucippa made a party for him. A wild rage[2] came over me, and my grief swells up with a flood of tears, on account of which I shall be a character in the tragedy. For I shall no longer see the sun rise. Thus shall I be fiercer than Medea and Phaedra.[3]

25. Ethical. From SOSIPATER to AXIOCHUS[1]

You recently buried your brother, they say, and you are profoundly stricken by an inconsolable grief. How am I to admire you as a wise man when you are overcome so violently by your emotions?

Death, as we commonly say,[2] is a sleep, longer to be sure than is usual, but quite short in the light of that day which is to come. The dead have departed from us for a brief pilgrimage, not for a long separation. Be resigned to the parting, while awaiting a reunion.[3] Don't afflict your soul with a craving for the flesh. For Plotinus too thought it "shameful that he was in a body",[4] so much did this mortal coil sadden that philosopher. For my sake, don't weep any more. Put limits on your grief.[5] Heal your feelings with sagacity. Be your own physician. You have the saying that cures:

Without anger and sadness, forgetfulness of all troubles.[6]

The Creator[7] never downgrades his work from better to worse. Let us leave to the dead that which is dead. For here life,[8] the most excellent product of the spirit, is afflicted by great disgraces. Indeed I would lament a birth rather than a death, since a birth is the beginning of weeping whereas death is the release from sorrows. Ignorance makes us cowards, and we are wary of death, not because it is evil, but because mortals do not know it. For nobody has brought back to us any knowledge of it; you complain about it constantly.[9] Then don't imitate Niobe, lest perhaps you too be thought to have consigned your human nature to a rock.[10]

26. Rustic. From THERISTRO to SPIRO

We are leaving for Etna, the mountain in Sicily, and saying farewell to Attica. For I never saw a land less fertile for the production of crops. Instead of pears,[1] it gave us myrtle; instead of barley, ivy.[2] Therefore, my first seeds having failed to sprout, I shall never again[3] sow that ungrateful soil. The farmer cannot endure hunger and a hostile army, nor can sailors contend with gales and thunder.

27. Love. From CECROPIS to DEXICRATES

The magnetic stone loves iron, they say,[1] and the more[2] it is united with its mate the livelier it is, according to report. For after[3] the stone is taken away from its partner, it loses its strength at once, and is deprived of its power.

Reactions of this kind, O Dexicrates, are present even in lifeless[4] objects. Why I am so deeply troubled[5] by the loss of your presence is harder to say than to feel, in my judgment. So may I sadden[6] those who sadden me, and become a Cupid's[7] dart, a[8] flash more impetuous than sea foam.

28. Ethical.[1] From HERACLIDES to ANTISTHENES

You have not yet overcome your anger toward me, O Antisthenes. On the contrary, you still feel bitter about me, and you hide your annoyance under the pretense of well-disposed words, as though preserving the glow of fire beneath the ashes.

Purge your heart of the remaining[2] hurt,[3] for this is ordained by our conversations. Otherwise you will be even more savage than the sea. For it lulls its ferocity and presents a mild appearance to sailors, when they soothe it with oil during exceptionally wild storms.[4]

29. Rustic. From LACHANO to PEGANO

Come to me tomorrow near the oil market. For I shall go into town and make a feast,[1] dearest friend, since I shall dedicate the firstlings of my flock to the Nymphs and to Pan.

At last the gods are gracious to us. My pails are full of milk. The ewes gave birth abundantly. The she-goats jump as though rejoicing at their good fortune. We have stopped struggling against poverty, that vicious and hostile beast[2] which clings to those who have it like a sore. It is an evil which intrudes itself very intimately, creating sluggishness and sadness, leading inevitably to misery and swiftly to grief, destructive of sleep and productive of anxiety, troublesome, leading to crime and disreputability, despicable, and inspiring no envy, since nobody wants to endure such an evil, not even if he were sentenced to suffer the insanity of Orestes.[3]

Therefore, leaving poverty to the poor, we shall establish ourselves in a different style of life by switching to high spirits together with pride.[4]

30. Love. From RHODINA to CALLIOPE

In the presence of my lovers, you disparage me[1] and mock my pregnancy,[2] my slim figure having become distended, and you even laugh out loud[3] at my body. On the other hand, you suppose that you have covered up your villainy. For you pluck out undeveloped fetuses, you wicked woman, and you have made abortions[4] more desirable than births. Smothering with powerful drugs the embryos alive in your womb, you have committed atrocities more abominable than African[5] Medea's. For although she had been a praiseworthy ally in her husband's struggles, his infidelity made her a child-killer.[6]

On the other hand, countless horrors are perpetrated by you, you little whore, for the sake of your beautiful appearance. Now finally stop hiding your cruelty and mocking our decency. Birth, in my judgment, is more humane than abortion. I want you to know, moreover, that the earth[7] too has been aroused by you and will not delay in imposing on you the punishment for destroying your offspring.

31. Ethical. From HEPHAESTIO to THALES

The peacock, a Persian bird, has acquired the arrogance of the Persians,[1] its beauty giving it a sense of greatness,[2] in which it seems to go beyond even its female consort.[3] It therefore puffs its feathers out like a crest and furnishes a magnificent spectacle to onlookers. When it opens out its round shapes, it resembles the heavenly display, with the eyes on its feathers duplicating the form of the stars. And this is the habit of Persian[4] birds in their zeal for excellence. For when they reveal their pulchritude, they do not begrudge it to painters.

You, on the other hand, sit on your writings, hide your accomplishments, and setting your works aside in obscurity, you disregard us who are deprived of so great a boon.[5] If, then, there is any ill-will in your making us sad,[6] what you are trying to do is unreasonable and quite far from your promise. However, if laziness was the cause of your delay, then you are on the same footing as the farmer who, having expended much sweat on the land, at the height of the summer gathers no grain.

32. Rustic. From POAS to AMPELIUS

Come, we shall wail together, as of yore.[1] The river has overflowed, pouring forth dreadful hardships over us. It has drenched the whole countryside and wantonly drowned the newly planted vines. But the disaster is greater inasmuch as the abomination refuses to recede from here. For it wants to tarry in the land and has made my field into its channel, a spectacle worthy of tears. Instead of vines, in our misery we shall plant fish. When the river so desires, we shall go prowling; when it so wishes, we shall suffer from hunger, a great favor having been conferred on me, as it seems.

Would that in the summer we did not pray at all to the clouds to bring rain to us in our drought! For thus we would never lose what we possess. For by itself the river is bad enough. But if the rains too are profuse,[2] they are more destructive than fire and no moderation restrains their roar.

33. Love. From GALATEA to THETIS

I praise your foresight and approve of it as very professional. For as if from some Pythian tripod[1] you predicted the future to me, and more penetratingly than the lynx[2] you examined deep secrets.

Callimachus has abandoned me. Rising on the swiftest wings of insolence, the wretch has flown away and fled from me, satiety having overcome his lust.

You often warned me: "O Galatea, don't trust oaths. Nothing comes more easily to a suitor than promises." For youths are intoxicated with erotic pleasures,[3] lose their good sense, and do and say whatever Cupid commands. For what they want to do is not under their own control. Skepticism is safer than trustfulness, and for the purposes of trickery a promise is a deceiver who engenders confidence.

34. Ethical. From THEMISTOCLES[1] to CHRYSIPPUS

Since wisdom is honored in fables too, now O Chrysippus, I shall tell you a tale not without significance.[2]

Once upon a time the birds went to Jupiter and acknowledged the Olympian as their ruler.[3] For the birds were troubled by the absence of government and, being deprived of that noble blessing, leadership, they therefore suffered from great disorder. Then Jupiter consented, his decision took effect, and he granted the suppliants a splendid gift, namely, their request for the appointment of a king.

He thereupon bade the birds go to ponds and springs and wash off their dirt, so that he entrusted to water the determination of the leadership. For, Jupiter's preference was for the display of beauty. Therefore the birds took a bath, then they come back at once[4] to Jupiter, and each displayed its own comeliness. But the magpie, being concerned about its misshapenness, falsified Nature's handiwork by decking out its own ugliness[5] with another's charm. The owl, however, detected the fraud and proved that the adornment was spurious. For, recognizing a feather as its own, the owl took it away as its own, setting an example to all the other birds that each should remove its own feather. And the magpie became a magpie once more.[6]

This fable, O Chrysippus, certainly[7] teaches a truth that expounds profound wisdom to us. For in like manner we humans possess nothing of our own here. Yet we who live but a short time, take pride[8] in our alien adornment. For when we are dead, we shall be deprived of what is not ours. Therefore despise wealth and the body, but take care of that immortal substance, your soul. For this is eternal and deathless, whereas those other things are perishable and belong to us briefly.

35. Rustic. From MYRONIDES to MOSCHIO

I lent my plow ox to Tycanias,[1] since he did not have a second as a mate for his yoke. But Tycanias also promised to let me have his bull, for I too had a defective herd, my very fine bull having died when that fierce epidemic raged in the pastures.[2] That honorable man, Tycanias, however, broke the agreement and as long as his purpose was being accomplished, he seemed to respect public opinion.[3] However, I bewail his cruel guile. For I have no plow oxen, he argues,[4] yet the plowmen's season has already passed. Therefore I shall organize a trial against Tycanias, and I shall summon the whole countryside to be judges, and he will be subject to a sentence for his chicanery. Moreover, I shall warn rogues to refrain from mischief by taking a lesson in decent behavior from the ruin of one man.

36. Love. From ERASMIUS to LYSISTRATUS

Cupids mock men, the winged gods of love making slaves of those who live beneath the moon, if we may believe the painters at all. Would that we might see the enemy himself! For thus the Cupids, who hurl the darts, would themselves be smitten. But the harm we suffer is greater because we do not even know the nature of our enemies.

I have become involved in just such an irrational passion. I am mad about Melanippe, Diodorus' daughter,[1] from afar,[2] though I have never seen the lady even in a dream, but have only heard from somebody that she is wonderfully charming.[3] I am pierced in my heart, having suffered no injury through my eyes, as was[4] usually the case, O Lysistratus. Now, however, my eyes have become ears[5] also, so powerful was the force of the Cupids. Then she is either a Fury[6] or an apparition of some sort — I do not know what is said about these matters, since there is not a single trustworthy witness to the truth. Nevertheless I am sick at heart, loving a girl I have not beheld.[7] And I realize that I have been blocked by an unforeseen[8] disorder. What was loved was unseen, and only a feeling of love was real.

37. Ethical. From EURYADES to CIMON

You promise much and do little, your tongue being more conspicuous than your deeds. But if you are renowned for the elegance of your diction, the artists wield greater power than your mouth does, since in their paintings they invent such things as Nature cannot produce. But if you think that with your promises you make your audience happy, you do cheer them up for a short while, but later you sadden them so much the more painfully. For even the most beautiful dreams do not give us as much pleasure when we are asleep as they sadden us when we wake up, since all hope vanishes[1] with sleep. Therefore let your actions agree with your rhetoric, lest you be hatefully regarded by your friends as a liar, and furnish your enemies with grounds for assailing you as indifferent to the truth.

38. Rustic. From TETTIGO to ORTYGO

You wretch, why in the world did you change[1] your clothes and let the partridges fly away?[2] Wine was your trouble. With wine, too, did Ulysses buy the eye of the Cyclops, as they say.[3] Therefore, unless you recover the birds,[4] together with you I shall jump off a cliff. For a boy to live a bad life is hard to bear. But if a son claims his grave sooner than his father does, that is more unendurable.

39. Love. From THETIS to ANAXARCHUS

You cannot love Thetis and Galatea at the same time.[1] For passions do not engage in struggle, since love is not divided.[2] Nor will you endure a twofold involvement. For just as the earth cannot be warmed by two suns, so one heart does not support two flames of love.

40. Ethical. From SOCRATES to PLATO

On the one hand, nobody is injured. On the other hand, everybody[1] spontaneously harms himself, since we are in control of our virtues and vices.

Philonides[2] took away your farm, which is external to you, and he did not sadden[3] your soul at all. Philip did you harm by taking possession of your ring, while in yourself you suffered no damage, for what we have acquired is not ours. The barbarians did away with your son; you suffered no fearful harm, since you did not acquire your child for all time. After a period when you did not have him, he was born to you recently, and once more he is not yours, since he did not subsist, but was created.[4] Therefore people do harm; they are not harmed.

I admired that Cyclops in Homer, for he says that nobody harms him while he is suffering an injury.[5] And this denial on the part of the shepherd was a declaration of the truth.

41. Rustic. From MARATHON to PEGANO

Fleeing from political disorders and the unbearable uproar of the city, I rented this farm with the hope of finding a change[1] in my spirit. But I have fallen upon even worse evils. For sometimes I have blight as my enemy, at other times locusts, and also hail now and then. Frost ruins my crops, like an implacable despot. And in my misery I donate my sweat to the winds. Alas, poor me! Where shall I turn? When I recall farm work, I cherish the city. When I encounter the hubbub in town, I love the country. What is not present seems[2] better than what is at hand. The only escape from my troubles is death, whether natural or, on the contrary, self-inflicted. Accordingly, with hanging as the cure for me, I feel awful, since it is stupid for unhappy people to choose[3] death.

42. Love. From PERICLES to ASPASIA

If you are looking for presents, you are not in love, since the gods of love are not influenced by gifts and instruct lovers to behave in the same way. Therefore, if you are in love, it is surely more befitting to give than to receive. But if you thirst for money and pretended, for the sake of wealth, to be in love, your feelings are belied by your tongue, which sells pleasures for gold to anyone who wants them.

43. Ethical. From DIOGENES to DEMONICUS

The eunuch Lydus, an artificial little woman, half a man,[1] whose nature conforms completely to no model, is despicable. For he is said to have a shameful tongue in every limb of his body. In imitation of Homer's Ulysses,[2] however, I am not susceptible to his darts. For if the female sex smites heroes, that is ineffectual,[3] to say something[4] terse and weighty[5] to you, in the manner of Diomedes.[6] It is surely befitting for eunuchs to bark and rave. For, being deprived of strength in their hands, they try to do everything with their tongues. My friends, however, emphasize[7] my forbearance in not subjecting that insolent rogue to punishment. For if in like manner the braying[8] had been done by a donkey,

I would of course never summon it to court. This surely is proclaimed with a discreet command.[9]

44. Rustic. From PRIAMIDES to CORYDO

Tomorrow I shall be at the feast. I have to prepare everything[1] for the wedding: beans, chickpeas, lots of dried figs, sweets, honey cake, and cookies. Bring your well-made shepherd's pipe and sing your very melodious[2] tunes, your knowledge of country music being most professional, O Corydo. For I have made up my mind to shift the bridal bed altogether in the direction of pleasure, heightened by the instrument's tones.

45. Love. From LEANDER to PYLADES

The gods of love distress me greatly. I am in love, but my beloved hates me. What shall I do, poor wretch that I am? The gods of love do not hold the balance even, as they weigh out bitter[1] tears for mankind. Then if they do an injustice, let them not be called gods at all. But if they do not belie their title, let them decide justly and allot sorrow to me according to what is fair.

46. Ethical. From DIOGENES to ARISTARCHUS

Alexander frightened Macedonia when he rode on his horse Bucephalus, who did not obey the reins at all, they say, nor was he calmed down by caressing hands. For he was an intractable beast, whose high spirit did not allow him to be ridden, an unapproachable evil, much feared by those around him. Then when it was his lot to have Alexander as his rider,[1] with his ferocity he mixed gentleness, as it were, exchanging loud neighing for docility, and he was seen to be tamed. For it was impermissible to oppose Alexander. Therefore, you too, O Aristarchus, pay[2] heed to Fortune, since it was not Alexander, but Fortune, that Bucephalus obeyed.

47. Rustic. From POIMNIO to ARNON

The udders of my ewes threaten to burst, and I am short of milk pails, I know not how. Therefore give me pails, while I supply you with milk as I exchange[1] small favors for generous gifts.

48. Love. From CHRYSOGONE to TERPANDER

Don't spurn a girl who abuses and also insults you. For lovers certainly receive pleasures and delights, yet they are often adorned[1] with blows and scars. But if you don't tolerate a scolding, neither will you gather the rose out of fear of the thorn.

49. Ethical. From LEONIDES to PERIANDER

Thetis' son [Achilles] had respect even for aged Priam and for his enemy's white hair,[1] and he gave back to the father his dead son, honoring Priam with a most calamitous gift. I marvel at Priam's recklessness, but I praise Achilles' kindness.

Be you also for me a grandson of Aeacus [Achilles], and pitying my white hair and my tears, give me back my son while he is alive. For I am just as unlucky as Priam. For since you are not a foe,[2] I would have communicated[3] with you about the boy, the letters of the message being written with tears, not ink. If you too, however, wish to be praised for your generosity, let your gift precede my request.[4] On the other hand, if you are not controlled by reason, but rage and grief are uppermost, you will rejoice, to be sure, for a short while. Yet you will be sorry longer when for your irrational anger you pay heavy penalties too.

50. Rustic. From CALAMO to SPIRO

If you wanted to be a farmer, stay away from the uproar of the town. But if you have a liking for courtroom attorneys and tribunals, throw away your hoe, pick up pen and paper, and plunge straight ahead, devil take you![1] For the farm community does not accept pettifoggers and those who constantly exclaim "Gentlemen of the jury."

51. Love. From RHODOCLEA to HYPSIPYLE

As I was strolling through Piraeus last night, I saw your lover with Chrysippa. Light was provided by a boy, and the arranger of the affair[1] was the old woman Habrotonon. But when I greeted the procuress, your lover urged that the incident be kept secret. Therefore don't believe him when he pledges [faithfulness] nor when he flatters you. For in both cases his tongue lies atrociously.

52. Ethical. From SOCRATES to CLEON

When wolves have surrounded a huge prey, after eating their fill they then become philosophical, as though they had the self-restraint of lambs,[1] and they exchange their beastly ways for a certain kindness, a full belly having taught the wolves excellent justice. They mingle with the sheep and respect them, until their belly subsides.[2]

You too, having taken on traits more ruthless[3] than the wolves', have a much greater lack of control of your greed. When an overabundance of gold bursts out of your coffers, you behave like the drunkards. For they become thirsty the more they stuff themselves with wine, and with drunkenness they beguile the peak of drunkenness. For they pass through the greatest ecstasy to the opposite mood, as the wine cheers the drunkards up and burns them out. Get rid of your unrestrained drunkenness, you profligate, lest you fall into the opposite state of affairs, being deprived by Lady Luck even of those things which she herself brought you. For by such punishments does she smite ingrates.

53. Rustic. From MINTHO to RHIZO

The Chrysippus River has carried away part of our land and attached it to yours. Because it does something stupid and silly, it is considered unjust.[1] If, however, your shoulders can't stand the weight of a lawsuit, don't accept gifts from rivers.[2] On the other hand, if you covet the property of others, you will soon be deprived of your own too, and you'll be sorry.[3]

54. Love. From MEDEA to JASON

Nothing intrigues men more, yet nothing is more cloying than the condition of love. Where are the floods of your tears that gushed forth at my feet? Whither have your thousand styles of speaking gone,[1] and the humility and submissiveness of your language? Such words, I am absolutely convinced, were never used by debtors in dealing with their creditors, nor by wounded prisoners under the control of their enemies.

No longer do you go without sleep night after night; you have given up your morning[2] serenades. You have cast aside[3] the innumerable messages, agreements, oaths, which you communicated to me[4] through match-making women. You have suddenly fallen for another girl, just as slumberers switch from one subject to another in their dreams without any interruption. I praise the painters because they depict the gods of love as winged, and by their art reshape reality and enhance the truth with their inventions.

55. Ethical. From PARMENIDES to CHRYSOSTHENES

To be awake without interruption is the mark of an immortal nature. A moderate amount of sleep, on the other hand, is our lot, as it should be, and the mark of a human being. But to keep on sleeping past the permissible limit suits the dead rather than the living. You, O Chrysosthenes, have lost the greatest part of your life. For you are always asleep and in this respect you have overstepped the bounds, like Homer's[1] Ulysses who, being about to leave the earth, uses sleep as an ocean, and sees neither the rising nor the setting sun.[2]

56. Rustic. From DAPHNO to AEGIRUS

Your fig trees spread their roots to my land because they do not wish to end up under your control, and by passing beyond your jurisdiction they surrender their fruit to me, since they have fled to my field of operations. Besides, this is the law of the farmers. Obey the ancient statutes,[1] you little old man.[2] But if you want to object to our customs, we shall throw you out of the Farmers' Association on the ground that you are a new and upstart lawgiver, and we shall banish you from our territory as an alien.

57. Love. From PYRRHIAS to PHILONIDES

If you are in love, do not find fault with the blemish in your beloved. For a lover's soul cannot fail to be blind, since the passion of lovers is irrepressible. If, on the other hand, you are not in love, why do you weep and moan and of your own free will create a commotion in yourself? Therefore you do wrong in both respects, since sometimes as a lover you are more than covetous, at other times as a critic you express an abhorrence.

58. Ethical. From DAMASCIUS to ANTIGONUS

If Socrates does not hold a lifelong pledge, do not engage him as your boy's teacher. On the other hand, let sons be deemed[1] pledged for life. For he who has been taught by Nature to be a father is especially qualified to be a teacher too, learning[2] through experience the conditions of procreation and the agonies of love.

59. Rustic. From CEPIAS to CORIANNUS

Be[1] my assistant, O Coriannus, by the noon hour. I shall enclose my land with fences[2] since we have wicked wayfarers, and I cannot contend with wild creatures and men at the same time: the hares ruin my vines, the caterpillars my vegetables. What, finally, shall I say about the moles? Surely they are a dreadful evil for the farmer, an unconquerable foe.[3] Therefore, work with me and take part in my labors. In such efforts I in turn shall cooperate with you. For this is how the ants likewise overcome their difficulties by their combined activities and achieve prodigious results.[4]

60. Love. From ANTHIA to ORION[1]

Everything is enslaved by love of women.[2] Lais is adored by Diogenes, Sostrate by Phrygius. Their philosophical skill has vanished; they have rejected[3] pure morals, forsaken heavenly virtue, and broken a lofty vow. Everything previously held by them in the strictest observance is discarded.

To me it seems an untimely[4] sport for an old man, wearing a venerable beard and exalted for his uprightness,[5] to consort with a young prostitute. I laugh out loud and can't hold back my jeers whenever I meet old men. For once upon a time they used to denounce love vociferously, and they declared lovers to be mad of their own free will, love being defined as a passion of an intemperate soul. A wise man endures everything. To hope for this is too presumptuous,[6] since Time and Chance can accomplish much.

61. Ethical. From SOSTRATUS to LYSISTRATUS

When I gave you much wonderful advice, I seemed to be weaving Penelope's shroud.[1] Here I am still starting with fabulous tales on the chance that you may benefit by hearing what I say.[2]

Flitting through the trees when they were sprouting their leaves, at the time of the heat the grasshopper chirped and enjoyed listening to its own music. The ant, however, followed the reapers, made its rounds about the threshing-floor, and buried its food in the hollows of the earth.[3] Finally the sun departed from the north, autumn passed, and winter raged[4] everywhere. The sea swallowed up the libations offered it to be calm. Sailors saluted their harbors as their saviors. The farmer sought refuge near his hearth,[5] as did the ant in the recesses of the earth and it had its food supply handy as the result of its exertions.

Then the grasshopper begged the hard-working ant for a share of its treasure. The ant, however, sent the grasshopper back to its singing,[6] laughing loudly at its idleness, and reproaching[7] it for its summer concerts. Accordingly, the grasshopper gained hunger from its song, while by its efforts the ant acquired delicacies.

This fable fits you, O Lysistratus.[8] Therefore get rid of your nonchalance,[9] for you will spoil your natural gifts if you fail to improve your strength[10] and the undeveloped health of your body by forgoing exercise as a matter of principle.

62. Rustic. From TETTIGO to PORPHYRIO

Corydo is a happy man and is favored by Fortune.¹ His vines are loaded with grapes; his pear-trees are full and eager to be picked; his olive-trees bend toward the ground, the mass of fruit crushing the heavily laden branches; and his meadows are lush.²

His wife, moreover, gives her husband additional reasons to rejoice. For he has so many children that by his fertility he seems³ to surpass both Danaus and Aegyptus. One child clings to the mother's breasts, another is weaned. Others crawl around, having not yet begun to stand up straight. Others babble and acquire their second set of teeth.⁴ Others walk vigorously and have attained steadiness on two feet.⁵ The differences in their height are arranged like the holes in a shepherd's pipe.

Yet you advised against my daughter's marrying Corydo, and you disparaged that tie⁶ as though abominating his humble ancestry. Woe is me, I have been cheated, I had little foresight. Noble birth is of no use to people, for all of them value nothing more than wealth.

63. Love. From CHRYSES to HEPHAESTIO

You are not in love with Diodota,¹ you have extinguished the embers of passion. For you have no feelings of jealousy concerning her when you see her keeping company with Lysistratus. Nothing provokes quarrels more than love does. How, then, are you in love if you endure such pains so calmly?

64. Ethical. From SOSTRATUS¹ to MELANIPPIDES

As teachers of boys and instructors in deportment, we firmly² control our pupils' misconduct by threats,³ since the youngsters are still unresponsive to reasoning and rules.⁴ For we frighten them by words rather than by whips.

On the other hand, I am astonished by your obduracy. For you are not afraid of being punished by the judges, nor do you pay any attention to words of warning. In your old age your attitude is less sensible than the youngsters'. Go to hell,⁵ if that is what you want. For if anybody can't be reformed by words and whips, putting up with his resentment when we are exhorting him⁶ is too irksome, and more disagreeable than cleaning out stable⁷ dung or draining the entire Atlantic Ocean with a cup.⁸

65. Rustic. From BUBALIO to CISSYBIUS

We are wronged by Gorgias' son, for the scoundrel trespasses¹ on horseback and covers up his misconduct by pretending to be hunting. For there are no hares among us, nor does the goat abound, nor the gazelle, nor the stag, nor any other game animal. For, the hares are kept down by warrens and snares, by which they are also trapped. The goats and gazelles are preyed upon by the lions hereabout. But after all why should I enumerate for you a thousand species of wild animals, which we too do not have?² We are kept down by our unmanliness,³ but the beasts by our astuteness.⁴ Now since you, O Cissybius, are related to the malefactor, give him useful⁵ advice, and let your words teach him to behave better.⁶ For I want you to know that he is going to be torn to pieces by my dogs if he encroaches on my fields hereafter. For, my bitch and her pups are already guarding my property, eager to get hold of soft human fat.

66. Love. From PEITHO to HIPPOLYTUS

Lovers pursue beauty, not virtue. For their lusts do not teach them restraint, but bodily beauty entices[1] their roving eye. Hence, if you love Rhodoclea for her character, your affection is not controlled by your carnal desires. For, virtuous thoughts are not affected by Love's dart.

67. Ethical. From ERATOSTHENES to AESCHINES

You swallow your promises as though they were vegetables, and the clacking of your teeth[1] seems to you to put an end to the matter. To those who accuse you, O most evil of men, your rebuttal is:

My tongue made the promise, but my mind made no promise.[2]

Don't you see[3] that a more severe form of punishment[4] awaits an unrestrained tongue? For whatever words we have misspoken, we shall suffer the penalty in deeds. By all means control your tongue. Avoid swearing, even to the truth. Does an oath perhaps seem a light matter to you? In fact it is weightier than any load. That is why Tantalus too was punished. For toward the gods his tongue was irreverent.[5]

68. Rustic. From SEUTLION[1] to CORIANNUS

I have finally caught that very mischievous little fox, and I am keeping that nuisance tied down in a network of ropes. I shall carry[2] her to the highway, O Coriannus, and call all the farmers together so that I may triumph publicly over the enemy. She will pay the penalty in front of the people for her many misdeeds by suffering a single punishment.

69. Love. From CALLIOPE to LAIS

Gorgias prides himself on his hair rolled up in a knot with a golden clasp,[1] while his cheeks are not yet darkened by whiskers. He is haughty in his handsomeness, and with his charm he entices us to swoon. But I shall depict old age, illness, and suffering on panels, which I shall put in front of the brute's door. For these are what he will have some day as adversaries of his good looks.[2]

70. Ethical. From PLATO to AXIOCHUS[1]

We control horses with bridles and whips. Sometimes we set a boat in motion by unfurling the sails, and at other times we hold it in check with anchors. That is how the tongue should be regulated, O Axiochus, now loaded with words, now lulled to silence.

71. Rustic. From RHODON to CYPARISSUS

The Lucanians are back again, it is said. Casting aside my scythe, I shall forge a spear and spike, and practice the art of war. For, the evil one does not let us rest. For me, summer is chillier than winter. For what is more horrible than war? I wept when spring drew near, so teeming with flowers is the field, so sweet the fragrance of myrtle in the land, so beautifully leafed out the plane tree. The crops are greening everywhere, all my plantings are ripening rapidly. Our enemies, however, harry us. For they love the sword better than the plough.

72. Love. From SOPATER to TELESILLA

Had Nature not mingled[1] satiety with the pleasures of love, the male sex would have been subordinated to the female,[2] O Telesilla. Don't you therefore develop a swelled head, you little whore! The flame of love has grown cold in me. For, the pain caused by Cupid's darts does not last forever.

73. Ethical. From PROCLUS to ARCHIMEDES

The octopus' habits are said to be ridiculous.[1] For they swallow their own tentacles and their unfortunate parts as[2] handy food. The wretched creatures, then, do away with the usefulness of their own members and are seen feasting on their own flesh.[3]

Don't you, O Xanthippus,[4] seem to be completely an octopus [in spirit]?[5] For my part, I think this is quite obvious. For you wrong your own father too cruelly, neither revering Nature nor heeding retribution, such as hell. But if you repent, the fading memory[6] of your past[7] misdeeds will intercede for you. On the other hand, if with implacable fervor you pursue a greedy course, the godhead will pay you back in equal measure. You will have sons who will be the spit and image of their father's knavery. Just so do the viper's offspring follow their mother's viciousness by tearing to shreds the belly that bore them and nourished them.[8]

74. Rustic. From ELAPHO to DORCON

Animal dung promotes fertility. Therefore give me the dung of your beasts, and you will receive fruit and vegetables in payment for the productivity. In addition, I shall reward your friendship with resounding thanks.

75. Love. From ARISTOXENUS to POLYXENE

Nothing lasts less than fun with a harlot.[1] The kisses of your lips are treacherous. For, a desire without roots[2] languishes quite rapidly. Henceforth[3] I shall adopt the self-restraint that is distasteful to many men. For it is said to be more enduring. By using a dowry, I'll enjoy[4] marriage. For to buy[5] a prostitute's faithfulness is very difficult.

76. Ethical. From DIOGENES to SOTION

A bit of fame,[1] that wee thing, seems to wise men to be[2] a dream, more monstrous and worthless than fabricated fables, fleeting, trivial, a plaything more insubstantial than echoes[3] and breezes. When absent, it saddens; but when it arrived, it caused greater grief. For it soon disappointed its pursuers.[4] Don't let Fortune's bubble delude you. For it mocks men just as it pleases. For, vanity of vanities is the business of man.

77. Rustic. From BUCOLIO to MYRONIDES

Your grandson, O Myronides, has misbehaved. With a red rag he caused great disarray[1] in my herd, and threw my favorite heifer down. This misfortune was treated by his pals as a joke. Don't give the boy double rations[2] any more. Even if well-nourished,[3] youths are unruly. And if you indulge them, their unruliness goes to extremes.

78. Love. From PERICLES to CALLIOPE

To win a maiden's love is quite difficult, in fact too difficult. My spirit is lackadaisical. What shall I do? For, pitiless Cupid has hurled his dart.[1] You be the judge[2] between Cupid and me, and give me a favorable decision. For, love is jealous. Even when it proposes love, it denies fulfillment.

79. Ethical. From ISOCRATES[1] to DIONYSIUS

Guards, standard bearers, heralds, and thrones raised on high cast a shadow over philosophy. This great departure from the virtues resulted. With your [change of] fortune, your character hasn't changed, has it?[2] You still have your thick skin.[3] For from the start you had a perishable nature.[4] Why then did this empty and insubstantial bit of fame inflate the muddy sack so much?[5] You are overcome by stark madness, you wretch, and have lost your understanding of Nature. In this way the wavering shifts of Fortune forced you to abandon your previous point of view and[6] pass beyond your reasonable aberration. Formerly, what was humble was, for you, exalted. Now, however, what is lowly and earth-bound is [for you] Fortune's loftiest peak.[7] Therefore[8] abandon your pretense of happiness. Flee fleeting Fortune. For by anticipating its treacherousness, you will not suffer from the sudden onset of its vicissitudes.

80. Rustic. From CROMYLO[1] to AMPELO

Nothing is more miserable than farming. We, poor wretches, are damaged even by the tyranny of the winds. The south wind ruins us. It made my grain bare, destroyed my vines, and spoiled my grapes. And I can't summon this assailant to court. I shall therefore throw away my scythe and sickle, don a shield, helmet, and sword, and become a soldier. By changing my trade, I shall cheat fate.

81. Love. From LEANDER to CALLICOMOS[1]

If anything[2] is dearer to men than gold, show me how[3] bliss is attained through more precious gifts. But if not, grant me your favors that much sooner. For you do not have the beauty of Danae nor are you more charming nor am I richer than Jupiter,[4] I who am buying your maidenhead with gold.

82. Ethical. From SOCRATES to ALCIBIADES

Even poetic fancies are full of all wisdom. Ulysses' comrades are said[1] to have stuffed their ears with wax when their ships bore them toward the lewd Sirens, while Ulysses was bound in chains, as though in an escape-proof prison.[2] Here the secrets of philosophy are introduced.[3] For in my opinion poetry invented the Sirens to stand for illicit pleasures. And I deeply admire Homer's coupling of the truth with his story. Together with what we hear, he mingled the truth, like a drink more soothing than nectar, as though he were diluting a very strong wine with water.[4] For in this way we are not deceived by the fictions nor are we made dizzy by an incomprehensible investigation.[5] Hence Homer symbolized the absence of sense experience by the wax, but philosophy by the chains. Accordingly, only Ulysses enjoyed[6] hearing the very sweet singing, and the chains checked his longing. For, the understanding of vices is the virtue of philosophy, and the curbing of pleasures is philosophy's combat trophy.

Life at present, O Antimachus,[7] is the mirror image of Ulysses' wanderings, and man in his misery is tossed about in a sea of troubles. The sounds of the Sirens' pleasures swirl[8] around us, and temptations[9] impinge on us like winds, now from this side and now from the other. We shall imitate Penelope's marriage,[10] therefore, by following philosophy as an unbreakable and divine chain of virtues.

83. Rustic. From ANTHINUS to AMPELINUS

The vintage is now approaching, and the sweet[1] grapes are full. Therefore, watch the public highway very carefully, and use your Cretan dog to help you. For, the wayfarer's hands are itchy, and quite ready to deprive the farmer of [what he has produced by] his sweat.

84. Love. From CHRYSIPPA to SOSIPATER

Haven't you, O Sosipater, been caught in love's snares through your passion for Anthusa?[1] It is the sign of a keen eye to fall in love with a beautiful girl. Don't groan because you have been overcome by pulchritude. For on account of your pains[2] your enjoyment will be greater. Love's tears are delightful. For, sorrow[3] is mingled with pleasure, and Cupid delights as he saddens. For the girdle of Venus is interwoven with a variety of feelings.

85. Ethical. From PLATO to DIONYSIUS

If you wish to overcome your sadness,[1] stroll through the tombstones and you will have the cure for your feelings. You will behold[2] man's greatest joys as in the end they take on the lightness of dust.[3]

Printed in Cracow in the house of Johannes Haller[1] in the year 1509 of our salvation.

BIBLIOGRAPHY

Czartoryski, Paweł, "Der Gebrauch der griechischen Sprache durch Copernicus und seine Übersetzung der Briefe des Theophylactos Simokattes", *Festschrift*, Copernicus Gymnasium Philippsburg, 1980
Epistolographers *Epistolae diversorum philosophorum, oratorum, rhetorum* (Venice: Aldus Manutius, 1499)
Epistolographi graeci, ed. Rudolf Hercher (Amsterdam: Hakkert, 1965, reprint of the Paris, 1873 edition)
Kytzler, Bernhard, *Erotische Briefe der griechischen Antike* (Munich, 1967)
Theophylactus Simocatta
    complete Greek works (Heidelberg, 1599)
    Letters, translated by Copernicus (Cracow, 1509; photofacsimile reprint, Warsaw, 1953)
        translated by Cimedoncius (Heidelberg, 1598)
    Natural Questions (Leiden, 1596, 1597)
        ed. with the *Letters*, by Jean-François Boissonade (Paris, 1835; reprint, New York/Hildesheim: Olms, 1976)
    Questioni naturali, ed. Lidia Massa Positano (Naples, 1953; 2nd ed., 1965; Collana di studi greci, XXIII–XXIV); tr. Luigi Torraca (Naples: Libreria Scientifica Editrice, 1966; Scrittori Bizantini tradotti, I)
Zanetto, Giuseppe, "La tradizione manoscritta delle Epistole di Teofilatto Simocatta," Accademia nazionale dei Lincei, Bollettino del Comitato per la preparazione dell'edizione nazionale dei classici greci e latini (Rome, 1976)

NOTES ON THE FRONT MATTER

[1] The god Mercury, grandson of Atlas, in addition to his other functions, presided over sage discourse. Hence, in Corvinus' poetic imagery, Mercury helps to produce Toruń's splendid men. The descent of Mercury from Atlas is recorded by Servius, the commentator on Vergil's *Aeneid* (VIII, 130): "Maia, daughter of Atlas, gave birth to Mercury"; *Servii ... in Vergilii carmina commentaria* (Hildesheim: Olms, 1961; reprint of 1881–1884 ed.), II, 219/3.

[2] See Introduction, end of Section V.

[3] Sigismund I, king of Poland from 1506 to 1548.

[4] Jadwiga (c. 1174–1243) duchess of Silesia, became that region's patron saint.

[5] The River Oława winds through Wrocław.

[6] Ovid, *Fasti*, I, 18. This work on the Roman calendar was originally dedicated by the poet to Augustus (*Tristia*, II, 551). But after that emperor's death in 14 A.D., Ovid addressed to Caesar Germanicus, the heir apparent and administrator of the eastern region, where the poet was then living in exile, a second dedication containing the line quoted by Copernicus. He had access to a copy of Ovid's complete works in the library of his Chapter (ZGAE, 5:378); that copy may have been part of the war booty carried off to Sweden in 1626.

NOTES ON THEOPHYLACTUS' *LETTERS*

Letter 1

[1] Copernicus' "The cricket is a musical being" (*Musicum animal cicada*) misunderstood Ὁ τέττιξ ὁ μουσικός (Ho tettix ho mousikos). For, Copernicus' formulation assumes that all crickets chirp, whereas Theophylactus' chirping cricket is a male. "The female cricket is mute," said Aelian, *On the Characteristics of Animals* (I, 20). This highly popular work, written about 200 A.D., was one of Theophylactus' favorite sources of information (or misinformation).

[2] See Introduction, p. 16.

[3] Here Copernicus omitted τοῖς ᾄσμασι (tois aismasi, in its songs), presumably because he felt this expression to be superfluous in the presence of "musical being," "starts to sing" and "more vociferous."

[4] See Introduction, p. 11.

[5] Copernicus' dictionary defined τερετίσματα (teretismata) as "strictly the cricket's songs, but the ancients used this term for lewd songs." Hence Copernicus did not use his dictionary's equivalent of τερετίζει (teretizei), and instead coined *teretisat*.

[6] Letter 25 refers to the philosopher Plotinus' feeling of "shame that he was in a body."

Letter 2

[1] Copernicus wrote *non ... honoravi* in the singular (I failed to honor), even though his Greek text had the plural form οὐ ... ἐτιμήσαμεν (ou etimesamen, we did not honor). Theophylactus often oscillated between the singular and plural forms of the first person, while referring to the same individual. Copernicus was aware of this stylistic peculiarity, although he may not have known that it was a result of Theophylactus' ardent adherence to the school of highly sophisticated writers who carefully cultivated a rhythmic form of prose.

Letter 3

[1] Copernicus mistranslated εὐπρέπεια (euprepeia, good looks) by *reverentia* (dignity) because he erroneously linked εὐπρέπεια with ῥυτίδων (rhytidon, wrinkles: wrinkled dignity). He failed to connect ῥυτίδων with ἐγγύς (engys), perhaps because he was not familiar with the construction of an adverb (ἐγγύς) governing the genitive case (ῥυτίδων).

[2] Copernicus' *deludens* (deceive) does not exactly match ἐκφαυλίζουσα (ekphaulizousa). This was equated in his dictionary with *vilipendo* (hold in contempt). The use of artificial cosmetics does indeed hold lovers in contempt (provided they are deceived). Rather than follow Theophylactus in skipping this proviso, Copernicus made it explicit by translating ἐκφαυλίζουσα by *deludens* instead of *vilipendens*.

[3] Copernicus mismatches κεκτημένη, a participle in the middle voice, with *acquisita*, as though this passive participle (having been acquired) were in the middle voice (having acquired for oneself). The middle voice, while quite common in Greek, does not exist in Latin.

Letter 4

[1] See Introduction, p. 9. Because Copernicus made Horus a person, he had to treat the imperative νομοθέτει (nomothetei, impose!) as though it were νομοθετεῖ (imposes, *dictavit*).

[2] As Copernicus' *processit* (proceeded) shows, he failed to connect συγκεχώρηται (synkechoretai) with συγχωρέω (synchoreo), which his dictionary equated with *concedo, permitto, indulgeo* (allow).

[3] Copernicus' dictionary treated ἀπειλέω (apeileo, threaten, *minor*) as though it were identical with ἀπειλέω (apeileo, leap). Under the influence of παλινδρομεῖ (palindromei, turned back), he made the wrong choice, *retraxit* (rolled back).

Letter 5

[1] See Introduction, p. 9.

[2] See Introduction, pp. 11–12.

[3] For ὡς ἔοικε (hos eoike) Copernicus wrote *ut contigit* (it so happened). Hence, it has been argued, he did not know the meaning of ὡς ἔοικε. Where this expression occurs in Letters 22 and 62, he omitted it. These two omissions have been adduced as further indications of his incomprehension of ὡς ἔοικε.

But Letter 22's τὸ τῆς τύχης, ὡς ἔοικεν, ἄδηλον (to tes tyches, hos eoiken, adelon; the unreliability of Chance, as it seems) makes Chance's unreliability merely apparent. That unreliability is more than merely apparent, however; it is real. Exercising his editorial judgment, therefore, Copernicus deleted ὡς ἔοικε from Letter 22.

Letter 62 describes a farmer as happy and favored by Fortune, as it seems (ὡς ἔοικε). His abundant crops and numerous children make his happiness and good fortune real rather than apparent. Once more Copernicus exercised his editorial discretion in deleting ὡς ἔοικε from Letter 62.

With regard to lovers' promises, Letter 33 says that skepticism seems (ἔοικεν) safer than trustfulness. Here ἔοικεν could not be omitted. However, by changing "seems" to "is" (*est*), Copernicus wrote: "Skepticism is safer than trustfulness." This change proves, not his ignorance of ἔοικεν, but his knowledge of the value of lovers' promises.

Letter 55 attributes uninterrupted wakefulness to the immortals, whereas a moderate amount of sleep is our lot, ὡς ἔοικε (as it should be; in Copernicus' translation, *ut decet*). This is one of the two equivalents of ἔοικε in his dictionary.

Here in Letter 5 none of these equivalents appealed to Copernicus as appropriate. The farmers are leaving because of the unacceptability of their gifts to their foes. In this causal explanation, the Greek text lacked a verb. In the absence thereof, ὡς ἔοικε implies that the gifts *seemed* unacceptable. But in that case, why leave? Perhaps other gifts would prove to be acceptable, and there would be no need to leave. In fact, however, the farmers are leaving. That is why Copernicus changed from ὡς ἔοικε (it seems) to *ut contigit* (it so happened).

[4] Here Copernicus translated κεκτημένοις (kektemenois) correctly by *habentibus* (having), by contrast with his erroneous use of *acquisita* in Letter 3 (see n. 3 thereon).

NOTES

Letter 6

¹ Copernicus linked *pretiosam*, a feminine adjective, with *imaginem*, a feminine noun (valuable portrait). His *imaginem* matches εἰκόνα (eikona), which was recorded in his dictionary without any indication of its gender. Copernicus may therefore have erroneously supposed εἰκόνα to be masculine. In that case he would have felt no qualms in linking it with χρυσοῦν (chrysoun) to form a "valuable portrait." But εἰκόνα, being feminine, would require χρυσῆν (chrysen). In Copernicus' Greek text, Τὸν χρυσοῦν Καλλικράτην (Ton chrysoun Kallikraten) forms a masculine unit (Callicrates the wealthy).

² Copernicus inadvertently omitted *mihi* (to me), which is needed in order to match οἶμαι (oimai, I believe).

³ Although Parrhasius won the highest renown as a painter of portraits, Helen of Troy was not mentioned in antiquity as one of his subjects. Her portrait by Zeuxis was mentioned by Pliny (*Natural History*, XXXV, 66), but Theophylactus' limited knowledge of Latin did not enable him to read a literary work in that language. As an imperial functionary, he may have been familiar with some of the official Latin titulature. But he did not even recognize *rex* (king) as a Latin word, assigning it instead to "the language of the barbarians"; L. M. Whitby, "Theophylact's Knowledge of Languages," *Byzantion*, 1982, 52: 425–428.

⁴ For διαποικῖλαι (diapoikilai, to blend), Copernicus wrote *commutasti* (you changed), parallel to *coegisti* (you compelled). He thereby showed that he did not recognize διαποικῖλαι as an infinitive, parallel to ψεύσασθαι (pseusasthai, to be deceptive). Although his dictionary did not contain the compound verb διαποικίλλω, it did include the simple verb ποικίλλω (poikillo), which it matched with *vario* and *distinguo* (diversify, decorate).

⁵ Copernicus mistranslated γραφέως (grapheos, painter) as though his Greek text read γραφῆς (graphes, painting, *picturam*).

Letter 7

¹ Up to this point Copernicus translated Letter 7 virtually word for word. But the next five words in his Greek test were grammatically defective. Hence he resorted to conveying what he took to be their essential meaning. In so doing, he omitted Theophylactus' "the mares perform an act of generosity," presumably "for whatever foal of the brood" is involved. Theophylactus may have relied on Aristotle's *History of Animals*, IX, 4:

> When horses pasture together, if a mare dies, they nurse one another's foals. And as a whole the equine species seems to be full of affection (611a 9–12).

Or Theophylactus' more immediate source may have been Aelian:

> Whenever mares leave their foals as orphans before they are weaned, the other mares feel pity and bring them up with their own foals (III, 8); foals deprived of their mothers are brought up on the milk of another mare (VI, 48).

² See Introduction, p. 11.

³ Copernicus' *habentes* (having) shows that he mistook συνεχῆ (syneche) as a participle, presumably connected somehow with the verb συνέχω (synecho, hold together). Yet his dictionary equated the adjective συνεχής (syneches) with *continuus*. Because he substituted the adverb μονόθεν (monothen) for the participle μονωθέν (monōthen), he could not see the intended link between this participle and ἔγγονα συνεχῆ (engona syneche, direct descendant).

⁴ See Introduction, p. 15.

Letter 8

¹ See Introduction, p. 16.

² Copernicus was not aware that the characterization μέγα κακόν (mega kakon, big nuisance, *magnum malum*), applied here to a jealous man, echoed the condemnation of a big book as a big nuisance. This famous judgment, expressed by the poet Callimachus about 260 B.C., was familiar to Theophylactus, because it was quoted by Athenaeus in the third century A.C. (after Christ) in his *Deipnosophistae* (Kitchen Philosophers), at the beginning of Book III.

Letter 9

¹ See Introduction, p. 14.

² Theophylactus' ἀπηθαλώθησαν (apeithalothesan) is a form of ἀπαιθαλόω, which was not included in Copernicus' dictionary. Hence he had to guess at its meaning, and he guessed wrong with *languebat* (grew faint). But the noun αἴθαλος (aithalos) is present in his dictionary, where it is equated with two words for soot and hot ashes. Had he recognized the link between this noun and his verb, he would have seen that ἀπηθα-

λώθησαν meant "burnt up," not "grew faint." In Letter 28 he matched αἰθάλη (aithalei) correctly with *favilla* (ashes).

³ According to the Greek text, "An untrustworthy thing (χρῆμα, chrema) is passion and love." But Copernicus overlooked his dictionary's distinction between the singular form χρῆμα (thing) and the plural form χρήματα (chremata, money). Hence he also made money untrustworthy, a sentiment not to be found in Theophylactus.

Letter 10

¹ For Copernicus' omission of Theophylactus' third example of Nature's universality, see Introduction, p. 14.

² Unable to find ἀπακριβόομαι (apakriboomai) in his dictionary, Copernicus mistranslated ἀπηκριβωμένον (apekribomenon, sparse) as though it were ἀποκρυπτόμενον (apokryptomenon, hidden, *abditum*). His dictionary equated ἀποκρύπτω (apokrypto) with *abscondo* (hide).

³ As the equivalent of ἀπονέμω, Copernicus' dictionary gave *distribuo* (impart). But he mistranslated ἀπένειμε (aponeime) as though it were in the passive voice and meant "deprived" (*abstulit*).

⁴ Here φιλόνεικον (philoneikon) was used in its pejorative sense, "contentious," as it was also in Letter 63, where Copernicus matched φιλονεικότερον (philoneikoteron) with *contentiosus* (nothing provokes quarrels more than love does). Yet here Copernicus departed from the pejorative sense, even though gold is at once described as that "from which man's worst misfortunes arise" (*per quod maxima mala hominibus oriuntur*). Despite this condemnation of gold as a source of trouble, Copernicus translated φιλόνεικον here in its good sense (*amicabile*, friendly). He was affected by the misunderstanding explained just above in note 3 as well as by the close of Letter 10, where gold is praised for stimulating man's best efforts. Copernicus' dictionary recorded only the pejorative sense of φιλόνεικον, not its good sense.

⁵ Copernicus' Greek text read ταυτόν, a misprint, which he silently emended to τοῦτον (touton, this, *hanc*). But when he linked his emendation with ψόγου (psogou, censure), he could not find a proper equivalent for the verb and therefore settled for *deplorasti*. Had he realized that what was wrong with ταυτόν was the absence of the sign for the breathing (ταὐτόν), he would have correctly understood Theophylactus' meaning: "You have the same impression as the owl." For in Letter 31, where ταὐτόν is again combined with πέπονθας (peponthas), Copernicus' familiarity with this expression is shown by his translation (*comparatus es*, you are on the same footing as ...).

⁶ Theophylactus goes far beyond Aristotle (*History of Animals*, XI, 1; 609a 9–10) and Aelian (III, 9), who say only that the owl does not see well by day.

⁷ Copernicus' defective Greek text read προσφόροις (prosphorois). Imagining this to mean something like ἀρχαῖς (archais), he resorted to *principibus et praepositis* (princes and potentates). These then "obtained obedience" from unspecified persons (*obedientiam praestitissent*), just as "the scepter of royal power received respect" in the preceding clause. But if προσφοραῖς (prosphorais) is the correct reading, Theophylactus meant: "the subject would be honored with magistracies and bounties."

⁸ In the 1509 edition Copernicus' *examinatur* was misprinted as *exanimatur*. For an explanation of this misprint see Letter 64, n. 3.

⁹ In his *Orations* Emperor Julian the Apostate (331–363) declared:

> They say that the Celts also have a river which is an incorruptible judge of offspring, and neither can the mothers persuade that river by their laments to hide and conceal their fault for them, nor the fathers who are afraid for their wives and sons in this trial, but it is an arbiter that never swerves or gives a false verdict (Second Oration, 81 D).

In his Letter 59 Julian identified the Celtic river as the Rhine.

Letter 11

¹ For χθές (chthes, yesterday) Copernicus mistakenly wrote *pridem* (long ago) instead of *pridie*. But later on, in Letters 14 and 20, he matched χθές correctly with *heri* (yesterday; the definition given to him by his dictionary). Yet with this improved knowledge, in his characteristic manner he did not return to Letter 11 to correct the error he had made in *pridem*.

² See Introduction, p. 12.

³ This gibberish is the result of a defective reading in Copernicus' Greek text and a misprint in the 1509 edition. Copernicus' dictionary matched πρόσοδον (prosodon) with three words. Two of these (*ingressus, accessus*; entry, approach) denote motion, and he chose a synonym (*aditum*). But his dictionary also offered *proventus* (increase), which he ignored on account of the flaw in his Greek text. This read πόροις (porois, forays),

whereas Theophylactus meant πόνοις (ponois, labors; profit from the labors of others). That profit was described by Theophylactus as unjust (ἄδικον, adikon). Copernicus' *iniustum* (unjust), however, was misprinted as *iustum* (just) in the 1509 edition. It therefore had Copernicus muttering: "You will learn that you do not have a just approach through intrusive forays," instead of Theophylactus' "You will learn not to acquire unjust profit from the labors of others." That is what a bee-keeper does when his bees invade his neighbor's meadows.

Letter 12

¹ See Introduction, p. 9.

² Copernicus omitted τυγχάνουσι πρὸς ἡμᾶς (tynchanousi pros hemas), thereby missing Theophylactus' point that the test would determine how the lovers "happen to feel about me."

³ Melpomene does not claim to be a beauty among women, just as the raven is not regarded as a handsome bird. But Chrysogone is so ugly that even in her absence Melpomene is grouped with the attractive women. In like manner, on the scale of comparative avian pulchritude, the raven benefits from comparison with the jackdaw.

Letter 13

¹ For instance, Aristotle, *History of Animals*, I, 1; IX, 1, 46; Aelian, II, 11; XIII, 25.

² The 1509 edition's *adiuratus* was a misprint of *admiratus*. In Copernicus' handwriting, *mi* and *iu* tend to look alike (for example, NCCW, I, folio 1ʳ, line 2: *hominum*; line 7: *apparentium*).

³ Here Copernicus was baffled by ἐκλελάκτικας (eklelaktikas, you kicked aside), since his dictionary did not include ἐκλακτίζω (eklaktizo, kick aside). But it did match λακτίζω (laktizo) with *calcitro* (kick). Copernicus failed to think back, however, from the compound verb ἐκλακτίζω to the simple verb λακτίζω. Hence he missed the rejection expressed by ἐκλελάκτικας. Had he realized that rejection was involved here, he would have continued with his favorite causal conjunction *enim* (for; you have rejected your father's knowledge, for you have wasted your life on gambling). Instead of *enim*, however, he wrote the adversative conjunction *At* (But). This shows that in trying to guess the meaning of ἐκλελάκτικας, he conceived of a transmission of the father's knowledge to the son: *suppeditasti* (you had access to your father's knowledge, yet you wasted your life). Copernicus' dictionary matched χορηγέω (choregeo) with *suppedito*, which it defined as "having the opportunity to do something" (*habeo facultatem ad aliquid faciendum*).

⁴ See Introduction, p. 10.

⁵ The 1509 edition's *quemvis* was a misprint of *quamvis*.

⁶ Plato (*Laws*, II, 653A): "With regard to reason and true, solid opinions, whoever possesses them even into his old age is fortunate."

⁷ Copernicus inadvertently omitted πάλαι (palai, old).

Letter 14

¹ See Introduction, p. 12.

² Misconstruing τὴν εὐθυμίαν (ten euthymian) as an adjective, Copernicus attached it to ᾠδήν (oiden) and emerged with a "sweet song" (*suave canticum*). According to Theophylactus, however, the boy "projects his contentment (εὐθυμίαν) as postprandial song." This became a lullaby (*somnum admodum invitans*) in the hands of Copernicus.

³ See Introduction, p. 12.

⁴ Copernicus' dictionary matched θρέμμα (thremma) not only with *pecus* (beast) but also with *nutrimentum* (fodder). He failed to recognize διέσπαρται (diespartai) as a passive form: the beasts have been scattered. Instead, he treated this verb as though it were in the active voice, *distraxit*: Damalus' son has scattered the fodder. Hence, for Copernicus, the unexpressed subject of ἐστιν (estin) is Damalus' son, who "is sluggish in haggling with a customer" (*ad circumventionem emptoris ignavus est*). Copernicus was seduced into this mistake about a customer because in the next sentence the boy "sold the fertilizer at a low price." Had Copernicus paid closer attention to inflectional endings, he might have seen that εὐάλωτα (eualota) does not fit into his scenario. This adjective can be combined only with θρέμματα: the beasts are easy prey for a marauder (πρὸς ἔφοδον, pros ephodon). This last expression could not be accommodated even in Copernicus' highly imaginative version.

⁵ See Introduction, p. 11.

⁶ See Introduction, p. 8.

⁷ "When the grub has increased in size in the ground, it becomes a nymph, and then they are tastiest, before the husk breaks.... At first the males are tastier, but after copulation, the females are, for they contain white eggs" (Aristotle, *History of Animals*, V, 30; 556b 6–8, 13–14). "I saw some people stringing crickets together and selling them for food" (Aelian, XII, 6).

⁸ See Introduction, pp. 8-9.
⁹ See Introduction, p. 12.
¹⁰ The 1509 edition's "quandem" was a misprint of *quandam*.
¹¹ Here Copernicus omitted ὁ κλεινίας (ho kleinias). If this were a personal name, it had not been assigned to anybody in this Letter. Did it perhaps designate Damalus' son? Since he had been discussed without being named, Copernicus could not be sure. On the other hand, the mystifying expression has been interpreted as a word coined by Theophylactus from κλεινός (kleinos, famous) to denote "the renowned" individual. If so, it would have been applied ironically to Damalus' son, like ὁ θεσπέσιος (ho thespesios, the marvelous) a few lines above. Copernicus of course could not find κλεινίας in his Greek dictionary, nor can we in ours.
¹² For "yield the rest of the land," see Introduction, p. 13.

Letter 15

¹ Copernicus' *densus* followed his dictionary's mistranslation of Theophylactus' λάσιος (lasios, hairy).

² Copernicus' *missurum* (about to release) mistook the tense of παρεάσασα (pareasasa, after having released).

Letter 16

¹ See Introduction, p. 17.

² In his dictionary Copernicus could not find ἐπινέμεις (epinemeis) in the active voice. But the middle voice was recorded there, and was matched with *depascor* (consume). Since this was inappropriate here, Copernicus translated as though the Greek text read ἐπινέεις (epineeis; equated in his dictionary with "spin"). Thus he arrived at *adijcis* ("you add" misfortune to misfortune, instead of "you switch from" misfortune to misfortune).

³ Here Copernicus omitted ἔπιδε (epide, look at), because he did not recognize this imperative form of the verb ἐπεῖδον (epeidon; Look at borrowing! It brings ...).

⁴ Copernicus inadvertently omitted αὐτομάτοις (automatois) from this reference to the growth of two heads where one of the hydra's heads had been cut off. In Letter 40 he translated αὐτομάτως (automatos) correctly by *ultro* (spontaneously).

⁵ Copernicus did not translate κἂν ταῖς καθ' ὕπνον φαντασίαις (kan tais kath'hypnon phantasiais, even in your fantasies while asleep). By matching κἂν (kan) with *quamvis* (although), Copernicus' dictionary implied that the contracted form κἂν was composed of καί + ἄν (kai + an). But the dictionary gave no indication that κἂν resulted from the contraction of καί + ἐν (kai + en, even + in). Being unfamiliar with this contraction, Copernicus simply omitted it. The omission left him with "sleep fantasies," which he did not see how to connect with "beware of borrowing." Hence he dropped the reference to "sleep fantasies," and replaced it by "in accordance with all sound thinking" (*secundum omnem cogitationem*).

Letter 17

¹ Here Copernicus understood ὅρον (horon) correctly; see Introduction, p. 9.

² Copernicus' *occupationem* (business) was intended to match χωρίον (chorion). This was equated by his dictionary with *ager, locus, praedium, villa*. Any one of these four terms would have brought Copernicus closer to Theophylactus' thought: The place near the top of the hill produced that evil beast for us. Finding no room for θηρίον (therion, beast) in his version, Copernicus simply omitted it. The place near the top of the hill was the courthouse.

³ By changing συνεκεκρότητο (synekekroteto) from the passive voice to the active (*vocavit*), Copernicus made Sostratus and Lopho codefendants, with Leucippus implicitly the plaintiff. Actually, Leucippus is the judge, while Sostratus and Lopho were opposing litigants.

⁴ Copernicus did not recognize Theophylactus' reminiscence of Euripides, *Helen*, line 1192: σὰς διέφθαρσαι φρένας (thy spirit distraught; passive verb, associated with an accusative of specification). Instead, Copernicus misconstrued Theophylactus' διέφθαρτο (diephtharto) as a form in the middle voice, governing a direct object (Leucippus corrupted the minds of all: *omnium mentes corrupit*). Yet Theophylactus' ὅλας τὰς φρένας (holas tas phrenas) can only mean the whole mind of one person, not the minds of all persons.

⁵ Here Copernicus correctly equated διέφθαρται (diephthartai) with *corrupta est*. Yet he did not turn back a few lines to διέφθαρτο and correct the error explained above in n. 4.

Letter 18

¹ Copernicus omitted "from the male" (τοῦ ἄρρενος, tou arrenos) as unnecessary.

NOTES

[2] Copernicus wrote *segmenta* (parts) for ψῆνας (psenas), a word not to be found in his dictionary. However, on sig. n 5ʳ, in the proper alphabetical order he inserted ψηνη -ης ἡ (psene, -es, he), matched by him with *rasura* (cutting). No such feminine noun ψηνη is included in the standard Greek dictionary of our time, which records only the masculine ψήν (psen, caprifying insect). Where Copernicus found Greek words not recorded in his dictionary has not yet been determined.

[3] Previously, in his *Natural Questions* (Chapter 4) Theophylactus had discussed the procedure adopted when palm trees of opposite sex were close together:

> In the case of the palm tree, a distinction is made between the sex of the females and that of the males. Accordingly, when the male and female are near each other, you may see the female lowering her branches like loving tresses toward the male and extending herself in her lust for intercourse. But since there are times when her branches do not reach the male, the orchardists perform a trick to promote the consummation. For they tie her reeds to the male. In this way the female attains a feeling of enjoyment. She rises from her curved posture and retracts her branches to their proper position, as though she had achieved intercourse with the male through the reeds.

[4] Copernicus' *per picturam* confuses the portrait with Theophylactus' painter of the portrait (διὰ ζωγράφου, dia zographou).

[5] Copernicus' "or" (*sive*) would correspond to the conjunction ἤ. But then, for him, πίναξιν (pinaxin, a dative plural form) would have to be on the same footing as γραφή (graphe, a nominative singular form), with both together forming the subject of the verb. Actually, Theophylactus' ἡ was not the conjunction, but the definite article: with its panels let the painting provide me with a view of your image.

Letter 19

[1] Here Theophylactus applies to material wealth Plato's pronouncement (*Phaedo*, 67B) that "the impure should not apprehend the pure" truth.

[2] Copernicus translated πεπίστευται (pepisteutai) by *creduntur* (is believed) because his dictionary matched πιστεύω (pisteuo) only with *credo* (believe). Here, however, "entrust," the other meaning of πιστεύω, was intended: the wealth is not yours, but it has been entrusted to you.

[3] Copernicus' *sitiens divitias* (thirst for riches) misconstrued λιμώττων πλουτεῖς (limotton plouteis, you starve amid your wealth). Hence, for Copernicus, the similarity between Chryses and Midas was greed. For Theophylactus, however, the analogy was starvation, since everything Midas started to eat turned into inedible gold. This was the hurt (τὴν κακίαν, ten kakian) suffered by Midas. But since it vanished in Copernicus' erroneous version, he omitted τὴν κακίαν.

Letter 20

[1] For ἀπαίσιον (apaision) Copernicus' dictionary offered three equivalents: *infaustus, indecens, execrabilis*. Yet he preferred *indignum* (unworthy) to these other terms, which connoted something more sinister than unworthiness.

[2] Copernicus' *indignante* (impatient) does not match ἀπρεποῦς (aprepous, disreputable; *indecens* in his dictionary). He has just called the messenger *indignum* (unworthy). Had he labeled this girl *indigna*, he would have hit closer to the mark. But perhaps he wanted to avoid repeating *indignum, indigna*.

Letter 21

[1] Copernicus omitted "if you perform a song" as unnecessary. Stopping the ears with wax, Homer's prescription against the Sirens' enchanting but dangerous songs (Letter 82), was Theophylactus' recommendation to those within hearing distance of the dirges.

Letter 22

[1] Copernicus translated ἐτετύφωτο (etetyphoto, puffed up with conceit) as though it were ἐτετύφλωτο (etetyphloto, *obcaecabatur*, blinded). He was led into this error by τυφλώττεις (typhlotteis, you are blind) near the end of Letter 22, and by τυφλώττουσι (typhlottousi, they are blind) in Letter 20.

[2] Copernicus failed to link the participle εἰωθώς (eiothos) with Alexander, who "was not in the habit of being enticed," according to Theophylactus. Instead, Copernicus mistakenly associated the participle with Chance. As a result, Copernicus' Chance "usually entices," an affirmative statement, by contrast with Theophylactus' "Alexander was *not* in the habit..." The negative οὐκ (ouk), having been erroneously detached from the participle by Copernicus, appeared to him to be somehow incomplete. He filled this supposed gap by inter-

polating *inconsiderantes* (the imprudent). Consequently he emerged with "Chance usually entices the imprudent" instead of Theophylactus' "Alexander was not in the habit of being enticed."

[3] Theophylactus follows Plutarch (*Life of Alexander*, Chapter 43, 690B; *On the Fortune or the Virtue of Alexander*, 332F) in having Alexander reach Darius only after he was already dead. Darius' condition was described by the perfect participle πεπτωκότα (peptokota, having fallen). Mistaking the tense, Copernicus translated by the present participle *cadentem* (falling). Yet his dictionary gave πέπτωκα (peptoka) as the perfect tense of πίπτω (pipto, fall; sig. d 2ᵛ), and in the third marginal entry on sig. C 5ᵛ he himself wrote: "πεπτωκα *cecidi a πιπτω*." Hence he knew that πέπτωκα, like its Latin equivalent *cecidi*, was in the perfect tense.

[4] Copernicus' dictionary gave ὁ ὑπήκοος (ho hypekoos) in the masculine singular as "the willing subject," sig. l 1ᵛ, Copernicus' *subditus ille*. But the neuter singular τὸ ὑπήκοον (to hypekoon), as in Theophylactus, was not given in the dictionary as equivalent to "the subjects," taken collectively.

[5] For Copernicus' omission of ὡς ἔοικεν, see Letter 5, n. 3.

[6] Copernicus did not recognize ἀγγελθέντων (angelthenton) as a passive form. Changing it to the active voice, he translated: "when many hailed him for his victory" (*pluribus ipsum victoria salutantibus*). But Theophylactus meant: "when very many successes were reported to him all at once."

[7] Failing to recognize μῖξον (mixon, mix!) as an imperative form of the verb, Copernicus translated it by *mixtum est* (is mixed), as though it were an indicative form. Hence Copernicus did not realize that Theophylactus' Alexander was urging the divinity to "mix some adversity too with things when they are good."

[8] According to Theophylactus, Alexander was wary of εὐεξίας (euexias), a condition not found in Copernicus' dictionary. Being therefore compelled to guess, he did not guess right ("prosperity"), and guessed wrong ("vagaries of Fortune," *fortunae allusiones*).

[9] Here Copernicus omitted: "and you will receive lessons more important than instruction" (γνώσεως, gnoseos). In the text's next sentence, which transfers the metaphorical setting from the classroom to the courtroom, γνώσεως recurs in its original meaning of a "trial." This was unfamiliar to Copernicus, whose dictionary matched γνῶσις only with *cognitio, scientia, notitia*. All three of these terms suggest "knowledge," and he chose *scientiae*, which makes absolutely no sense there. Here, evidently discouraged, he omitted the entire clause.

[10] Copernicus wrote *mulctarum* (fines) because his dictionary matched πλημμέλημα (plemmelema) with the "levy derived from those condemned in a debt proceeding" (*proventus qui ex damnatis in aere procedit*).

Letter 23

[1] In Theophylactus, this accusation was addressed directly to Milo (you have harmed, ἐλυμήνω, elymeno). But Copernicus did not recognize this form of the verb as belonging to the second person. Instead, his translation shifted to the third person (*laesit*, he has harmed), without indicating who the third person might be. Letter 23, however, essentially concerns only two persons.

[2] See Introduction, p. 12.

[3] Copernicus put *pessime* (you scoundrel) here at the end of the question, instead of at the end of the conditional clause in the following sentence, where his Greek text placed παμπόνηρε (pamponere).

Letter 24

[1] Copernicus omitted ἐρεβοδιφῶντες (erebodiphontes, prying in the Nether Darkness) because his dictionary did not contain this echo of Aristophanes, *Clouds*, line 192. On the other hand, his dictionary did include ἔρεβος (erebos, Nether Darkness) and also διφάω (diphao, pry). But he failed to see that ἐρεβοδιφῶντες was a compound of ἔρεβος + διφάω.

[2] To match σκηπτός (skeptos, thunderbolt), Copernicus wrote *fulmen*, which was misprinted in the 1509 edition as *flumen* (river).

[3] Medea, the principal character in Euripides' play of that name, killed her brother and her own children. Phaedra, in Euripides' *Hippolytus*, made a false accusation against Hippolytus and thereby set off a train of events that culminated in his death and her suicide.

Letter 25

[1] Letter 25 was detached from the rest of Theophylactus' *Letters* in Paris manuscript 1389, which regarded the addressee as that Axiochus who is an interlocutor in the pseudo-Platonic dialog of that name dealing with the problem of death; see Introduction, n. 49.

[2] "Sleep... the brother of Death"; the "twins, Sleep and Death" (Homer, *Iliad*, XIV, 231; XVI, 672, 682).

[3] See Introduction, p. 15.

[4] Theophylactus is echoing the first sentence of Porphyry's *Life of Plotinus* (c. 204-270), the founder of the

Neoplatonist school of philosophy. Where ἐδόκει (edokei) appears in Copernicus' Greek text, Porphyry (234–c. 301) wrote ἐῴκει (eoikei).

⁵ See Introduction, p. 9. Here again, as in Letter 4, Copernicus confused the imperative νομοθέτει with the indicative νομοθετεῖ (*legem tulit*).

⁶ Copernicus was not aware that Theophylactus was quoting from Homer's *Odyssey* (IV, 220–221) the reference to the "drug ... that, banishing pain and anger, brings forgetfulness of all troubles." For Theophylactus, the healing drug was "reason" (λόγον, logon), which Copernicus misinterpreted as merely the quoted "saying" (*verbum*). In his dictionary, on sig. ? 7ᵛ, in the margin he wrote "νηπενθής ὅ *sine tristitia*." This Latin equivalent (*sine tristitia*) he used here in his translation. But the source from which he obtained νηπενθής misled him into believing that this adjective is a noun, as is shown by his ὅ (which, incidentally, should have no accent).

⁷ Here again, as in Letter 4, Copernicus missed the significance of δημιουργός (demiourgos) as the Creator; see Introduction, p. 9.

⁸ Copernicus inserted "life" (*vita*), for without this clarification his readers might not have understood Theophylactus' "treasure of the soul" (τὸ βασιλικὸν τῆς ψυχῆς, to basilikon tes psyches). Because Copernicus felt it necessary to insert *vita*, he changed θνητοῖς, θνητά (thnetois, thneta, mortal) to *mortuis, mortua* (dead).

⁹ A defect in Copernicus' Greek text caused him to misunderstand this sentence. It began, in his text, with οὐ (ou, not), instead of with οὗ (hou, that of which: that of which anyone lacks knowledge, about that he complains readily). "He complains" is κατηγορεῖ (kategorei) in the active voice. But the same form would occur in the middle and passive voice as "you complain." No wonder poor Copernicus became confused! In order to arrive at his result, he had to alter the meaning of ἀφήρηται (apheiretai, has been deprived) to "brought back" (*retulit*). He also interpolated *nobis* (to us), for which there was no equivalent in his Greek text. By these various detours he arrived at: "For nobody has brought back to us any knowledge of it [death]; you complain about it constantly" (*Nullus enim nobis scientiam eius retulit, hanc protinus accusas*). On the other hand, Theophylactus meant: "Anyone who has been deprived of knowledge about anything, complains about it very readily."

¹⁰ Niobe, weeping for the loss of her children, was turned into stone, according to a well-known myth.

Letter 26

¹ See Introduction, p. 11.
² See Introduction, p. 15.
³ Here Copernicus translated αὖθις (authis) correctly by *denuo* (again). Yet he did not return to Letter 25 to rectify his mistranslation of αὖθις there by *statim* (at once; see Introduction, p. 15).

Letter 27

¹ One such author was Achilles Tatius (I, 17):

> The magnetic stone loves the iron. If it merely sees it and touches it, it attracts it toward itself as though it had an internal yearning.

² In Copernicus' Greek text τοῦτο (touto) was defective. He tried to remedy this defect by *tanto magis ... quanto magis* (the more united, the livelier). A manuscript he did not see had a better remedy: it lives as long as (τοσοῦτον, tosouton) it is united.

³ Copernicus' *Postquam* (after) is not the best correlative with *continuo* (at once). His *Postquam* translated ὁπηνίκα (hopenika), which was not to be found in his dictionary. However, he wrote ὁπηνίκα in the margin (sig. a 1ᵛ), and equated it with *quando* (when, whenever) there, and also in Letter 28. *Quando* would have been a better correlative here with *continuo* ("whenever ... at once," rather than "after ... at once").

⁴ As pointed out in the Introduction (p. 22), the 1509 edition misprinted *mammatis* (breasted) for Copernicus' *inanimatis* (lifeless). In his handwriting, *inani-* could easily be misread as *mam-* like *min/nim* (Letter 10, n. 8), and *mn/min* (Letter 84, n. 1).

⁵ Here Copernicus omitted φίλτατε (philtate, O dearest one). This omission led him to link ἐπὶ τοσοῦτον (epi tosouton) with πάθοιμι (pathoimi), now directly before this phrase, instead of with ἀπολιμπανομένη (apolimpanomene), right after it. As a result, Copernicus missed the question: "What would happen to me if I were deprived of your presence for so long?"

⁶ Having missed the question translated at the end of n. 5, just above, Copernicus did not realize that the answer to this question begins with οὕτω λυπήσαιμι (houto lypesaimi): Under those conditions I would sadden (those who are already sad). Instead of this intensification of the feeling of grief, Copernicus imagined a mutual interaction: I make those sad who make me sad (*contristem contristantes me*). But in order to establish this mutuality, he had to interpolate *me*, for which there was of course no warrant in his Greek text.

⁷ Copernicus' *cupidinis* (Cupid's) missed the parallel between τοῖς ἐρῶσι (tois erosi, those in love) and τοὺς λυπήσαντας (tous lypesantas, those who are sad). Again, an intensification of feeling is intended. Just as those who are already sad would become sadder, so those in love would become more passionate.

⁸ Here Copernicus omitted ἀφροδίσιος (aphrodisios, passionate). Later on, in Letter 72, he translated ἀφροδισίοις (aphrodisiois) by *venereis* (amatory). Here, however, he must have felt ἀφροδίσιος to be just another of Theophylactus' superfluities since it is juxtaposed with *cupidinis iaculum* (a lover's dart).

Letter 28

¹ Without acknowledgment of Theophylactus' authorship, Letter 28 was used as the example of a conciliatory letter in a treatise on the various kinds of letters (*De forma epistolari*, misattributed to Proclus or Libanius; Hercher, p. 12/no. 36). Letter 28 was shortened and modified to suit the compiler's taste.

² Copernicus mistakenly linked λοιπόν (loipon), which is not a feminine form, with τὴν λύμην (ten lymen), both of which are feminine. Hence he did not realize that Theophylactus, using λοιπόν as an adverb here, meant: "It is time to purge your heart." Copernicus' previous error with regard to λοιπόν is discussed in the Introduction, p. 13.

³ Copernicus mistranslated λύμην (lymen, hurt) by *tristitiam* (sorrow), his dictionary's equivalent for λύπη (lype). While looking for λύμη, his eye must have come to rest at λύπη. It may have stopped there because he had just used *tristitiam* (correctly) for ἀνίαν (anian). Whereas the Aldine edition together with its family *b* of manuscripts read λύμην, λύπην appeared in family *c* and in two additional Paris manuscripts (Zanetto, p. 71/17–18).

⁴ Theophylactus, *Natural Questions*, chapter 7: "Sailors, I hear, manage to calm the sea by spreading oil on it, contriving in this way to tranquilize it when it is turbulent." As the author of his own *Natural Questions*, Theophylactus may have been familiar with a similar work, *Causes of Natural Phenomena*, by Plutarch, who asks (No. 12): "Why does the sea become clear and calm when oil is spread over it? Does the wind slipping over the smoothness so caused produce no impact or swell?" In *History of Technology*, eds. A. Rupert Hall and Norman Smith (London, 1978), John C. Scott, "The Historical Development of Theories of Wave-Calming using Oil," gives our author's name as "Theophilus Simocrate" (p. 166/8).

Letter 29

¹ Copernicus translated δαιτυμών (daitymon) by *conviva* (table companion), just as his dictionary did. But he may have known Homer's *Odyssey*, IV, 621–624:

> Table companions (δαιτυμόνες, daitymones) came to the palace of the godlike king, driving their sheep, bringing wine, that joy of men, and bread sent them by their wives wearing beautiful headbands. Thus did they prepare the feast in the halls.

In Letter 73 Copernicus translated δαιτυμόνες (daitymones) by *epulari* (feasting).

² Copernicus was surely unaware that here Theophylactus was quoting line 78 of the *Farmer*, a play by Menander that has been only partially preserved.

³ In Aeschylus' *Libation Bearers* (lines 1048–1062) Orestes goes mad after avenging the murder of his father by killing his mother and her lover.

⁴ Copernicus wrote *animo* (pride) because his Greek text had the wrong reading ψυχῆς (psyches). It did not occur to him that the correct reading was τύχης (tyches, fortune): together with the [change in our] fortune.

Letter 30

¹ Copernicus found διασύρεις (diasyreis) equated in his dictionary with *detraho* (disparage). But the construction governed by διασύρω was not indicated in his dictionary. Hence, he did not realize that it is a transitive verb, like σκώπτεις (skopteis, mock), to which it is joined by καί (and), with both these verbs governing the same object (ἡμᾶς, hemas, me; you disparage and mock me). By separating the two verbs and treating διασύρεις as though it were intransitive, Copernicus emerged with "You disparage my lovers" (*Amatores mihi detrahis*). But Ἐπὶ τῶν ἐραστῶν (epi ton eraston) means "In the presence of my lovers."

² Copernicus' Greek text read τὸ καὶ τό (to kai to, this and that), which he translated faithfully by *passim* (you mock at me for this and that). This vague expression, wholly inappropriate in this context, was a scribal error for τοκετῷ (toketoi, pregnancy). Although τοκετῷ and τὸ καὶ τό do not look alike, they sound almost the same. A scribe who was producing a new manuscript by listening to the finished work being read aloud might easily imagine he heard τὸ καὶ τό rather than τοκετῷ. This substitution, on the other hand, would hardly have been made by an amanuensis who was making a fresh copy by looking at a written manuscript.

³ Copernicus did not recognize σφριγῶν (sphrigon) and διαπαίζουσα (diapaizousa) as participles associated

with Calliope. Instead of having her laugh out loud, he turned σφριγῶν into a noun (*virescentiam*) and linked διαπαίζουσα with Rhodina. She "mocks her body's development" in Copernicus' mistranslation (*virescentiam corporis defraudanti*). Toward the end of Letter 30, however, where διαπαίζουσα is repeated, Copernicus linked it correctly with Calliope.

[4] Here Copernicus omitted ἢ τοὺς τοκετοὺς αἱρετωτέρας (e tous toketous hairetoteras, more desirable than births).

[5] See Introduction, p. 8.

[6] See Letter 24, n. 3.

[7] Copernicus' *terram* (earth) matched his Greek text's γῆν (gen). But τὴν δίκην (ten diken, Justice) was probably the correct reading.

Letter 31

[1] See Introduction, p. 8.

[2] Copernicus omitted "and supereminence" (καὶ ὑπέρογκον, kai hyperonkon), presumably because he felt this addition to be a characteristic exuberance on the part of Theophylactus.

[3] Instead of *femineam*, the 1509 edition has *pulchritudinem* (beauty). Its presence here, between *pulchritudine* in the line above and *pulchritudinem* two lines below, is evidently due to dittography, a common enough fault in Copernicus.

[4] Again Copernicus mistook "Persian" for "industrious." Contributing to his confusion was the contrast between the peacock's activity and the laziness (*pigritia*) of the writer discussed in the following paragraph.

[5] Thales, the earliest Greek philosopher, left no writings.

[6] See Introduction, p. 16.

Letter 32

[1] In this echo of Aristophanes, *Knights* (line 9), Copernicus failed to recognize γερόντιον (geronion) as Poas' address to Ampelius as "you little old man." Instead, Copernicus mistakenly linked γερόντιον with ξυναυλίαν (xynaulian) as though it were γερόντειαν (*consonantiam vetulam*, old harmony).

[2] In Copernicus' Greek text, the misprint πλουτήσαιεν (ploutesaien), a plural form, misled him into regarding "the rains" as its subject. He persisted in this error despite ἐστιν (estin) and χαλινοῖ (chalinoi), both of which are singular forms. The subject of all three of these verbs is ποταμός (potamos, river): "But if the river is also flooded (πλουτήσειεν, plouteseien) with rains, it is more destructive than fire, and does not keep its wildness within bounds."

Letter 33

[1] Herbert W. Parke and D. E. W. Wormell, *The Delphic Oracle* (Oxford, 1956), I, 24.

[2] Copernicus wrote *lynce* (lynx, the animal), whereas his Greek text read λυγκέως (Lynkeos, of Linceus, the mythical hero). For the animal, the corresponding genitive form would be λυγκός (lynkos). Both the hero and the animal were reputedly sharp-sighted. Theophylactus is quoting Aristophanes' *Plutus*, line 210.

[3] Copernicus mistakenly linked ἐρωτικῶν (erotikon, a genitive form) with οἱ νέοι (hoi neoi, the young, a nominative form) instead of with the genitive ἡδονῶν (hedonon, pleasures). Hence he emerged with "For, young lovers are intoxicated with pleasures" (*amantes enim iuvenes voluptatibus inebriantur*).

Letter 34

[1] This famous name was misprinted "Themistodes" in the 1509 edition. In Copernicus' handwriting, *cl* might easily be mistaken for *d* (for example, NCCW, I, folio 1ʳ, line 10). See Introduction, n. 49.

[2] This tale is an elaboration of a fable in Aesop; see Ben Edwin Perry, *Aesopica* (Urbana: University of Illinois Press, 1952), p. 361, no. 101.

[3] Copernicus' Greek text was defective: Ὀλύμπιον ἐπρεσβεύοντο ἡγεμόνα (Olympion epresbeuonto hegemona). He tried to solve this puzzle by having the birds "acknowledge the Olympian as their ruler" (*Olympium praeficiebant ducem*). His attempted solution, however, is not in agreement with the rest of the fable, since the birds did not acknowledge Jupiter as their ruler. Instead, they asked him to appoint a king over them. "He granted the suppliants a splendid gift, namely, their request for the appointment of a king" (*dedit supplicantibus magnum donum, regalis videlicet dignitatis postulationem*). In his translation of this sentence, Copernicus had the solution of the puzzle in his hands. But of course he was not aware that in the archetype of the manuscript from which the Aldine edition was printed ἡγεμόνα was followed by σφίσιν ἐγχειρίσαι. These two words had inadvertently been omitted at an early stage of the copying and recopying process, and were later restored or replaced by an equivalent expression, but not in the manuscript on which the Aldine edition was based; see Zanetto, pp. 68–69.

⁴ Copernicus' *denuo* (again) mistranslated εὐθύς (euthys, at once) as though it were αὖθις (authis, again).
⁵ See Introduction, p. 16.
⁶ Here Copernicus translated αὖθις (authis) correctly by *denuo* (again), as he had in Letter 26. Yet once more he did not go back to Letter 25 to rectify his mistranslation of αὖθις there by *statim* (at once).
⁷ The 1509 edition misprinted "admo-modum." Copernicus wrote *admodum* (certainly) to match καθάπερ ὕπαρ (kathaper hypar, in reality), which seemed to him redundant in combination with *veritatem admodum proclamat* (certainly teaches a truth).
⁸ To match σεμνυνόμεθα (semnynometha, we take pride), the 1509 edition printed *privabimur* (we shall be deprived). It printed *privabimur* again, at the end of the same line. The first *privabimur* is surely an example of the common printer's error known as anticipatory dittography. Instead of the first *privabimur*, Copernicus may have written *gloriamur*.

Letter 35

¹ Copernicus wrote "Cicanias" four times in this Letter. His Greek text, however, read "Tycanias," a name made up by Theophylactus from "tykane" (a threshing instrument). This substitution of C for T was not a printer's error. For in Copernicus' handwriting these capital letters do not look at all alike, as may be seen from the enlargements in NCCW, I, Plate XII.
² Copernicus wrote *pascuis* (pastures), even though θρέμμασιν (thremmasin) was matched in his dictionary with *pecus* (cattle) and *nutrimentum* (fodder).
³ Copernicus translated τὰ δόξαντα (ta doxanta) by *opiniones*, which was replaced in Baranowski's edition by *conditiones*. This gratuitous emendation misled PII, 88, n. 57, to conclude that τὰ δόξαντα means *das Beschlossene* (the agreement), which Copernicus has already expressed as *pacta*. Here, however, τὰ δόξαντα has its usual meaning, public opinion, correctly rendered as *opiniones* by Copernicus.
⁴ Copernicus inserted *inquit* (he argues) for the purpose of making clear that Myronides' lack of oxen was Tycanias' justification for breaking his promise. Nothing in Copernicus' Greek matches *inquit*. He introduced it in order to clarify this passage, which is regarded by the experts as hopelessly incomprehensible. The main difficulty is ἄντλην, a sequence of letters suggesting a misprinted form of ἀντλίαν (antlian, filth). This was equated in Copernicus' dictionary with *sentina*. He assumed that *sentina* was in turn another misprint, for *sententia* (statement). But when he put "statement" in his translation, he felt the need of making clear to his readers whose statement this was. That is why he departed from his Greek text by inserting *inquit*. He had no way of knowing that, instead of ἄντλην, the correct reading was ἀπάτην (apaten, guile).

Letter 36

¹ Copernicus translated ἀπόγονον (apogonon, daughter) by *natam*. The 1509 edition, however, repeating "or" from "Diod*or*o," printed *ornatam* (painted), thereby transforming the girl's father into her painter.
² Copernicus' dictionary did not include ἐκτόπως (ektopos, from afar). He translated this word in its derivative sense (*plane*, quite).
³ See Introduction, p. 10.
⁴ Copernicus wrote *consueverunt* (they were accustomed), referring to Erasmius' eyes. Copernicus did not recognize εἰώθειν (eiothein) as a verbal form in the first person singular (as was my usual experience), just as he misunderstood εἰωθώς (Letter 22, n. 2).
⁵ By missing Melanippe's musical accomplishments, Copernicus made it hard for his readers to understand why Erasmius' "eyes have become ears also."
⁶ Copernicus omitted τὴν θεωρίαν ἠγνόηκα (ten theorian egnoeka, this is a sight I have not seen). This unusual insertion of a short independent sentence inside a longer independent sentence evidently puzzled him.
⁷ See Introduction, p. 10.
⁸ For πανικῆς (panikes, phantom), Copernicus wrote *repentina* (unforeseen), which was only the first word in his dictionary's lengthy explanation. On the other hand, for μετειληφέναι (meteilephenai) it offered him five perfectly satisfactory equivalents. But in order to locate them, he would have had to know that he had to look for μεταλαμβάνω (metalambano, share). Instead, he wrote *praeventum ... esse* (have been blocked), perhaps under the influence of φθάνω (phthano, anticipate).

Letter 37

¹ Copernicus mistranslated συνανίπτανται (synaniptantai, fly off together) as though it were συναπονίπτονται (synaponiptontai, washed away, *abluitur*). His dictionary did not include the compound verbs συνανίπταμαι (synaniptamai, fly off together) and ἀνίπταμαι (aniptamai, fly away). But it did match the simple

verb ἵπταμαι (hiptamai) with *volo* (fly). Instead of using this base, Copernicus focused on νίπτω (nipto, wash) and conjured up a non-existent compound συναναvίπτω.

Letter 38

[1] Copernicus' *detraxisti* (take off) mistranslated διήλλαξας (diellaxas, changed), as though it were a form of διαλοξεύω (dialoxeuo, cast aside).

[2] Instead of *emisisti* (you released), the 1509 edition misprinted *emisti* (you bought), presumably under the influence of *emisse* two lines below.

[3] *Odyssey*, IX, 345–359.

[4] At this point Copernicus' Greek text read: "O Tettigo." But the heading over Letter 38 names Tettigo as the writer, not the addressee. Hence Copernicus omitted "O Tettigo" rather than reverse the order of the names in the heading, and thereby make Tettigo the addressee.

Letter 39

[1] Copernicus inserted *simul* (at the same time) in order to make quite plain Letter 39's message: only one love affair at a time.

[2] The 1509 edition misprinted *patitur* (permits) for *partitur* (is divided).

Letter 40

[1] Letter 40 begins with the paradox Ἀδικεῖται μὲν οὐδείς, ἀδικοῦσι δὲ πάντες (Adikeitai men oudeis, adikousi de pantes; On the one hand, nobody is injured; on the other hand, everybody does harm). The second part of the paradox (ἀδικοῦσι δὲ πάντες) was omitted by Copernicus.

[2] Leonides, in Copernicus' Greek text (Letter 49 is ascribed to a Leonides). Letter 57 is addressed to a Philonides.

[3] Copernicus mistranslated ἐλυμήνατο (elymenato, hurt) as though it were ἐλυπήσατο (elypesato, saddened, *contristavit*). Yet in Letter 23 he had matched ἐλυμήνω (elymeno) with *laesit* (harmed), and λυπήσαιμι, λυπήσαντες (lypesaimi, lypesantes) with *contristem, contristantes* (sadden, Letter 27). In Letter 28, however, he confused λύμην (lymen, hurt) with λύπην (lypen, sorrow), the same sort of error he repeated here (see Letter 28, n. 3).

[4] Copernicus' "since he does not exist, but did exist" (*cum non sit, sed fuit*) mistakenly changed the tense of ἦν (en) from imperfect to present, and altogether missed Theophylactus' distinction between the eternal subsistence of the immortals and the temporary existence of mortals.

[5] *Odyssey*, IX, 408.

Letter 41

[1] Copernicus used *alterationem*, which was labeled a word coined by him (PII, 94/n. 70). But it occurs nearly a thousand years earlier in Boethius' *Commentary on Porphyry*, IV, and *Commentary on Aristotle's Categories*, IV; Migne, *Patrologia latina*, 64:118 B, D; 290A.

[2] Copernicus preferred "seems" (*videtur*) to ἐστι (esti, is): what is not present may seem better than what is at hand, without actually being so.

[3] Copernicus translated αἰωρεῖν (aiorein) by *assumant* (choose) as though his text read αἱρεῖν (hairein), which his dictionary equated with *capio* (take). It also matched αἰωρέω (aioreo) with *suspendo* (hang). But the highly unusual expression αἰωρεῖν τὸν θάνατον (aiorein ton thanaton, die by hanging) evidently baffled Copernicus. Many later students of Theophylactus reacted in the same way to this expression.

Letter 43

[1] See Introduction, p. 15.

[2] A sorceress tells Ulysses: "A spirit impervious to enchantment resides in your breast" (*Odyssey*, X, 329).

[3] For the word ἀδρανές (adranes, ineffectual), Copernicus mistakenly substituted ἀνδρεῖον (andreion), which his dictionary translated as *virilis* (manly), by contrast with the eunuch "half a man."

[4] Copernicus mistranslated τί (ti) by "why" (*quid*), because he did not notice that the acute accent was transferred from the following word to τι (ti), the enclitic indefinite pronoun.

[5] Copernicus' dictionary offered him seven equivalents for σοβαρός (sobaros). He used first, *arrogans*, in the comparative form *arrogantius*, to match σοβαρώτερον (sobaroteron). Yet Diogenes would hardly characterize his own utterance as arrogant.

[6] In Homer's *Iliad*, V, 348–351, Diomedes warns the goddess Venus to stay away from battle.

[7] Copernicus transferred the adjective πολλήν (pollen, great), from "forbearance" to the verb. Hence,

Copernicus' *multum praedicant* (emphasize) is twice removed from κατηγοροῦσιν (kategorousin, censure). Whereas Theophylactus meant: "My friends censure me for my great forbearance," Copernicus emerged with the opposite attitude: "My friends loudly proclaim my forbearance."

⁸ Copernicus' Greek text printed εἰλακτίσειεν without separating the conditional conjunction εἰ (ei, if) from the verb λακτίσειεν (laktiseien, kicking). Of course he could not find any εἰλακτίζω in his dictionary. But he did come across ὑλακτέω (hylakteo), equated with *latro* (bray). He therefore wrote *latrasset*, as though translating ὑλακτίσειεν (hylaktiseien) instead of λακτίσειεν. Through no fault of his own, Copernicus failed to recognize the reference to Socrates. Hence he did not realize that Theophylactus was recalling the anecdote in Diogenes Laertius (II, 21), about Socrates, who was often physically attacked by his opponents in an argument, yet did not respond in like manner. Once, "when he was kicked, and someone expressed surprise that he did not react, he said: 'Had a donkey kicked me, would I have sued him?'"

⁹ See Introduction, p. 9.

Letter 44

¹ According to the Greek text, however, "everything has been prepared."

² Copernicus had to guess how to characterize these tunes, since ὑπερύμνους (hyperhymnous) in his Greek text is a nonexistent form. The most closely similar word ὑπερυμνήτους (hyperymnetous) would mean "highly extolled."

Letter 45

¹ Copernicus wrote *tristes* (bitter), as though his Greek text read ἄνια (ania) instead of ἄνισα (anisa, unfair).

Letter 46

¹ "Only Alexander mounted this Bucephalus, because it deemed all the other riders unworthy" (Arrian, *Anabasis of Alexander*, V, 19, 5).

² Unfamiliar with ἔσο (eso) as a rare imperative form of the verb "to be," Copernicus translated it by *eris*, as though it were the future tense "you will be" (heedful; ἔσει, esei).

Letter 47

¹ For ἀμειβόμενος (ameibomenos) Copernicus wrote *recepturus* (receive), thereby making Poimnio a rather ungracious participant in this transaction.

Letter 48

¹ Copernicus' dictionary did not contain the verb ὡραΐζονται (horaizontai, adorned). But it included the related adjective ὡραῖος (horaios, beautiful) and noun ὡραιότης (horaiotes, beauty). Hence he could not have been unaware of the meaning of ὡραΐζονται. Turning it into its opposite *deformantur* (are disfigured), he may have felt, would bring Theophylactus' somewhat strained expression closer to his readers' experience of life.

Letter 49

¹ See Introduction, p. 15.

² Copernicus' *pueri hostis* mistakenly linked ὑπὲρ παιδός (hyper paidos) with the preceding word πολέμιον (polemion, foe of the boy), instead of with the following word ἐπρεσβευσάμην (epresbeusamen, communicated about the boy).

³ Copernicus' *mitto* (I am sending) missed the conditional construction ἄν ... ἐπρεσβευσάμην (I would have communicated).

⁴ By erroneously substituting the factual statement "I am sending" for Theophylactus' conditional construction "I would have communicated," Copernicus created an inconsistency with the indicated order of precedence. For, the gift cannot precede the request, if Leonides is sending the request (as Copernicus' mistaken *mitto* declares). The gift can precede the request only if Leonides does not actually send the request, but "would have" done so.

Letter 50

¹ Here Copernicus omitted γερόντιον (gerontion, you little old man, you!), even though his dictionary matched it with *vetulus*. Previously he had misconstrued γερόντιον (Letter 32, n. 1).

Letter 51

¹ Copernicus' *nuntia amicitiae* (arranger of the friendship) mistook φιλία (philia) as the noun (friendship)

instead of the adjective φίλια (friendly). Because the old woman at the head of the procession (πρόπομπος, propompos) was friendly, Rhodoclea greeted her. The connection between the old woman's friendliness and Rhodoclea's greeting is broken by Copernicus' mistranslation of φιλία/φίλια.

Letter 52

[1] Copernicus omitted τὸ ἦθος (to ethos, as their characteristic).
[2] Aelian, IV, 15.
[3] In the 1509 edition *prudentiores* was misprinted for *impudentiores*, matching ἀναιδέστερον (anaidesteron).

Letter 53

[1] Copernicus did the best he could with his defective Greek text: ἀλλότριον ἐπαγγέλλεται (allotrion epangelletai, *iniquum perhibetur*). He had no way of knowing that an unpublished manuscript (Paris 3047) contained what Theophylactus had actually written: ἀλλοτρίοις γὰρ δώροις τὸ φιλότιμον ἐπαγγέλλεται (allotriois gar dorois to philotimon epangelletai, the river is considered generous because it gives away what belongs to others). Copernicus' text apparently lost a whole line: -οις γὰρ δώροις τὸ φιλότιμ-.
[2] Theophylactus' δῶρα ποταμῶν (dora potamon) echoes Herodotus' famous description of Egypt as δῶρον τοῦ ποταμοῦ (doron tou potamou, a gift of the River Nile; II, 5).
[3] At this point Letter 53 in the Aldine edition stops, because that is where its (lost) prototype stopped. But other families of Theophylactus manuscripts continue with: "and be punished by the arbitrator's decisions" (Zanetto, p. 77/1-2).

Letter 54

[1] As a translation of διέπτησαν (dieptesan, flown away), Copernicus' *pervenerunt* is not quite adequate since it fails to suggest the idea of flight. He did not connect διέπτησαν with διαπέτομαι or διίπταμαι (diapetomai, diiptamai, fly away). Neither of these verbs was included in his dictionary. He similarly failed to link συνανίπτανται with συνανίπταμαι (which is likewise missing from his dictionary; Letter 37, n. 1).
[2] See Introduction, p. 10.
[3] See Introduction, p. 12.
[4] The preceding Note 3 refers to a misunderstanding which caused Copernicus to omit μοι (moi, to me) from his translation. He thereby misdirected Jason's messages, agreements, and oaths from Medea, the writer of Letter 54, to "another girl."

Letter 55

[1] See Introduction, p. 14.
[2] In the *Odyssey*, Ulysses is ordered to communicate with the dead, and for this purpose his ship "came to the limits of [the world], the deep-eddying Ocean," where the sun never appears, whether rising or setting (XI, 13–19). The unconventional comparison of uninterrupted sleep with "an ocean" baffled Copernicus. Two datives (ocean, sleep) are associated with χρώμενος (chromenos, using). Choosing "ocean" as the instrumental dative and "sleep" as the secondary dative, Copernicus mistranslated: Ulysses "uses the ocean as some sleeping" [thing; *oceano tamquam dormiente quodam utitur*]. For the Aldine edition's χρώμενος, Boissonade (p. 65/1, p. 294/n. 8) substituted Paris 2991's νηχόμενος (swimming). This was repeated by Hercher (p. 779/26), whose translation was *somni cuidam oceano innatans* (swimming as though in an ocean of dreams). But Ulysses left the earth by alertly crossing Ocean in a boat, not by swimming.

Letter 56

[1] "If a tree located near a boundary sends its roots into a neighbor's land too, it becomes common property" (*Corpus iuris civilis*, II, 1 :31).
[2] Here Copernicus translated γερόντιον correctly, by contrast with his error in Letter 32 (n. 1) and his omission in Letter 50 (n. 1).

Letter 58

[1] Failing to recognize νοείσθωσαν (noeisthosan) as an imperative form of the verb, Copernicus mistranslated it as though it were in the indicative mood (sons are deemed, *existimantur*). They are to be deemed pledged for life to their parents. For, every father is naturally his own son's teacher, better even than Socrates, reputedly the greatest of all Greek teachers.
[2] See Introduction, p. 11.

Letter 59

¹ Making the same mistake here as in Letter 46 (n. 2), Copernicus mistranslated ἔσο (eso, Be!) as though it were ἔσει (esei, you will be).

² Copernicus wrote *saepibus* (fences), because αἱμασιᾷ (haimasiai, a stone wall) was so mistranslated in his dictionary.

³ As Copernicus' *bellum* shows, he mistook πολέμιον (polemion, foe) for πόλεμον (polemon, war). Later on, however, in Letters 68 and 71 he matched πολέμιον, πολέμιοι correctly with *hoste, hostes* (foe, foes). Yet with his improved knowledge he did not come back here to Letter 59 in order to correct his confusion of πολέμιον with πόλεμον.

⁴ The cooperative labor of the ants was depicted by Aelian (II, 25).

Letter 60

¹ Copernicus failed to recognize that "Brion" in his Greek text was a less suitable reading than "Orion" (in Paris 2991).

² See Introduction, p. 10.

³ Here Copernicus matched παρώσαντο (parosanto) correctly with *dereliquerunt* (rejected). Previously, in Letter 54, he had misunderstood παρώσω (paroso), because the compound verb παρωθέω (parotheo) was not included in his dictionary. The simple verb ὠθέω (otheo = *expello*), however, did appear there. When Copernicus grasped the connection between the simple verb and its compound, he translated the latter properly here, but did not turn back to rectify his error in Letter 54.

⁴ For ἄωρον (aoron) Copernicus' dictionary offered three equivalents: *deformis, indecens, intempestivus*. Using none of these three, he wrote *despectus* (contemptible), which is quite close in meaning to the first two. The last one, however, is what Theophylactus had in mind: the combination old man/young prostitute is untimely (*intempestivus*).

⁵ Copernicus' dictionary did not give him ὑπερωφρυωμένον (hyperophryomenon). Neither does our modern dictionary. But he reduced the compound verb to its simple form ὀφρυόομαι (ophryóomai). This he wrote in the margin of his dictionary opposite ὀφρῦς (ophrys, eyebrow) and ὀφρυόεις (ophryoeis), which his dictionary translated as *altus* (high). Hence he chose *laudatum* (exalted). He thereby missed the pejorative meaning conveyed by Theophylactus' compound verb, "supercilious in his uprightness."

⁶ Copernicus missed the connection between this conclusion and its basis: since Time and Chance can accomplish much (*multa enim tempus et fortuna potest*), why is it too presumptuous to hope for this (*id autem sperare nimis est praesumptuosum*)? The hope in question is that a "wise man endures everything" (*omnia ferre sapientem*). In this context, "everything" includes abstention from sex in old age: to hope [for this (abstention)] is much more reasonable since Time and Chance can accomplish much. Copernicus' dictionary matched ἔμφρων (emphron) with *prudens* (reasonable). Yet his *nimis ... praesumptuosum* did not come at all close to ἐμφρονέστερον (emphronesteron, more reasonable).

Letter 61

¹ Penelope, Ulysses' wife, pretending to know her husband was dead, asked her suitors to wait until she finished weaving a shroud. "And then by day she wove the massive cloth, but unraveled it by night" (Homer, *Odyssey*, II, 104–105).

² Perry, *Aesopica*, p. 475, no. 373.

³ See Introduction, p. 15.

⁴ Failing to recognize ἀπήντα (apenta, came upon [the earth]), Copernicus resorted to *ubique* (everywhere), as though the word were ἀπάντῃ (hapantei). In addition, he inserted *saeviebat* (raged), as though some form of ἀπηνής (apenes, rough) were present. Copernicus committed a similar error in his Latin translation of the Pythagorean letter which he included in his holograph of the *Revolutions* (NCCW, I, fol. 11ᵛ–12ᵛ). There he matched ἀπαντῶσι (apantosi, those whom we meet) with *omnibus* (everybody; fol. 11ᵛ/3 up), as though ἅπασι (hapasi) were the reading in his Greek text.

⁵ For ἀλέαν (alean, hearth), Copernicus repeated *aream*, which he had just used, four lines above, for ἅλω (halo, threshing-floor, *area*).

⁶ Copernicus changed Theophylactus' "The ant sent the singer away from her doors."

⁷ Instead of ἀνεμίμνησκε (anemimneske, reminded), Copernicus wrote *objiciebat* (reproached).

⁸ Here Copernicus omitted "For, being lazy, you are more miserable than a feverish patient, and you devour double rations to no purpose."

⁹ Here Copernicus omitted "dear friend" (Ὦ βέλτιστε, O beltiste).

NOTES

¹⁰ Copernicus omitted τηλικαύτην (telikauten, [which is] so great), perhaps because he felt it to be inconsistent with ἀκόσμητον (akosmeton), which he translated by *incultam* (undeveloped).

Letter 62

¹ See Letter 5, n. 3.

² "His threshing-floor is adequate for his νελοαῖς furrows" was omitted by Copernicus, since these Greek letters make no sense. Perhaps Corydo's furrows were "freshly watered" (νεολούταις, neoloutais).

³ Copernicus inserted *videatur* (he seems) in order to protect Theophylactus from his own incautiousness. In Greek mythology, fifty daughters were fathered by Danaus, and fifty sons by Aegyptus, the eponymous hero of Egypt.

⁴ See Introduction, p. 10.

⁵ Copernicus' dictionary did not include βούπαιδες (boupaides, big boys; "others have attained the status of big boys"). Consequently Copernicus had to make a guess. Instead of παῖς (pais, child), he thought of πούς (pous, foot). As for the prefix βου-, he did not see its connection with βοῦς (bous, bull), and imagined a link with Latin *bi-*. In this way he emerged with "two feet" instead of "big boys."

⁶ Copernicus' *genus* (ancestry) does not match ἀγχιστείαν (anchisteian, tie), which refers back to the proposed marriage. On the other hand, Copernicus' *genus* mistakenly looks forward to *ignobilitatem* (humble ancestry). When he encountered ἀγχιστεία again in Letter 65, he matched it correctly with *affinitas* (relation), the definition offered him by his dictionary.

Letter 63

¹ Copernicus omitted ἔτι (eti, any more).

Letter 64

¹ In the Greek text, this name was Socrates. Copernicus apparently repeated Sostratus from Letter 61.

² Copernicus' Greek text read ἀπαραιτήτοις (aparaitetois, inflexible) as a description of the boys. But they are frightened by words, it went right on to say. Since frightened boys are not inflexible, Copernicus omitted that word, an incorrect reading found in the family of manuscripts which included the manuscript underlying the Aldine edition. The correct reading ἀπαραίτητοι (aparaitetoi), applied to the teachers, not to their pupils, occurs in a different family of manuscripts (Zanetto, p. 76/14 up −12 up).

³ Copernicus' *minis* (threats, corresponding to ἀπειλαῖς) was misprinted in the 1509 edition as *nimis* (too much). In Copernicus' handwriting *nim-* and *min-* are almost indistinguishable (for example, NCCW, I, folio 1ʳ/16: *nimirum*). This near identity accounts also for the misprint *exanimatur/examinatur* (Letter 10, n. 8).

⁴ Since the boys are frightened by words, Copernicus made them responsive (*susceptivi*), exactly the opposite of ἀνεπίδεκτοι (anepidektoi, unresponsive, a word not found in his dictionary). This reversal forced Copernicus to omit "still" (ἔτι, eti), because it implies that the boys will become responsive only when they are older. Furthermore, Theophylactus' boys are still unresponsive to "reasoning and rules" (φρονήσεως καὶ νόμων, phroneseos kai nomon). By contrast, Copernicus made his responsive boys amenable to "rebukes and warnings" (*correctionum, admonitionum*).

⁵ The 1509 edition's *barbarum* (barbarian) was a misprint for *barathrum*, which Copernicus took in this transliterated form from his dictionary. He omitted τοίνυν (toinyn, therefore).

⁶ Copernicus did the best he could with his defective Greek text. He did not see that if κακίας (kakias) were substituted for κακίαν (kakian), the faulty syntax would be mended (... persuading him to abstain from wrongdoing).

⁷ Copernicus omitted αὐγείου (augeiou, Augeas'). In his dictionary, on sig. F 4ᵛ, in the margin he wrote Αυγειον in the proper alphabetical order. His capitalization of the initial letter signifies his awareness that this was the name of a person. By the same token, his omission of this personal name from his translation shows his unfamiliarity with the fable of Augeas, the droppings of whose thirty thousand cattle stayed in place for thirty years.

⁸ The 1509 edition enclosed *cotila* (cup) in parentheses, a printing device which appears nowhere else in the edition. In his dictionary, below κοτύλη (kotyle) Copernicus wrote *acetabuli mensura* (measuring cup for vinegar).

Letter 65

¹ Copernicus' *transgreditur* (trespasses) does not correspond to συνωρικεύεται (synorikeuetai). This word was not to be found in this form in his dictionary. On sig. h 7ʳ, however, in the margin he wrote συνωρις (synoris), without indicating its meaning (a pair of horses) or its equivalence with ξυνωρίς (xynoris), which his dictionary defined as *iugum mulorum* (a team of mules). Copernicus was unaware that Letter 65's συνωρικεύ-

εται ... ἱππαζόμενος (hippazomenos) echoes Aristophanes' *Clouds*, where a rich long-haired youth ἱππάζεταί τε καὶ συνωρικεύεται (rides a horse and also drives a pair of horses; line 15).

[2] Copernicus' *privamur* (lack) took the place of συλλιμώττομεν (syllimottomen, jointly starve), which he could not find in his dictionary, nor can we in ours. Yet λιμός (*fames*, hunger) and λιμώττο (*famesco*, starve) were both in his dictionary. Hence he could easily have surmised the meaning of συλλιμώττομεν. Nevertheless, *privamur* seemed more appropriate to him in this context: the pretended hunter has no reason to trespass because there is a lack of game, whether well-nourished or famished. There is no other suggestion in Letter 65 of hunger in Bubalio's household.

[3] Mistakenly saying that the Aldine edition reads ἀνυδρία (anydria, drought), PII, 113, n. 97, declares that Copernicus nevertheless used ἀνανδρία (anandria, unmanliness), the reading of the later editions, but makes no suggestion how Copernicus could have foreseen what editors would do after his death. He had no need of such clairvoyance, however, since the Aldine reading is ἀνανδρία (anandria) in agreement with the later editions.

[4] See Introduction, p. 11.

[5] Copernicus mistranslated συμφέρον (sympheron, useful) as though it were σύμφορον (symphoron, companion, *sodalem*, his synonym of his dictionary's *comitem*).

[6] Copernicus wrote: "correct him with whatever words you can" (*quibuscumque potes verbis corrige*), because his Greek text was hopelessly corrupt: τούτον instead of τοῦτον, σά instead of σοί, and an intrusive σέ.

Letter 66

[1] Matching δελεάζουσιν (deleazousin), Copernicus' *inescant* was misprinted "investant" in the 1509 edition.

Letter 67

[1] See Introduction, p. 10.

[2] Quoted from Euripides, *Hippolytus*, line 612.

[3] Copernicus failed to recognize εἰδώς (eidos, knowing) as a participial form of οἶδα (oida, know). In his dictionary he found εἴδομαι (eidomai, seem). This middle or passive verb he turned into a transitive verb (see), with εἰδώς as a (non-existent) indicative form. In addition, he changed the affirmation "not knowing" into a question ("Don't you see?").

[4] Copernicus' *maiorem supplicii poenam* (punishment greater than the penalty) is nonsense due to his dictionary's inadequate definition of πλημμέλημα (plemmelema; see Letter 22, n. 10). That definition could be adopted in Letter 22, but here πλημμέλημα has its ordinary meaning (fault: punishment greater than the fault).

[5] Quoted from Euripides, *Orestes*, line 10.

Letter 68

[1] "Seutlion" was misprinted as "Senthon" in the 1509 edition. In Copernicus' handwriting, *u* and *n* are practically indistinguishable (for instance, NCCW, I, folio 69^r/17: *thuribuli*); the close resemblance of *li* and *h* is exemplified by *Solis* and *Trianguli* (NCCW, I, Plate XII).

[2] See Introduction, p. 11.

Letter 69

[1] In the shape of a cricket.

[2] See Introduction, p. 16.

Letter 70

[1] See Introduction, n. 49.

Letter 72

[1] Copernicus omitted "joyless" (ἀτερπῆ, aterpe). For the contrast between joyless satiety and the pleasures of love, ἀτερπῆ is so essential that its omission by Copernicus seems inadvertent rather than deliberate.

[2] See Introduction, p. 10.

Letter 73

[1] See Introduction, p. 11.

[2] Here Copernicus omitted ἀληθῶς (alethos, really).

[3] The myth of the hungry octopus eating its own tentacles occurs twice in Aelian (I, 27; XIV, 26). Yet

more than five centuries before Aelian, Aristotle had said that the octopus "is often found with its tentacles removed, these having been eaten by the conger-eel" (*History of Animals*, VIII, 2; 591a 5).

⁴ See Introduction, p. 16.

⁵ Copernicus omitted τὴν γνώμην (ten gnomen, in spirit). This omission is linked with his misunderstanding of τοῖς ὅλοις (tois holois, completely) as "everybody" (*omnibus*). This in turn is connected with his emphasis on Proclus' attitude: For my part, I think ... (*equidem ... puto*).

⁶ See Introduction, p. 15.

⁷ Copernicus failed to recognize Theophylactus' παρῳχηκότων (paroichekoton) as a participial form of his dictionary's παροίχομαι (paroichomai, pass).

⁸ According to Aristotle (*History of Animals*, V, 34),

> the little vipers are brought forth in membranes which burst open in three days. Sometimes, however, they themselves, eating their way through, get out (558a 29–30).

Although Aristotle did not say so explicitly, he clearly meant that the young viper ate its way through the membrane. Aelian, however, has the newborn viper "gnawing through its mother's belly," even though he reports (I, 24; XV, 16) the denial of that behavior by Theophrastus, Aristotle's successor as head of the Lyceum.

Letter 75

¹ In his dictionary Copernicus found ἑταίρα (hetaira) matched with *amica* (sweetheart) and *meretrix* (prostitute). Here he used *amatoriis*, but in the closing sentence of Letter 75 he shifted to *meretricariam*.

² In order to match the rare word ἀφύτευτος (aphyteutos, not planted), Copernicus coined *incomplantatum* as the opposite of *complantatum*, a familiar late Latin expression meaning "planted" or "planted together."

³ Here Copernicus recognized the adverbial use of λοιπόν (loipon, henceforth, *deinceps*). In Letter 14, he had misconstrued λοιπόν substantively on account of the proximity of the genitive τῶν ἀγρῶν (ton agron), which induced him to write *residuum agrorum* (the rest of the land; see Introduction, p. 13).

⁴ See Introduction, p. 10.

⁵ Copernicus' *emere* (buy) was misprinted in the 1509 edition as *eruere* (uproot). In his handwriting *m* and *ru* sometimes tend to look alike (for example, NCCW, I, folio 6ʳ/2 up).

Letter 76

¹ As a diminutive of *gloria* (fame), *gloriuncula* was coined by Copernicus to match δοξάριον (doxarion).

² Here Copernicus omitted ὕπαρ (hypar). His dictionary translated it by *re vera, vere* (really), so that it seemed to him hardly suitable in combination with ἐνύπνιον (enypnion, *somnium*, dream), a departure from reality. Previously he suppressed ὕπαρ as redundant (Letter 34, n. 7).

³ See Introduction, p. 11. Copernicus mistakenly linked ἤχων (echon, echoes) with παίγνιον (paignion, plaything) instead of with καὶ πνευμάτων (kai pneumaton, and breezes).

⁴ Here Copernicus omitted "and before materializing, it rushed toward non-existence."

Letter 77

¹ Copernicus could not find the compound verb διαταράττω (diataratto, cause great disarray) in his dictionary. But had he looked up the simple verb ταράττω (taratto), he would have seen it matched with *turbo* (disturb). Instead, he resorted to *terruit* (scared).

² For χοινίκοιν (choinikoin), Copernicus' dictionary offered him two equivalents: *semodium* (half-bushel) and *pedicae genus* (a kind of shackle). He mistakenly chose the latter. One choinix was the normal daily ration.

³ For σφριγῶντες (sphrigontes), Copernicus' dictionary offered him *iuvenesco, vigeo* (thrive). But, feeling himself bound by his wrong choice of "shackles" (*pedicis*), he substituted *corripiantur* (restrained) for "thriving".

Letter 78

¹ "For Cupid is pitiless when he hurls his dart" would have been closer to the Greek text.

² The 1509 edition misprinted *indices* (show) instead of *iudices* (judge), matching δίκαζε (dikaze). In Copernicus' handwriting, *n* and *u* often look alike (for example, NCCW, I, folio 1ʳ/4: *sunt*), as was pointed out in Letter 68, n. 1.

Letter 79

¹ Copernicus was not alone in failing to realize the fictitious character of Letter 79; see Introduction, n. 49.

² Copernicus' *Numne* turned "You haven't changed your character" into the equivalent question. He could not discern any systematic mark of interrogation in the Aldine *Epistolographers*. The semicolon had appeared in the Greek manuscripts as early as the eighth or ninth century to serve as the question mark, and it was so used in Aldine Greek texts later than the *Epistolographers*.

³ Copernicus' intention to translate θύλακον (thylakon) literally as *saccum* (bag) was thwarted by the 1509 edition's misprint *sacrum* (rite). In Copernicus' handwriting, *c* and *r* tend to look alike (for example, NCCW, I, folio 2ʳ, line 7 up: *occultetur*). His dictionary equated θύλακος only with "bread basket" (*vas ad panem ferendum*) and did not indicate its metaphorical use for a person (like our "bag of bones" or "windbag"). Copernicus' choice of *saccum* was influenced by the occurrence of ἀσκόν (askon, sack) a few lines below.

⁴ For σύστασιν (systasin), Copernicus' dictionary offered only *commendatio* (recommendation) in Latin, and συμφωνία (symphonia, harmony) and πλῆθος (plethos, multitude) in Greek. But none of these three equivalents of σύστασις suited the present context. Copernicus therefore chose *consistentiam* as at least an etymological equivalent of σύστασις.

⁵ Copernicus could not find διαφυσάω (diaphysao, inflate) in his dictionary. Hence he mistook διεφύσησε (diephysese, did inflate) for διέφυσε (diephyse, did cling), for which he wrote *adhaesit*. Since he missed the idea of inflation, he omitted κενόν (kenon, empty), κοῦφον (kouphon, insubstantial) and τοσοῦτον (tosouton, so much). For his coining of *gloriuncula* (bit of fame), see Letter 76, n. 1.

⁶ Copernicus omitted παρεσκεύασαν (pareskeuasan), which his dictionary matched with *praeparo, suborno* (prepare, incite). He must have felt the difference between ἠνάγκασαν (enagkasan, *coegerunt*, forced) and παρεσκεύασαν to be so slight that by omitting the second main verb he could streamline Theophylactus' overblown style.

⁷ See Introduction, p. 9.

⁸ Presumably Copernicus intended to translate οὐκοῦν (oukoun) by *Itaque*, not *Ita* (the reading of the 1509 edition: in this way).

Letter 80

¹ The name was misprinted with an initial E in the 1509 edition. In Copernicus' handwriting, capital C and capital E are unmistakably different (NCCW, I, Plates IX, XII).

Letter 81

¹ This name was misprinted in the 1509 edition as "Calliconius," another example of the confusion between *m* and *ni* in Copernicus' handwriting (Letter 64, n. 3).

² Copernicus' *quid* shows that he translated τι (ti, anything) instead of his Greek text's τοι (toi), the enclitic particle often found in a conditional clause.

³ Closer to the Greek would have been: "show me and I shall..."

⁴ According to the fable Danae was locked up in a tower, where Jupiter could approach her only by turning himself into a shower of gold.

Letter 82

¹ See Introduction, p. 14.

² These two devices for counteracting the lure of the Sirens are recommended in the *Odyssey*, XII, 48–54.

³ Copernicus' *Introducuntur* (are introduced) does not really match Μυθολογεῖται (Mythologeitai, conveyed in mythical form).

⁴ Copernicus omitted the identification of water with fiction (ψεῦδος, pseudos) and wine with truth (ἀληθείᾳ, aletheiai).

⁵ Copernicus omitted "of reality" (τῶν ὄντων, ton onton).

⁶ Copernicus' *delectavit* (enjoyed) resulted from his confusion of τερατεύεται (terateuetai, [Homer] relates) with τέρπεται (terpetai, enjoys). Yet in Letter 54 he had correctly matched τερατεύονται (terateuontai) by *repraesentant*.

⁷ See Introduction, p. 16.

⁸ See Introduction, p. 10.

⁹ The adjective ἐξαίσια (exaisia, portentous) belongs with ἠχήματα (echemata, sounds). These two words stand so far apart, however, that Copernicus missed the connection between them, and mistranslated ἐξαίσια as though it were a noun (*illecebrae*, temptations).

¹⁰ Although Copernicus' Greek text read ὁμόζυγα (homozyga, husband), he wrote *coniugium* (marriage), not *coniungem* (spouse, the definition in his dictionary). After twenty years' absence Penelope's husband, Ulysses, finally returned, having committed his fair share of infidelities. On the other hand, Penelope's faithfulness

NOTES

is the main theme of Homer's *Odyssey*. Hence, the marriage rather than the husband seemed to Copernicus to be the better model for philosophy.

Letter 83

[1] Copernicus mistranslated the noun γλεύκους (gleukous, of new wine) as though it were a form of the adjective γλυκύς (glykys, sweet, *dulcis*). Contributing to this difficulty was the defective entry in his dictionary, which misprinted γλῆκος for γλεῦκος. On sig. H 1ᵛ, in the margin he wrote the correction γλευκος (without accent or final letter, as was the common practice of the time).

Letter 84

[1] In Copernicus' Greek text, Sosipater was told: "You have not been caught in love's snares." Yet this statement does not jibe with Sosipater's being "overcome by pulchritude." Hence Copernicus turned the opening of Letter 84 into a question. But his *numne* (Haven't?) was misprinted in the 1509 edition as *numine* (by divine will), and his *es* (have you been?) by *est* (has he been?). The first of these misprints resembles the misprint explained in Letter 64, n. 3.

[2] Copernicus' *maiorem ... laborum fruitionem habebis* (you will have enjoyment greater than your pains) shows that he did not recognize the causal genitive in πόνων (ponon, on account of your pains).

[3] See Introduction, p. 16.

Letter 85

[1] A stroll through a cemetery was prescribed "to overcome sadness" (λύπης ... κρατεῖν, lypes kratein). Copernicus changed the mood, however, to "being overcome by sadness" (*tristitia ... teneri*).

[2] Copernicus changed the future tense ἐπόψει (epopsei) to the imperative (*aspice*, Behold!).

[3] In Letter 4 Copernicus had matched *ulterius* (further) with περαιτέρω (peraitero). Here in Letter 85, however, he missed the construction of the adverb περαιτέρω governing the genitive (κόνεως, koneos, dust), just as he had done in a similar case in Letter 3 (see Letter 3, n. 1). Here, instead of linking περαιτέρω with κόνεως (beyond the grave), he connected κόνεως with φύσημα (physema; lightness of dust). In this way Copernicus emerged with: "in the end (*tandem*) they take on the lightness of dust," whereas Theophylactus meant: "beyond the grave they become hollow." In his dictionary Copernicus could not find φύσημα, and for the same reason he misinterpreted διεφύσησε (diephysese) in Letter 79 (see Letter 79, n. 5). Yet the simple verb φυσάω (physao) was matched there with *inflo, spiro* (puff up, blow).

Colophon

[1] Even Haller's name was misprinted in the 1509 edition (Halle*s*).

COPERNICUS
MINOR ASTRONOMICAL WRITINGS

COPERNICUS' *COMMENTARIOLUS*

INTRODUCTION

The earth is a planet in motion. This basic truth was proclaimed and rationally elaborated, for the first time in mankind's long history, by the little treatise commonly called Copernicus' *Commentariolus*. That title was not bestowed on it by him. In fact, he did not give it any title. Nor did he attach his name to it. He had good reason to conceal his authorship. Since the earth is a planet, it is in heaven along with the other planets. Since the earth is in heaven, the old distinction between earth and heaven is dissolved. Nobody on earth has to wait for death in order to go to heaven. Every human being is already in heaven at the instant of birth. The familiar exhortation to lead a moral life in order to ascend to heaven is bound to sound silly to anyone who understands the theological implications of the Copernican astronomy and wants to lead a moral life for secular reasons.

As a canon of Varmia, as a Roman Catholic ecclesiastic, Copernicus had no desire to weaken belief in the cosmological underpinnings of his own religion. On the other hand, as an innovative astronomer, he was profoundly convinced that the traditional view of the earth as stationary was wrong. Living in an age of savage sectarian strife, he knew only too well what fate awaited those accused of spreading disbelief in hallowed dogma. Yet his momentous discovery, or re-discovery, would not remain quietly locked up in his own brain. He therefore committed it to paper. And what then? It is not true that he "concealed his theories" throughout "the entire period that had elapsed since his first discovery of the heliocentric theory."[1] He neither "concealed his theories" nor did he have them printed. Instead, he chose a middle course. Avoiding complete silence on one side, and unrestricted publication on the other side, he distributed handwritten copies to a few trusted professional friends.

> He had as friends ... Cracow astronomers, formerly his fellow-students, with whom he corresponded about eclipses and observations of eclipses,

according to his earliest well-informed biographer.[2]

One of these handwritten copies found its way into the possession of a professor at the University of Cracow, Matthew of Miechów (1457–1523), who completed the inventory of his library on 1 May 1514. An entry in that inventory reads as follows:

> A manuscript of six leaves expounding the theory of an author who asserts that the earth moves while the sun stands still.[3]

This description unmistakably fits Copernicus' *Commentariolus* with respect to both its length and its essential contents. The entry also shows that the manuscript circulated without the author's name and without any title. Matthew of Miechów's last will and testament disposed

---

[1] Sw 1974, p. 191; afterwards, Sw (1976, p. 116) found out about the "readers to whom Copernicus sent" *Commentariolus*.

[2] Simon Starowolski, *Hekatontas* (Venice, 1627), p. 161; StC XXI, 15.

[3] Leszek Hajdukiewicz, *Biblioteka Macieja z Miechowa* (Wrocław, 1960), p. 218, no. 189: *sexternus Theorice asserentis Terram moveri, Solem vero quiescere*.

of the wooden box containing his copy of *Commentariolus*. With this clue as a guide, the writer of the recent study of Matthew's library searched long and hard and unsuccessfully for the present whereabouts of Matthew's copy of *Commentariolus*, which he concluded no longer exists.[4]

By the same token no trace has ever been found of Copernicus' original draft of *Commentariolus*. Nevertheless it is possible to form a reasonable conjecture concerning its fate. Toward the end of his life Copernicus acquired his only disciple, George Joachim Rheticus (1514–1574). Arriving in Frombork with an armful of valuable books for his new master, Rheticus undoubtedly received gifts in exchange. These apparently included *Commentariolus* as written by Copernicus' own hand. Whereas Rheticus bequeathed his unfinished trigonometrical works and Copernicus' autograph manuscript of *Revolutions* to his younger collaborator, he left the remainder of his library to a fellow-physician, Thaddeus Hájek (1525–1600). The following year, on 1 November 1575, at the ceremonies celebrating the crowning of the king of the Romans Rudolph II at Regensburg, Hájek, who was the emperor's personal physician, met the illustrious Danish astronomer Tycho Brahe (1546–1601). A book about the nova of 1572, a subject dear to the heart of Brahe, had just been published by Hájek. While talking to Brahe, Hájek found out that he had a high regard for Copernicus, and therefore gave him Copernicus' draft of *Commentariolus*, which had formed part of Rheticus' legacy to Hájek. The latter could not have chosen a more suitable person to whom to transmit this precious document. For, as Brahe related in a published work,

> A certain little treatise (*tractatulus*) by Copernicus, concerning the hypotheses which he formulated, was presented to me in handwritten form some time ago at Regensburg by that most distinguished man Thaddeus Hájek, who has long been my friend. Subsequently, I sent the treatise to certain other astronomers in Germany. I mention this fact to enable the persons into whose hands the manuscript comes to know its provenience.[5]

The manuscript owes not only its distribution to Brahe but also the title it now bears. For when Hájek turned it over to Brahe, telling him that its author was Copernicus (as he had learned from Rheticus), the work still lacked a title. This deficiency was made good by Brahe when he had his staff prepare copies to be sent to other astronomers. Those copies displayed the following heading:

Nicolai Copernici[6] *de hypothesibus motuum caelestium a se constitutis commentariolus*
Nicholas Copernicus' little treatise on the hypotheses formulated by himself for the heavenly motions.

In a speech delivered to the entire faculty and student body of the University of Copenhagen early in September 1574, Brahe spoke of Ptolemy's "hypotheses formulated by himself" (*hypotheses ab ipso constitutas*) and Copernicus' "hypotheses formulated differently" (*hypothesibus aliter constitutis*).[7] The speech's expressions *hypotheses ab ipso constitutas* and *hypothesibus*

---

[4] Ibid., p. 99.

[5] Brahe, *Astronomiae instauratae progymnasmata*, part 2, in TB, II, 428/34–40: not in Brahe's *De nova ... stella ... anno ... 1572*, as in Paul Moraux, "Copernic et Aristote," p. 234, n. 4, *Platon et Aristote à la renaissance*, XVIe Colloque international de Tours (Paris, 1976). Brahe wrote and published *De nova ... stella* in 1573, two and a half years before he received the *Commentariolus* from Hájek.

[6] Misreported as "Coppernici" by MCV, I, 5/1, followed by Lindhagen's title page and PII, 184/1.

[7] TB, I, 149/24–25.

aliter constitutis resemble the title's *hypothesibus ... a se constitutis* so closely as to justify the conjecture that the title was devised by Brahe in order to facilitate citation of the document whose provenience he wanted astronomers to know. In the title he called Copernicus' little work *commentariolus*, instead of which he used the synonym *tractatulus* later on, when he wrote his published reference to it without having the manuscript before him.

Brahe's title spoke of Copernicus' "hypotheses," a term never used in *Commentariolus* and recently mistranslated as "models."[8] In Copernicus' later years, when the expression *hypothesis* became part of his vocabulary, he used it in *Revolutions* mainly as a synonym for "principle" and "assumption." The word *constitutis* in Brahe's title has also been mistranslated, as though it meant "invented."[9] But when Brahe spoke to his Copenhagen audience about Ptolemy's hypotheses *ab ipso constitutas*, he and they knew that Ptolemy's hypotheses had not been "invented by himself." As for Copernicus' hypotheses *aliter constitutis*, they could hardly have been "invented differently." Brahe's admiration for Copernicus did not go so far as to give him credit for having "invented" the hypotheses formulated in *Commentariolus*, which disclaims originality with regard to its most famous idea, the motion of the earth.

Among the specialists to whom Brahe sent copies of *Commentariolus* (as it came to be known after Brahe's christening) was Henry Brucaeus (Van den Brock, 1530–1593), professor of astronomy at the University of Rostock. While Brahe was a student there from 1566 to 1568, he became quite friendly with Brock, and the two kept up an active correspondence after Brahe left Rostock. Hence, when Brahe drew up the list of astronomers in Germany who were to receive copies of *Commentariolus*, he made sure that Brock was included.

Brock himself did nothing with the copy of *Commentariolus* which he received from Brahe. For he was a staunch anti-Copernican, and on 12 June 1584 he wrote to Brahe:

In propounding the motion of the earth, Copernicus did not need such elaborate hypotheses, which are undoubtedly false.[10]

But the picture changed with the matriculation of Duncan Liddel (1561–1613) at the University of Rostock in October 1585.[11] Nominally Liddel was merely a student (although he was already thoroughly familiar with the Copernican astronomy) while Brock was an (anti-Copernican) professor. But "although Brock was an excellent astronomer, he learned the Copernican system from this teacher" (Liddel).[12] Brock's copy of *Commentariolus* has perished. Fortunately, however, he let Liddel make his own copy of it. This Liddel did by interleaving his copy of *Commentariolus* in his copy of the second edition of *Revolutions*. Liddel finished this operation on 2 November 1585, shortly after he had matriculated at Rostock. In later years, when he returned to his native Aberdeen, he donated his library to what is now the University of Aberdeen. There, in his copy of *Revolutions*, his manuscript of *Commentariolus* was recently discovered.

---

[8] Sw 1973, pp. 423, 433. According to Sw 1976, p. 119, "... the word hypothesis [has] the fundamental meaning in planetary theory, 'model,' which is of course its meaning in the title of the *Commentariolus*." But *Commentariolus*' title was not written by Copernicus.

[9] Sw 1973, p. 423.

[10] TB, VII, 85/23–25.

[11] *Die Matrikel der Universität Rostock*, eds. Adolph Hofmeister and Ernst Schäfer (Rostock/Schwerin, 1889–1922), II, 216. Liddel was born in 1561, not 1551 (as in JHA, 1973, 4:124).

[12] Duncan Liddel, *Ars medica* (Hamburg, 1607–1608, 1617, 1628), letter from Johannes Caselius to John Craig, dated 1 May 1607.

Our two other manuscripts of *Commentariolus* are quite different from the Aberdeen manuscript, which we shall call "A." For, the other two manuscripts ("V" for Vienna, and "S" for Stockholm) were written by professional scribes. It was their duty to reproduce their prototype as precisely as possible. Liddel, by contrast, was not a scribe, but a scholar, "the first in Germany to teach the theories of the heavenly motions according to the hypotheses of Ptolemy and Copernicus at the same time."[13] Unlike a mere copyist, Liddel did not feel obligated to produce an exact replica of his prototype, which Brock had received from Brahe. On the contrary, Liddel departed from his prototype whenever he so pleased. Although he surely recognized that *Commentariolus*' style was quite compact, he nevertheless strove for even tighter compactness by omitting words which in his judgment could be deleted without impairing the clarity of the exposition. Less frequently he altered *Commentariolus*' vocabulary in line with the development of ideas during the three-quarters of a century between the composition of *Commentariolus* and the production of A.[14] Elsewhere he sought to improve *Commentariolus*' Latin style. All in all, then, A is not the best guide to the text of *Commentariolus* as it left Copernicus' hands.

A's prototype was the copy of *Commentariolus* which Brock had received from Brahe. This, as was said above, has not survived. But it had formed part of the batch of copies sent by Brahe "to certain other astronomers in Germany." Hence A's prototype, which is lost, resembled S, a surviving member of Brahe's batch. S survived because it was combined with that copy of the second edition of *Revolutions* which once belonged to the distinguished Polish astronomer Jan Hewelke (1611–1687), who latinized his surname as Hevelius and acquired this copy of *Revolutions*, together with *Commentariolus*, in 1659. To whom did Brahe send S? From whom did Hewelke obtain S? How and when did S pass to the library of the Swedish Royal Academy of Sciences in Stockholm? Further research will be required to answer these questions.

Meanwhile, the affinity between S and A's prototype is revealed by the omission of *ex. 25. partibus constituto semidiametro orbis* in both S and A's prototype.[15] These six missing words follow *orbis* and end with *orbis*. Evidently a scribe's eye jumped from an occurrence of *orbis* to its next occurrence, overlooking in the process the six omitted words. These were left out in Brahe's headquarters, which dispatched S and A's prototype when the batch of copies of *Commentariolus* was sent "to certain other astronomers in Germany."

Brahe's chief assistant, the Danish astronomer Christian Sørensen Longberg (1562–1647, who latinized his surname as Longomontanus), first joined Brahe in 1589, several years after the batch of copies of *Commentariolus* had been sent out. While Longberg was with Brahe, a copy of *Commentariolus* was made for him. But when Longberg was preparing to depart from Brahe's headquarters in Benátky,[16] on 18 July 1600 he left his copy of *Commentariolus* as a memento[17] for his friend Joannes Eriksen, another of Brahe's assistants. This Longberg-

---

[13] Ibid.

[14] See n. 39 in the commentary. A differs so much from S and V that A cannot be grouped with the copies of *Commentariolus* sent by Brahe "to certain other astronomers in Germany." Yet it has been surmised (JHA, 1973, 4:125) that "in Rostock in 1585 Liddel (then resident in Frankfurt) was one of those ... to whom Tycho Brahe made his copy of the *Commentariolus* available." If Liddel was "then resident in Frankfurt," how could Brahe have made the *Commentariolus* available to him in Rostock?

[15] See n. 165 in the commentary.

[16] Longberg's *Benachia* does not mean Prague, as Sw mistakenly thinks (1973, p. 431).

[17] Longberg's μνιμόσιον does not mean a "dedication," as Sw mistakenly thinks (1973, p. 431). A dedication is written by an author. The author of *Commentariolus* was Copernicus, not Longberg, who was the possessor of a copy of *Commentariolus*.

Eriksen copy later passed, with the rest of Brahe's library, into the Austrian National Library in Vienna. There it lay unnoticed for centuries until it was found by Maximilian Curtze (1837–1903), who published it in MCV, 1878, I, 5–17. Three years later, (Carl) Arvid Lindhagen (1856–1926) published S in *Bihang till Kungliga Svenska Vetenskapsakademiens Handlingar*, 1881, VI, no. 12, pp. 3–15. The following year Curtze collated V and S in MCV, 1882, IV, 5–9. A collation of A with V and S is included in the notes accompanying the present translation of *Commentariolus*.

A sometimes differs from S while agreeing with V. Yet A never saw V, whereas A's prototype resembled S. Where S is wrong and V is right, A's zeal in correcting errors in his prototype brought A into agreement with V and disagreement with S. These situations indicate, not any dependence of A on V, but rather A's independence with regard to his prototype, which was akin to S.

Although V and S were both produced in Brahe's headquarters, they were not made in the same way. For, V was written by a scribe who was listening to the prototype being read aloud to him. That is why V wrote *ac si* when the reader said *axi*.[18] These two combinations of letters sound alike, but they do not look alike. Only a scribe who was relying on his sense of hearing would write *ac si* for *axi*. On the other hand, only a scribe who was relying on his sense of vision would jump from one occurrence of the word *orbis* to the next occurrence of *orbis*, while omitting six words between the two occurrences.[19] This is what was done by S, but it was not done by V. Hence, unlike V, S was written by a scribe who was looking at the prototype with his own eyes, not listening to a reader. S was sent abroad, while V remained with Longberg-Eriksen at Brahe's headquarters. A's omission of these six words shows that A's prototype was akin to S, but not to V.

S was not written throughout by a single hand. In line 15 of S' first sheet the first hand is abruptly replaced by a second hand, which continues to the end of *Commentariolus*. A major difference between the two hands is their treatment of the letter *u/v*. Above that letter, in both its vocalic and consonantal forms, the first hand places a small *u*, or the remnants thereof, in the characteristically German style of the period. By contrast, the second hand does not accord *u/v* any special distinguishing mark.

Since V and S were produced in two different ways, it is not surprising that they differ from each other even though they are both descended from the same progenitor, the (lost) Hájek–Brahe MS of *Commentariolus*. But where V and S both show the same erroneous reading, that defect was already present in the Hájek–Brahe MS. In one instance such a defect may be traced back to Copernicus himself.[20] Hence the original draft or autograph may have passed from Copernicus to Rheticus to Hájek to Brahe.

Copernicus wrote *Commentariolus* before 1 May 1514.[21] How long before that date? Not during his student years after 1491 at the University of Cracow, whose professors showed not the slightest interest in geokineticism. Nor about 1500, when Copernicus in Rome "lectured on astronomy before a large audience of students and a throng of great men and experts in this branch of knowledge,"[22] whose silence about what Copernicus said proves that he did not mention geokineticism, then a highly controversial idea which would surely have provoked a host of heated reactions from so articulate an audience. Nor before the middle

---

[18] See n. 145 in the commentary.
[19] See n. 165 in the commentary.
[20] See n. 188 in the commentary.
[21] See n. 3, above.
[22] TCT, p. 111.

of 1508, when Corvinus referred to the "alternating movements"[23] of Copernicus' sun, which became motionless in *Commentariolus*. Hence we may safely conclude that Copernicus wrote *Commentariolus* at some time between the latter half of 1508 and early 1514. Late in 1510 he finally decided not to try to succeed his uncle as bishop of Varmia, and instead to devote his best energies to astronomy.[24] For he had caught a glimpse of his new geokinetic, heliostatic universe, the glimpse that found its first expression in *Commentariolus*.

As a general rule, *Commentariolus* does not indicate from what sources it drew its information. Ptolemy, the greatest ancient astronomer, was not yet directly available, since the Greek text of *Syntaxis* was first published in 1538, and the earliest printing of a Latin translation was completed on 10 January 1515.[25] *Epitome*, however, was published in Venice in 1496, the year of Copernicus' arrival in Italy; *Commentariolus*' use of *Epitome*, III, 2, and V, 22, is evident.[26] So is *Commentariolus*' use of Giorgio Valla's *Seek and Avoid* as well as of Pliny's *Natural History*.[27] What other sources were tapped by *Commentariolus* will have to be determined by further research.

Copernicus sent out a few copies of *Commentariolus*, as we saw above. Perhaps an unpleasant rebuff made him decide against disclosing *Revolutions*, which he was already planning while he was writing *Commentariolus*.[28] In later years, while working on *Revolutions*, he said nothing about *Commentariolus*, since it expressed views which were contradicted or abandoned in *Revolutions*.[29] For all that Copernicus did about it, the world would never have known about the existence of *Commentariolus*. The same may be said about Rheticus (who presumably received Copernicus' original draft) and Hájek, who acquired it from Rheticus and passed it on to Brahe. Thanks to Brahe, however, *Commentariolus* survived, since all three of our manuscripts are descended in one way or another from the Copernicus–Rheticus–Hájek–Brahe MS. Brahe's enthusiastic diffusion of *Commentariolus* evoked only two responses, Liddel's and Longberg's. Thereafter a curtain of silence descended for two and three-quarter centuries until Curtze revived interest in *Commentariolus*, an interest that has continued unabated until the present day.

---

[23] See above, Theophylactus, Introduction, section V, last paragraph.

[24] TCT, pp. 334–335.

[25] This translation is listed by Sw (1973, pp. 425–426) as a possible source for *Commentariolus*. How can a work dated 10 January 1515 have been a source for a work completed before 1 May 1514?

[26] See nn. 92, 140 in the commentary.

[27] See nn. 48, 200 in the commentary. For a useful chronology of Valla's life and works, see Gianna Gardenal, pp. 93–97, in Vittore Branca, ed., *Giorgio Valla tra scienza e sapienza* (Florence: Olschki, 1981; Civiltà veneziana Saggi 28).

[28] See n. 35 in the commentary.

[29] See nn. 110, 169, 173 in the commentary.

COMMENTARIOLUS

Our predecessors assumed, I observe, a large number of celestial spheres[1] mainly for the purpose of explaining the planets' apparent motion by the principle of uniformity. For they thought it altogether absurd that a heavenly body, which is perfectly spherical, should not always move uniformly.[2] By[3] connecting and combining uniform motions in various ways, they had seen,[4] they could make any body appear to move to any position.

Callippus and Eudoxus,[5] who tried to achieve this result by means of concentric circles,[6] could not thereby[7] account for all the planetary movements, not merely the apparent revolutions of those bodies but also their ascent, as it seems to us, at some times and descent at others, [a pattern] entirely incompatible with [the principle of] concentricity. Therefore for this purpose it seemed better to employ eccentrics and epicycles, [a system] which most scholars finally accepted.[8]

Yet the widespread [planetary theories], advanced[9] by Ptolemy and most other [astronomers], although consistent with the numerical [data],[10] seemed likewise to present no small difficulty. For these theories were not adequate[11] unless they also conceived[12] certain equalizing circles,[13] which made the planet appear to move at all times with uniform velocity[14] neither on its deferent sphere nor about its own [epicycle's] center. Hence this sort of notion seemed neither sufficiently absolute nor sufficiently pleasing to the mind.[15]

Therefore, having become aware of these [defects], I often considered whether there could perhaps be found a more reasonable arrangement of circles,[16] from which every apparent irregularity would be derived while everything in itself would move uniformly, as is required[17] by the rule of perfect motion. After I had attacked this very difficult and almost insoluble problem, the suggestion at length came to me how it could be solved with fewer and[18] much more suitable[19] constructions than were formerly put forward, if some postulates (which are called axioms)[20] were granted me. They follow in this order.[21]

POSTULATES

1. There is no one center of all the celestial orbs or spheres.[22]
2. The center of the earth is the center, not of the universe,[23] but only of gravity and of the lunar sphere.
3. All the spheres encircle the sun,[24] which is as it were in the middle of them all,[25] so that the center of the universe is near the sun.[26]
4. The ratio of the earth's distance from the sun to the height of the firmament is so much smaller than the ratio of the earth's radius to its distance from the sun that the distance between the earth and the sun is imperceptible in comparison with the loftiness of the firmament.[27]
5. Whatever motion appears in the firmament is due, not to it, but to the earth. Accordingly, the earth together with the circumjacent elements performs a complete rotation on its fixed poles in a daily motion, while the firmament and highest heaven abide[28] unchanged.[29]
6. What appear to us as motions of the sun are due, not to its motion, but to the motion of the earth and our sphere, with which we revolve about the sun[30]

as[31] [we would with] any other planet.[32] The earth has, then, more than one motion.

7. What appears in the planets as [the alternation of] retrograde and[33] direct motion is due, not to their motion, but to the earth's.[34] The motion of the earth alone, therefore, suffices [to explain] so many apparent irregularities in the heaven.

Having thus propounded the foregoing postulates, I shall endeavor briefly to show to what extent the uniformity of the motions can be saved in a systematic way. Here, however, the mathematical demonstrations intended for my larger work[35] should be omitted for brevity's sake, in my judgment. Nevertheless, in the explanation of the circles themselves I shall set down here the lengths of the spheres' radii. From these anybody familiar with mathematics will readily perceive how excellently this arrangement of circles[36] agrees with the numerical data and observations.

Accordingly, lest anybody suppose that, with the Pythagoreans, I have asserted the earth's motion gratuitously,[37] he will find strong evidence here too in my exposition of the circles.[38] For, the principal arguments by which the natural philosophers attempt to establish[39] the immobility of the earth rest for the most part on appearances. All these arguments are the first to collapse here, since I undermine the earth's immobility as likewise due to an appearance.

THE ORDER OF THE SPHERES

The celestial spheres[40] embrace[41] one another in the following order. The highest is the immovable sphere of the fixed stars, which contains and[42] gives position to all things. Beneath it is Saturn's, which Jupiter's follows,[43] then Mars'. Below Mars' is the sphere on which we revolve; then Venus'; last is Mercury's. The lunar sphere, however, revolves around the center of the earth[44] and moves with it[45] like an epicycle.[46] In the same order also, one sphere surpasses another in speed of revolution, according as they measure out greater or smaller expanses of circles. Thus Saturn's period ends[47] in the thirtieth year, Jupiter's in the twelfth, Mars' in the third,[48] and the earth's with the annual revolution; Venus completes its revolution in the ninth month,[49] Mercury in the third.

THE APPARENT MOTIONS OF THE SUN

The earth has three motions. First, it revolves annually in a Grand Orb[50] about the sun[51] in the order of the signs, always describing equal arcs in equal times.[52] From the Grand Orb's center to the sun's center the distance is $1/25$[53] of the Grand Orb's radius. This Orb's radius is assumed[54] to have a length imperceptible in comparison with the height of the firmament. Consequently the sun appears to revolve with this motion, as if the earth lay in the center of the universe. This [appearance], however,[55] is caused not by the sun's motion but rather by the earth's.[56] Thus, for example, when the earth is in the Goat, the sun is seen diametrically opposite in the Crab, and so on. Moreover, on account of the aforementioned distance of the sun from the Orb's center, this apparent motion of the sun will be nonuniform, the maximum inequality being $2 1/6°$.[57]

The sun's direction with reference to the Orb's center is invariably toward a point of the firmament about 10° west of the more brilliant bright star in the head of the Twins.[58] Therefore, when the earth is opposite this point,[59] with the Orb's center lying between them,[60] the sun is then seen at its greatest distance [from the earth].[61] By this Orb not only is the earth[62] revolved, but also whatever else is associated with the lunar sphere.

The earth's second motion is the daily rotation. This is in the highest degree peculiar to the earth,[63] which it turns on its poles in the order of the signs, that is, eastward. On account of this rotation the entire universe[64] appears to revolve[65] with enormous speed.[66] Thus does the earth rotate together with its circumjacent[67] waters and nearby air.

The third is the motion in declination. For, the axis of the daily rotation is not parallel to the Grand Orb's axis, but is inclined [to it at an angle that intercepts] a portion of a circumference, in our time[68] about 23 $1/2$°. Therefore, while the earth's center always remains[69] in the plane of the ecliptic, that is, in the circumference of a circle of the Grand Orb, the earth's poles rotate,[70] both of them describing small[71] circles about centers [lying on a line that moves] parallel to the Grand Orb's axis. The period[72] of this motion also is a year,[73] but not quite, being nearly[74] equal to the Grand Orb's [revolution]. The Grand Orb's axis, however, being invariant[75] with regard to the firmament, is directed toward what are called the poles of the ecliptic. The poles[76] of the daily rotation would always be fixed in like manner at the same points of the heavens by the motion in declination combined with the Orb's motion, if their periods were exactly equal. Now with the long passage[77] of time it has become clear that this alignment of the earth changes with regard to the configuration of the firmament. Hence it is the common[78] opinion that the firmament itself has several motions. But even though the principle[79] involved is not yet[80] sufficiently[81] understood, it is less surprising that all these phenomena can occur[82] on account of the earth's motion.[83] I am not prepared to state to what its poles are attached. I am of course aware that in more mundane matters a magnetized[84] iron needle[85] always points[86] toward a single spot in the universe.[87] It has[88] nevertheless seemed a better view to ascribe the phenomena to a sphere, whose turning[89] governs the movements of the poles. This sphere must doubtless be sublunar.

EQUAL MOTION SHOULD BE MEASURED NOT BY THE EQUINOXES BUT BY[90] THE FIXED STARS

Accordingly, since the equinoxes and the other cardinal points of the universe shift considerably, whoever attempts to derive from them the equal length of the annual revolution necessarily falls into error. Besides,[91] different[92] determinations of this length were made in different ages on the basis of many observations. Hipparchus computed it as 365 $1/4$ days,[93] and al-Battani the Chaldean as $365^d\ 5^h 46^m$,[94] that is,[95] 13 $3/5^m$ or [13]$1/3^m$ less than Ptolemy.[96] Hispalensis, on the other hand, increased al-Battani's length[97] by the 20th[98] part of an hour, since he determined the[99] tropical year as[100] $365^d\ 5^h\ 49^m$.[101]

Lest these differences should seem to have arisen from[102] errors of observation, [let me say that] if anyone will study the details carefully, he will find that the discrepancy has always corresponded to the shift in the equinoxes. For when the cardinal points of the universe moved[103] 1° in 100 years, as was found in Ptolemy's time, the length of the year was then what Ptolemy himself reported. When however in the following centuries they moved with greater rapidity in opposition to lesser motions,[104] the year became shorter by as much as[105] the cardinal points' displacement increased. For by their swifter recurrence[106] they encountered the annual motion in a shorter time. Therefore the derivation of the equal length of the year from the fixed stars is more accurate. Thus I used the Spike of the Virgin and discovered that the year has always[107] been 365 days,[108] 6 hours, and about $1/6$ hour, the value also found in ancient Egypt.[109] The same reasoning must be employed also with the other motions of the heavenly bodies because their apsides, which are likewise fixed in the firmament,[110] with their true testimony make manifest the laws of the motions as well as heaven itself.

THE MOON

The moon, on the other hand,[111] seems to me to have four motions in addition[112] to the annual revolution which has been mentioned.[113] For on its deferent sphere it revolves once a month about the center of the earth in the order of the signs.[114] That deferent in fact carries what is commonly called[115] the epicycle of the first anomaly or of the argument, but I call the first or larger epicycle. In its upper portion this[116] [larger epicycle] revolves in the direction opposite to the deferent's in a period of a little more than a month. A second epicycle is attached to this larger epicycle, by which it is carried around. Lastly, as the moon clings to this second epicycle, it completes[117] two revolutions a month in the direction opposite to the first epicycle's. As a result, whenever the larger epicycle's center touches the line drawn from the Grand Orb's center through the earth's center (I call this line the Grand Orb's diameter),[118] the moon is then nearest to the larger epicycle's center. This occurs around the new and full moon. But contrariwise at the quadratures, halfway between new and full moon, the moon is most remote [from the larger epicycle's center]. In length, the larger epicycle's radius[119] is to its deferent sphere's radius as $1\ 1/18 : 10$,[120] and to the smaller epicycle's radius as $4\ 3/4 : 1$.[121]

By reason of these [arrangements], therefore,[122] the moon appears to be fast at some times, and at other times slow, as well as to drop down and climb higher. Into the first anomaly[123] the smaller epicycle's motion introduces two irregularities.[124] For it withdraws the moon from uniform motion on the larger epicycle's circumference, the maximum inequality being 12[125] $1/4°$ of a circumference of corresponding[126] size or diameter;[127] and[128] it also brings the larger epicycle's center at times farther from [the moon], at times nearer [to it], within the limits of the [smaller epicycle's] radius.[129] Therefore, since for this reason the moon describes unequal peripheries of circles[130] around the larger epicycle's center, it happens that the first anomaly undergoes complicated variations. Thus, the greatest variation of this kind does not exceed 4°56' near conjunctions and oppositions to the sun,[131] but in the quadratures[132] it increases to[133] 7°[134] 36'.

Those, however, who believe that this[135] [variation] is caused by an eccentric circle improperly treated the motion on that circle as nonuniform[136] and, in addition, they fell into two manifest[137] errors. For, the consequence by mathematical reasoning is that[138] in the quadratures, when the moon drops down[139] to the lowest part of the epicycle, it would appear nearly four times greater (if the entire [disk] were luminous) than when new and full,[140] unless they also irrationally claim that the size of its body increases and decreases.[141] So too, because the earth's size is sensible in comparison with its distance [from the moon], the eccentric makes the parallax increase very greatly near the quadratures. But if anyone investigates rather carefully, he will find that both [the apparent size and parallax] differ very little in the quadratures as compared with new and full moon, and accordingly he will not lightly doubt that my theory is the truer.

Indeed, with these three motions in longitude,[142] the moon passes through the points[143] of its motion in latitude.[144] The epicycles' axes are parallel to the deferent's axis,[145] and therefore the moon does not[146] move away from the [plane of the] deferent. But the deferent's axis is inclined to the axis of the Grand Orb or ecliptic and therefore makes the moon move out of the plane of the ecliptic. Thus the deferent's axis is inclined at an angle which subtends[147] 5°[148] of the circumference of a circle. Its poles revolve[149] [around centers lying on a line that moves] parallel to the ecliptic's axis,[150] in nearly the same manner as was explained regarding declination.[151] Also[152] in the present case they move in the reverse order of the signs but[153] much more slowly, with a revolution being completed in the nineteenth year.[154] It is the common opinion that this [motion] takes place in a higher sphere, to which the poles are attached as they revolve in the manner[155] described. Such[156] a fabric[157] of motions,[158] then, does the moon[159] seem to have.[160]

THE THREE OUTER PLANETS
SATURN—JUPITER—MARS

Saturn, Jupiter, and Mars have a similar system of motions, since their spheres completely enclose the Grand Orb associated with the year and revolve in the order of the signs[161] around its center as their common center. But Saturn's sphere completes its revolution in the thirtieth year, Jupiter's in the twelfth year, and Mars' in the twenty-ninth month,[162] just as if these revolutions are delayed[163] by the spheres' size.[164] For if the Grand Orb's radius is divided into 25 units, the radius of Mars' sphere[165] will be 38,[166] Jupiter's 130[167] $5/12$, and Saturn's 230[168] $5/6$.[169] By "radius" [of the sphere] I mean the distance from the sphere's[170] center to the center of the first[171] epicycle.[172]

For, each [deferent sphere] has two epicycles.[173] One of these carries the other, in much the same way as was explained in the case of the moon.[174] The arrangement, however, is different. For, the first epicycle revolves in the direction opposite to the deferent sphere's, the periods of both being equal. On the other hand, the second[175] epicycle, revolving in the direction opposite to the first's with twice the velocity, carries the planet around. As a result, whenever the second epicycle is at its greatest distance from the deferent sphere's center, or again at its nearest approach thereto, the planet is then at its closest to the [first] epicycle's center; but it is at its greatest distance therefrom whenever [the second epicycle is] at a quadrant's distance [from the two positions just mentioned and] halfway [between them].[176] Therefore, through the combination of these motions of the deferent sphere and epicycles, as well as the commensurability of their periods, it happens that these withdrawals and approaches occupy absolutely[177] fixed places of their own[178] in the firmament. [These planets] constantly adhere[179] to unchanging patterns of motion throughout, so that their apsides are immovable: Saturn's, near the star described as being above the Archer's elbow;[180] Jupiter's, 8°[181] east of the star called the end of the Lion's tail;[182] and Mars', 6 1/2° west of the Lion's heart.[183]

The sizes[184] of their epicycles are as follows. In those units of which the Grand Orb's radius was taken to be 25 [25ᵖ 0ᵐ],[185] the radius of Saturn's first[186] epicycle consists of 19ᵖ 41ᵐ, while the second epicycle's radius has 6ᵖ 34ᵐ. Similarly in the case of Jupiter, the first epicycle has[187] a radius of 10ᵖ 6ᵐ; the second, 3ᵖ 22ᵐ. As for Mars, its first [epicycle's radius is] 5ᵖ 34ᵐ; its second [epicycle's radius is 1ᵖ][188] 51ᵐ. Thus, the first [epicycle's] radius is throughout three times greater than the second [epicycle's radius].

Now the irregularity imposed by the epicycles' motion[189] upon the deferent sphere's motion has usually been called the "first anomaly" which, as I said, adheres[190] throughout[191] to unchanging boundaries in the firmament. For there is a second anomaly, in which the planet is seen sometimes to retrograde and often to become stationary. This second anomaly happens by reason of the motion, not of the planet, but of the earth[192] as it changes[193] its observational position on the Grand Orb. For since the earth's speed surpasses the motion of the planet, the line of sight directed toward the firmament regresses,[194] and the earth more than neutralizes the planet's motion.[195] This regression peaks at the time when the earth is nearest to the planet, that is, when it comes between the sun and the planet at the planet's evening rising.[196] On the other hand, about the time when the planet is setting in the evening or rising in the morning, the earth advances the line of sight in the forward direction. But when the line of sight is moving in the direction opposite to the planet's and at an equal rate, the planet seems[197] to stand still because the opposite motions neutralize each other[198] in this way. This generally[199] happens when the angle at the earth between the sun and the planet is about 120°.[200] In all these planets, however, the lower the sphere by which the planet is moved, the greater is this anomaly. Hence in Saturn it is smaller than in Jupiter, and again[201] greatest in Mars, in accordance with the ratio of the Grand Orb's radius to their radii. This anomaly peaks for each of them at the time when the planet is seen along a line of sight tangent to the Grand Orb's circumference.[202] For us, at any rate, these three planets do indeed wander about[203] [in longitude].

In latitude, on the other hand, their digression is twofold. [In the first place,] while the epicycles' circumferences remain in a single plane with their deferent, the planets deviate[204] from the ecliptic[205] in accordance with the axes' inclinations. These do not revolve,[206] as in the case of the moon,[207] but are always directed[208] toward the same region of the heavens. Therefore the intersections of the circles (the deferent's and the ecliptic, these intersections being[209] called the "nodes") also occupy eternal places in the firmament. Thus, the node where the ascent toward the north begins is, for Saturn, 8 1/2° east of the star said to be in the head of the eastern Twin;[210] for Jupiter, 4° west of that[211] star; and[212] for Mars, 6 1/2°[213] west of Vergiliae.[214] Hence in these and the diametrically opposite[215] [nodes] a planet has no latitude.

On the other hand, its maximum[216] latitude, which occurs when these planets are at a quadrant's distance[217] [from the nodes], is quite variable.[218] For, the inclination[219] of the axes and[220] circles seems, as it were, to be attached to those nodes, while oscillating [around them]. In fact, it peaks at the time when the earth is nearest to the planet, that is, when the planet is rising in the evening. For then the axis' inclination is $2\,2/3°$ for Saturn, $1\,2/3°$ for Jupiter, and $1\,5/6°$ [221] for Mars. On the other hand,[222] near evening setting and morning rising, the earth being then[223] at its greatest distance [from the planet], this inclination decreases for Saturn and Jupiter by $5/12°$, but for Mars by $1\,2/3°$. Thus this variation is most notable in the greatest latitudes, and for any[224] latitude it diminishes as the planet's[225] distance from the node lessens, so that the variation increases and decreases[226] in phase with the latitude.

In the second place,[227] it happens that the earth's motion on the Grand Orb causes the apparent latitudes to change for us.[228] Thus, [the earth's] approach toward and[229] withdrawal from [the planet] increase and[230] decrease the angles of the apparent latitude, as mathematical reasoning requires.[231] If in fact this oscillating motion[232] occurs along a straight line, it is nevertheless possible for a motion of this kind to be produced by a combination of two spheres.[233] Although these are concentric,[234] [the higher] one carries around the other one's poles, which are inclined. In addition, the lower sphere makes the poles of the deferent sphere bearing the epicycles revolve with twice the velocity of the upper sphere and in the opposite direction. The deferent's poles are also inclined, their inclination away from the poles of the sphere halfway[235] above being equal to the inclination of this sphere's poles away from those of the highest sphere. So much for Saturn, Jupiter, and Mars as well as[236] the spheres which enclose the earth.

VENUS

What is enclosed within the Grand Orb's embrace, that is, Venus' and Mercury's motions, remains to be investigated. To begin with, Venus has a system of circles[237] closely resembling that of the outer planets, but the motions are executed differently. As was said above,[238] Venus' deferent sphere[239] completes its revolution in the ninth month, which is likewise the period of the larger epicycle. Their composite motion brings the smaller epicycle back in a constant relation to the firmament everywhere, and establishes the higher apse at the point toward which I said[240] the sun is directed. On the other hand, the smaller epicycle's period, while incommensurable with the other two,[241] is commensurable[242] with the Grand Orb's motion: in one revolution of the Orb, the smaller epicycle completes two revolutions. As a result, whenever the earth is in the diameter drawn through the apse, the planet is then[243] nearest to the larger epicycle's center, and farthest from it [when the earth is] on the perpendicular [to the line of apsides] at a quadrant's distance from them. [This arrangement] closely resembles the way in which the moon's smaller epicycle in its aspects is related to the sun.[244] But the radii of the Grand Orb and of Venus' deferent sphere have the ratio $25^p:18^p$;[245] the larger epicycle, $3/4^p$; and the smaller, $1/4^p$.[246]

Venus too is sometimes seen to retrograde, particularly when it is nearest to the earth. Its regression occurs for a reason that in a certain way is like the reason for the outer planets' regression, but is its opposite. For their regression occurs because the earth's motion is faster [than theirs], but in this[247] case because it is slower; moreover, in their case the earth's sphere is enclosed, whereas in this case it does the enclosing. Hence Venus is never[248] in opposition to the sun, since the earth cannot come between them. On the contrary, it turns back within fixed elongations to either side of the sun. These are determined by tangents to the circumference drawn from the earth's center, and never exceed 48° in our observations. This in substance is the motion by[249] which Venus is carried around in longitude.

Its latitude also changes for a twofold reason. For Venus too has[250] the axis of its deferent sphere inclined, at an angle of 2 1/2°;[251] and the node whence it turns north is in its apse. But although in itself this inclination is one and the same, the digression arising from it appears to us[252] as twofold. For when the earth faces either node of Venus,[253] these digressions are seen on perpendiculars [to the nodal plane] above and below it, and are termed the "reflexions." The deferent sphere's natural inclinations, which are[254] called the "declinations," appear[255] when the earth is at a quadrant's distance [from the nodal line], but[256] they are the same.[257] In all the other positions [of the earth], however, both latitudes mingle and are combined:[258] the larger one prevails over the other, as they augment and eliminate each other by their likeness and[259] difference.

But the axis' inclination is the following. It has an oscillating[260] motion hinged, not on the nodes as in the case of the outer planets,[261] but on certain other points that revolve by performing annual revolutions of their own with reference to the planet. As a result,[262] whenever the earth faces an apse of Venus, at that time the oscillation[263] peaks, and this [affects] the planet itself, no matter what part of its deferent it is in then. Consequently, if the planet is then in an apse or its diametric[264] opposite, it will not completely lack latitude even though it is then in the nodes. From these [peak positions], however, this oscillation decreases until the earth moves through a quadrant of a circle away from the aforesaid [apsidal] location and, their motions being similar,[265] until the peak[266] point of this deviation has moved an equal distance away[267] from the planet, when absolutely no trace of this deviation is found.[268] The deviational swing continues[269] uninterrupted, with that initial [peak] point dropping from north to south and moving as far away from the planet as the earth moves away from the apse. The planet is thereby brought to the region which had previously been south. Now, however, by the law of opposition, it becomes north [and remains so] until the planet again reaches[270] the peak when a semicircle[271] of the libration is completed. Here the deviation becomes maximal once more, being similar[272] [in sign] and equal to its initial [value]. Thus, finally, the remaining semicircle is traversed in the same way [as the first]. Consequently this latitude, which is usually called the "deviation," never becomes southern. Here too it seems reasonable that these phenomena are produced by two concentric spheres with oblique axes, as[273] I explained in the case of the outer planets.[274]

MERCURY

Of all the orbits in the heaven, however, the most remarkable is that of Mercury, which traverses almost untraceable[275] paths so that it cannot be easily studied. There is the further difficulty that it generally follows a course invisible in the sun's rays and is observable on very few days. Yet Mercury too will be understood, provided that it is investigated by someone of superior ability.[276]

For Mercury[277] too, as for Venus, two epicycles revolving on their[278] deferent sphere will be suitable. For, as in the case of Venus, the larger epicycle has the same period as its deferent sphere,[279] and fixes the position of Mercury's apse[280] at $14\ 1/2°$[281] east of the Spike of the Virgin. The smaller epicycle, on the other hand, revolves with twice the [earth's] speed. But by contrast with the principle governing Venus, in every position of the earth passing over[282] Mercury's [higher] apse or facing it from the opposite direction, the planet is farthest from the larger epicycle's center, and nearest to it [when the earth is] at a quadrant's distance [from the apsidal line].[283] As I said,[284] Mercury's deferent sphere completes its revolution in the third month, that is, in 88 days. Its radius contains $9\ 2/5^p$, of the 25^p which I have assumed[285] for the Grand Orb's radius.[286] Of these units, the first epicycle takes $1^p\ 41^m$, while[287] the second epicycle takes one-third as much,[288] that is, about 34^m.[289]

But this combination of circles is not sufficient here,[290] by contrast with the other [planets].[291] For when the earth passes through the aforementioned positions with respect to the apse, the planet seems to move along a far smaller periphery, and again, when the earth is at a quadrant's distance [from the apsidal line], along an even[292] larger periphery, than is consistent with the aforesaid system of circles.[293] Yet no other perceptible[294] longitudinal irregularity is produced by this [disparity].

Its occurrence, consequently, is reasonably explained by a certain approach [of the planet] toward and withdrawal from the deferent's center[295] along a straight line. This oscillation must be caused by two interlocking[296] small spheres, whose axes are parallel to the deferent's axis. At the same time the center of the larger epicycle, or of this whole [epicyclic structure], is exactly[297] as far away from the center of the small sphere[298] which without any gap contains[299] [the epicycle's center] as the center of this [inner sphere] is from the center of the outer [small sphere]. This distance has been found[300] to be $0^p\ 14$[301] $1/2^m$, the universal measure I have used[302] being $25^p\ 0^m$. In addition, the outer small sphere's motion[303] performs two revolutions in the course of a year,[304] while the inner one completes four revolutions in the same time with twice the speed in the opposite direction. For by this composite motion the centers[305] of the larger epicycle are carried along a straight line, just as I explained with regard to the oscillating latitudes.[306] In this manner, therefore, when the earth is in the aforementioned positions with respect to the apse, the larger epicycle's center is nearest[307] to the deferent's center, but farthest[308] [from it when the earth is] at a quadrant's distance [from the apsidal line]. However, [when the earth is] at the midpoints, that is, 45° from these [four points, just mentioned], the larger epicycle's center joins[309] the outer small sphere's center, and both [these centers] coincide. This approach-and-withdrawal amounts to $0^p\ 29^m$ of the aforementioned units. And this ends the discussion of Mercury's motion in longitude.

In latitude it does not differ from[310] Venus, except that it is always in the opposite region.[311] For where Venus turns north, Mercury heads south. But its deferent sphere is inclined to the ecliptic at an angle of 7°. Here too[312] there is a deviation, but[313] it is always southern and never[314] exceeds $3/4°$. Otherwise, what was said about Venus' latitudes[315] may[316] be recalled here too,[317] to avoid frequent repetition of the same statements.

Thus,[318] Mercury runs on seven circles in all;[319] Venus, on five;[320] the earth, on three,[321] and around it [322] the moon on four;[323] finally, Mars, Jupiter, and Saturn on five each.[324] Thus altogether, therefore, 34[325] circles[326] suffice to explain the entire structure of the universe and the entire ballet of the planets.[327]

BIBLIOGRAPHY

AIHS   *Archives internationales d'histoire des sciences*
          Aiton, Eric John, "Celestial Spheres and Circles," *History of Science*, 1981, *19*:75–114
CSE   "Copernicus' Spheres and Epicycles," AIHS, 1975, *25*:82–92
Sw   Swerdlow, Noel M.
Sw 1973   "The Derivation and First Draft of Copernicus' Planetary Theory; a Translation of the *Commentariolus* with Commentary," *Proceedings of the American Philosophical Society*, *117*:423–512
Sw 1974   "The Holograph of the *Revolutions* and the Chronology of its Composition," JHA, *5*:186–198
Sw 1975   "On Copernicus' Theory of Precession," at pp. 49–98 in *The Copernican Achievement*, ed. R. S. Westman (Berkeley, 1975), reviewed in *Polish Review*, 1976, *21*:225–235
Sw 1976   "Pseudodoxia Copernicana," AIHS, *26*:108–158
U   University of Uppsala, Sweden, folio 15^r of Copernicus' notes in his copy of the 1492 edition of the *Alfonsine Tables*

NOTES

[1] Here *Commentariolus* writes *orbium caelestium* for the "celestial spheres." The presence of these two words in the title of Copernicus' later and longer work *De revolutionibus orbium caelestium* (cited hereafter as *Revolutions*) was condemned by some scholars as an alien intrusion. However, they overlooked the occurrence of *orbium caelestium* in the unquestionably authentic text of *Revolutions* (NCCW, I, fol. 8^r/1; NCOO, II, 5/8-9). They were also unacquainted with the appearance of *orbium caelestium* here at the very beginning of *Commentariolus*, since they did not even know about its existence. As the use of *orbium caelestium* in both *Commentariolus* and *Revolutions* shows, this expression was a permanent part of Copernicus' professional vocabulary. See nn. 22, 40, below.

[2] V, A: *aeque*; S: *aequaliter*. Although V and S are descended from the same prototype, they sometimes differ because V was written by a scribe who was listening to a reader, whereas S was written by a scribe who was looking at his prototype with his own eyes. A's prototype was akin to S, but not to V. A's replacement of *aequaliter* by *aeque* was part of A's constant striving for more intense brevity. The resulting agreement between V and A is fortuitous, and does not indicate any affinity between V and A. In Copernicus' mind, uniform motion (*aeque* or *aequaliter moveri*) implied (1) uniform velocity, traversing equal distances in equal times, and (2) circular motion, always equally distant from its own center.

[3] V, S, A: *et*. V's *et* was mistaken for *etiam* by MCV, I, 5/9, because the long tail of the letter *q* in the line above was misread as a stroke over *et*, which would have made it an abbreviation of *etiam* (for a similar misreading, see n. 84, below). The context requires *ex*, not *et*. The presence of the erroneous reading *et* in all our manuscripts indicates that this defect was a blemish in Brahe's MS too.

[4] V: *adverterant*; S, A: *animadverterant*. Writing while listening, V omitted the first two syllables.

[5] The source from which *Commentariolus* obtained information about these two ancient Greek astronomers failed to make clear that it was Eudoxus who originated the concentric system, while Callippus improved upon his predecessor's scheme.

[6] Discussions of Eudoxus and Callippus regularly referred to their use of concentric *spheres*. By contrast, *Commentariolus*' choice of *concentricos circulos* here shows that concentric spheres were less dominant in Copernicus' thinking than were the circles traced on the surfaces of those spheres by moving bodies or points. See n. 326, below.

[7] V, S: *et*, corrected to *ex* by A. The presence of *et* in both V and S indicates that this faulty reading was a blemish in Brahe's MS too.

[8] Here S' second hand begins to write and continues to the end of *Commentariolus*.

[9] V: *fuerunt*; S, A: *fuere*.

[10] V, S: *quamquam ad numerum responderent*, omitted by A in a constant effort to make *Commentariolus* even more compact.

[11] V: *sufficiebant*; S, A: *sufficiebat*. S omitted the customary supralinear stroke over the *a* to indicate the elision of *n*, and A failed to correct this oversight in his prototype. The plural form *sufficiebant* is required to match the other plural forms *prodita fuere (fuerunt)*, *responderent*, and *videbantur*.

[12] V: *imaginarentur*; S, A: *imaginentur*. S omitted the medial syllable *-ar-*, which is needed to maintain the proper sequence of tenses. A failed to correct this defect in his prototype, which was akin to S.

[13] *Commentariolus'* *aequantes quosdam circulos* contrasts with *Revolutions'* avoidance of the term "equant" (except in V, 25); see NCCW, II, 417, 429.

[14] Sw (1973, p. 434) mistranslates: "the planet never moves with uniform velocity." But Ptolemy's equant made the planet move with uniform velocity at all times (*aequali semper velocitate sidus moveri*). Its velocity was uniform at all times, however, on its equant, but not on its deferent sphere nor about its own epicycle's center at a distance from the equant's center.

[15] Copernicus rejected the equant because it made the planet move with uniform velocity about a point from which its distance was not constant, thereby violating the second part of Copernicus' rule of perfect motion (see n. 2, above).

[16] *Commentariolus* looked for an arrangement of circles (*modus circulorum*), not spheres; see n. 326, below.

[17] V: *possit*; S, A: *poscit*. V's incorrect reading simply echoes *possit*, three lines above.

[18] V, A: *ac*; S: *et*. Instead of following his prototype, A simply repeated *ac* in the line above, thereby coming into fortuitous agreement with V.

[19] V: conventioribus; S, A: *convenientioribus*. Writing while listening, V omitted the medial syllables -*ien*-, a mess ignored by MCV, I, 6/8.

[20] The postulates are "incorrectly called 'axioms,'" says Sw (1973, p. 437). Having committed the sort of gross grammatical errors of which Sw is flagrantly guilty (see nn. 120, 141, below), he has the impertinence to call Copernicus' Latin incorrect. Over a space of a dozen years Copernicus attended three outstanding universities (Cracow, Bologna, Padua), where the language of instruction was Latin. Almost everything Copernicus ever wrote was in Latin, including the world-renowned *Revolutions*, which initiated modern astronomy, if not all modern science. To see such a giant lectured about the meaning of the word "axiom" is a pathetic spectacle.

What is wrong with *Commentariolus'* calling postulates "axioms"? "They are hardly self-evident," says Sw (1973, p. 437), as though the word "axiom" always denoted a self-evident proposition. But in Aristotle's logic "the axioms are the primary propositions from which a proof proceeds" (*Posterior Analytics*, I, 10; 76b 14–15). Copernicus followed Aristotle's logic, not Sw's.

A work preserved in Nuremberg Stadtbibliothek MS Cent VI, 12, fol. 1^r–66^v, is entitled *Almagesti minoris libri VI* (Ernst Zinner, *Leben und Wirken des Johannes Müller von Königsberg genannt Regiomontanus*, 2nd ed. [Osnabrück, 1968], pp. 75, 315). Wishing to shorten this title, Sw omits *libri VI* (Six Books). The remaining words (*Almagesti minoris*) are in the genitive case. If they are to serve as a short title, they must be cited in the nominative case. Trying hard to do so, Sw emerges with "Almagestum minor" (1973, pp. 425–426, 512). But *Almagestum*, being of the neuter gender, requires *minus* (neuter), not *minor* (masculine or feminine). Thus, that learned astronomer Giovanni Battista Riccioli entitled his valuable work *Almagestum novum* (neuter), not *novus* (masculine) or *nova* (feminine).

[21] "The introduction of these postulates at this point appears unmotivated" to Sw (1973, p. 437), because Sw misunderstands their nature as the primary propositions from which *Commentariolus* proceeds. It has just rejected the two most important previous theories: concentrics, and eccentrics with equant. By means of its postulates, it now lays the foundation for the exposition of its own system. In short, it introduces the postulates exactly where they logically belong. Although no other place would have been suitable for them, and no better place is suggested by Sw, he condemns *Commentariolus'* indispensable arrangement as "this flaw," which is "intelligible if one considers that *Commentariolus* may well have been written in haste with no revision" (1973, p. 437). Actually, in its admirable compactness, without a superfluous word, *Commentariolus* gives every sign of being the end product of much reflection, careful planning, and superb organization.

[22] Postulate 1 is directed against the principle of concentricity, which required all the celestial spheres to share a common center. Again, *orbium caelestium* is *Commentariolus'* expression for the celestial spheres; see n. 1, above.

[23] Postulate 2 is directed against the Eudoxan system of concentrics and the Ptolemaic system of eccentrics, since both of these systems identified the center of the earth with the center of the universe.

[24] V: *solent*; S, A: *solem*. Writing while listening, V linked *solem* with the initial consonant of the following word *tamquam*, thus emerging with *solent*, which does not fit this context.

[25] V, S: *omnium*, omitted by A as superfluous, although actually it is helpful.

[26] Postulate 3 enunciates the principal difference between *Commentariolus* and its geokinetic predecessors, the Pythagoreans, from whom it dissociates itself soon after Postulate 7. In the center of their universe the Pythagoreans placed an imaginary Fire, around which their non-central sun revolved.

[27] Postulate 4 forestalls the objection that the absence of annual stellar parallax disproves the annual motion of the earth, as proclaimed by Postulate 6. Postulate 4 makes the stars in the firmament so remote that whatever shift in the position of any star is produced by the earth's yearly revolution around the sun, that shift cannot be detected by the unaided eye.

[28] V, S, A: *permanente*. V's *permanente* was misreported as *perveniente* by MCV, I, 6.

[29] Sw (1973, p. 436) mistranslates: "the sphere of the fixed stars remains immovable and the outermost heaven," as though *Commentariolus* were merely reiterating Ptolemy's thesis that the sphere of the fixed stars forms the outermost heaven. But far from repeating Ptolemy, Postulate 5 contradicts him. For, Ptolemy's outermost heaven of the fixed stars moved, whereas *Commentariolus*' "firmament and highest heaven abide unchanged" (*firmamento immobili permanente ac ultimo caelo*). Sw tries to support his mistranslation by saying that *Commentariolus*' "sphere of the fixed stars is the outermost heaven because a sphere of diurnal rotation is no longer necessary" (1973, p. 438). But a sphere of diurnal rotation is always necessary, since diurnal rotation is an astronomical fact. For *Commentariolus*, that necessary sphere of diurnal rotation was the earth. For Ptolemy, the diurnal rotation had been performed by the starry sphere, which was his outermost heaven. Ptolemy's outermost heaven moved, whereas *Commentariolus*' did not.

[30] V: *circa ☉ volvimur*; S, A: *circumvolvimur*. In Brahe's MS the astronomical sign ☉ for the sun may have been followed by a supralinear stroke for the letter *m* to stand for *solem*. If so, this combination may have been misinterpreted by S as belonging with *circa* to form *circum-*.

[31] V, S: *seu*, corrected to *ceu* by A. The agreement of V and S shows that the defective reading *seu* was already present in Brahe's MS.

[32] V, S, A: *aliquo alio sidere*. The agreement of all three manuscripts with regard to this sound reading renders unnecessary the emendation *aliquod aliud sidus* proposed by MCV, I, 6/31, and adopted by PII, 186/30.

[33] V, S: *ac*; A: *et*, perhaps repeating *et* two lines above.

[34] The formulation of Postulate 7 is open to the objection that the planets' direct motion is due in part to their own motion. For, *quod apparet in erraticis retrocessio ac progressus* means that the alternation of the planets' apparent retrograde and forward directions is due to the earth's motion. Postulate 7 reduces the planetary loops to the rank of mere appearances or optical illusions. Yet, according to Sw (1973, p. 437), "Copernicus has raised no objection to Ptolemy's representation of the second anomaly," that is, the apparent planetary loops.

[35] *Commentariolus*' reference to *maiori volumini* shows that Copernicus was already planning *Revolutions*. Sw asks:

> What is the "larger book" he [Copernicus] refers to? There is no reason to believe it was to be anything like *Revolutions*. ... I believe that the sort of book Copernicus was contemplating when he wrote *Commentariolus* would have consisted of geometrical demonstrations of the equivalence of Ptolemy's and his own models (1973, p. 439).

Some anti-Copernicans have asserted the equivalence of the Ptolemaic and Copernican astronomies. Copernicus himself, on the other hand, always emphasized the crucial differences between these two incompatible systems: for him, the earth was a heavenly body in motion, whereas for Ptolemy it had been a non-heavenly body at rest.

The weird notion that Copernicus ever contemplated demonstrating the equivalence of the Ptolemaic and Copernican systems was soon abandoned by Sw, who shifted from an inaccurate statement to the following obscurantist statement:

> What kind of larger book Copernicus had in mind when he wrote this [reference to *Revolutions* in *Commentariolus*] is by no means clear (1974, p. 188).

Here Sw also contended that "there is, in fact, no evidence whatsoever for" the conclusion that *Commentariolus* referred to *Revolutions*. But there are in fact two excellent pieces of evidence for this conclusion. First, *Commentariolus* refers to a larger work; Copernicus wrote only one larger work, *Revolutions*. Secondly, *Commentariolus* refers to the "mathematical demonstrations intended" (*mathematicas demonstrationes ... destinatas*) for the larger work. Many mathematical demonstrations are included in Copernicus' only larger work, *Revolutions*.

In connection with his unreasonable effort to deny that *Commentariolus* refers to *Revolutions*, Sw says:

> in all probability *Revolutions* was entirely a work of the 1530's, and I can see no evidence for giving any part of it an earlier date (1974, p. 194).

In June 1542, when Copernicus wrote the Preface to *Revolutions*, he said that a friend

> urgently requested me to publish this volume (*librum hunc*) and finally permit it to appear after being buried among my papers and lying concealed not merely until the ninth year but by now the fourth period of nine years (NCCW, II, 3/35–37).

A "fourth period of nine years" (*quartum novennium*) begins after 27 years. Subtraction of 27 from 1542 leaves 1515 as the latest year in which "this volume" began to lie concealed among Copernicus' papers. He plainly says that he concealed "this volume" (*librum hunc, Revolutions*). This famous utterance is falsified in two different ways by Sw (1973, p. 423; 1974, p. 191): first, Sw counts the 27 years "from the time that the theory of the earth's motion first occurred to Copernicus"; secondly, Sw misquotes Copernicus as saying that he "concealed his theories," whereas Copernicus explicitly specifies that he concealed "this volume." Nobody, not even Sw, knows when "the theory of the earth's motion first occurred to Copernicus." But this first occurrence took place in or before 1514. By 1515, at the latest, *Revolutions* began to be concealed by Copernicus. Several parts of *Revolutions* were demonstrably written at various times between 1515 and 1541. The evidence that some sections were written before 1530 — evidence that Sw says he cannot see — is overwhelming. *Commentariolus*' reference to *Revolutions* in or before 1514 is unmistakable to any reader not obsessed by unhistorical prejudices. "What kind of larger book Copernicus had in mind when he wrote" *Commentariolus* is "by no means clear" to Sw, nor to anyone else, since Copernicus never had any kind of larger book in mind other than *Revolutions*.

[36] *Commentariolus* claims excellent agreement between the observations and its "arrangement of circles" (*circulorum compositio*), not spheres; see n. 326, below.

[37] Here Copernicus proclaims his awareness that he was not the first to realize that the earth is a body in motion: his geokinetic predecessors were the Pythagoreans (he still knew nothing about Aristarchus of Samos). According to his source of information about the Pythagoreans, they based their geokineticism on reasoning he regarded as unconvincing. He accepted their conclusion that the earth moves, while rejecting their supporting arguments. Copernicus "accuses them ... of not knowing what they were talking about," says Sw (1973, p. 439). This is Sw's opinion of the Pythagoreans, which he foists on Copernicus, as he does so often elsewhere.

[38] *Commentariolus* promises evidence for the earth's motion *in circulorum declaratione*, "in its exposition of the circles," not spheres (see n. 326, below). Sw asks:

> Is this promise kept? Not explicitly, for nowhere does Copernicus point out that any particular apparent motion ... is evidence or proof of the earth's motion (1973, p. 439).

One *Commentariolus* passage after another proves how faithfully and explicitly it kept its promise to show that a particular apparent motion outside the earth was evidence of the earth's real motion (see nn. 56, 66, 83, 192, below).

[39] V, S: *astruere*; A: *asseverasse*. A in 1585 regarded this attempt to deny the earth's mobility as a relic of the remote past, whereas three-quarters of a century earlier *Commentariolus* correctly treated this effort as still a living force to be overcome.

[40] Once more *orbes caelestes* is *Commentariolus*' expression for the celestial spheres (see nn. 1, 22, above).

[41] V, S, A: *complectuntur*. V's *complectuntur* was misreported by MCV, I, 7, as *complectunt*. But the upward flourish at the end of *complectunt* turns it into the deponent form, just like *sequitur*, two lines below.

[42] V, S: *et*; A: *ac*, presumably in order to avoid repeating *et* three words before.

[43] All three manuscripts (V, S, A) have Mars' sphere follow directly below Saturn's, omitting Jupiter's. This omission was therefore a blemish in Brahe's MS too. Even A failed to fill this gap. In S it was filled by a third hand above the line.

[44] The lunar sphere revolves around the center of the earth, while the earth's sphere revolves around the sun (Postulate 6). That is why Postulate 1 said: "There is no one center of all the celestial spheres."

[45] V, S: *ea*; A: *eo*, thereby changing the reference from the earth (feminine) to the center (neuter) of the earth.

[46] Since the lunar sphere revolves around the center of the earth and moves with it like an epicycle, the moon periodically penetrates into and emerges out of the earth's sphere. This is therefore not a solid sphere. Yet Sw (1976, p. 110) insists that Copernicus' spheres were solid. The expressions *sphaera solida* and *orbis solidus* were used by other astronomers, but never by Copernicus.

[47] V, A: *restituitur*; S: *restituuntur*. For his prototype's plural form *restituuntur*, A substituted *restituitur* in order to match the singular form *peragit* at the end of this sentence. A's alertness brought him into fortuitous agreement with V.

[48] The number for Mars is missing in both V and S, and therefore was missing in their prototype, Brahe's MS. By contrast with the omission of Jupiter's sphere (see n. 43, above), A noticed this gap, and filled it with "second" (*biennio*), which A took from *Revolutions*, I, 10 (NCCW, I, fol. 9^v/3 up). But A overlooked *Commentariolus*' later section on "The Three Outer Planets," where Mars' period is given as ending in the 29th month (see n. 162, below). This error is found in Giorgio Valla's *Seek and Avoid*, XVI, 1: *Mars ... duobus annis et quinque mensibus suum orbem conficit* (sig. bb 4^v/12-13), a translation of Cleomedes, I, 3, 17, into Latin. "Copernicus ... could immediately compute the correct period," 23rd month, from the *Alfonsine Tables*, says Sw (1973, p. 441). But Copernicus did not do so. Yet Sw insists: "Copernicus derived his parameters by simple extraction from the *Alfonsine Tables*," which "is the source of the numerical parameters in ... *Commentariolus*"

(1973, pp. 425, 426). After finishing *Commentariolus*, Copernicus improved his knowledge of Mars' period, which he later gave correctly in *Revolutions*, V, 1 (NCCW, II, 228–229, 234–235).

[49] Again, by computation from the *Alfonsine Tables* Copernicus could have learned that Venus completes its revolution in about 225 days, that is, in the eighth month, not the ninth. Then once more the *Alfonsine Tables* is not "the source of the numerical parameters in ... *Commentariolus*," despite Sw (1973, p. 426). After finishing *Commentariolus*, Copernicus improved his knowledge of Venus' period, which he later gave correctly in *Revolutions*, V, 1 (NCCW, II, 228–229, 236–237).

[50] Here *Commentariolus* introduces *orbis magnus* (Grand Orb) as the new heliocentric designation of the earth's real annual revolution around the sun. With regard to the term *orbis magnus*, which became the distinctive feature of the innovative vocabulary of the Copernican astronomy, Sw remarks:

> Kepler ... clearly does not know why [Copernicus used the expression *orbis magnus*] when he guesses, both in the *Cosmographic Mystery* and in the *Epitome of Copernican Astronomy*, that the earth's orbit is called the Great Sphere because it has so many uses (1973, p. 442).

In the *Cosmographic Mystery* Kepler explains that *orbis magnus* is so "called on account of its multiple uses" (GW, I, 20 = VIII, 36). This definitive explanation (which Sw calls a "guess") shows that Kepler clearly did know why *Commentariolus* introduced *orbis magnus*. In Sw's second "guess" (*Epitome of Copernican Astronomy*) Kepler expands his previous explanation that *orbis magnus*

> is the name applied by Copernicus to the earth's true orbit (*orbita*) about the sun. This orbit is located in the space between the orbit of Mars outside it and the orbit of Venus within it; and he calls it "Grand" not on account of its size, since the circular orbits of the outer planets are much larger, but on account of its extraordinary usefulness in saving the apparent motions of not only the sun but also all the primary planets (GW VII, 403/5–9).

In these thoroughly informed and absolutely categorical affirmations by Kepler, Sw sees only lack of knowledge and guesswork.

[51] V: *sol*; S: *solem*; A: ☉. Writing while listening, V missed *solem*'s ending *-em* under the influence of the initial syllable *am-* of *ambiens*, the following word.

[52] V: *In tribus*; S: *temporibus*; A: *in temporibus*. A restored *in*, which was missing in his prototype. V's amusing error *tribus* (three) was due to his reader's misinterpretation of a contracted form of *temporibus* (t̄ribus).

[53] V, A: *parte*, omitted by S. A's restoration of *parte* brought A into fortuitous agreement with V.

[54] In Postulate 4, above.

[55] V: *tamen*; S, A: *autem*. Which of these two acceptable readings was present in Brahe's MS?

[56] This is the first of the passages mentioned in n. 38, above, in which *Commentariolus* keeps its promise to show that the earth's real motion produces a corresponding apparent motion outside the earth — in this instance, the sun's apparent annual orbit.

[57] V, S: *duobus gradibus et sextante unius*. Constantly striving for greater compactness, A omitted *unius*. In Figure 1 let C denote the center of the Grand Orb, and S the sun. Join CS, and extend it to meet the Grand Orb in both directions at E_0, E_0. Draw SE perpendicular to SC. Join EC. As the earth E revolves around the

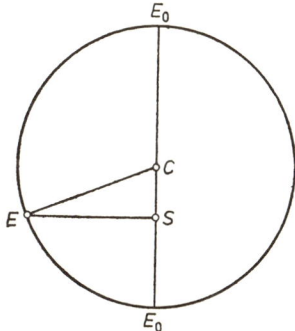

Fig. 1. Earth's nonuniform revolution around the sun
C — center of Grand Orb, E — earth, E_0, E_0 — earth's apsidal line, S — sun

Grand Orb, ⊰ CES measures the constantly changing difference (inequality) between two directions: (1) EC, from the earth to C, the center of the Grand Orb; (2) ES, from the earth to S, the sun. When E reaches E_0, E_0 (the line of apsides), ⊰ CES = 0. When E reaches E_{max} (as in Figure 1), ⊰ CES attains its maximum value. This is given in *Commentariolus* as 2° 10′. It also gives $CS = 1/_{25} \times CE_0$. The correspondence between a maximum inequality of 2° 10′ and an eccentricity of $1/_{25}$ is not precise, but rather a convenient approximation. For if $CE_0 = 100,000$, then $CS = 4000$. In *Revolutions*' Table of the Straight Lines Subtended in a Circle (NCCW, II, 32/19–20), the sine of 2° 20′ is 4071, and of 2° 10′ is 3781, the difference being 290, or 29 for 1′, so that 4000 is the sine of 2° 17′ 33″ ≃ 2° 10′.

[58] Pollux, the somewhat less brilliant star being Castor. *Commentariolus* always defines celestial places by referring to nearby stars, since these "give position to all things" (according to "The Order of the Spheres"). Previously, celestial longitudes had been reckoned from the vernal equinox, regarded as constant. But "the equinoxes ... shift considerably," *Commentariolus* will maintain at the beginning of its next section. Hence the vernal equinox can no longer be used as the starting point from which to measure celestial longitudes.

[59] V: *toto*; S, A: *loco*. V's reader mistook the abbreviation of *loco* (*lō*) for *toto* because the short horizontal stroke over the *o* extended a little too far to the left, thereby making the *l* look like a *t* (*to*, a common abbreviation for *toto*).

[60] V, A: *mediante*; S: immediante. A corrected the faulty reading (immediante) in his prototype, and thereby came into fortuitous agreement with V. *Commentariolus*' *mediante* here resembles *mediantibus* later on.

[61] The solar apogee is therefore about 10° west of Pollux. This star's celestial longitude, like that of the other stars, was subject to correction for precession, the amount of which is not precisely defined in *Commentariolus*. Hence its value for the longitude of the solar apogee is uncertain.

[62] V, S, A: *terram*. V's *terram* was misreported by MCV, I, 8/19, as *terra*, which is the required reading. The erroneous reading *terram* (*terrā*), between *nō* (*non*) and *solū* (*solum*) appears in both V and S; hence, it was already present in Brahe's MS.

[63] Sw (1973, p. 444) mistranslates *sibi maxime proprius* by "most certainly belongs to it," as though the earth's second motion belongs to it more certainly than the first and third motions do. The second motion is peculiar to the earth, which is the only heavenly body that performs a rotation in *Commentariolus*.

[64] V: *motus*; S, A: *mundus*. V's incorrect reading *motus* was simply an echo of the preceding word *totus*.

[65] V, A: circumvagi; S: circumagi. Writing while listening, in place of *circumagi* V incorrectly substituted circumvagi under the influence of *voragine* just before it and *videtur* just after it. The required form of *circumvagor* would have been circumvagari (like *pervagari*, in the first sentence of the section on "The Moon"), not circumvagi. In like manner, a little later on, V wrote the non-existent form circumvaguntur instead of *circumaguntur*, which A has correctly (see n. 70, below).

[66] This apparent daily rotation of the heavens, caused by the real diurnal rotation of the earth, is the second phenomenon adduced by *Commentariolus* in keeping the promise mentioned in n. 38, above.

[67] V, S, A: *circumfluis*. This correct reading, found in all three manuscripts, is equally applicable to both *aqua* and *aere*. The emendation *circumflua*, which would be restricted to *aqua*, is printed by Sw (1973, p. 444) as though it were his own, although it was proposed a century ago (MCV, I, 8). Nevertheless, *Commentariolus*' *circumfluis* extends also to *vicino aere*. That "nearby air", whose "motion, acquired from the earth by proximity, shares without resistance in its unceasing rotation," differs from "the uppermost belt of air," which "is unaffected by the earth's motion on account of its great distance from the earth" (NCCW, II, 16/25–34).

[68] V: *circulo*; S, A: *saeculo*. Writing while listening, V mistook the correct reading *saeculo* for *circulo*, which he heard more often while producing his copy of *Commentariolus*.

[69] V: *supermanente*; S, A: *semper manente*. Writing while listening, V mistook the correct reading *semper manente* for *supermanente*.

[70] V: circumvaguntur, misreported by MCV, I, 9/2, as *circumvagantur*, which would have been the proper form of *circumvagor*. The reading of S, A, however, is *circumaguntur*, which is the proper form of *circumago*. V's circumvaguntur sounds like a delayed echo of V's circumvagi (see n. 65, above).

[71] V, S: *parvos*; A: *parvulos*, emphasizing the small size of these circles as compared with the circle traversed by the earth's center.

[72] V, S, A: *complet*. V's *complet* was misreported as *complent* by MCV, I, 9/4, which ignored the scribe's deletion of the letter *n*.

[73] V: *annuos*, misreported by MCV, I, 9/4, as *annuas*, the required feminine form; S, A: *annuas*. V wrote the masculine form *annuos* under the influence of the preceding word *motus*, a masculine form.

[74] V, S: *paene*, changed by A to *fere*, which may be merely an inadvertent repetition of *fere*, seven words earlier.

[75] V, S: *immutabilem*, changed by A to *immobilem*, an inferior reading.

[76] V: *polus*, misreported as *polos* by MCV, I, 9/8; S, A: *polos*. V's mistaken *polus* was affected by *complexus*, the preceding word.

[77] V: *tractum*; S, A: *tractu*. V's incorrect reading *tractum* was affected by the following word *deprehensum*.

[78] V, S: *plerisque*, omitted by A, although *Commentariolus* included it in order to emphasize the widespread acceptance of this opinion.

[79] V, S, A: *lege*. V's *lege* was misreported as *loge* by MCV, I, 9.

[80] V: *nedum*; S, A: *nondum*. V's incorrect reading *nedum* was influenced by the preceding word *lege*.

[81] V, S: *satis*, omitted by A, striving for greater compactness.

[82] V: *fieri*, omitted by S, A. Their omission of *fieri*, which is indispensable, shows that A's prototype was akin to S and, like S, was defective in this respect.

[83] This third motion of the earth is proposed by *Commentariolus* to account for the precession of the equinoxes, which had previously been explained by having the sphere of the stars move. Like the earth's annual revolution and daily rotation, its axial motion too was intended by *Commentariolus* to produce phenomena observed outside the earth, and therefore to be the third motion fulfilling the promise mentioned in n. 38, above.

[84] V, S, A: *magnete*. V's *magnete* was misreported as *magnetem* by MCV (I, 9), which misread the long tail of the letter *q* in the line above as a horizontal stroke over the final *-e* in *magnete*. For a similar misreading, see n. 3, above.

[85] V, A: *virgula*; S: *ungula*, an understandable confusion of *vir-*, as then written, with *un-*, on the part of a scribe copying a manuscript at which he was looking with his own eyes. A corrected the faulty reading in his prototype, and thereby came into fortuitous agreement with V.

[86] V, S: *nitatur*; A: *nutatur*, an unsuccessful attempt by A to improve *Commentariolus*' terminology where no improvement is required. A's *nutatur* may have been suggested by the incorrect reading *nutum* (discussed in n. 89, below). In any case A's *nutatur* would eliminate *nitatur*'s implied idea of striving or effort, which was an integral part of Copernicus' concept of gravity too: "For my part I believe that gravity is nothing but a certain natural desire" (*appetentiam*; NCCW, II, 18/5–6).

[87] Just as the compass needle points north, the combined effect of the earth's first and third motions in *Commentariolus* keeps the axis of the earth's second motion pointed toward the region of the pole star.

[88] V: *est*, missing in S, A. A followed his prototype and failed to insert *est*, although it is needed. The omission of the common word *est* is understandable on the part of a scribe who was copying a manuscript at which he was looking with his own eyes.

[89] V: *motum*; S, A: *nutum*. A scribe copying a manuscript at which he was looking with his own eyes might understandably confuse *mo-*, as then written, with *nu-*. A's acceptance of *nutum* may have suggested the emendation *nutatur* (discussed in n. 86, above).

[90] V, S: *sed stellas*; A: *sed ad stellas*, improving *Commentariolus*' Latin style.

[91] V, A: *etiam*; S: *et*. A improved an inferior reading in his prototype and thereby came into fortuitous agreement with V. The common abbreviation of *etiam* (*et* surmounted by a curved stroke) does not appear anywhere in S. Brahe's MS had *etiam*, as is shown by V. Hence, the curved stroke over *et* fell out when A's prototype was being made from Brahe's MS. The meaning of *etiam* is "besides," not "for," as it is mistranslated by Sw (1973, p. 451). The length of the year should not be measured from the equinoxes, according to *Commentariolus*, for two reasons: (1) the equinoxes shift; (2) in addition (*etiam*), such measurements have yielded different results.

[92] V: *diversa*, omitted by S, A. A's omission of *diversa*, which is indispensable, shows that A's prototype was akin to S. The presence of *diversa* in V indicates that Brahe's MS also had *diversa*, which Sw mistranslates by "irregular" (1973, p. 451). The determination was different (*diversa*) in different (*diversis*) ages. *Commentariolus*' *diversis*, *diversa* echo *Diversi diversas* at the beginning of *Epitome*, III, 2.

[93] This historically incorrect statement was taken by *Commentariolus* from *Epitome*, III, 2. *Revolutions* is better informed than *Commentariolus*, but still wrong:

> Hipparchus ... declared that the $1/4^d$ lacked a small fraction. This was later established as $1/300^d$ by Ptolemy (III, 13; NCCW, II, 144/22–25).

Yet PS (III, 1) quotes Hipparchus' statement (in a work that has not come down to us) that the year is 365 $1/4^d$ "less about $1/300^d$."

[94] V, S, A: *46*. V's *46* was misreported as "r6" by MCV, I, 10.

[95] V: *hoc*; S, A: *hoc est*. The presence of *est* in S, A shows that Brahe's MS also had *est*, a very common word understandably omitted by a scribe who was writing while listening.

[96] According to *Epitome*, III, 2,

Hipparchus and Ptolemy declared the year to be the return of the sun to an equinoctial or solstitial point. Hence the amount of time from the sun's entry into the autumnal equinoctial point until its next entry into the same point is said to be the length of the year.

From this linkage of the names of Hipparchus and Ptolemy, *Commentariolus* inferred that they accepted the same length of the year. Hence *Commentariolus* ascribed also to Ptolemy the erroneous length 365 $1/4^d$, thereby overlooking *Epitome*, III, 2's correct statement of Ptolemy's value ($365\ 1/4^d - 1/300^d = 365^d\ 5^h\ 55^m\ 12^s$).

The length of the year assigned by *Epitome* to al-Battani ($365^d\ 5^h\ 46^m + 2/5^m$) was less than $365^d\ 5^h\ 59\ 5/5^m$ ($= 365\ 1/4^d$) by $13\ 3/5^m$. Why $13\ 1/3^m$ was introduced by *Commentariolus* as an alternative has not yet been made clear. Perhaps a source available to *Commentariolus* gave $2/3^m$ (instead of $2/5^m$) as the fraction by which al-Battani's year exceeded $365^d\ 5^h\ 46^m$.

Sw's "admittedly drastic restoration" of the text (1973, p. 452) assumes that *Commentariolus* would express $4/5$ ($= 0.800$) by resorting to $1/2 + 1/3$ ($= 0.833$) immediately after writing simply, clearly, directly, and exactly *3 quintis* ($= 3/5$).

[97] V, S: *huic*, corrected by A to *hunc (annum)*. The presence of *huic* in both V and S indicates that Brahe's MS also had this defective reading.

[98] V, S: *vigesima*; A: *25ta*. A changed *Commentariolus*' $1/20^h = 3^m$ to $1/25^h = 2\ 2/5^m$ as the amount by which Hispalensis increased al-Battani's length. *Commentariolus*, evidently content to ignore fractions of a minute, treated these lengths as $365^d\ 5^h\ 49^m$ and $365^d\ 5^h\ 46^m$, with a difference of $3^m = 1/20^h$. On the other hand, A insisted on taking account of the fractions. Thus, for A, al-Battani's length was $365^d\ 5^h\ 46\ 3/5^m$, so that Hispalensis' year, $365^d\ 5^h\ 49^m$, was $2\ 2/5^m = 1/25^h$ longer than al-Battani's.

[99] V: *49 minutis annum*; S: *49 m. in annum*; A: *49' annum*. If A's prototype agreed with S, A recognized the impropriety of *in*, which A deleted.

[100] V, A: *ex*, omitted by S. If A's prototype agreed with S, A recognized the necessity of inserting *ex* and thereby came into fortuitous agreement with V.

[101] *Commentariolus*' Hispalensis (the man from Hispalis = Seville) was formerly misidentified with Isidore of Seville (MCV, I, 10; PII, 191) and with Jabir ibn Aflah (called al-Ishbili, Arabic for "the man from Seville"). Neither of these older writers from Seville, however, made the year $365^d\ 5^h\ 49^m$ long, as did *Commentariolus*' Hispalensis. He was finally identified correctly by Ludwik Antoni Birkenmajer, *Stromata Copernicana* (Cracow, 1924), p. 353, as Alfonso de Corduba Hispalensis. His *Almanach perpetuum* was published in Venice on 15 July 1502, when Copernicus was in residence at the University of Padua, then under the control of Venice, some twenty miles away. At the very outset of his *Almanach* (sig. a 1^v), Hispalensis gave the number of days in the year as $365\ 1/4^d - 11^m$ (*365. et quartum minus undecim minutis horae*) = $365^d\ 5^h\ 49^m$, precisely the value assigned to Hispalensis by *Commentariolus*.

This obviously correct identification of *Commentariolus*' Hispalensis with Alfonso de Corduba Hispalensis was universally accepted without the slightest demur until Sw pretended that *Commentariolus*' last value ($365^d\ 5^h\ 49^m$) "is that of the *Alfonsine Tables*," and also "is simply a rounding" of the *Alfonsine Tables*' year (1973, p. 452, and n. 2). Copernicus owned a copy of the second edition (Venice, 1492) of the *Alfonsine Tables*. But Sw carefully avoids specifying where it gave $365^d\ 5^h\ 49^m$ as the length of the year, or any length that could be so rounded. Later versions of the *Alfonsine Tables* were modified by their editors. Half a century after Copernicus' edition, and some three decades after he wrote *Commentariolus*, the *Alfonsine Tables* were said to subtract from $365^d\ 6^h$, not 11^m (as *Commentariolus*' Hispalensis did), but "about $10^m\ 44^s$, which is a little more than $1/6^h$" (Peurbach, *Theoricae novae planetarum*, ed. Reinhold [Wittenberg, 1542], sig. e 4^v–5^r: *Annum enim faciunt [Alphonsini] 365 dierum cum quadrante minus 10 scrupulis 44 secundis fere, id quod paulo plus est sextante unius horae*). Less than a decade after publishing this statement about the *Alfonsine Tables*, Erasmus Reinhold (1511–1553) finally brought out his own *Prussian Tables*, which were widely consulted for a long time after they were first printed in 1551. Reinhold did not find $365^d\ 5^h\ 49^m$ as the length of the year in the *Alfonsine Tables*. But *Commentariolus* did find $365^d\ 5^h\ 49^m$ as the length of the year in Alfonso de Corduba Hispalensis.

Not content to claim that *Commentariolus* could find in the *Alfonsine Tables* what was not there, Sw sought to link *Commentariolus* with the *Alfonsine Tables* in still another way. *Commentariolus* quoted from Hispalensis, the man from Seville. Sw asks: "Could *Hispalensis* in fact be a misreading of *Hispaniensis*?" (1973, p. 452). To lend plausibility to the possibility of such a misreading, Sw remarks that *Hispaniensis* "would be written *Hispāiensis*" (1973, p. 451). But all three manuscripts of *Commentariolus* read *Hispalensis*, not *Hispaniensis* nor *Hispāiensis*. Besides, who could have misread what? *Commentariolus* quoted from a book published by Hispalensis, whose name was transmitted faithfully by all three manuscripts of *Commentariolus*. From the paleographical point of view, therefore, Sw's emendation has absolutely no basis whatever.

Why, then, did Sw propose it? "It is likely that *Hispalensis* (*Hispaniensis*?) refers to Alfonso X who was, after all, a Spaniard" (1973, p. 452). But when Copernicus refers to a Spaniard, he calls him *Hispanus*, not *His-*

paniensis (*Revolutions*, III, 2, 6, 16; NCOO, II, 118/37, 126/20, 157/2). Besides, who ever called Alfonso X "*Hispaniensis*"? Certainly not Copernicus' edition of the *Alfonsine Tables*. It mentioned its royal and imperial patron four times: once, as "king of Castile," and three times as "king of the Romans and of Castile," referring to Alfonso's partially successful candidacy in the inconclusive election of the Holy Roman Emperor in 1257 (sig. A 4^r, B 1^r, a 1^r, k 6^r). Copernicus' *Alfonsine Tables* never referred to Alfonso X as *Hispaniensis*, as though he were an ordinary Spanish commoner, instead of king of Castile and Leon. Copernicus' *Alfonsine Tables* never referred to Alfonso X as king of Spain, a title that did not even exist constitutionally in his time. Historically as well as paleographically, there is not the slightest foundation for Sw's proposal to substitute Alfonso X for Alfonso de Corduba Hispalensis as *Commentariolus*' Hispalensis.

Incidently, to use Sw's idiosyncratic spelling (1973, pp. 425, 430, 431, 433; like "accidently," 1973, p. 460), *Commentariolus* never refers to Alfonso X nor to his *Alfonsine Tables*. Neither does *Revolutions*.

Although *Commentariolus* names four astronomers in this paragraph, it mentions only three lengths of the year, since it treats Hipparchus' and Ptolemy's as identical. These three lengths are different (not "irregular," as in Sw, 1973, p. 451; see n. 92, above): $365^d\ 6^h$; $365^d\ 5^h\ 46^m$ (plus an uncertain fraction of a minute); $365^d\ 5^h\ 49^m$. These different lengths were determined in different ages: Hipparchus'–Ptolemy's, in Greek antiquity; al-Battani's, in the Muslim Middle Ages; and Hispalensis', by a Christian contemporary. Between the first two, the difference was a decrease of about 13^m; between the last two, an increase of 3^m.

[102] V, A: *ex*; S: *a*, followed by an obliterated letter, the combination being misreported by Lindhagen (p. 8/28) as *ex*, the required reading. If A's prototype agreed with S, then A's correction of the defective reading brought A into fortuitous agreement with V.

[103] V: *mutabant*; S, A: *mutabantur*. V wrote *mutabat* with a curved stroke over the final syllable to indicate the elision of *-n-*, while forgetting to attach to the final *-t* the flourish needed to convert *mutabat* to the passive voice.

[104] Sw (1973, p. 451) mistranslates *motibus inferioribus obviantes* by "since they are opposite in direction to the lower motions." The nature of these supposed "lower motions" is not explained in Sw's lengthy comment on this passage. Lesser or slower motions were invoked by *Commentariolus* to explain why the equinoxes moved faster in some centuries than in others, for the rate of that precessional motion was (mistakenly) believed by Copernicus to be variable. Since the precession is always westward, its direction cannot affect its rate, despite Sw's mistranslation.

[105] Sw's "as" (1973, p. 451) does not adequately match *tanto ... quanto*. The length of the year decreased by "as much as" the displacement increased.

[106] V: *occursu*, omitted by S, A. The presence of *occursu* in V shows that this was the reading in Brahe's MS too. The omission of *occursu* in S was inadvertent. A failed to recognize the necessity of inserting some such term as *occursu* in his prototype.

[107] V, S: *semper*, omitted by A. By this omission A weakened *Commentariolus*' contrast between the ever changing tropical year and the always constant sidereal year. *Commentariolus* did not explain exactly how it used the Spike of the Virgin to arrive at this astonishingly accurate determination of the length of the sidereal year.

[108] V, A: *dierum*; S: *diebus*. If A's prototype, which was akin to S, also had *diebus*, then this was emended by A to *dierum* for the sake of consistency with the genitive forms (*horarum*, *sextantis*).

[109] *Commentariolus* took this misstatement from *Epitome* (III, 2), which said that the "oldest of the Egyptians" found the length of the sidereal year to be $365\ 1/4^d + 1/130^d$. The fraction $1/130^d \simeq 11^m$, which was rounded by *Commentariolus* to $\sim 1/6^h \simeq 10^m$. The ancient Egyptians had three calendars: (1) lunar, with an intercalary month; then (2) solar, with 365 days, requiring (3) a revised lunar calendar. It was not until they were conquered by the Macedonians, Greeks, and Romans that the Egyptians acquired the leap year. The distinction between the tropical and sidereal years lay far beyond the reach of ancient Egyptian astronomy (DSB, XV, 706–709).

[110] *Commentariolus* still accepted the traditional teaching that the apsides maintained a constant position among the stars. Copernicus' subsequent discovery that the apsides move with reference to the stars was announced in *Revolutions*, which therefore advanced beyond *Commentariolus* in this important respect. Copernicus' later silence about *Commentariolus* may have been due to his feeling that, in this feature, as well as in others, his earlier, briefer treatise had been superseded (see nn. 169, 173, below).

[111] V, S: *vero*, omitted by A. *Commentariolus*' contrast between the apsides' fixed position, so strongly emphasized at the end of the preceding section, and the moon's multiple mobility was abandoned in A's drive for greater compactness.

[112] V, S, A: *praeter*. V's *praeter* was misreported as *prefer* by MCV, I, 10, which mistook the first *e*'s flourish (signifying the diphthong *ae*) as the tail of the letter *f*, where in fact the letter *t* stands.

[113] In "The Order of the Spheres."

[114] Here V's folio 59 (according to the older numeration) ends. At the bottom of fol. 59ᵛ a librarian's pencil noted parenthetically: "fol. 60 is missing." The missing section is preserved, however, in S and A. Sw (1973, p. 431) calls V "the first manuscript of the *Commentariolus*." If the first manuscript had a big gap, how was that gap filled by S and A? Actually, A's prototype was written earlier than V.

[115] S: *quem vocant*, omitted by A. In striving for greater compactness, A weakened *Commentariolus*' contrast between the conventional terminology (*quem vocant epicyclum primae diversitatis sive argumenti*) and its own more direct labels (*nos vero primum sive maiorem*).

[116] S, A: *anni*, which is clearly wrong because the first lunar epicycle has no connection with the year. The attractive emendation *qui* was proposed by the late Ryszard Gansiniec (1888–1958). But while *u* and *n* were often confused, would a copyist misread *q* as *an*? Perhaps we shall obtain the correct reading if V's missing fol. 60 is ever recovered.

[117] S: *perficit*; A: *proficit*, an unnecessary emendation.

[118] S, A: *diametrum*. This correct reading is erroneously emended to *semidiametrum* by Sw (1973, p. 455). But "diameter" is necessary, since in Figure 2 the larger epicycle's center (LEC_2) periodically touches the diameter in question at a distance from the earth's center (EC) that exceeds the semidiameter or radius ($LEC_2 - GOC$). In contemporary astronomy the line $GOC - EC$ is now called "the direction of the mean sun," a term not found in *Commentariolus*. Instead, it uses the expressions "line drawn from the Grand Orb's center through the earth's center" or "Grand Orb's diameter." CSE (p. 83) explained that "mean sun" is Sw's "label for what Copernicus calls the center of the earth's annual orbit." Sw exclaims that CSE "objects strenuously to my calling the center of the earth's orbit the mean sun" (1976, p. 111). For the sake of readers not thoroughly familiar with the difference between the technical vocabularies of early modern and contemporary astronomy, CSE noted the equivalence of "center of the earth's orbit" with "mean sun." In this purely lexicographical remark, Sw professes to see a strenuous objection to himself.

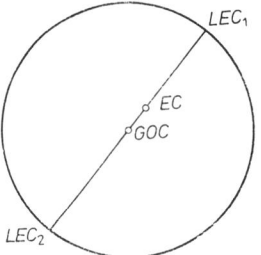

Fig. 2. Grand Orb's diameter
GOC — Grand Orb's center, EC — earth's center, $LEC_{1,2}$ — larger epicycle's center

[119] S, A: *diametri*. Since this erroneous reading occurs in both S and A, A's prototype was akin to S, and A did not correct the error. The emendation *semidiametri* (TCT, 1st ed., 1939, p. 69) is printed by Sw (1973, p. 455) as though it were his own.

S, A: *diametri et*; this obtrusive *et* was correctly deleted by MCV, IV, 6/14, 7. After *et*, S made a false start in spelling *epicicli*, the next word: S wrote the letters *epicl* and then deleted them.

[120] S: *10 partem ... cum 18a unius particulae*; A: *10 partem ... cum 18a unius particula*. A's *particula* is undoubtedly an inadvertent error, since in the corresponding passage in the section on "Venus" A wrote *dodrantem ... unius particulae*. Sw (1973, p. 455) replaces *10 partem* (tenth part) by *10am unius partis*. This would-be emendation leaves the adjective *10am* (*decimam*) without a noun to modify, since *partem* (which it modifies in both our manuscripts) is discarded by Sw. Furthermore, *Commentariolus' 10 partem de semidiametro* (misprinted "semiametro" in MCV, IV, 6/15, and PII, 193/3) *orbis sui deferentis* is precise: tenth part of the radius of its deferent sphere. By contrast, Sw's *10am unius partis* (tenth of one part [of the deferent's radius]) is unacceptably imprecise, since it does not specify the size of the one part. Then, instead of *Commentariolus' cum 18a unius particulae* (together with an eighteenth of one small [tenth] part [of the deferent's radius]), Sw proposes *cum 18am decem particularum*. Sw thereby combines the preposition *cum* with the accusative case, instead of the ablative, the sort of error committed by beginners wrestling with the rudiments of Latin grammar. Sw also transfers *unius* from this complex fraction's second section (where it was put by *Commentariolus*) to the first section. In the second section, in place of this transposed *unius*, Sw writes *decem*, for which there is no warrant in S and A; where Sw's *decem* comes from will soon be evident. Sw also changes *particulae* to *particularum*, thereby emerging with an "eighteenth of ten particles." Are these ten particles (*particularum*)

of the same size as the previously mentioned one part (*partis*) of unspecified size? Sw's "tenth of one part ... with an eighteenth of ten particles" is offered, forsooth, as an emendation of *Commentariolus*' "tenth part of the deferent's radius plus an eighteenth of one [such tenth] part," an absolutely clear and precise expression that requires no emendation whatsoever.

Commentariolus' complex fraction was couched essentially in words. Elsewhere, however, Copernicus wrote it numerically, in two slightly different but equivalent forms. For this purpose he used one of the blank sheets he had inserted by a binder when two of his books were bound together. That sheet may be designated "U," because the composite bound volume in question is now preserved in the library of the University of Uppsala, Sweden.

As may be seen in the facsimile (NCCW, IV, Plate XXXII), in the form of a fraction U records the ratio of the lengths of the "radii of the moon's deferent and larger epicycle," in that order. The number 10, written as the numerator of the fraction, is associated with the deferent; it does not belong with the epicycle, where Sw's *decem* (10) would put it, and it has nothing to do with Sw's particles (*particularum*). The denominator of the fraction is presented as a double-digit value. Its first digit appears as a dot and the number one (.1), signifying one-tenth of the deferent's radius. The second digit is represented by a short diagonal stroke, followed by $\frac{1}{18}$, signifying one-eighteenth of the preceding one-tenth. In order to avoid the inconvenience of fractions, larger units were used in those days. Thus, raise the deferent's radius 10 to 60. Then the larger epicycle's radius becomes ($1/10 \times 60 =$) 6 in the first column, and ($1/18 \times 1/10 \times 60 =$) $1/3$ in the second column, or $6\,1/3$ when the deferent's radius = 60.

Just below the first fractional value of the epicycle's radius, U's bottom line has a second fractional value. This time, the denominator omits the dot and the short diagonal stroke. This is the value verbalized in *Commentariolus*:

> *Quantitas autem semidiametri epicycli maioris continet decimam partem de semidiametro orbis sui deferentis cum decima octava unius particulae.* The length of the larger epicycle's radius contains a tenth part of the radius of its deferent sphere, together with an eighteenth part of one [such tenth] part.

Thus again, if the deferent's radius = 60, the larger epicycle's radius = ($1/10 \times 60 =$) 6; ($1/18 \times 1/10 \times 60 = 1/3$; $6 + 1/3 = 6\,1/3$; $\frac{6\,1/3}{60} = \frac{1\,1/18}{10}$.

[121] S: *quinquies dempta una parte ipsius*; A: *quinquies dempta una ipsius parte*. A's alteration of the word order means little, for A often made such changes for stylistic reasons. But A drew a thick diagonal stroke through *parte*, perhaps to indicate dissatisfaction with its lack of precision: the larger epicycle's radius contains the smaller epicycle's radius five times minus one "part" of the smaller epicycle's radius. The correct emendation, *quarta* for *parte*, was made by MCV, IV, 6/17, on the basis of U's *epi[cyclus] a ad b* $19/4$ (epicycle *a*: epicycle *b* = 19:4 = [$5 - 1/4$]:1).

[122] S: *igitur*, omitted by A, striving for greater compactness.

[123] The "first lunar anomaly" denotes the variation in the moon's distance from the center of the earth. S, A: *prima ... diversitate*, a reading incompatible with the syntax of this sentence. For that reason *primae ... diversitati* was proposed in TCT, 1st ed., 1939, p. 70, an emendation which is printed by Sw (1973, p. 455) as though it were his own.

[124] S: *variationum*; A: *variationem*. Before A was discovered, S' *variationum* was corrected by TCT, 1st ed., 1939, p. 70, to *variationem*, an emendation which is printed by Sw (1973, p. 455) as though it were his own.

[125] S, A: *17*, corrected by TCT (1st ed., 1939, p. 70) to *12*, an emendation which is printed by Sw (1973, p. 455) as though it were his own. In Figure 3 the moon M is carried on the surface or circumference of the smaller epicycle, whose center (SEC) rotates on the larger epicycle's circumference. Join the larger epicycle's center (LEC) with SEC, and extend this line of centers to the moon in M_1. When the moon is in M_1 or M_2, it may be regarded as moving uniformly, together with SEC, along the larger epicycle's circumference. In all its other positions on the smaller epicycle, however, the moon deviates from uniform motion along the larger epicycle's circumference. This deviation reaches its maximum when the moon is in M_3 or M_4, where the tangents drawn from LEC touch the smaller epicycle. $\angle SEC - LEC - M_3 (M_4)$ measures the maximum deviation. The size of this angle is regulated by the ratio $SEC - M_3 (M_4) : SEC - LEC = 1 : 4\,3/4 = 21{,}053 : 100{,}000$. Then, according to *Revolutions*' Table of the Straight Lines Subtended in a Circle (NCCW, II, 32/42 - 33/8), the angle = $12°9'$, so that *Commentariolus*' $12\,1/4°$ is an approximation.

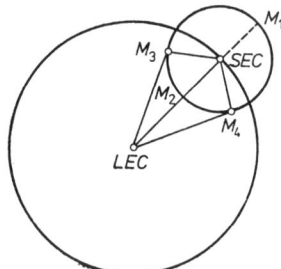

Fig. 3. Moon's variable distance from the center of its larger epicycle
M — moon, *SEC* — smaller epicycle's center, *LEC* — larger epicycle's center

[126] S, A: *respondentes*, corrected to *respondentis* by MCV, IV, 6/23.

[127] S, A: *diametri*. This perfectly sound reading is emended by Sw (1973, p. 455) to *semidiametri*. Why?

[128] S, A: *cum*. S' *cum* was misreported as *eum* by Lindhagen (p. 9/30), who was followed by MCV, IV, 6/23, and PII, 193/11. Ignoring TCT's restoration of *cum* (1st ed., 1939, p. 71) instead of *eum*, Sw makes this misreading worse by changing it to *eam*, which forces Sw also to substitute *a centro* for *centrum* (1973, p. 455). Since the letters *c* and *e* were written very much alike in S, Lindhagen's misreporting of *cum* as *eum* is understandable. But in the presence of TCT's correct reading *cum*, how is Sw's change to *eam* (and of *centrum* to *a centro*) to be understood?

[129] In Figure 3 the distance from *LEC*, the larger epicycle's center, to the moon varies. The mean distance may be increased or decreased within the limits of the smaller epicycle's radius. Sw (1973, p. 455) translates:

> it [the motion of the smaller epicycle] also sometimes draws the moon away from and sometimes brings it toward the center of the larger epicycle by an amount equal to the length of its semidiameter.

The semidiameter of what? The larger epicycle? The moon? The smaller epicycle?

[130] *Commentariolus' inaequales circulorum ambitus* is mistranslated by Sw (1973, p. 455) as "irregular circumferences of circles." How can a circle's circumference be irregular? By definition, a circumference's regularity consists of the equidistance of each of its points from the center of the circle. *Commentariolus* refers to "unequal peripheries of circles." When two circles have unequal peripheries, their diameters are unequal in length, but each circle is perfectly regular. In explaining this feature of Copernicus' lunar theory, Michael Maestlin (1550–1631), who introduced Kepler to Copernican astronomy, wrote:

> For the new and full moon, the epicycle of the apparent motion is *NIO*, but for the half moon *PLQ*. ... The first epicycle becomes nearer to the earth and farther away ... because the apparent epicycle, as though composed of two, increases and decreases in size ("The Dimensions of the Heavenly Orbs and Spheres, according to the *Prussian Tables*, in Conformity with the Thinking of Nicholas Copernicus," reprinted in GW, I, 134/19–27).

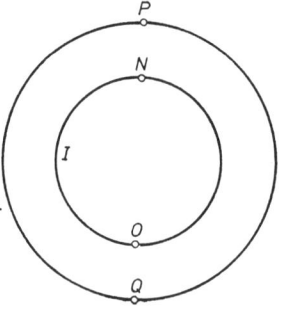

Fig. 4. Maestlin's diagram for Copernicus' lunar theory (somewhat simplified)
NIO — apparent epicycle at new and full moon, *PLQ* — apparent epicycle at half moon

Commentariolus' "peripheries of circles" (*circulorum ambitus*, *NIO* and *PLQ* in Figure 4) are "unequal" (*inequales*). They are not "irregular circumferences of circles," despite Sw.

[131] The new moon is in conjunction with the sun, while the full moon is in opposition to the sun.

NOTES

[132] The half-moon is in one of the two quadratures.

[133] S: *ad*, omitted by A, perhaps inadvertently.

[134] S, A: *6°*, corrected by TCT, 1st ed., 1939, p. 71, to *7°*, an emendation which is printed by Sw (1973, p. 455) as though it were his own; see Figure 5.

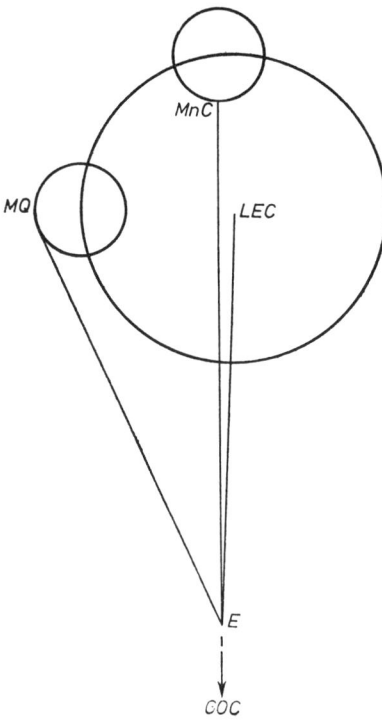

Fig. 5. Moon's variable angular displacement from the circumference of its larger epicycle
E — earth, LEC — larger epicycle's center, GOC — Grand Orb's center, MnC — moon near conjunction, MQ — moon at quadrature,
$\angle (MNC - E - LEC)_{max} = 4°56'$, $\angle MQ - E - LEC = 7°36'$

[135] S: *hoc*, changed to the plural *haec* by A.

[136] The Ptolemaic lunar theory placed the moon M on an epicycle whose center EpC moved on the circumference of an eccentric circle Ec (Figure 6). While moving uniformly with respect to the earth E, the epicycle's center EpC moved nonuniformly with regard to the eccentric's center EcC. This objection to Ptolemaic lunar theory recalls the discarding of the entire Ptolemaic system at the beginning of *Commentariolus* on the ground that it permitted nonuniform motions.

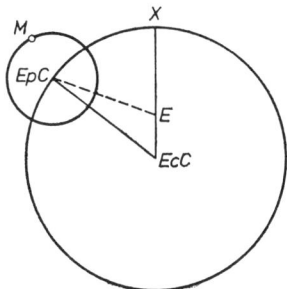

Fig. 6. Ptolemaic Lunar Theory
M — moon, EpC — epicycle's center, EcC — eccentric's center, E — earth, $\angle EpC - E - X$ increases uniformly, $\angle EpC - EcC - X$ increases nonuniformly

[137] S: *manifestos*, changed to the adverb *manifeste* by A.

[138] S, A: *cum*, perhaps under the influence of *dum* four words later. MCV, IV, 6/32, correctly emended to *quod*.

103

[139] Sw (1973, p. 461) mistranslates *dependet* as "rests." The moon is never at rest.

[140] This discrepancy between Ptolemaic lunar theory and observed phenomena was known to *Commentariolus* from *Epitome*, V, 22:

> But it is noteworthy that the moon does not appear so large at quadrature, when it is in the epicycle's perigee, whereas, if its entire disk were visible, it should appear four times its apparent size at opposition, when it is in the epicycle's apogee.

According to Sw, *Commentariolus* remarks

> that the only way of compensating for this large variation would be to maintain, improbably, that the body of the moon itself undergoes a bimonthly variation in its true size (1973, pp. 462–463).

Commentariolus does not remark that this is the "only" way. Nor does it indicate any awareness that a Muslim astronomer had previously assigned to the moon two epicycles and a concentric deferent in order to overcome this gross defect in the Ptolemaic lunar theory. *Commentariolus'* silence with respect to Ibn al-Shatir (c. 1305–c. 1375) contrasts sharply with its explicit divergence from its Pythagorean predecessors who also had espoused the motion of the earth. It was not Copernicus' habit to claim that he was the first to introduce any astronomical innovation. Nor did he systematically acknowledge what he owed to his predecessors. Nevertheless, *Commentariolus'* ratio of the larger epicycle's radius to the deferent's radius is $1\ ^{1}/_{18}:10$, and to the smaller epicycle's radius is $4\ ^{3}/_{4}:1$, whereas Shatir's corresponding ratios are about $1\ ^{1}/_{10}:10$ and $4\ ^{5}/_{8}:1$ (DSB, XII, 359–360). Moreover, Shatir's earth was stationary, and his sun was a moving planet equipped with an epicycle. By contrast, *Commentariolus'* sun was stationary without any epicycle, and its earth was a moving planet. No evidence has been adduced that Copernicus ever heard of Shatir or his lunar theory. *Commentariolus'* knowledge of Islamic astronomers was derived from *Epitome*, which never mentions Shatir.

[141] S: *diminutionem magis sui corporis et temerarie asserit*; A: *diminutionem corporis sui magis temerarie quispiam asserit*. Forgetting that it attributed the eccentric to "Those who believe" (*Qui ... arbitrantur*) in the plural, *Commentariolus* inconsistently shifts here to the singular *asserit*. A emended by interpolating *quispiam* ("somebody claims"). Sw's emendation (1973, p. 461) is *asseritur*, a passive form in the indicative mood, whose subject must be in the nominative case. But *diminutionem* is in the accusative case. Despite his incompetence in Latin grammar, Sw undertakes to emend a Latin text. A deleted *et*, because it becomes awkward once *quispiam* is interpolated. On the other hand, if *asserit* is replaced by *asserunt*, *et* shows that *Commentariolus* is still thinking about the advocates of the eccentric. MCV, IV, 6/35, changed *magis* to *magnitudinis*, observing that *magis* lacked only a supralinear stroke to become *magnitudinis*.

[142] S, A: *longitudinum*, the genitive plural for the moon's three motions in longitude (deferent, larger epicycle, smaller epicycle), matching the genitive singular *latitudinis* for the moon's one motion in latitude. S' *longitudinum* was pardonably misreported by Lindhagen (p. 10/11) as *longitudinem*. In reprinting Lindhagen's reading, MCV, IV, 7/41, felt it necessary to interpolate [*in*] before *longitudinem*, and PII, 194/7, did the same. When correctly read, however, the passage is perfectly sound and requires no interpolation. Nevertheless, Sw suspects "textual corruption, possibly the loss of a full line" (1973, p. 464).

[143] S: *puncta*; A: "*punctus*," obviously a slip of the pen, perhaps influenced by *motus*, two words later.

[144] Here V's folio 61ʳ begins, closing the gap due to the loss of V's folio 60 (see n. 114, above).

[145] V: *ac si*; S, A: *axi*. V's *ac si* is the kind of error that is made by a scribe who is listening to the reading aloud of a prototype rather than looking at it with his own eyes.

[146] V, S, A: *nullam*. V's *nullam* was misreported by MCV, I, 11/1, as *nullum*. V did not always close the top of the letter *a*.

[147] V: *superatenduntur*; S: *superadtenduntur*; A: *semper attenduntur*, an infelicitous emendation of an awkward reading in his prototype. At first MCV incorrectly emended to *superadduntur* (I, 11/6), but later shifted to *supratenduntur* (IV, 7/47).

[148] Where *Commentariolus* repeated the traditional value of 5° for the moon's latitude, A wrote in the margin: "Tycho makes it 5° 20' at the present time."

[149] V, S, A: *circumferuntur*. V's *circumferuntur* was misreported as *circumferantur* by MCV, IV, 7.

[150] V, S, A: *in aequidistantia axis*. For *axis*, the unanimous and sound reading of all three manuscripts, Sw substitutes *axi* (1973, p. 464).

[151] In "The Apparent Motions of the Sun," the third motion.

[152] V, S, A: *Sed* (But). See the following note.

[153] V, S, A: *et* (and). *Sed* (n. 152) and *et*, being interchanged in both V and S, were interchanged in Brahe's MS too. This interchange was pointed out by Fritz Rossmann, *Nikolaus Kopernikus Erster Entwurf seines Weltsystems* (Darmstadt, 1966; reprint of 1948 ed.), p. 45. Rossmann's obviously correct and indispensable emen-

dation is spurned by Sw, who condemns Rossmann's "accompanying notes" (like those of all Sw's predecessors) as "in all cases superficial, non-technical, and frequently erroneous" (1973, p. 432). There were people like Sw in the time of Gottfried Wilhelm Leibniz (1646–1716), who once said about them: *qui spernunt, non intelligunt* (those who despise, fail to understand; *Die philosophischen Schriften*, VII [Hildesheim, 1961; reprint of 1890 ed.], 481).

[154] The axis of the moon's deferent, being inclined to the axis of the ecliptic, "makes the moon move out of the plane of the ecliptic." The points of intersection (the lunar nodes) revolve as the poles of the moon's deferent revolve. The direction of this revolution is the "reverse order of the signs" or westward. This is likewise the direction of the earth's third motion. But, having dismissed Rossmann's emendation (see the preceding note), Sw remarks that Copernicus

> contrasts the regression of the lunar nodes with the motion of the inclination in such a way that it seems as though he did not realize that the motion of the inclination must be opposite to the annual motion (1973, p. 449).

Incidentally, Sw insists on the label "motion of the inclination" for what *Commentariolus* called *motus declinationis*. Sw complains that *motus declinationis* "is not very clear." It was so clear to Copernicus that he retained it in *Revolutions*, I, 11 (NCOO, II, 22/27, 24/15). With regard to *Commentariolus*' account of the motion of the lunar deferent's axis, Sw remarks:

> The odd thing about Copernicus' description is his statement that, in contrast to the motion of the inclination, [the deferent's axis] moves opposite to the order of the signs. ... The specification of direction implies ... that Copernicus thinks the motion of the inclination takes place in the order of the signs, which would be incorrect (1973, p. 465).

> ... in the *Commentariolus* ... the earth's axis is carried about in the wrong direction (1975, p. 52).

If *Commentariolus* incorrectly directed the "motion of the inclination" in the order of the signs, then the first and third motions it attributed to the earth would both be eastward. In that case, these two motions in combination would not keep the poles of the earth's daily rotation at (nearly) the same points of the heavens. But *Commentariolus* explicitly says:

> The poles of the daily rotation would always be fixed ... at the same points of the heavens by the motion in declination combined with the [Grand] Orb's motion ("The Apparent Motions of the Sun," last paragraph; see n. 87, above).

The combination of the Grand Orb's motion (earth's first motion) with the motion in declination (earth's third motion) could keep the poles of the daily rotation (earth's second motion) fixed only if the first and third motions took place in opposite directions. Since the first motion is described by *Commentariolus* as occurring "in the order of the signs," that is, eastward, the third motion must be westward. Yet Sw concludes:

> when Copernicus wrote the *Commentariolus* he really did not know that the motion of the inclination must be opposite to the order of the signs. He did not understand his own precession model (1973, p. 465).

Commentariolus explained the precession of the equinoxes by introducing, for the first time in the history o the human race, a motion of the earth's axis. The direction of that motion could readily be inferred by any normal reader. *Revolutions* stated the direction explicitly:

> The third motion ... occurs in the reverse order of the signs, that is, in the direction opposite to that of the motion of the center. These two motions are opposite in direction (NCCW, II, 23/17–20).

What is called the "motion of the center" here is termed the [Grand] "Orb's motion" in *Commentariolus*. Both there and in *Revolutions* the motion in declination is opposite in direction to the Grand Orb's motion or motion of the center. Had Rossmann's emendation been understood by Sw, he could have printed it as though it were his own, just as he misappropriates the emendations mentioned in nn. 67, 119, 123, 124, 125, 134, 166, 221, 252, 266. Sw's own attempts at emendation are discussed in nn. 96, 101, 118, 120, 127, 128, 141, 142, 150, 155, 177, 194, 219, 224, 231, 257, 258, 282, 297, 305.

Commentariolus asserts that the poles of the lunar deferent revolve "much more slowly" (*longe tardiore motu*) than the earth's motion in declination. For, the latter completes its revolutions in a little less than a year (*annuas fere complet revolutiones*), while the former waits until the nineteenth year for one revolution (*ad unam revolutionem decimum nonum annum expectet*). *Commentariolus*' assertion shows, according to Sw, "that at this point Copernicus has forgotten that the earth is also revolving around the sun" (1973, p. 465). How does the assertion that once in nearly nineteen years is much slower than once in nearly one year show that here Copernicus forgot his greatest contribution to astronomy — the earth's annual revolution around the sun? How did Sw arrive at his weird conclusion?

Side by side Sw draws two diagrams of his own design depicting the regression of the lunar nodes. The first diagram illustrates what Sw calls *Commentariolus*' "simple model"; the second diagram presents what Sw labels the "correct model." From the first diagram Sw omits the earth's orbital revolution around the sun, an omission which Sw attributes to *Commentariolus*. Forgetting to include the earth's orbital motion, Sw blames Copernicus for his own forgetfulness, and then charges that the great astronomer "was confused about the model for the regression of the lunar nodes" (1973, p. 465).

The lunar nodes complete a regression "in the nineteenth year," or about 19° a year. Therefore they move "much more slowly" than the motion in declination, which "completes its revolutions [of 360°] in a little less than a year." Sw, however, insists that "the motion of the axis of the lunar orbit is much *faster* [Sw's italics], not slower, than the motion of the inclination." "The motion of the inclination" (heretofore Sw's term for *Commentariolus*' motion in declination) has without any warning suddenly been identified with "the period of the precession of the equinoxes," which "takes thousands of years" because it averages about 50" a year. This is the amount contributed annually by *Commentariolus*' motion in declination to the long-term period of the precession of the equinoxes. The regression of the lunar nodes (about 19° a year) is faster than the rate of precession (about 50" a year) but much slower than the motion in declination (about 360° a year). Sw imagines that he has found an error in *Commentariolus*. The only error here is Sw's use of the same term for two such disparate quantities as 360° a year and 50" a year.

[155] V, S, A: *modum*. This perfectly sound reading, found in all three manuscripts, is emended to *motum* by Sw (1973, p. 464). Why?

[156] V: *Tandem*; S, A: *Talem*. V's incorrect reading was due to a slight slip in oral communication.

[157] V: *fabrica*; S, A: *fabricam*. V omitted the required stroke over the final -a.

[158] V: *motum*; S, A: *motuum*. V forgot to repeat the letter -u.

[159] V, A: *luna*; S: *lunam*. For *lunam* in his prototype A substituted *luna*, although *lunam* is an acceptable reading. A's agreement with V is fortuitous.

[160] V, S, A: *habere*. V's *habere* was misreported as *haberi* by MCV, IV, 7, thereby making V's performance in this last sentence of the section on "The Moon" look worse than it actually was. The sentence contains only seven words, and V wrote three of them incorrectly. Evidently, V or his reader was in a great hurry to finish this section.

[161] V, A: *signorum*, omitted by S. A supplied this indispensable word, which was omitted in his prototype. A's agreement with V is fortuitous.

[162] V, S, A: 29. *mense*; see n. 48, above.

[163] V: *remoretur*; S: *remoratur*; A: *remoraretur*. V's reading is correct. A's prototype had the wrong mood in *remoratur*. By emending it to *remoraretur*, A made *Commentariolus*' conditional statement contrary to fact (just as if these revolutions were delayed by the spheres' size, but the revolutions are not so delayed). *Commentariolus*' conditional statement is not contrary to fact, while A's emendation is contrary to *Commentariolus*.

[164] Sw (1973, p. 466), misinterpreting *magnitudo* (size) as "massiveness," contends that *Commentariolus*

> fell back on the possibility that the sheer massiveness of the spheres retarded the planetary motions. There is no mention of this theory in *Revolutions*, and Copernicus obviously does not take it very seriously here. Nevertheless, the very mention of the possibility that the spheres have some kind of mass is further evidence that Copernicus had no doubt that the motions of the planets were controlled by solid spheres.

But *Commentariolus* is talking about actualities, not possibilities. The periods of the three outer planets are actually related to the size of their orbits: the larger the orbit, the longer the period. No mass or massiveness is attributed to the planetary spheres by *Commentariolus* here or anywhere else (see n. 184, below).

[165] V: *semidiametro magni orbis ex .25. partibus constituto semidiametro orbis Martii 30 partes obtinebit*; S: *semidiametro magni orbis Martii, 30 partes obtinebit*; A: *semidiameter magni orbis Martii 30 partes obtinet*. S jumped from *orbis* following *magni* to *Martii*. Such a leap from a word (*orbis*) to the next appearance of the same word sometimes occurs also in modern printing. Like S, A too omitted *ex .25. partibus constituto semidiametro orbis*. This omission shows that A's prototype resembled S in this respect. However, A could not repeat his prototype's *semidiametro*, which is made ungrammatical by the omission of *ex ... orbis*. Hence, A changed *semidiametro* to *semidiameter*. In like manner A substituted *obtinet* for his prototype's *obtinebit*. For as a future tense, *obtinebit* is suitable in conjunction with *semidiametro ... constituto*. The omission of this ablative absolute, however, makes the future tense *obtinebit* less acceptable than the present tense *obtinet*. V's second *semidiametro* is an erroneous

duplication. At its first occurrence *semidiametro* is divided at the end of a line, with *-metro* appearing at the beginning of the following line. At its second occurrence the word is again divided in exactly the same way, and V repeated *-metro* from the line above, instead of writing *-meter* as required by the context. MCV incorrectly emended the second *semidiametro* to *semidiametros* (I, 11/20; IV, 7).

[166] V, S, A: *30*. TCT (1st ed., 1939, p. 74) changed *30* to *38*, an emendation which is printed by Sw (1973, p. 466) as though it were his own. The ratio of the radius of Mars' sphere to the Grand Orb's radius is given in *Revolutions*, V, 19, as $10,000 : \sim 6580 \simeq 38 : 25$ (NCCW, II, 269/21–22), in agreement with *Commentariolus*. U (lines 1, 10) gives 6583 and "about 38" (NCCW, IV, Pl. XXXII). Since V, S, A all show the scribe's substitution of 0 for 8, the erroneous value *30* was already present in Brahe's MS.

[167] V, S, A: *130*. S' value was misreported as *230* by MCV, IV, 7, confusing Jupiter's *130* with Saturn's *230*. U's value for Jupiter (NCCW, IV, Pl. XXXII/12) was misreported as *30*, instead of *130*, by PII (195/7 up, 211/17).

[168] V, S, A: *230*. V's value was reported by MCV, I, 11/21, as *236*. Originally the scribe did write the third digit as 6 (perhaps under the influence of the approaching *sextantem*). But later the upper portion of the curved stroke of 6 was almost completely erased, with the lower loop remaining intact as 0. MCV, IV, 7, mistakenly assigned to Jupiter S' Saturn value, which is *230*, not *236* (as misreported by Sw, 1973, p. 466).

[169] V, S, A: *sextantem*. The occurrence of *sextantem* ($= 1/6$) in all three manuscripts shows that this erroneous substitution for *dextantem* ($= 5/6$) was already present in Brahe's MS. The fraction was expressed, not verbally (as in the manuscripts), but numerically as $5/6$ in U (NCCW, IV, Pl. XXXII/13). U's $5/6$ was misreported as $5/8$ by PII, 195/6 up. The ratio $230\ 5/6 : 25$ is confirmed by U's line 5, where 1083 is given for Saturn:

$$10,000 : 1083 = 230.84 : 25 \simeq 230\ 5/6 : 25.$$

U (line 13) places "Saturn's radius $230\ 5/6$" below the heading in lines 8–9: "Ratio of the celestial spheres to an eccentricity of 25 units." This 25-unit eccentricity, in the lower half of U, denotes the Grand Orb's radius, that is, the distance of the orbiting earth from its orbit's center. Because this center is itself at a distance from the sun, the center of the universe, the Grand Orb is an eccentric. This eccentric's radius, divided into 25 units, is used throughout *Commentariolus* as the fundamental constant for measuring the sizes of the other planets' spheres.

On the other hand, the upper half of U records a different scale that was not adopted in *Commentariolus*. This scale is not preceded by any explanatory heading. Its first word ("eccentricity") denotes the Grand Orb's radius as a variable part of 10,000, arbitrarily adopted as the length of the radius of an outer planet's sphere. Thus, in its upper half, U (line 5) gives 1083 as Saturn's eccentricity, while line 13 (in U's lower half) gives Saturn's radius as $230\ 5/6$. These are simply two different ways of expressing the same proportion since, as we just saw,

$$10,000 : 1083 \simeq 230\ 5/6 : 25.$$

But the term "eccentricity" in the first ratio is a variable, relating the Grand Orb's radius to 10,000 as the constant length of the radius of a planetary sphere. On the other hand, in the second ratio, the term "eccentricity" is a constant, 25, denoting the length of the eccentric Grand Orb's radius as a yardstick for measuring the length of the radii of the other planets' spheres.

Commentariolus adopted the second ratio because Copernicus then believed the distance between the Grand Orb's center and the center of the universe to be constant. He later learned that this distance is variable rather than constant. This realization may have induced him to shift to the first ratio in *Revolutions*, and to avoid mentioning *Commentariolus* (see n. 110, above).

The meaning of the term "eccentricity" in U is misunderstood by Sw, who claims that the numbers in the upper half of U

> directly give the proportion of the radius of the epicycle to the radius of the eccentric where the radius of the eccentric is 10,000 (1973, p. 471).

But these numbers in the upper half of U have nothing whatever to do with any epicycle. They relate the length of the eccentric Grand Orb's radius as a variable to the constant length (10,000) of the radius of a planet's eccentric sphere. Sw acknowledges: "My entire analysis hangs on this one word" (*eccentricitas*, 1973, p. 478). Since Sw misunderstands the meaning of this one word "eccentricity," Sw's entire analysis is self-condemned.

[170] V: *orbis*, omitted by S, A. The presence of this indispensable word in V and its omission in S, A is another indication that A's prototype resembled S.

[171] V: '/.; S: '\.; A: /—/. S' symbol for "one" was misinterpreted as the abbreviation for *id est* by Lindhagen (p. 10/34), who was corrected by MCV, IV, 8.

[172] This definition explains that the radius of an outer planet's sphere extends from the sphere's center to the first epicycle's center, and no farther. *Commentariolus*' definition is misconstrued by Sw as a "distinction."

Without specifying what two entities would be distinguished from each other in this supposed distinction, Sw proceeds:

> If he [Copernicus] were talking only about circles and epicycles, this distinction [what distinction?] would be unnecessary since the center of the epicycle is obviously located at the circumference of the circle on which it moves. With solid spheres, however, the radius of the solid sphere itself must extend to the outer edge of the epicycle, or better, to the outer surface of the epicyclic sphere. The arrangement is shown in Figure 12. The radius of the solid sphere is $\overline{ST}$, and the epicyclic sphere with center F is tangent to the larger sphere at T and presumably rotates on an axis attached to the solid sphere at M and N. Hence Copernicus must specify that he is calling $\overline{SF}$, not $\overline{ST}$, the semidiameter of the sphere (1973, pp. 466–467).

Commentariolus itself supplied no diagrams. Sw's Figure 12 is our Figure 7. Although Sw himself has just said that "the radius of the solid sphere is $\overline{ST}$," he concludes that "Copernicus must specify that he is calling $\overline{SF}$, not $\overline{ST}$, the semidiameter of the sphere." A sphere's semidiameter or radius, however, must extend all the way from the sphere's center to its surface. Any line, such as $\overline{SF}$, which falls short of the full distance from a sphere's center to its surface cannot be called a semidiameter or radius by Copernicus or by anybody else. Any discussion leading to the conclusion that a non-radius must be called a radius is self-condemned. This verdict applies with just as much force to a two-dimensional circle as to a three-dimensional sphere.

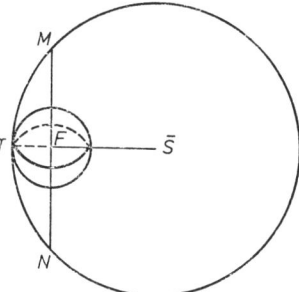

Fig. 7 (Sw's Fig. 12). Swerdlow's misplacement of an outer planet's first epicycle

[173] For the three outer planets, *Commentariolus*' two epicycles (on a concentric deferent) were superseded by a single epicycle (on an eccentric deferent) in *Revolutions*. This change may have been partly responsible for Copernicus' later silence about *Commentariolus* (see nn. 110, 169, above).

[174] *Commentariolus*' moon clings to the surface or circumference of its second epicycle. The center of this smaller epicycle is carried around on the surface or circumference of the moon's first or larger epicycle, as is shown in Figure 3, above. The relation of the moon and of both its epicycles to its deferent sphere is indicated in Figure 8, which reproduces the essential features of Sw's Figure 5 (1973, p. 456). For purposes of compari-

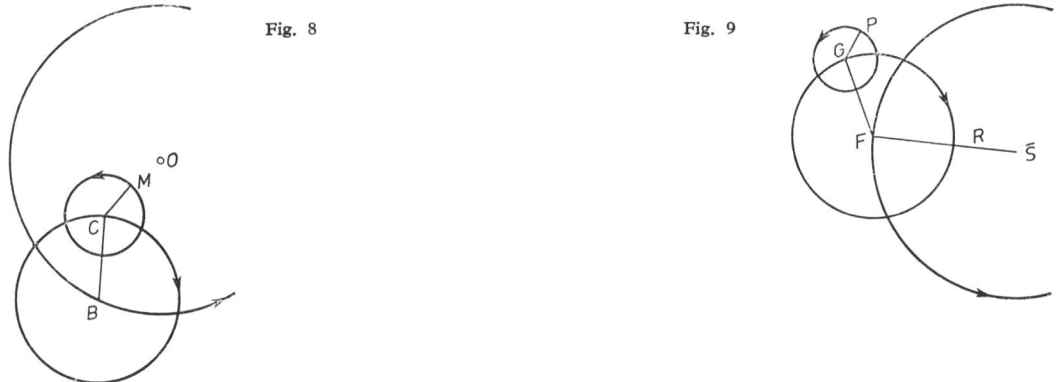

Fig. 8. (Sw's Fig. 5, simplified). Moon's deferent sphere and two epicycles
M — moon, O — earth, B — center of first epicycle, C — center of second epicycle
Fig. 9 (Sw's Fig. 14, simplified). Outer planet's deferent sphere and two epicycles
I — planet, $\overline{S}$ — center of Grand Orb, F — center of the first epicycle, R — radius of the deferent sphere, G — center of the second epicycle

son, Figure 9 reproduces the essential features of Sw's Figure 14 (1973, p. 468). Here the radius R of the planet's deferent sphere extends all the way from $\overline{S}$, the center of the Grand Orb, to F, the center of the planet's first epicycle. Here the radius R does not stop short of the deferent's surface, as $\overline{SF}$ does in Figure 7 (Sw's Figure 12). There the first epicycle is wholly inside the deferent sphere and tangent to it. Here, by contrast, the first epicycle's center F is located on the deferent's surface or circumference, not somewhere inside the deferent. Therefore, the first epicycle is not entirely contained within the deferent sphere. Nor is it true that "the second epicyclic sphere is entirely contained within the first in the same way" (Sw 1973, p. 467). Sw proclaims that "the intersection of spheres is not permitted." This is Sw's own proclamation. Nothing remotely resembling it is found in *Commentariolus* or in *Revolutions*. Sw's proclamation is refuted by his own Figures 5 and 14 (our Figures 8 and 9), where the second epicyclic sphere constantly passes freely in and out of the first epicyclic sphere and the deferent sphere. By the same token, the first epicyclic sphere constantly passes freely in and out of the deferent sphere.

Sw's proclamation that "the intersection of spheres is not permitted" is connected with Sw's misconception that *Commentariolus*' spheres possess "sheer massiveness" (see n. 164, above). On the contrary, they penetrate one another with the greatest of ease, and are penetrable, in this respect resembling light.

[175] V: *alter*; S, A: *altero*, presumably under the influence of the following word *vero*. Once more, A's erroneous reading reveals the affiliation of A's prototype with S.

[176] V: *in quadrantis* (should be *quadrantibus*) *autem mediantibus remotissimum*, omitted in S, A. A's omission of these five words again demonstrates the affinity of A's prototype with S. A's shift to *mediante* (see n. 60, above) is justified by *mediantibus* here.

[177] V, A: *maxime*; S: *maximo*. A corrected the erroneous reading *maximo* in his prototype, and thereby came into fortuitous agreement with V. Emending to *maximae*, Sw mistranslates:

> The farthest distances and closest approaches of this kind retain positions fixed with respect to the sphere of the fixed stars (1973, p. 466).

But it is not only "the farthest distances and closest approaches" that are fixed, since the planets

> continuously observe relations in [these] motions that are everywhere fixed [with respect to the sphere of the fixed stars]

as Sw's rendering continues. *Commentariolus* emphasizes that the withdrawals and approaches are absolutely (*maxime*) fixed everywhere (*ubique*).

[178] V: *sibi sub*; S, A: *si sub*. A's prototype agreed with S' incorrect reading, which may have lost the final syllable of *sibi* under the influence of *sub*, the following word.

[179] V, S, A: *observant*. V's *observant* was misreported as *observent* by MCV, I, 12/7.

[180] *Commentariolus* places Saturn's higher apse (or apogee) near this star in the Archer. Hence, the planet's apogee has almost the same celestial longitude as this star. It is described here as *stellam quae super cubitum esse dicitur Sagittatoris*, and as *quae est super cubitum dextrum ... Sagictarii* in Copernicus' edition of the *Alfonsine Tables*. There (sig. c 4ʳ/6) the star's celestial longitude is given as 4 signs, 41° 58′. Such a sign evidently differs from the ordinary sign, which has only 30°. In fact, two kinds of sign are distinguished in the first Canon or Proposition of this edition:

> A degree ... is the sixtieth part of a physical sign, six of which make a circle or revolution; or [a degree is] the thirtieth part of a common sign, twelve of which make a circle or revolution. ... Hence, when 60° have accumulated, they are replaced in these *Tables* more often by one physical sign, although in some of the tables included herein, if 30° have accumulated, they are replaced by one common sign (*signum commune*), as will be clear to whoever uses [these *Tables*, sig. A 5ʳ].

In the case of the Archer's elbow, the physical sign is intended rather than the common sign, so that our star's celestial longitude = $4 \times 60°$ (= 240°) + 41° 58′ = 281° 58′.

The modern designation of the Archer's elbow is "h² Sagittarii." Using the notation "λ_A Saturn" for the longitude of Saturn's apogee, Sw (1973, pp. 479–480) says:

> We now compare the tropical longitudes of the apogees from the *Alfonsine Tables* for the supposed epoch of the Alfonsine star catalog, Era Alfonso = June 0, 1252, with the longitudes of these stars in the catalog.
>
> h² Sagittarii — λ_A Saturn = 4,41; 58°–4,10 ;39° = 31 ;19° west.

Sw's "4,41 ;58°" = 4 signs, 41° 58' = 281° 58' = the celestial longitude of the Archer's elbow in the Alfonsine star catalog (1492 ed., sig. c 4ʳ/6). By the same token, Sw's "4,10; 39°" = 4 signs, 10° 39' = 250° 39' = the Alfonsine longitude of Saturn's apogee (according to Sw, who does not indicate his source for this numerical value).

In Copernicus' 1492 edition of the *Alfonsine Tables* (sig. b 1ᵛ), however, the table entitled "Radices of Saturn's apogee at the eras given here, without the motion of the eighth sphere," locates Saturn's apogee for the era of Alfonso at 4 2 35 20 41 0 = 4 signs, 2° 35' 20'' 41''' 0'''' = 242° 35' (without the negligible smaller fractions). Copernicus' edition of the *Alfonsine Tables* therefore placed Saturn's apogee at 242° 35', and Archer's elbow at 281° 58'; the planet's apogee was accordingly (281° 58' − 242° 35' =) 39° 23' west of the star. By contrast, *Commentariolus* located Saturn's apogee near (*circa*) the star, not 39° 23' away from it. Clearly, *Commentariolus* did not take both of these celestial positions from the *Alfonsine Tables*.

Sw, however, clinging to his unsupported belief in *Commentariolus*' dependence on the *Alfonsine Tables*, says (1973, p. 480) that

> since such a longitude for Saturn's apogee is unheard of, I would guess that Copernicus made an error of $1^s = 30°$ in noting its position.

By Sw's unexplained calculation, the difference between the two positions is 31° 19', so that an error of 30° would narrow the gap to 1° 19'. So small a separation would fit *Commentariolus*' description of Saturn's apogee as near the Archer's elbow. But Sw's "$1^s = 30°$" is wrong, since here $1^s = 60°$. Moreover, although Sw maintains that "such a longitude for Saturn's apogee is unheard of," Sw carefully avoids equating that "unheard of" longitude with a precise number of degrees, minutes, and seconds. When Sw "guess[es] that Copernicus made an error of $1^s = 30°$," it is Sw who made an error, or more than one.

Saturn's apogee had been put at 224° 10' in Ptolemy's *Syntaxis* (XI, 8); at 233° 46' in *Epitome*, XI, 14 (sig. n 1ʳ/5-7); at 240° 5' in the *Toledan Tables* (*Osiris*, 1968, *15* :45); at 242° 35' in the *Alfonsine Tables* (1492 edition); and at 252° 45' 25'' in a Cracow annotation of Campanus of Novara's *Theory of the Planets* (F. S. Benjamin and G. J. Toomer, *Campanus* [Madison, 1971], p. 301). That annotation is dated 1473, the year of Copernicus' birth. About four decades later, when *Commentariolus* was written, such steadily increasing values for the longitude of Saturn's apogee may have brought it somewhere near 264° 50', the celestial longitude of Archer's elbow in Valla's *Seek and Avoid* (sig. ee 1ʳ/25). It was not Copernicus' habit to specify his sources in any systematic way. But if Saturn's apogee was located near 264° 50' in some (still unidentified) source, it may have been the basis of *Commentariolus*' location of the planet's apogee near Archer's elbow.

The continued eastward displacements of Saturn's apogee over the centuries may have looked like improved determinations to Copernicus when he was writing *Commentariolus*. Later, however, after realizing that Ptolemy was wrong in regarding the planetary apsides as fixed in relation to the stars, Copernicus made his important discovery that the planetary apsides constantly undergo a slight shift among the stars. Saturn's apogee in *Commentariolus* may therefore have been, not the error imagined by Sw, but an essential step in Copernicus' progress toward his discovery of the secular movement of the planetary apsides. This advance over Ptolemy's doctrine of the fixity of the planetary apsides should be pondered by those who never tire of repeating the old falsehood that Copernicus did nothing but paraphrase Ptolemy.

[181] V: g̅r̅ (*gradibus*); S: *igitur*; A: i̅g̅r̅ (*igitur*). A's prototype agreed with S' mistaken reading.

[182] This star's modern name is β Leonis.

[183] This star, formerly called Basiliscus or Regulus, is now known as α Leonis.

[184] *Magnitudines* has exactly the same meaning here as in the passage discussed in n. 164, above. In both cases "size" is intended, without any reference whatsoever to "massiveness."

[185] In "The Three Outer Planets," 3rd sentence.

[186] V, S, A: *primum*, corrected to *primi* by MCV, I, 12/13. The presence of the erroneous reading *primum* in all three manuscripts indicates that this blemish marred Brahe's MS too.

[187] V, S: *continent*; A: *continerent*. A's shift to the subjunctive mood is inconsistent with the indicative mood in the corresponding verbs *sunt, constat, habet, est*.

[188] V, S, A: *secundus minut. 51*. The omission of *partis 1* in all three manuscripts indicates that this defect was a blemish in Brahe's MS too. In fact, this fault goes all the way back to Copernicus himself, since U (line 11) also omits the number for the whole unit and gives only 51^m for the radius of Mars' second epicycle (NCCW, IV, Pl. XXXII). MCV, I, 12/18, correctly inserted the missing *partis 1*, which is required because "the first [epicycle's] radius is throughout three times greater than the second [epicycle's radius]." For the radius of Mars' first epicycle, U (line 10) gives $5^p\ 34\ 1/2^m$ (not 34^m, as in Sw 1973, p. 429), and $5^p\ 34\ 1/2^m = 3 \times (1^p\ 51^m\ 30^s) \simeq 1^p\ 51^m$.

[189] V: *motum*; S, A: *motus*. V's incorrect reading may be due to the influence of *motum* three words later.

[190] V, S, A: *observant*, correctly emended to *observat* by the late Ryszard Gansiniec.

[191] Here *ubique* recalls *ubique* in the passage discussed in n. 177, above.

[192] This is the fourth passage in which *Commentariolus* keeps its promise to show that an apparent motion outside the earth (in this case the observed planetary loops) provides evidence for the earth's real motion (see n. 38, above).

[193] V, S: *variantis*; A's *variātib.[us]* is undoubtedly a slip of the pen.

[194] V, S, A: *radio visuali ad firmamenti a(d)spectum obviante*. Against the unanimous reading of all three manuscripts, Sw (1973, p. 480) emends *firmamenti* to *firmamentum*, and translates:

> the line of sight [passing through the planet] to the fixed stars is moving in the direction opposite to [the motion of] the position of the observer.

In "Equal Motion Should Be Measured Not by the Equinoxes," however, *Commentariolus* combines *obviare* with the dative case (not the accusative case) in *motibus inferioribus obviantes* (see n. 104, above). In "The Apparent Motions of the Sun", *positionem ad faciem firmamenti mutari* resembles *radio visuali ad firmamenti aspectum obviante* here.

[195] V, A: *motum*; S: *nutum*. By correcting this incorrect reading in his prototype, A came into fortuitous agreement with V.

[196] V: *ortu*; S, A: *ortus*. A's prototype agreed with S' incorrect reading *ortus*, which may have been affected by *sidus* four words earlier.

[197] V, A: *videntur*; S: *videtur*. If A's prototype agreed with S' correct reading *videtur*, A should not have shifted to the plural form *videntur*, since this verb's implied subject *sidus* occurs seven times in this paragraph in the singular number. V's mistaken *videntur* may have been affected by the plural forms in the next two words, *adversis motibus*.

[198] V, S: *invicem*, omitted by A, presumably because it was felt to be superfluous in the presence of the reflexive *se ... perimentibus*.

[199] V: *quod plerumque*, omitted by S, A. This omission is a further indication that A's prototype was akin to S.

[200] This imprecise statement was taken by *Commentariolus* from Pliny's *Natural History* (II, 59):

> In the trine aspect, that is, at 120° from the sun, the three outer planets have their morning stations, which are called the first stations ... and again at 120°, approaching from the other direction, they have their evening stations, which are called the second stations.

Far more advanced than Pliny's trine aspect for both stations of all the outer planets was a double theorem of Apollonius. That great Greek geometer had demonstrated an ingenious method of determining the stations, and the length of the retrograde arc between them, for each of the three outer planets moving either on an epicycle or on an eccentric. Apollonius' theorem was so highly admired by Ptolemy that he accorded it the rare distinction of being incorporated in his *Syntaxis* (XII, 1), whence it was transferred to *Epitome*, XII, 1–2. Other sections of *Epitome* (III, 2; V, 22) were adopted by *Commentariolus* (see nn. 92, 140, above). But such partial use of *Epitome* by no means indicates familiarity with all the rest of that lengthy treatise. In particular, had *Epitome*'s sophisticated treatment of the stationary points been known to Copernicus while he was writing *Commentariolus*, Pliny's simple trine aspect would surely have been discarded by him at that time as it was a quarter-century later in *Revolutions* (V, 35–36). *Commentariolus*' reliance on Pliny's dubious generalization proves that Copernicus was then unfamiliar with Apollonius' theorem as expounded in *Epitome*, XII.

According to Sw, however, "Copernicus' derivation of his planetary theory" in *Commentariolus* comes from *Epitome*, XII, 1–2:

> The eccentric model for the second anomaly [is] mentioned briefly by Ptolemy in *Syntaxis*, XII, 1 ...; it is this alternate model that leads directly to the heliocentric theory. ... Ptolemy had said that the eccentric representation of the second anomaly was usable only for the superior planets, but in *Epitome*, XII, 2, Regiomontanus describes an equivalent eccentric model for the inferior planets. ... I believe that Copernicus arrived at the heliocentric theory after a careful investigation of these two propositions in Book XII of the *Epitome* (1973, p. 471).

> ... Copernicus was investigating an alternative eccentric model of the second anomaly that was described in detail by Regiomontanus in Book XII of the *Epitome*. ... The model leads directly to the heliocentric theory, although its two forms for the superior and inferior planets lead respectively to the Tychonic and Copernican theories (1973, p. 425).

Copernicus' derivation of his theory rests upon the eccentric model of the second anomaly and therefore upon these two propositions in the *Epitome*. In this way Regiomontanus provided the foundation of Copernicus' great discovery. It is even possible that, had Regiomontanus not written his detailed description of the eccentric model, Copernicus would never have developed the heliocentric theory (1973, p. 472).

Regiomontanus ... was, through these two propositions, virtually handing it [the heliocentric theory] to any taker (1973, p. 476).

If "to any taker," why not to Regiomontanus himself? If "the eccentric model for the second anomaly mentioned briefly by Ptolemy ... leads directly to the heliocentric theory," why did it not lead Ptolemy directly to that theory? Why did it not lead Ptolemy's predecessor Apollonius to the heliocentric theory? Why did Tycho Brahe reject Copernicus' heliocentrism?

These four outstanding intellects, separated by immense stretches of time and space, all agreed in regarding the earth as not a heavenly body, as no more interchangeable with a heavenly body than a dead elephant is interchangeable with a firefly on the wing. But after those who would listen had learned from Copernicus that the earth is a heavenly body, a planet in motion, the category of primary planets was revised to include the earth and exclude the sun and moon. Rearrangement of the six planets in their new category became the order of the day. Kepler, a confirmed Copernican, produced a successful reform of the Copernican astronomy, by contrast with the failure of Tycho Brahe, who steadfastly rejected Copernicus' mobile earth. Nevertheless, when Brahe addressed the entire faculty and student body of the University of Copenhagen early in September 1574, he declared:

In our time Nicholas Copernicus may not undeservedly be called a second Ptolemy. Through observations made by himself he discovered certain gaps in Ptolemy, and he concluded that the hypotheses formulated by Ptolemy admit something unsuitable in violation of the axioms of mathematics. Moreover, he found the Alfonsine computations in disagreement with the motions of the heavens. Therefore, with wonderful intellectual acumen he formulated different hypotheses. He restored the science of the heavenly motions in such a way that nobody before him reasoned more accurately about the movements of the heavenly bodies (TB, I, 149/22–30).

This is what the greatest astronomer of the second half of the sixteenth century thought of the greatest astronomer of the first half of that century.

Brahe's high regard for Copernicus' astronomical abilities is not shared by Sw. Falsely claiming that *Commentariolus* gave a "false description of the apparent motion of Mercury," Sw argues that Copernicus

may well have copied the Mercury model from some other treatise without fully understanding its relation to Mercury's apparent motion (1973, p. 429).

When Copernicus wrote *Commentariolus* he really did not know that the motion of the inclination must be opposite to the order of the signs. He did not understand his own precession model ... and he was confused about the model for the regression of the lunar nodes (1973, p. 465).

One may seriously wonder whether he [Copernicus] understood the fundamental properties of his model for the first anomaly [of the planets; 1973, p. 469].

Copernicus had few observations and could use fewer. ... He was also impeded by his lack of technical proficiency. ... Even if Copernicus knew what nature showed, he could not have chosen [*sic*] to represent it for sheer lack of mathematical originality. It was fortunate that he had Ptolemy to represent, otherwise he would have had nothing (1973, p. 486).

But that is almost the only thing he can claim for the model [of Venus' motion in latitude] with any truth. In other respects it is a mess, and I am disappointed that he did not have the sense to get rid of it in *Revolutions* (1973, p. 499).

The foregoing assessments provided the basis for CSE's conclusion (p. 83) that Sw "holds a very low opinion of Copernicus' mental ability." CSE's conclusion was ironically characterized by Sw (1976, p. 108) as a "delicate remark." Neither delicate nor indelicate, CSE's remark was correct, and its correctness was not disputed by Sw 1976. Instead, Sw 1976 chose to refrain from referring to those Sw 1973 passages which fully justified CSE's conclusion and were not disavowed by Sw 1976.

If Sw 1973 was right in saying that Copernicus copied from others without full understanding; that he did not understand one of his own models, was confused about another, and did not understand the fundamental properties of a third; that he lacked technical proficiency and mathematical originality; and that he made a mess of Venus' latitude theory and did not have the sense to get rid of it, how could such a puny intelligence derive the heliocentric astronomy from Apollonius'-Ptolemy's-Regiomontanus' theorem, from which three such giants took only a method for ascertaining the planetary stations?

Uniform circular motion (as defined in n. 2, above) prevails throughout Apollonius'-Ptolemy's-Regiomontanus' theorem, which admits no equant. But the "planetary theories advanced by Ptolemy and most other astronomers" were rejected by *Commentariolus* because they

> conceived certain equalizing (*equantes*) circles. ... This sort of notion seemed neither sufficiently absolute nor sufficiently pleasing to the mind. Therefore, having become aware of these defects, I often considered whether there could perhaps be found a more reasonable arrangement of circles, from which every apparent irregularity would be derived while everything in itself would move uniformly, as is required by the rule of perfect motion. After I had attacked this very difficult and almost insoluble problem, the suggestion at length came to me how it could be solved with fewer and much more suitable constructions than were formerly put forward (paragraphs 3–4).

This candid explanation that Copernicus' geokinetic heliocentrism was motivated by his rejection of the equant was repeated in *Revolutions*, V, 2, which condemned the ancient astronomers for admitting that "a circular motion can be uniform with respect to an extraneous center not its own" in the case of the planets and the moon. Copernicus then explicitly declared:

> These and similar situations gave me the occasion to consider the motion of the earth and other ways of preserving uniform motion and the principles of the science (NCCW, II, 240/26–27, 30–32).

By Copernicus' own account, therefore, his geokinetic planetary theory was derived from his rejection of the equant, which plays no part in Apollonius'-Ptolemy's-Regiomontanus' theorem. Hence, *Commentariolus*' planetary theory was not derived from Apollonius'-Ptolemy's theorem as presented in Regiomontanus' *Epitome*.

Commentariolus' planetary theory mounted two epicycles on a deferent whose center was also the center of the universe. Apollonius'-Ptolemy's-Regiomontanus' theorem, on the other hand, placed only one epicycle on its concentric deferent, and no epicycle on its eccentric deferent. Therefore, *Commentariolus*' planetary theory was not derived from Apollonius'-Ptolemy's theorem as presented in Regiomontanus' *Epitome*. Ibn al-Shatir had used a biepicyclic concentric. But he is not mentioned in *Epitome*, nor is there any evidence that Copernicus ever heard of him and his biepicyclic concentric (see n. 140, above).

Whereas *Commentariolus* was written without any awareness of Apollonius' theorem, this was later used in *Revolutions* to locate the planetary stations and to determine the length of the retrograde arc between them (V, 35–36). Apollonius is mentioned there by name as having enunciated this theorem regarding the stations (NCCW, II, 302/32). But no acknowledgment that Apollonius was the source of the geokinetic astronomy was ever made by Copernicus there or anywhere else. On the contrary, he explicitly says with regard to Apollonius' theorem:

> Although it conforms to the hypothesis of a stationary earth, nevertheless it is compatible also with my principles based on the mobility of the earth, and therefore I too shall use it (NCCW, II, 302/33–35).

Sw's attempted derivation of *Commentariolus*' planetary system from Apollonius' theorem is a pretentious castle of sand that is washed away by the high tides of Copernicus' own emphasis on the equant's impropriety, *Commentariolus*' acceptance of Pliny's trine aspect in ignorance of Apollonius, and *Commentariolus*' use of a biepicyclic concentric rather than a monoepicyclic concentric or an eccentric-without-epicycle.

Sw (1973, p. 468) says: "It was the bisected eccentricity that ... Copernicus found objectionable." Copernicus did not object to bisection. Quadrisection (the bisection of a bisection) was *Commentariolus*' prescription for the outer planets: "The first [epicycle's] radius is throughout three times greater than the second [epicycle's] radius]." Here the sum of the epicyclic radii is quadrisected in order to obtain the length of the second epicycle's radius; and since the deferent spheres are concentric, no eccentricity is involved. What Copernicus found objectionable was, not any bisected eccentricity, but the equant. In Figure 10, the equant Eq is just as far from the eccentric's center CE in one direction as the center of the universe CU is in the opposite direction:

$$Eq\text{--}CE = CE\text{--}CU$$

Application of the term "bisected eccentricity" to the case of the equant requires defining "eccentricity" as the distance from the equant to the universe's center, a distance bisected by the eccentric's center. Since Copernicus discarded the equant, its distance from any other point was not what he meant by "eccentricity."

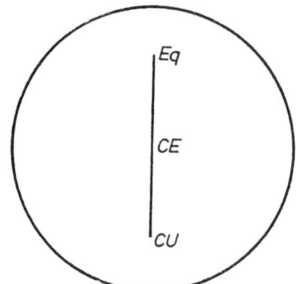

Fig. 10. Equant
Eq — equant, *CE* — center of eccentric, *CU* — center of the universe

[201] V: *versus*; S, A: *rursus*. The mistaken reading *versus* resulted from a slip in oral communication between V and his reader.

[202] As *Commentariolus'* reference to the Grand Orb's circumference (*circumferentiam*) shows, here the Grand Orb is regarded as a two-dimensional circle rather than a three-dimensional sphere; see n. 326, below.

[203] V, A: *pererrant*; S: *percurrunt*. Did A's prototype diverge from S here, or did A try to improve on his prototype here, as he often did elsewhere? If the latter, then A's agreement with V is fortuitous.

[204] V: *declinant*; S: *declinent*. Originally A wrote *declinent*. Then, recognizing the error in this reading, A turned the second *e* into an *a*. Hence A's prototype had the same faulty reading *declinent* as S.

[205] Sw (1973, p. 482) translates: "Since the circumferences of the epicycles remain in the same plane, they incline from [the plane of] the ecliptic...." *Commentariolus*, however, does not say that the inclination to the ecliptic is *caused* by the fact that both epicycles remain in the same plane.

[206] V, S, A: *circumducibiles*. V's *circumducibiles* was misreported as *circumducibile* and incorrectly emended to *circumducibilem* by MCV, I, 13/18.

[207] See the last paragraph of the section on "The Moon."

[208] V, S, A: *vergentes*. V's *vergentes* was misreported by MCV, I, 13/19, as *vergentem*. This misreading was linked with the erroneous emendation *circumducibilem* mentioned in n. 206, above.

[209] V, S, A: *quas*. V's *quas* was misreported by MCV, I, 13/20, as *quos*. V did not always close the top of the letter *a*.

[210] Pollux, previously identified as "the more brilliant bright star in the head of the Twins" (see n. 58, above).

[211] V, S: *ante eam ipsam stellam*. A omitted *ipsam*, presumably because it was felt to be expendable in the drive for greater compactness.

[212] V, S: *autem*; A: *vero*, perhaps on account of the next word *Vergiliae*.

[213] V: *part. 6. et medio*; S: *part. 6 1/2*; A: 6 1/2°. V's *medio* should be *media*, in the feminine form, as in the location of Saturn's node, three lines above: *part. 8 et media*.

[214] Sw (1973, p. 488) comments:

> The *Vergiliae* are the Pleiades, and it is not certain which of the four stars in the group listed in
> the Alfonsine catalog Copernicus intends by this vague designation.

The designation *Vergiliae*, however, does not appear in the star catalog included in the edition of the *Alfonsine Tables* owned by Copernicus. But in his own star catalog Copernicus wrote *Vergiliae* in the right margin alongside Taurus 30 (NCCW, I, fol. 59ʳ). *Commentariolus* always identifies a celestial position by reference to an individual star, never to a group of stars or constellation.

[215] V, S: *positis*, omitted by A, who wrote *In his igitur ac e diametro* twice, each time omitting *positis*. When A reached *positis* in his prototype, he paused momentarily to consider the advisability of omitting it with a view to greater compactness. Then, having decided to omit it, A forgot that he had already written the six preceding words, and proceeded to write them again.

[216] V, S, A: *maximam*. V's *maximam* was misreported by MCV, I, 13, as *maximum* because V did not always close the top of his letter *a*.

[217] V: *quae in his in quadraturis contingit*; S: *quae in his in his in quadraturis contingit*; A: *his in quadraturis*. A's deletion of *quae in* and *contingit* as expendable may have been prompted by his prototype's repetition of *in his*, if it shared that defect with S.

NOTES

[218] V, S, A: *diversam*. V's *diversam* was misreported by MCV, I, 13, as *diversum* because V did not always close the top of his letter *a*.

[219] V, S, A: *inclinare*. Since this erroneous reading is found in all three manuscripts, it was already present in Brahe's MS. MCV, I, 13/27, correctly replaced *inclinare* by *inclinatio*, which is needed as the subject of *videtur* and *fit*. The intrusion of *inclinare* may have been due to anticipation of *instare*, five words later. Sw (1973, p. 483) emends *instare* to *instabilis*, which Sw translates as "unsteady." Yet eight lines below, in Sw's own translation (1973, p. 483), the inclination's irregularity "increases and decreases uniformly with the latitude," which is not unsteady.

[220] V: *circulorumque*; S, A: *circulorum quae*. A failed to correct *quae* in his prototype, which shared this incorrect reading with S.

[221] V, S, A: *dextante*. V's *dextante* was misreported by MCV, I, 13/31, as *sextante*. Mistakenly assigning *sextante* to both V and S, Sw (1973, p. 483) claims *dextante* as his own emendation, although it is present in all three manuscripts. PII, 197/21, printed *sextante* ($= 1/_6$), making Mars' maximum latitude $1\ 1/_6°$, which *Commentariolus*' next sentence says decreases by $1\ 2/_3°$.

[222] V: *E contrario autem*; S: *E contra vero*; A: *E contra autem*. A substituted *autem* for *vero* in his prototype, whereas previously A had substituted *vero* for *autem* (see n. 212, above).

[223] V, A: *tunc*; S: *habent*. Was S' strange reading somehow affected by the following word *absistente*? This was changed to *absente* by A, immediately after he replaced *habent* by *tunc*. In this way A came into agreement with V's *tunc*, but not with V's *absistente*.

[224] V, S: *alicui*, which is misreported by Sw (1973, p. 483) as *aliqui*. For the sake of increased brevity A omitted *alicui*, which Sw unnecessarily emends as *alibi*.

[225] V, S: *sidus*, omitted by A with a view to greater compactness.

[226] V: *deficiens*; S, A: *decrescens*. At first S wrote *def-* before deleting these three letters and shifting to *decrescens*.

[227] V, S, A: *etiam*. V's *etiam* was misreported as *et* by MCV, I, 14/1, who overlooked the curved stroke over V's *et*.

[228] V: *visib. nostris*; S, A: *visibiles nobis*. Because V mistakenly wrote *nostris* instead of *nobis*, V's abbreviation *visib.* was expanded by MCV, I, 14/2, to *visibus*. This, however, should be *visibiles*, to go with the preceding word *latitudines*, matching *visibilis latitudinis* in the following clause. MCV's erroneous reading *visibus nostris* was followed by PII, 197/28.

[229] V: *vel*; S, A: *et*.

[230] V: *vel*; S, A: *et*. Although V's *vel* is consistent with the *vel* reported in the preceding note, the sequence *et ... et* in S, A is the usual expression in other Copernican passages of this nature.

[231] V, S, A all put a period after *exposcit*, and then start a new sentence. This punctuation is printed by Sw (1973, p. 483) as though it were an emendation. The comma after *exposcit* in MCV, I, 14/4, and PII, 197/30, was an unwarranted departure from the manuscripts.

[232] V: *liberationis*; S, A: *librationis*. A scribe listening to a prototype being read aloud might understandably think he was hearing *liberationis* (freedom) instead of the far less familiar technical term *librationis*.

[233] The dominant tradition in the theory of motion held that circular and rectilinear motion were different in kind. Circular motion, deemed appropriate to the celestial region extending from the moon outward to the stars, was produced by the rotation of a sphere. Rectilinear motion, on the other hand, was confined to the region below the moon, where it was produced by a push or a pull. The earliest demonstration that motion in a straight line could be produced by two interlocking spheres was provided by Nasir al Din (1201-1274), known as Tusi from his birthplace Tus in Persia (NCCW, II, 358, 385). His ingenious device, christened a "Tusi couple," is what *Commentariolus* now proceeds to describe. Tusi wrote partly in Persian, mostly in Arabic. Neither language was known to Copernicus. Exactly how he found out about Tusi's couple is still an unsolved problem. Whatever the line of transmission from Tusi to Copernicus may turn out to have been, *Commentariolus*' use of two intertwined spheres to produce a rectilinear motion connected with the latitudinal oscillations of the three outer planets was momentous. For it entailed the introduction of motion in a straight line into the heavens, from which rectilinear motion had been rigorously excluded by being confined to the infralunar region. *Commentariolus*' earth, however, having become a planet, was in the heavens. Hence rectilinear motion, a common enough occurrence on the earth, could become a heavenly phenomenon as well.

[234] Sw takes "concentric" to mean that these two librational spheres are both centered on the Grand Orb's center ("concentric" to $\bar{S}$," 1973, p. 489; see Figure 11). Sw explains:

> The higher [librational sphere] with pole F ... carries the pole G of the lower [librational sphere]. G in turn ... carries H, the pole of the planet's sphere. This will make H librate on either side of F (1973, p. 489).

Yet Sw's Figure 32 (simplified in our Figure 11) places H, the pole of the planet's deferent sphere, on the periphery of the *higher* librational sphere, whose pole is F. But *Commentariolus* says: "the *lower* sphere makes the poles of the deferent sphere ... revolve." Furthermore, "the higher [librational sphere] ... carries the pole G of the lower" (librational sphere; Sw 1973, p. 489). Yet Sw's Figure 32 does not place G on the periphery of the

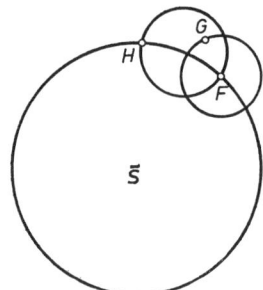

Fig. 11 (Sw's Fig. 32, simplified). Outer planet's latitudinal librations
$\bar{S}$ — mean sun = Grand Orb's center, F — pole of higher librational sphere, G — pole of lower librational sphere, H — pole of planet's deferent sphere

higher sphere. Sw also fails to explain how the two librational spheres can be "concentric to $\bar{S}$" while they "make H librate on either side of F." Sw translates *cum sint concentrici* by "Since these [librational spheres] are concentric, [the higher] one carries around the inclined poles of the other ..." (1973, p. 483). How can the polar interlocking of these spheres be *caused* by their concentricity? Concentric spheres must be interlocked at their common center, but they need not be interlocked at their poles.

Commentariolus' interlocking spheres produce a motion along a straight line. Such rectilinear motion results from a Tusi couple functioning in a plane. But if the couple consists of spheres, not circles, the resulting motion takes place along an arc, not a straight line. To denote the members of its Tusi couple, *Commentariolus* uses the term *orbis* (three times). Yet *Commentariolus* regards these *orbes* as functioning in a plane, in other words, as circles. According to Sw (1973, p. 432), "In the *Commentariolus orbis* always means sphere." But the *orbes* constituting *Commentariolus*' Tusi couple were treated by *Commentariolus* as circles, not spheres.

[235] V, S: *mediate*; A: *mediante*. V's *mediate*, misreported as *mediale*, was incorrectly emended to *immediate* by MCV, I, 14/9, followed by PII, 198/2. A was so eager to emend *mediate* to *mediante* that he wrote *orbis mediante* twice. The correct reading *mediate* shows that *Commentariolus* assumed the radii of the two librational spheres to be equal.

[236] V, A: *ac*, S: *in*. If A's prototype shared S' reading *in*, then A's emendation of his prototype brought A adventitiously into agreement with V.

[237] Venus has a system of circles (*circulorum compaginem*) closely resembling that of the outer planets (*persimilem ... qualem illi superiores*). To these four planets, *Commentariolus* assigns circles, not spheres; see n. 326, below.

[238] In "The Order of the Spheres."

[239] V, A: *orbis*; S: *orbes*. By emending the incorrect reading *orbes* in his prototype, A came into fortuitous agreement with V.

[240] In "The Apparent Motions of the Sun"; see n. 61, above.

[241] This incommensurability differentiates Venus' motions from those of an outer planet. For, the latter's smaller epicycle revolves in half the time required by its larger epicycle and deferent sphere, so that all three of these motions are mutually commensurable.

[242] V, S, A: *imparitatem*, correctly emended to *paritatem* by Sw (1973, p. 490). The presence of *imparitatem* in all three manuscripts indicates that Brahe's MS too had this defective reading, which may be an echo of *impares*, eight words earlier.

[243] V: *ad absidem diametro porrecta fuerit, sidus tunc centro*; S: *ad absidem diametro porrecta fuerit sidus tunc centrum*; A: *ad absidem diametro tunc centrum*. A's acceptance of the incorrect reading *centrum* induced him to omit *porrecta fuerit sidus*. The omission of *fuerit* left A without a verb for *tellus*. Moreover, the omission of *sidus* makes the center of A's larger epicycle perform the approach and withdrawal, whereas *Commentariolus*' planet (*sidus*) approaches the larger epicycle's center and withdraws from it.

[244] The moon's smaller or second epicycle carries the moon around through its successive phases, which are determined by its aspects or spatial relations to the sun.

[245] V, S, A: *10*, disagreeing with *18* in U's line 14 (U's upper half omits Venus; NCCW, IV, Pl. XXXII).

The presence of the incorrect reading *10* in all three manuscripts indicates that this defect was already present in Brahe's MS. In V a later hand added an upper loop to the zero, thereby changing the incorrect reading *10* to the correct reading *18*, just as the incorrect reading *236* was changed to the correct reading *230* for the length of the radius of Saturn's sphere (see n. 168, above).

[246] Alongside the *10* altered to *18* in V, the later hand mentioned in the preceding note wrote in the left margin: "larger epicycle 1.48; smaller, 0.36." These values are the sexagesimal equivalents of *Commentariolus*' ratio $^3/_4 : ^1/_4$, as is indicated by the rest of the marginal note: in units "of which the radius is 60." This "60" was misreported as "1.60" by MCV, I, 15, through a confusion of the marginal note's abbreviation (ē) for *est* with "1." MCV then changed its own misreading "1.60" to "60," which it applied to the Grand Orb's radius, giving the proportions

$$60:25 = 1\ ^4/_5 : ^3/_4 = ^3/_5 : ^1/_4$$

MCV's string, however, omits the radius of Venus' deferent. When the deferent is added to the marginal note's sexagesimal transformation, the proportions of the radii of the Grand Orb and Venus' three circles become

$$60^p : 43^p\ 12' : 1^p\ 48' : 0^p\ 36'$$

MCV's misreading (1.60 for 60) was followed by PII, which made matters worse by assigning $^2/_3$ (instead of $^3/_4$) to the larger epicycle's radius (p. 198/7 up).

[247] V, S: *hic*; A: *his*. A wished to improve *Commentariolus*' Latin style by matching *his* here with *illis* six words earlier, while keeping the pair *hic ... illic* four words and eight words later.

[248] V, S: *Quapropter nec umquam*; A: *Quare nec utrique*. A changed *Quapropter* to *Quare* with a view to brevity. For *unquam* (*unq̃*) in his prototype, A mistakenly substituted *utrique* (*utriq̃*).

[249] V, S, A: *quo*. V's *quo* was misreported as *qua* by MCV, I, 15/9, followed by PII, 199/9. Sw (1973, p. 493) mistakenly reports *quo* as found only in S.

[250] Sw (1973, p. 494) mistranslates *habet* as "keeps": Venus "keeps the axis of its sphere inclined at an angle of 2 $^1/_2$°." Yet, a little later on, Sw translates: "The inclination of the axis is as follows. It has a movable libration" (1973, p. 497). Since the axis "has a movable libration," it is not true that Venus "keeps the axis of its sphere inclined at an angle of 2 $^1/_2$°."

[251] V: *ij :s*; S, A: *5*. S and A's prototype omitted the whole number 2, leaving the abbreviation *s* for *semissis* (= $^1/_2$), which could be misinterpreted as the number 5. This reading made Lindhagen so uneasy, however, that he appended to his article a facsimile of the manuscript sheet in question in order to permit the reader to decide the matter for himself.

[252] V, S: *non*. Recognizing that *non* was wrong, A wrote something which he then deleted so effectively that it is illegible. Sw (1973, p. 494) prints the correct emendation *nobis* as though it were his own, although it was introduced by Müller, as was pointed out by TCT, 1st ed., 1939, p. 83.

[253] Sw mistranslates: "For when the earth moves into either of the nodes of Venus ..." (1973, p. 494). The earth never moves into a node of Venus.

[254] V, S: *et has*, omitted by A, striving for greater compactness.

[255] V, S: *apparent*; A: *apparentes*. While interchanging the order of the words *naturales apparent*, A mistakenly added the final syllable *-es* of *naturales* to *apparent*.

[256] V, S: *autem*; V: *vero*, unpleasantly repeating *vero* four words earlier.

[257] V, S, A: *eaedem*, which Sw (1973, p. 494) emends to *eadem*, identified with the earth. But in four similar passages *Commentariolus* does not use *eadem* (or its equivalent) in this way. Thus, with regard to the variable distance between Venus and its larger epicycle's center, *Commentariolus* writes:

> *quandocumque tellus in linea ad absidem diametro porrecta fuerit, sidus tunc centro maioris epicycli proximum sit, et in transverso quadrantum remotissimum.*

Again, with reference to the variable distance between the centers of Mercury's larger epicycle and deferent sphere, *Commentariolus* similarly writes:

> *in memoratis ad absidem telluris sitibus centrum epicycli maioris centro orbis proximum est, in quadraturis autem remotissimum.*

Earlier, in explaining the moon's approach toward and withdrawal from its larger epicycle's center, *Commentariolus* likewise writes:

> *quandocumque centrum epicycli maioris contingit lineam a centro orbis magni transeuntem per centrum terrae ... tunc luna sit ad centrum maioris epicycli proxima, quod quidem circa novam et plenam lunam accidit, et e contra in quadraturis mediantibus iisdem remotissima.*

Finally, to describe how an outer planet approaches toward and withdraws from the center of its first epicycle, *Commentariolus* also writes:

> alter [*epicyclus*] ... *circumagit sidus adeo ut, quandocumque sit in summa a centro orbis distantia vel rursus in maxima vicinitate, tunc sidus sit centro epicycli quam proximum, in quadrantibus autem mediantibus remotissimum.*

In these passages *Commentariolus* describes what happens when anything in motion is midway between two of its previously mentioned positions 180° apart (*in transverso quadrantum, in quadraturis, in quadraturis mediantibus, in quadrantibus ... mediantibus*). In none of these passages does *Commentariolus* refer again to the moving entity (*tellus, centrum epicycli maioris* [twice], *alter* [*epicyclus*]) when it is at these midpoints. Hence *eaedem*, the reading of all three manuscripts, should be retained, while Sw's emendation *eadem* should be rejected.

[258] V, S, A: *permixtae*, which Sw (1973, p. 494) emends to *permixte*. Sw then translates: "both latitudes are mixed together in a disorderly way." *Commentariolus*, however, says that Venus' reflexions and declinations mingle and are combined in a perfectly orderly way.

[259] V, S, A: *ac*. V's *ac* was misreported as *et* by MCV, I, 15/19, followed by PII, 199/19.

[260] V: *liberationem*; S, A: *librationem* (see n. 232, above).

[261] See the penultimate paragraph of "The Outer Planets."

[262] V, A: *unde*; S: *undeque*, under the influence of the following word *quandocumque*. By correcting this error in his prototype, A came into fortuitous agreement with V.

[263] V: *liberationis*; S, A: *librationis* (see nn. 232, 260, above).

[264] V, A: *diameter*; S: *diam*[*etrali*]*ter*. V was emended to *e diametro* by MCV, I, 15/26 (followed by PII, 199/26) before S' correct reading was made known. Like V's reader, A misinterpreted the contraction *diamter* in his prototype as the syntactically unacceptable nominative singular *diameter*.

[265] V: *simi*(*li*) *tutidinem*; realizing that his pen had slipped, V erased the letters within the parenthesis (*li*), although he should have erased *ti*; V's reading was misreported as *similitudine* by MCV, I, 15/19.

S: *similitudinem*; its presence in both V and S indicates that this defective reading was already present in Brahe's MS.

A: *similitudine*, correcting the error in his prototype, which was akin to S.

[266] V, S, A: *maxime*, corrected by Müller to *maximae*. This emendation is printed by Sw (1973, p. 497) as though it were his own, although it was credited to Müller by TCT, 1st ed., 1939, p. 84.

[267] V, S, A: *destiterit*. S' *destiterit* was misreported as "distiterit" by Lindhagen (p. 14/7), followed by MCV, IV, 8, and Sw 1973, p. 497.

[268] V: *usque* (not in S, A). Gansiniec emended *usque* to *usquam* (anywhere). However, here *Commentariolus* is referring, not to the deviation anywhere, but only to its disappearance when the earth is 90° away from Venus' apse-line, and the point of maximum deviation is 90° away from Venus. V's *usque* comes at the end of line 5 on folio 64ᵛ, four lines below the first line, which ends with *quoadusque*. Hence, V's *usque* may be nothing more than an inadvertent dittography.

[269] V: *continuata*; S, A: *continuato*. V's *continuata* was misreported as *continuato* by MCV, I, 15/31-32, followed by Sw 1973, p. 497. In PII, 200/3, *continuatio* is a misprint.

[270] V: *perveniat*, omitted by S, A, perhaps under the influence of *peracto* three words later.

[271] V, S, A: *circulo*, corrected to *semicirculo* by Sw (1973, p. 497). The presence of the wrong reading *circulo* in all three manuscripts indicates that Brahe's MS too contained this defect.

[272] V, S: *et primae simul et aequalis*; A: *et primae similis et aequalis*. The presence of *simul* in both V and S indicates that this was the reading in Brahe's MS also. But A recognized that the second *et* in his prototype was incompatible with *simul*, which he correctly changed to *similis*.

[273] V: *sic*; S, A: *sicut*. V's incorrect reading *sic* was misreported as *sicut*, the correct reading, by MCV, I, 16/7.

[274] In the last paragraph of "The Three Outer Planets."

[275] V: *investigabiles*; S, A: *impervestigabiles*. Listening to the prototype being read aloud, V reduced the more intensive form *impervestigabiles* to the briefer form *investigabiles*.

[276] V, S: *altiori ingenio quispiam*; A: *ingenio quispiam*. A omitted *altiori*, which might be misinterpreted as a description by Copernicus of himself. Since he never claimed to be endowed with superior ability, *Commentariolus' quispiam* (someone) meant "someone else," perhaps a contemporary or later astronomer for whom *Commentariolus* cleared the way.

[277] V: *hinc*; S, A: *huic*. V's incorrect reading was the result of imperfect communication from mouth to ear to hand.

[278] V, S: *suo*, omitted by A, steadily striving for greater compactness.

NOTES

²⁷⁹ See "Venus," 1st paragraph.

²⁸⁰ V, S, A: *absides*. V's *absides* was misreported by MCV, I, 16/17, as *absidis*, the required reading.

²⁸¹ V, S: *medium*; A: *30′*. V's *medium* was misreported by MCV, I, 16/17, as *medio*, the reading which would be required to remove the inconsistency in case between *Commentariolus' gradibus* (ablative) and *medium* (not ablative).

²⁸² V: *supervenit*; S, A: *superaverit*. These two acceptable readings, which require no emendation, are emended by Sw (1973, p. 500), who introduces *supervidet*, a verb never used by Copernicus. Moreover, it would eliminate the manuscripts' explicit reference to the earth's physical passage over Mercury's higher apse. As between *supervenit* and *superaverit*, the latter is undoubtedly the better reading.

²⁸³ By contrast, Venus is nearest to its larger epicycle's center whenever the earth is on Venus' apsidal line, and when the earth is at a quadrant's distance from the apsidal line, Venus is farthest away from its larger epicycle's center.

²⁸⁴ In "The Order of the Spheres." V, S: *duximus*; A: *diximus*. The presence of *duximus* in both V and S indicates that Brahe's MS too showed this faulty reading, which was correctly changed to *diximus* by A. V's *duximus* was misreported as *diximus* by MCV, I, 16/22.

²⁸⁵ Near the beginning of "The Three Outer Planets."

²⁸⁶ V, S: *semidiametrum*; A: *semidiameter*. Near his faulty *semidiameter* A wrote the marginal note which is reported in n. 289, below.

²⁸⁷ V, S: *autem*, omitted by A in his striving for greater compactness.

²⁸⁸ V, A: *eius*; S: *etiam*, under the influence of the preceding word *tertiam*. If A's prototype shared S' faulty reading *etiam*, then A's emendation *eius* brought A into fortuitous agreement with V.

²⁸⁹ In the margin A converted *Commentariolus'* sexagesimal values into approximately equivalent decimal values:

But with the [Grand] Orb's radius = 100^p, Mercury's radius will be 38^p; and the first epicycle's, 7^p or $6\ 1/2^p$; the second epicycle's, $2\ 1/6^p$.

A's values are roundings of $37\ 3/5^p$, $6^p\ 44^m = 6.73^p$, $2^p 16^m = 2.267^p$.

²⁹⁰ V, S: *hic*, omitted by A in his constant striving for greater compactness.

²⁹¹ To Mercury, as well as the other planets, *Commentariolus* ascribes a combination of circles (*circulorum concursus*), not spheres; see n. 326, below.

²⁹² V, S: *etiam*; A: *iam*, apparently through inadvertence.

²⁹³ Mercury has a system of circles (*ratio circulorum*), not spheres; see n. 326, below. *Commentariolus'* system for the three outer planets and Venus consists of a concentric deferent surmounted by two epicycles. Each of these four planets traces an orbit regulated by a combination of circles which are invariant in dimension. This rigid pattern, however, does not suit Mercury, because that planet seems to move along arcs belonging to circles of various sizes. This idea is put forward hesitantly by *Commentariolus* as a pattern that Mercury seems (*apparet*) to follow. The word *apparet* tones this statement down from a categorical affirmation ("follows") to a tentative suggestion ("seems to follow"). In saying that Mercury seems to move along arcs of unequal circles, *Commentariolus* seeks to justify its treatment of Mercury as different from the other four planets.

Misunderstanding *Commentariolus' apparet* as undertaking to give a complete description of Mercury's apparent motion, Sw exclaims:

> The statement that Mercury "appears" to move in a smaller orbit when the earth is in the apsidal line and in a larger orbit when the earth is 90° from the apsidal line is utter nonsense as a description of the apparent motion of Mercury. No one — not ... even Copernicus in *Revolutions* — gives such a description of Mercury's apparent motion (1973, p. 504).

After writing *Commentariolus*, Copernicus improved his understanding of Mercury's motions, which he treated much better in *Revolutions*. There "Mercury does not always describe the same circular circumference. On the contrary ... it traces an exceedingly varying circuit"; Mercury "describes around F, the center of its orbit, unequal circles"; "here the planet [Mercury] describes a larger circle than it does there" (NCCW, II, 280/20-22, 282/32-33, 283/35-36). In the theory of Mercury's longitude, *Revolutions* went beyond *Commentariolus* by taking into account the compensatory effect of visual distance on the size of Mercury's observed pathway. Thus in Maestlin's diagram (somewhat simplified in our Figure 12) A is the center of the universe, $BCSD$ is the earth's orbit, and E is the center of FG, a small eccentric circle carrying the center of Mercury's expanding and contracting epicycle:

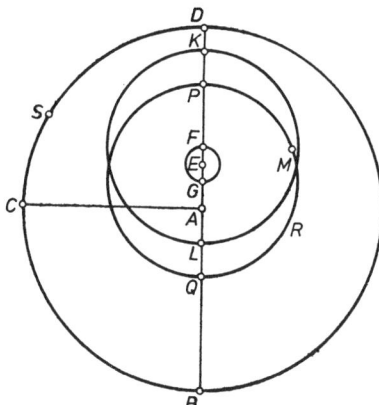

Fig. 12. Maestlin's diagram for Mercury's motion in longitude (somewhat simplified)

Whenever the earth is in the [apsidal] line of Mercury's apogee B or perigee D ... Mercury is found at its epicycle's extremities K or L, nearest to [E], the eccentric's center. Hence in that situation the circle of Mercury's path is KML, the smallest possible of all. ... But if the earth, for instance at C, is midway between the apsides, the eccentric's center drops down to G, the point nearest to A, the center of the universe. ... Mercury, however, ascends to the epicycle's outermost point P or Q, and describes PRQ, the most extensive circle of its pathway. ... When the earth is in B ... Mercury's circle, KML, is the smallest and looks the smallest, because the earth is at its greatest distance from Mercury. When [the earth is] in ... D, however, [Mercury's] circle [KML] is still minimal, but looks bigger because it is at its closest [to the earth]. When [the earth] at C [is] on the perpendicular [to the apsidal line, Mercury's] circle [PRQ] is the largest, to be sure, but looks no bigger because it is at a greater distance from the earth. On the other hand [when the earth is] near S, a third of a circumference from B, or a sixth from D, ... according to Copernicus [Mercury's circle] looks the biggest ... because that result is achieved in that position by the mutual balancing of distance and true size (Maestlin, in GW, I, 143/17–47).

[294] V, S, A: *percipiatur*. V's reading was misreported by MCV, I, 17, which mistook the passive form *percipiat[ur]*, indicated by the flourish attached to the *t*, for the plural form *percipiant*, which would have been indicated by a supralinear stroke placed over the *a* and not attached to the *t*.

[295] V: *accessum quendam et recessum centro*; S: *accessum quendam et recessum a centro*; A: *accessum quendam centro*. V's *centro* was misreported as *centri* by MCV, I, 17/3, followed by PII, 201/7. A's omission of *et recessum a* achieves compactness by doing violence to an expression often used by Copernicus.

[296] V, S, A: *circumdata*, corrected to *circumdatis* by MCV, I, 17/4. The presence of *circumdata* in all three manuscripts indicates that this defective reading was a blemish in Brahe's MS too.

[297] V, S, A: *asse*, incorrectly emended to *axe* by MCV, I, 17/5, followed by PII, 201/10. Calling *asse* "meaningless," Sw thought it "was perhaps the end of a divided word and could as well have been -*isse* or -*esse*," left over when "a line or so ... dropped out of the text" (1973, pp. 503, 505). This alleged gap, leaving -*asse*, -*isse*, or -*esse*, vanishes from Sw's second thoughts (1976, p. 157). There Sw objects to the translation of *asse* by "exactly" (TCT, 1st ed., 1939, p. 87) on two grounds. First, *Commentariolus* reads *asse* instead of *ex asse* or *in asse* (a "very frequent expression," *Thesaurus linguae latinae*, II, 747). But the Greek word ὁλοκλήρου (holoklerou, completely) was equated with *asse* preceded by no preposition (*Corpus glossariorum latinorum*, II, 23/55). Secondly, Sw contends that "(entirely) does not fit the context," which is given by Sw as *sive totius illius asse tantum distet*. The actual context, however, is *asse tantum distet ... quantum* (the first distance is exactly equal to the second distance).

[298] V: *orbiculi*; S, A: *orbiculum*. A failed to correct the incorrect reading *orbiculum* in his prototype, which was akin to S.

[299] Sw's translation "connecting" (1973, p. 503) does not match *continentis*. *Commentariolus* means that the larger epicycle's center is located on the surface of the inner small sphere.

[300] The passive voice (*repertum est*, has been found) implies that $0^p\ 14\ 1/2^m$ as the length of the radius of the inner small sphere was found by someone else before *Commentariolus* was written. Recent research has thrown a flood of new light on the development of post-Ptolemaic planetary theory prior to Copernicus, but has not yet succeeded in identifying the source of *Commentariolus*' $14\ 1/2^m$ as the length of the radius of the small sphere.

NOTES

[301] V, S, A: *14*. V's *14* was misreported as *24* by MCV, I, 17/7, followed by PII, 201/12, and JHA, 1973, 4:126/12 up. Sw (1973, p. 503) mistakenly reports *14* as given only by S.

[302] Here the deponent verb *mensi sumus* says that Copernicus himself assigned $25^p\ 0^m$ to the length of the radius of the earth's orbit as the universal measure. By contrast, the passive voice *repertum est* implies that $0^p\ 14\frac{1}{2}^m$ for the length of the radius of Mercury's inner small sphere was found by someone else (see n. 300, above).

[303] V: *motur*; S, A: *motus*. V's incorrect reading was corrected to *motus* by MCV, I, 17/9.

[304] Here *anno vertente* is not used in its technical sense of "tropical year," as defined in *Revolutions*, III, 13:

> I must therefore distinguish the seasonal year from the sidereal year. For I term (*vocamus*) that year "natural" or "seasonal" which marks the four annual seasons for us, but that year "sidereal" which returns to one of the fixed stars. However, the natural year, which they (*vocant*) also call "tropical" (*vertentem*) is nonuniform (NCCW, II, 144/13–16; NCOO, II, 142/8–12).

This nonuniform year is called by Copernicus himself *annus naturalis* or *annus temporalis*, while he ascribes *annus vertens* to others. He himself uses *anno solis vertente* (*Revolutions*, I, Introduction) to mean "year," without distinguishing between the tropical and sidereal varieties (NCOO, I, 8/6–7). On the other hand, *Revolutions*' Preface says:

> astronomers ... are so uncertain about the motion of the sun and moon that they cannot even establish and observe the constant length *vertentis anni*.

Since constant length differentiates the sidereal year from the tropical year, here *vertentis anni* means the sidereal, not the tropical, year. In short, *annus vertens* is used by Copernicus to mean "sidereal year" or "tropical year" or just plain "year." The last is the sense intended in our passage. Instead of its *anno vertente*, the corresponding passage in *Revolutions*, V, 25, uses *annuo spatio telluris*, which is equally applicable to both the tropical year and the sidereal year (NCOO, II, 288/17). Mismatching our passage's *annus vertens* with "tropical year," Sw writes about

> Copernicus slipping and referring to the outer small sphere's completing two revolutions in a tropical year (*in anno vertente*) when of course he means a sidereal year (1973, p. 505).

[305] V, S, A: *centro*. This erroneous reading, which is under the influence of the preceding word *composito*, was present in Brahe's MS, since it is found in all three manuscripts. The correct emendation to *centra* was made by MCV, I, 17/11. Sw (1973, p. 503) emends to *centrum*, which entails substituting *Praefertur* for *Praeferuntur*, the unanimous reading of all three manuscripts. These two paleographically implausible emendations are made by Sw in order to restrict the larger epicycle to a single center. However, since this center is not fixed, but slides back and forth along a straight line, *Commentariolus* naturally regards this epicycle's various positions as having various centers.

[306] In "The Three Outer Planets," last paragraph.

[307] V: *proximus*; S, A: *proximum*. V's erroneous *proximus* is just below *-bus*, the final syllable of the divided word *sitibus*.

[308] V, S, A: *remotiss*. V's abbreviation was mistakenly expanded to the masculine form *remotissimus* by MCV, I, 17/14, which had failed to notice that V's erroneous *proximus* had to be changed to *proximum* (see the preceding note). PII, 201/19, printed *remotissimus*, inconsistent with *proximum*, the correct neuter form.

[309] V, S: *applicat*; A: *se applicat*, in an effort to improve *Commentariolus*' Latin style by making the verb reflexive.

[310] V, S: *vero haud secus facit quam*; A: *autem facit haud secus ac*. A's feeling that he was not bound inflexibly to his prototype is clearly shown by his replacement of *vero* by *autem*, and of *quam* by *ac*, as well as by his transposition of *facit* to precede, instead of following, *haud secus*.

[311] V: *tractum*; S, A: *tractu*. V's *tractum* was misreported as the correct reading *tractu* by MCV, I, 17/20, who ignored the supralinear stroke over the *u*.

[312] V, S: *quoque*, omitted by A in order to prevent it from being erroneously linked with the following words "always southern" (*semper australis*). Yet Sw translates:

> The deviation, which in the case of Mercury is also always southern, never exceeds $3/4°$ (1973, p. 509).

To what does Sw's "also" look back? Not to the previously mentioned latitude, for there *Commentariolus*' compressed statement "where Venus turns north, Mercury heads south" must be supplemented by the converse:

"and where Venus turns south, Mercury heads north." Thus, this latitude of Mercury is not "always southern." Nor can Sw's "also" look back to Venus' deviation. For, as Sw himself translates *Commentariolus*' statement about Venus: "this latitude, which is commonly called the 'deviation,' never becomes southern" (1973, p. 497). As Sw himself remarks, "Venus is only moved north by the deviation, never south" (1973, p. 499). Hence, there is nothing to which Sw's "also" can look back. It was precisely in order to avoid this erroneous linkage of *quoque* ("also") with *semper australis* ("always southern") that A omitted *quoque* altogether. But its presence in both V and S indicates that it belongs in *Commentariolus*. It must be linked, however, with the preceding word *hic*: "here too [as in Venus] there is a deviation." But Mercury's deviation is always southern, whereas Venus' deviation is never southern.

[313] V: *autem*, omitted by S, A.

[314] V, S: *numquam*, toned down to *non* by A.

[315] V: *latitudines*; S, A: *latitudinem*, an inferior reading.

[316] V: *conveniet*; S, A: *convenit*, the better reading.

[317] V, S: *quoque*, omitted by A for the sake of greater compactness.

[318] V, S: *Sicque*, shortened to *Sic* by A.

[319] Deferent sphere, 2 epicycles, 2 longitudinal small spheres, 2 latitudinal small spheres.

[320] Deferent sphere, 2 epicycles, 2 latitudinal small spheres.

[321] Annual revolution, daily rotation, axial rotation.

[322] V, S, A: *eam*. V's *eam* was misreported as *eum* by MCV, I, 17. V did not always close the top of the letter *a*.

[323] Deferent sphere, 2 epicycles, nodal sphere.

[324] Deferent sphere, 2 epicycles, 2 latitudinal spheres.

[325] $34 = 7+5+3+4+(3 \times 5)$. Aristotle (*Metaphysics*, XII, 8) had put the number of celestial spheres at 55. That is why *Commentariolus* (¶4) emphasized that the problem of the planets "could be solved with fewer ... constructions than were formerly put forward."

[326] "To explain the entire structure of the universe, 34 circles suffice." In this triumphant conclusion *Commentariolus* referred to circles (*circuli*), not spheres (*orbes*). In like manner it recalled the concentric circles (*circulos*) of Eudoxus and Callippus, whose concentric spheres had been discussed since Aristotle. *Commentariolus* sought

> a more reasonable arrangement of circles (*circulorum*), from which every apparent irregularity would be derived while everything in itself would move uniformly.
>
> In the explanation of the circles (*circulorum*) themselves I shall set down here the lengths of the spheres' radii.
>
> Anybody familiar with mathematics will readily perceive how excellently this arrangement of circles (*circulorum*) agrees with the numerical data and observations.

The earth's motion will receive powerful proof in *Commentariolus*' "exposition of the circles (*circulorum*)." The latitude of the three outer planets varies with the oscillating "inclination of their axes and circles (*circulorum*)." "In its system of circles (*circulorum*) Venus closely resembles the outer planets." Whereas two epicycles mounted on a fixed concentric deferent suffice for Venus and the three outer planets, "this combination of circles (*circulorum*) is not sufficient" for Mercury. Clearly, when Copernicus wrote *Commentariolus*, his mind focused on the circles rather than on their associated spheres.

According to Sw, "Copernicus had no doubt that the motions of the planets were controlled by solid spheres" (1973, p. 466). After hearing that Copernicus did have such a doubt, Sw remarked:

> Now it is odd that if Copernicus wished to do without an assumption as universal and fundamental as solid spheres he would not say so. ...
>
> That he does not deny the existence of solid spheres really constitutes by itself a certain amount of evidence that he accepted the common assumption of his times. ... His acceptance of solid spheres was as complete as that of any other astronomer before Tycho (1976, p. 114).

In like manner Sw referred to

> the assumption, shared by all astronomers of Copernicus' time, that all planetary motions are controlled by the rotation of solid spheres (1976, p. 120).

Sw's undocumented assertion that "all astronomers of Copernicus' time" assumed "that all planetary motions are controlled by the rotation of solid spheres" ignores the widespread denial of the existence of such spheres. That denial was brought sharply to Copernicus' attention by a close friend who helped to overcome the astronomer's reluctance to publish *Revolutions*. In personal conversations with Copernicus, this friend recalled a Muslim philosopher's judgment that "epicycles and eccentrics could not possibly exist in the realm of nature," and that "the Ptolemaic astronomy is nothing, as far as existence is concerned" (TCT, p. 194). "But it is in agreement with the computations, though not with reality" was the rest of Ibn Rushd's (Averroes') judgment in his *Commentary on Aristotle's Metaphysics* (Book XII, summa 2, chapter 4, comment 45).

Ibn Rushd's judgment was echoed by a renowned lecturer at the University of Paris, Jean Buridan (c. 1295–c. 1358), whose *Questions on Aristotle's Metaphysics* declared:

> Up to the present no way has been found to explain ... planetary phenomena except by assuming eccentrics or epicycles. ... This way of assuming or imagining epicycles and eccentrics is quite valid for computing and knowing the positions of the planets and their configurations with respect to one another and to us. Nothing more than this is sought by the astronomers. Accordingly, they are permitted to use such figments of the imagination even though these do not really exist (ed. Paris, 1518, sig. AA 1^r).

Buridan's ideas spread eastward across Europe and became very popular at the University of Cracow. The Jagiellonian Library there possessed manuscript copies of eighteen of his writings, as well as eleven different commentaries on Aristotle's *Metaphysics*. These referred to Ibn Rushd's and Buridan's arguments against the validity of epicycles and eccentrics. It was at the University of Cracow that Copernicus learned the fundamentals of astronomy. The curriculum had shortly before been enriched by the introduction of Peurbach's *New Theory of the Planets*, concerning which Sw remarks:

> Peurbach illustrates and describes in detail the (purely Ptolemaic) solid sphere models. ... In the later fifteenth and sixteenth centuries, the *New Theory* was widely used as a school text to be read, possibly with a commentary (1976, p. 116).

Such a commentary was the vehicle by which Peurbach's *New Theory* was brought into the curriculum at the University of Cracow by its foremost astronomer, Albert of Brudzewo (1445/1446–1495). In his *Commentary on Peurbach's New Theory of the Planets*, Brudzewo dealt with Peurbach's eccentric spheres:

> No mortal man knows whether these eccentrics really exist in the spheres of the planets, unless we admit (as some people claim) that the eccentrics, like the epicycles, are made manifest by the revelation of spirits. If we reject this claim, then the eccentrics are devised solely by the imagination of the astronomer (ed. L. A. Birkenmajer [Cracow, 1900], p. 26).

Brudzewo (p. 27) also quoted with approval from Richard of Wallingford (Part I, Proposition 10, Corollary 3):

> Among the heavenly bodies there are no such eccentrics or epicycles as are devised by the astronomical imagination for its own use. No educated person could regard them as probable. Without such imaginative astronomical constructions, however, no systematic science of the motion of the stars can be established which would so pinpoint their positions at any moment as to be in accord with what we see (cf. J. D. North, *Richard of Wallingford* [Oxford, 1976], I, 278).

Hence it is not true that Richard of Wallingford "leaves the description of the physical properties of the planets to Albumasar," as in Robert S. Westman, "The Astronomer's Role in the Sixteenth Century," *History of Science*, 1980, *18*:138, n. 34.

When Copernicus was in his most impressionable years, the dominant doctrine at the University of Cracow looked upon eccentrics, epicycles, and the rest of the celestial machinery as "imaginative astronomical constructions," existentially improbable yet professionally indispensable. Cracow alone refutes Sw's unsupported pronouncement about "the assumption, shared by all astronomers of Copernicus' time, that all planetary motions are controlled by the rotation of solid spheres" (1976, p. 120).

This assumption, mistakenly described by Sw "as universal and fundamental," was not shared by all astronomers of Copernicus' time, and certainly not by Copernicus himself.

> That he does not deny the existence of solid spheres really constitutes by itself a certain amount of evidence that he accepted the common assumption of his times

is a fallacious inference of Sw, who misunderstands Copernicus' scientific strategy. The astronomer neither asserts nor denies the existence of solid spheres. His failure to deny no more implies acceptance than his failure to assert implies rejection. His silence rather indicates his unwillingness to express an opinion in public about a controversial issue which was not absolutely clear to him.

Why do the planets keep on moving? In Copernicus' time they were not believed to be moving by themselves. On the contrary, the visible planets were thought of as being moved by invisible movers to which they were firmly attached. What was the nature of these unseen movers? *Commentariolus* and *Revolutions*, destined for publication, preserved a discreet silence. In a private letter, however, Copernicus told his correspondent:

> First we learn that the planets' apparent motions are nonuniform; subsequently we conclude that there are epicycles, eccentrics, or other circles (*circulos*) by which the planets are carried in this way (p. 146, below).

The circles which moved Copernicus' planets were of course associated with spheres (*orbes*). But the spheres mentioned in *Revolutions*, I, 4, 10, are gratuitously converted into "solid spheres" by Sw (1976, p. 129). Moreover, Sw professes to cite

> the testimony of Tycho ... in full accordance with Copernicus' remarks, that Copernicus took the existence of "solid and real" spheres for granted (1976, p. 158).

What Tycho Brahe actually wrote with regard to a planetary sphere, however, was: "Let this be solid and real as Copernicus also seems to have believed" (Sw's translation, 1976, p. 130). Far from testifying categorically that Copernicus believed in solid spheres, Brahe advisedly adopted the prudent expression "seems to have believed." Brahe knew not only *Commentariolus*, but *Revolutions* as well. On the basis of his close study of these two works, Brahe concluded that Copernicus "seems to have believed" in solid spheres. Brahe's cautious statement is perverted by Sw into "the testimony of Tycho ... in full accordance with Copernicus' remarks, that Copernicus took the existence of 'solid ...' spheres for granted." Tycho gave no such testimony, nor did Copernicus make any such remarks.

Although Copernicus never said that he believed in solid spheres, Kepler momentarily thought otherwise. His *New Astronomy*, in referring to a solid sphere (*orbis solidus*), stated parenthetically: "which Copernicus believes in" (GW, III, 73/27–28). Kepler never repeated this mistake in any of his voluminous writings.

The great Muslim scientist Ibn al-Haytham wrote an astronomical work in Arabic which was translated into Latin around 1300. As thus translated, we are told by Sw (1976, p. 118), "Ibn al-Haytham explains that the common noun *falak* (in the Latin translation *orbis*)

> is applied to all round magnitudes whether they are spherical bodies or spherical surfaces, or surfaces or circumferences of circles. However, in the description of planetary models the meaning of *falak* (*orbis*) is restricted to 'sphere'."

The foregoing excerpt begins as an actual quotation, since the Latin translation of Ibn al-Haytham reads as follows:

> *Orbis est nomen commune, et dicitur super omnes quantitates rotundas sive sit corpus spericum, sive superficies sperica, sive superficies circuli aut circumferencialis circuli* (José Maria Millas Vallicrosa, *Las traducciones orientales en los manuscritos de la Biblioteca Catedral de Toledo* [Madrid, 1942], p. 287).

On the other hand, the closing sentence in the excerpt (However ... "sphere") is not in Ibn al-Haytham, as translated. Sw's homogenized presentation does not indicate where Ibn al-Haytham leaves off and Sw begins. The unsuspecting reader is left with the mistaken impression that it was Ibn al-Haytham who said: "in the description of planetary models the meaning of ... *orbis* is restricted to 'sphere'." That restriction was imposed, not by Ibn al-Haytham, but by Sw.

Ibn al-Haytham's astronomical work was recently translated from the Arabic into English. According to the translator,

> Ibn al-Haytham uses, in a discussion of a planetary model, the word *falak* to refer to the circle produced by the intersection of a plane and a spherical surface. Swerdlow commits another error when he claims: "In the Latin translation the predominant word for these spheres is *orbis*, which is occasionally replaced by *sphaera*." In fact, the two Latin terms are used consistently to translate two different Arabic terms. In the translation ... used by Swerdlow, *orbis* is used for *falak*, whereas *sphaera* is reserved for *kurah*, a term which unambiguously refers to the solid sphere. ... In conclusion the only term used by Ibn al-Haytham which refers exclusively to solid spheres is *kurah*. The much more common term *falak* is used for both two-dimensional and three-dimensional figures (Tzvi Langermann, "A Note on the Use of the Term Orbis (Falak) in Ibn al-Haytham," AIHS, 1982, *32*:113).

Rosen ... claims that Copernicus' references to spheres can be taken as a fiction with no physical existence says Sw (1976, p. 129) without adducing any citation to support this misrepresentation. By the same token, with regard to the assumption by other astronomers of "uniformly rotating spheres in the heavens as the physical basis of planetary theory," according to Sw (1976, p. 109) "Rosen ... maintains that this was not the case" for Copernicus. Again, no citation is adduced by Sw in support of this misrepresentation. Before Sw was born, TCT plainly said:

> Copernicus used these spheres (*orbes*) throughout his work. He avoided taking sides in the controversy over the question whether the spheres were imaginary or real, whether, that is, they were simply a mathematical means of representing the planetary motions and a convenient geometrical basis for computing the apparent paths, or whether they really had a physical existence in space and like a piece of machinery produced the observed phenomena. But whether the planets were carried by material balls or hoops or by imaginary spheres or circles through a medium of whatever type, the resultant computation of the actual planetary courses was the same. From Copernicus' language it sometimes appears that he regarded the planet as attached to a three-dimensional sphere; but more often a two-dimensional great circle of the sphere was the geometrical figure to which he affixed the planet. For astronomical, as opposed to cosmological or astrophysical, theory it was a matter of indifference whether a planet was thought to be attached to a sphere or to a great circle thereof (1st ed., 1939, pp. 11–12).

TCT emphasized Copernicus' "repeated shift from sphere to circle and back again" (1st ed., 1939, p. 13). Nearly four decades later, Sw agrees: "Copernicus moves back and forth between referring to spheres and circles" (1976, p. 121). Whatever their nature may have been, however, Copernicus' *orbes* were not solid spheres.

While Copernicus was a student in the University of Padua, then under the control of Venice, in that city a famous publisher in 1501 issued GV, which mentioned

> certain persons, and especially the Aristotelians, who, prompted by their belief in opposition to the opinion of the foremost astronomers, ... maintain that there are no eccentric, no epicyclic globes because, they say, the true and solid bodies of the planets cannot be transported by circles and lines that are drawn and lack body, so that a body cannot be coerced by an incorporeal body, nor are even epicycles to be called bodies lest there seem to be unoccupied [space] in heaven (sig. gg 7/20 up–15 up).

This assault on the physical reality of eccentrics and epicycles came at the end of GV's Latin version of Proclus' *Hypotyposis*. "Version" is a more appropriate term than "translation" to describe the relation between GV and Proclus, who said:

> In their eagerness to display the motions of the heavenly [bodies] as uniform, the professional astronomers unwittingly portrayed the bodies' very essence as nonuniform and shot through with variations. For with regard to the eccentrics and epicycles which they keep talking about, what shall we say? Are these merely imagined, or do they also have physical reality in their spheres, in which they are fastened? For if they are merely imagined, the astronomers unknowingly shifted from physical bodies to mathematical concepts, and derived the causes of natural motions from entities that do not exist in nature.
>
> I shall add that the astronomers would be illogical also in describing the motions. For, the motions are in accordance with our concepts. Hence, the heavenly bodies imagined on these [eccentrics and epicycles] do not in reality move nonuniformly. And if the eccentrics and epicycles also exist physically, their connection with the spheres in which these circles exist is obliterated by the astronomers. They make these circles move independently, and the spheres independently. The circles do not even move like one another, but in opposite directions. Moreover, their distances from one another are confused, since sometimes they come together and lie in one plane, while at other times they diverge and intersect one another. Hence there will be all sorts of divisions and coalescences and separations of the heavenly bodies.
>
> Moreover, the treatment of these contrived hypotheses also seems haphazard. For in each hypothesis why is the eccentric as it is, either stationary or mobile? Why is the epicycle as it is, with the planet being moved either according to or contrary to the order of the signs [of the zodiac]? And what are the reasons for those planes and divergences? I mean the true reasons, those which definitely put an end to all exertion by the mind when it understands them. No answer of any sort is forthcoming. Instead, actually proceeding backwards, the astronomers do not derive

systematic conclusions from the hypotheses, as the other disciplines do. On the contrary, from the conclusions they try to form the hypotheses, from which they should have inferred these conclusions. And in the process they do not even appear to have provided what was possible.

Yet this much should be known, that these are simpler than all [the other] hypotheses and more suitable for divine bodies, and that they have been devised to find out how the planets move, which really move as they seem to move, the purpose being to bring within reach the measure of their motions (Ch. VII, 50-58).

As the head of the Neoplatonic Academy in Athens, and as a prolific and widely read author, Proclus (410-485) influenced the attitude of many thinkers. His questioning of the physical reality of eccentrics and epicycles, while conceding their professional usefulness, helped to shape the outlook of Ibn Rushd, Maimonides, Richard of Wallingford, Buridan, Brudzewo (a professor at the University of Cracow), and Copernicus (a contemporary student at the University of Cracow).

According to Sw, "it is odd that if Copernicus wished to do without an assumption as universal and fundamental as solid spheres he would not say so." It is odd that in the face of the foregoing overwhelming (and intentionally limited) evidence to the contrary, anybody should speak of "an assumption as universal... as solid spheres."

Since Sw's solid spheres formed no part of Copernicus' cosmos, why did his planets and planetary spheres keep going round and round? As a university student, he had read in Aristotle's *Heavens*, II, 7, 289a 14-16 the decision

> to make each of the heavenly bodies out of the substance in which it happens to have its motion since, as I said, there is something which by nature moves in a circle.

That is why in his *Revolutions*, I, 4 (NCCW, II, 10/29-33) Copernicus said:

> The motion of the heavenly bodies is circular, since the motion appropriate to a sphere is rotation in a circle. By this very act the sphere expresses its form as the simplest body, wherein neither beginning nor end can be found, nor can the one be distinguished from the other, while the sphere itself traverses the same points to return upon itself.

[327] Copernicus derived "ballet of the planets" from Martianus Capella (I, 18), mentioned by name in *Revolutions*, I, 10 (NCCW II, 20/6).

COPERNICUS' *LETTER AGAINST WERNER*

INTRODUCTION

JOHANN WERNER

Johann Werner, a clergyman with a natural bent toward the mathematical disciplines, was born in Nuremberg on 14 February 1468.[1] At the age of 16, he entered the University of Ingolstadt on 21 September 1484,[2] but dropped out without obtaining a degree. A friendly mathematician and astronomer, however, arranged to have an ecclesiastical benefice conferred on him on 22 November 1492.[3] The income therefrom enabled him in the following year to go to Rome, where he was ordained a priest.[4] He was still in Rome when he observed an eclipse of the moon on 18 January 1497.[5] Several months later, after much study but again without receiving any academic degree, he departed from Rome, leaving behind the small library he had accumulated.[6] His offer to buy books and maps in Florence for a well-known Nuremberg chronicler was accepted on 16 June 1497.[7] He then returned to his birthplace, where as a confirmed astrologer he cast a horoscope at about 9:39 P.M., 8 November 1497,[8] and celebrated his first mass on 29 April 1498.[9]

A bishop who was interested in astronomy and mathematics spent the summer of 1501, from May to mid-September, in Nuremberg. Almost every day he invited Werner to discuss mathematical questions with him.[10] Presumably at his suggestion, the empress herself on 24 July 1503 intervened energetically on Werner's behalf with Nuremberg's municipal authorities, strongly urging them to provide him with a suitable post. With her own hand she added at the end of the official communication: "I want you to give me satisfaction in this matter."[11] With such powerful support Werner did not have long to wait. In the course

---

[1] Leonhard Krentzheim, *Chronologia* (Görlitz, 1577), II, 337. Without offering any reason for changing 1468 as the year of Werner's birth, and indeed without indicating any awareness that he was altering Krentzheim's date, previously unchallenged, Schottenloher (pp. 147–148) assigned Werner's birth to 1466. The date 1468 was verified in the British Library's copy of Krentzheim by my former student, Mrs. Theano N. Paschalides.

[2] *Die Matrikel der Ludwig-Maximilians-Universität Ingolstadt-Landshut-München*, ed. Götz von Polnitz (Munich, 1937–1941), I, 134.

[3] Kressel, p. 289.

[4] Kist.

[5] Werner's 1514 collection of essays, sig. d 1^r, as quoted in *Abhandlungen zur Geschichte der mathematischen Wissenschaften*, 1907, 24:1 (cited hereafter as "Abh"), p. 150, n. 3.

[6] Austrian National Library, Vienna, Codex 10650, fol. 87^r: *ea bibliothecula, quam Romae olim habueram*, as quoted by Schottenloher, p. 149, n. 2.

[7] Hermann Grauert, "Dante in Deutschland," *Historisch-politische Blätter für das katholische Deutschland*, 1897, 120:339–340.

[8] This is the date given in line 1 of Kressel's Plate 1 (facing p. 288), which presents a photocopy of the first seven lines of the folio cited in n. 6, above. On the other hand, the notation in *Tabulae codicum manu scriptorum praeter graecos et orientales in Bibliotheca palatina vindobonensi asservatorum*, VI (Vienna, 1873), 217, is: "written on 14 February 1498." Is this the date when the horoscope was copied? Schottenloher's assertion (p. 149, n. 2) that Vienna Codex 10650, fol. 87, is in Werner's own handwriting should be tested against Werner's autograph, *Historical Journal*, cited in n. 17, below.

[9] Kist.

[10] Karl Morneweg, *Johann von Dalberg* (Heidelberg, 1887), pp. 286–287, 330.

[11] Kressel, pp. 289–290.

of the next few months he was appointed to a secular priesthood in a suburb of Nuremberg. Later he became a vicar of St. John's, which was then also in a suburb, and he was assigned to its infirmary (originally a leprosarium). This was the position which he occupied for the rest of his life.

Its duties were not onerous enough to keep him from his beloved intellectual pursuits.[12] His talents made a deep impression on a mathematician (who was also the Imperial Historiographer) when he visited Nuremberg in the spring and summer of 1512.[13] On 19 December 1513 the Historiographer elicited from the Holy Roman Emperor a copyright for a series of works which Werner hoped to publish. This copyright, in which the emperor calls Werner "our chaplain," was then printed in the front matter preceding Werner's first collection of mathematical studies, published in Nuremberg on 4 November 1514.[14] In the Dedication of one of these studies to a Nuremberg patrician, Werner thanked him

> for lending me no small sum of money to enable me to publish more opportunely the little works [containing] my thoughts and discoveries.[15]

The number of copies printed in 1514 exceeded the demand, and the unsold surplus was later modified in an expanded posthumous edition (Ingolstadt, 1533).[16] Werner did not live to see his second collection of mathematical studies in print (Nuremberg, 1522). For in that year, at the age of 54, he succumbed to the plague. His (unpublished) autograph *Historical Journal from 1506 to 1521* breaks off abruptly on 7 August 1521.[17] While others were free to flee the stricken city, Werner's official duties required him to attend to the sick and dying. A handwritten entry on the last sheet of the copy of his 1522 collection in the State Library in Munich notes: "While this copy was in press, the author of the work died."[18]

Before (and even after) the discovery of this entry, which definitely places Werner's death in 1522, his demise was dated in 1528 by confusion with another Johann Werner. Carl Christian Hirsch's *Lebensbeschreibungen aller Herren Geistlichen* (Nuremberg, 1756) presented the biographies of all the Nuremberg clergymen since the Lutheran Reformation. In connection with St. John's, Hirsch listed (II, 467) a "Johann Werner," who "arrived in 1525, died or left in 1528." Our Johann Werner, however, describes himself as "vicar or rector of the chapel of St. John the Baptist and St. John the Evangelist in Nuremberg" at the beginning of his *Historical Journal from 1506 to 1521*. Hence, Hirsch's Johann Werner, who arrived in 1525 and departed (or died) in 1528, is surely not our Johann Werner, who is recorded as a clergyman in St. John's as early as 1 October 1508.[19]

The correctness of the entry in the Munich copy of Werner's 1522 collection of essays is confirmed by the *Great Burial Register of St. Sebald 1517–1572*, an unpublished manuscript (No. 6277) preserved in the Germanic Museum in Nuremberg. On folio 15^r this *Register* records the burials in 1522 of 25 persons between 12 March and 11 June. No. 20

---

[12] Werner's autograph description of his obligations and possessions was published by Herold, pp. 106–108. Werner locates St. John's *in suburbio Nurembergensi* (p. 106/last line).

[13] Kressel, p. 298.

[14] G. W. Panzer, *Annales typographici*, VII (Nuremberg, 1799), p. 454, no. 104.

[15] Werner's Dedication was reprinted in Briefwechsel, II, 476–481, with the quoted passage at 479/7–9.

[16] Ernst Zinner, *Geschichte und Bibliographie der astronomischen Literatur in Deutschland zur Zeit der Renaissance*, 2nd ed. (Stuttgart, 1964), p. 148, no. 1019; p. 179, no. 1516.

[17] Bachmann, p. 331.

[18] MK, p. 433, n. 5.

[19] Kist; Kressel, p. 290.

is "Hans Werner, vicar at St. John's." This compact designation (*vikary zu S. Johanns*) matches our Johann (or Hans) Werner's somewhat fuller self-description as *vicarius ... cappellae beatorum Johannis Baptistae et Johannis Evangelistae* at the beginning of his *Historical Journal*. As No. 20 in the list of 25 burials, our Johann Werner may have died toward the end of May 1522. The statement that he died "before 5 May 1522"[20] arose from a confusion with 5 May 1523, the day on which the processing of Werner's (lost) will was terminated.[21]

Nearly a quarter-century after Werner's death, his contribution to the nascent science of meteorology, containing his weather observations during the years 1513–1520, was published posthumously.[22] Less fortunate was his translation of Euclid's *Elements* into German, the first in that language. With Werner's aforementioned creditor acting as intermediary, a Nuremberg manufacturer of cannon, interested in the education of his son, arranged to have Werner

> translate into the German language the 15 Books of Euclid. To make all the theorems more intelligible, however, Werner was required to provide a clear example for every theorem. For every Book he was accordingly given 10 gulden.[23] But neither the trustee nor the heirs know the whereabouts of these Books.

So reported the author of *Information about Nuremberg Artists and Craftsmen*, which was dedicated on 16 October 1547.[24] Fifteen years later, when Books I–VI of Euclid were published in German translation, the title page claimed that the "like had never before been seen in the German language."[25]

The fate of two other works by Werner is linked with Copernicus' *Sides and Angles of Triangles*,[26] which was published in Wittenberg in 1542[27] by Copernicus' disciple, Rheticus. He dedicated this little work to a Nuremberg mathematician George Hartmann (1489–1564), with the intention, as Rheticus explained to Hartmann,

> of stimulating you to publish whatever you have of this nature, whether ancient or recent.[28]

After Werner's death in 1522, Hartmann had rescued from oblivion Werner's manuscripts on *Spherical Triangles* and *Meteoroscopes* (as certain types of observational instruments were called in those days). But instead of responding to Rheticus' stimulation by publishing these two manuscripts, Hartmann turned them over to Rheticus when he arrived in Nuremberg.

---

[20] Kist; Bachmann, pp. 317, 331.

[21] Herold, p. 108; Briefwechsel, I, 437, n. 13; Kressel, p. 302. Werner was still alive on 11 January 1522, when he dated the preface of his 1522 collection of essays.

[22] *Canones ... complectentes praecepta et observationes de mutatione aurae*, ed. Johann Schöner (Nuremberg, 1546).

[23] The total was therefore 150 gulden, rather than "100 taler," as in DSB, XIV, 274, or "15 Gulden," as in Ernst Zinner, "Die fränkische Sternkunde im 11. bis 16. Jahrhundert," p. 27/20, in *XXVII. Bericht der naturforschenden Gesellschaft in Bamberg* (1934).

[24] Johann Neudörfer, *Nachrichten von Künstlern und Werkleuten in Nürnberg*, ed. G. W. K. Lochner (Vienna, 1875), p. 48; Quellenschriften für Kunstgeschichte und Kunsttechnik des Mittelalters und der Renaissance, X; reprint Osnabrück: Zeller, 1970.

[25] *Die sechs erste Bücher Euclidis* (Basel, 1562).

[26] *De lateribus et angulis triangulorum*, "one of the finest works in the history of the subject, marked out by its lucid derivations of the essential formulae," says John David North, "Werner, Apian, Blagrave, and the Meteoroscope," *British Journal for the History of Science*, 1966–1967, *3*:59.

[27] North errs in saying (p. 59) that it was "published posthumously," since Copernicus died on 24 May 1543.

[28] Rheticus' Dedication was reprinted by Burmeister, III, 45–47, with our passage at p. 46/6 up–5 up.

For at the beginning of May 1542, Rheticus took a leave of absence from his teaching post at the University of Wittenberg in order to go to Nuremberg to supervise the printing of Copernicus' *Revolutions*. This task was so absorbing that while it lasted Rheticus could do nothing with the Werner manuscripts. Afterwards, when his desk was clear, he was troubled by a doubt about their authenticity, since he knew that Werner's

> first and second meteoroscopes had been published by certain persons as their own. For, as I later heard from George [Hartmann], they had expounded the theory and construction of the device which they had examined in his home. ... While uncertain whether our [Werner] or they are the authors of the *Meteoroscopes*, I see nothing further appearing on this subject. I therefore considered that Werner's *Meteoroscopes* should not be withheld any longer, especially since he discussed two additional meteoroscopes, far superior to the first [two].[29]

Rheticus discreetly refers to "certain persons" (*quidam*, in the plural). In reality, he had in mind only the editor of the 1533 re-issue of Werner's 1514 collection of essays, an editor whose plagiarizing proclivities are understood better today. After receiving the Werner manuscripts from Hartmann in 1542, Rheticus waited to see whether anything further on the subject would appear in print. After fifteen years, during which nothing was published, in 1557 Rheticus finally overcame his doubt about the authenticity of the manuscripts, and determined to go ahead in the conviction that Werner was their true author.

At that time Rheticus was living in Cracow, then the capital of the kingdom of Poland. A Cracow printer undertook to issue Werner's two treatises, "now published for the first time, under the supervision and editorship of George Joachim Rheticus." But instead of Werner's two treatises, only the title page,[30] the royal and imperial copyright, and Rheticus' Introduction were actually printed. On further reflection Rheticus decided not to publish the treatises which Werner, "prevented by death, could not polish and perfect to the degree required by his subject matter."[31] A copy of the truncated publication is preserved in the library of the University of Cracow. In the lower right-hand corner of the title page, Jan Brożek (1585-1652), a professor at the University of Cracow and indefatigable collector of documents pertaining to Copernicus and Copernicus' disciple Rheticus, wrote:

> Only this Introduction [was] printed in Cracow. He [Rheticus] had planned to send the rest of the work to Germany, as I learned from a letter written in Rheticus' own handwriting to Wolf.[32] I do not yet know whether [the rest] was sent and printed.[33]

---

[29] Rheticus' Introduction to Werner's *Spherical Triangles and Meteoroscopes* (Cracow, 1557), sig. a 4ᵛ/2-12; facsimile in Abh; reprint in Burmeister, III, 132-140, with a translation into German at III, 140-150, and our passage at 136/3 up-137/6.

[30] North errs (p. 62) in saying that "Rheticus set up a few pages in type, around the year 1557."

[31] Rheticus' Introduction, Abh, sig. a 4ᵛ/15-16; Burmeister, III, 137/10-11.

[32] Rheticus' letter to Caspar Wolf in Zürich has not been found. Presumably Brożek had access to it because it had been brought back to Cracow by Anthony Schneeberger (1530-1581), a doctor from Zürich who taught medicine at the University of Cracow and was a close friend of Rheticus. When the latter left Cracow for Hungary, he gave a medical book to Schneeberger (L. A. Birkenmajer, *Stromata Copernicana*, p. 368; Burmeister, II, 43, no. 15). Another book which passed from Rheticus to Schneeberger later became the property of Brożek (Birkenmajer, *Stromata*, p. 367). After Schneeberger's death, part of his library passed into the hands of a Cracow professor who was on good terms with Brożek and showed him many letters of interest to him (Hilfstein, StC XXI, 32).

[33] Facsimile, Abh, title page.

Rheticus' correspondent, Caspar Wolf (1525–1601), was the municipal physician of Zürich, where a posthumous edition of Conrad Gesner's *Bibliotheca universalis* was published in 1574 under the title *Bibliotheca instituta et collecta*. At page 424, the editor stated:

> George Joachim Rheticus published Werner's four Books on triangles and six Books on meteoroscopes (unless I am mistaken).

The editor[34] was mistaken, because he was putting into print what he had learned from Rheticus' (lost) letter to Wolf, written while Rheticus still intended to publish Werner's two manuscripts. Evidently Rheticus did not inform his Zürich correspondent(s) that in the end he decided not to send "the rest of the work to Germany." As a result, Werner's treatises were not published in the sixteenth century. Indeed, it was only in the present century that they were finally printed (in Abh).

In his Introduction to his abortive edition of Werner, Rheticus says absolutely nothing about Copernicus' *Letter against Werner*, to which Werner could not reply because he died before it was written. Rheticus had good reason not to mention Copernicus' *Letter against Werner*, which denounced Werner as "deranged" and "foolish," and his theory as a "hallucination."[35] On the other hand, Rheticus' Introduction said about Werner's discussion of spherical triangles:

> No expert could add anything that would not be superfluous, nor remove anything without harming the entire work, since nothing in it is superfluous. ... In what he did, he carries out his plan so wisely that with his skillful learning he easily surpasses the writings of all his predecessors in this field.[36]

Into this resounding paean of praise, Copernicus' *Letter against Werner* could only have interjected a discordant note.

WAPOWSKI AND COPERNICUS' *LETTER AGAINST WERNER*

That discordant note was provoked by Werner's "Motion of the Eighth Sphere," at sig. k 1^r–v 4^r in his second collection of essays (Nuremberg, 1522). Werner's "Motion of the Eighth Sphere" came to the notice of Bernard Wapowski (c. 1475–1535), the founder of Polish scientific cartography.[37] Nowadays a book like Werner's second collection of essays would be reviewed in appropriate scientific periodicals. But these had not yet come into existence. In their absence Wapowski sought the opinion of his old friend Copernicus, who had been his fellow-student at the University of Cracow.

A year and a half after Copernicus, in the summer semester of 1493 Wapowski was admitted to the University of Cracow, as is indicated by the official entry: *Bernhardus Stanislai de*

---

[34] Josias Simler (1530–1576), not Wolf (as in Abh, p. 158, and Burmeister, I, 159). Rheticus first met Gesner when they were both young fellow-students in Zürich (Burmeister, I, 20). Two decades later Rheticus studied medicine with Gesner for a short while, and contributed his "Recent Invention" to one of Gesner's publications (ibid., I, 98; II, 71, no. 33).

[35] *Letter against Werner*, near nn. 93, 182, below.

[36] Abh, sig. a 5/3–8; Burmeister, III, 137/9 up–4 up.

[37] Karol Buczek, "Bernard Wapowski, der Gründer der polnischen Kartographie," *Comptes rendus du Congrès international de géographie Varsovie 1934*, IV (Warsaw, 1938), 61–62.

Radochonycze dioc. Cracoviensis[38] (Bernard, son of Stanislaus, from Radochońce in the diocese of Cracow). In accordance with the usual practice at that time, no surname was mentioned. Hence for a while Wapowski was confused with another Cracow student whose given name was also Bernard. This Bernard was awarded a bachelor's degree in 1493, and two years later a master's degree.[39] But Bernard Wapowski left the University of Cracow without having received any degree, just like Copernicus. Again like Copernicus, Wapowski went on to study law at the University of Bologna, which awarded him a doctorate in canon and civil law on 6 February 1505.[40] From Bologna, in the spring of 1505 he moved to Rome, where he joined the staff of the Polish Embassy. He was still in Rome on 3 February 1511, when he wrote with his own hand in the official register of a religious order that he had become a member.[41] After returning to his native country, he settled in Cracow and became a secretary of the king of Poland.

Knowing that Copernicus in far-off Frombork was not in touch with foreign cultural centers and developments, as he himself was, Wapowski sent Copernicus a copy of Werner's "Motion of the Eighth Sphere," remarking that it was widely praised, and requesting the astronomer's judgment.

What Wapowski sent Copernicus was not a complete copy of Werner's 1522 collection of essays, but only "The Motion of the Eighth Sphere" at sig. k 1^r–v 4^r. For, the preceding essays dealt, not with any astronomical problem, but with three purely mathematical topics: conic sections, duplication of the cube, division of a sphere in a given ratio. It has been asserted that Copernicus must have possessed a complete copy of Werner's 1522 collection of essays.[42] But there is no positive evidence to support that assertion, while its contrary is implied in Copernicus' acknowledgment to Wapowski at the beginning of the *Letter against Werner*:

> Some time ago, my dear Bernard, you sent me a little treatise (*opusculum*) on "The Motion of the Eighth Sphere," published by Johann Werner of Nuremberg. Your Reverence stated that the work was widely praised and asked me to give you my opinion of it.

Copernicus says nothing about having received Werner's entire collection of essays, but confines his acknowledgment to "The Motion of the Eighth Sphere."

When did Werner's essay (or 1522 collection of essays) reach Wapowski in Cracow? When did the royal secretary send "The Motion of the Eighth Sphere" to Frombork? Our only clue is Copernicus' use of an imprecise word at the outset of his reply, as just quoted:

---

[38] Album, II, 24.

[39] PI¹, 149; PI², 221; PII, 172.

[40] *Primus liber secretus juris pontificii ab anno 1377 ad annum 1528*, fol. 180, as cited in MK, p. 457, from the manuscript in Archivio di Stato, Bologna. The law degrees awarded from 1378 to 1450 were included in *Il "Liber secretus iuris caesarei" dell'Università di Bologna* (Bologna, 1938–1942), ed. Albano Sorbelli. While Wapowski was studying law at the University of Bologna, he did not join the German Nation there. Its statutes admitted students "whose native language is German, even though their home is elsewhere." Hence Copernicus was admitted on 6 January 1497 (1496, according to the local calendar; *Acta nationis germanicae universitatis bononiensis*, eds. Ernst Friedländer, Carlo Malagola [Berlin, 1887], p. 248/43; language stipulation, p. 4/6). There is no documentary evidence that Wapowski arrived in Bologna before Copernicus left without a law degree in the autumn of 1500 (TCT, p. 326). Hence there is no tangible basis for the assertion (StC XVIII, 167, n. 3) that Copernicus and Wapowski were fellow-students at the University of Bologna as they had been at the University of Cracow.

[41] PI², 221.

[42] Z, p. 156. A complete copy of Werner's 1522 collection of essays was presented by Rheticus to his former teacher A. P. Gasser, when he visited his birthplace in April 1539 (MK, p. 584). That copy is now in the Jagiello-

> Some time ago (*pridem*), my dear Bernard, you sent me a little treatise on "The Motion of the Eighth Sphere."

Evidently Copernicus did not dash off an instant reply. Instead, on 3 June 1524 he sent Wapowski a carefully considered report, which has come to be known in the literature as his *Letter against Werner*.[43]

Cast in the form of a private letter to Wapowski, it was not intended for publication. Nevertheless it has been called an "open letter," "intended for the public, as is shown by its content and form," with the author "permitting the recipient to give it wider distribution."[44] But nothing in the *Letter against Werner* authorizes further distribution. Nothing in its form shows that it was an open letter, intended for the public. Had that been Copernicus' purpose, he would surely have refrained from using such harsh language about Werner, whose work had been widely praised by others, as he was informed by Wapowski. Copernicus' *Letter against Werner* was a private communication to Wapowski, not intended for the general public.

Did Copernicus depart from his usual practice and keep a copy of his *Letter against Werner*? For if Werner ever saw it and chose to reply, the exact wording of the original would be indispensable for the purpose of drafting a rebuttal. In the event, this precaution proved to be unnecessary. For about two years before Copernicus wrote the *Letter against Werner*, the latter had died, a fact of which Copernicus was unaware.[45]

For his part, Wapowski deemed the scientific content of the *Letter against Werner* to be too important to be shut up in his personal files. For in the first place, Copernicus' *Letter against Werner* made a notable contribution to the emerging discipline of chronology by correcting Werner's woeful error of eleven years in dating Ptolemy's catalog of fixed stars, a conspicuous landmark in the history of astronomy. Secondly, the *Letter against Werner* insisted on technical terms being defined and used with precision: in a recurring periodic nonuniform motion, the mean velocity cannot also be the slowest, although Werner would have it both ways. Lastly, the *Letter against Werner* maintained the primacy of fact over theory. Having constructed a theory in conflict with ancient observations, Werner concluded that the observations were wrong. On the contrary, Copernicus replied, the theory is wrong, being inconsistent with itself to boot. With regard to our attitude toward the ancient scientists, Copernicus writes that we must

> hold fast to their observations, bequeathed like a legacy. But if anyone, holding fast to his own view, thinks that they are untrustworthy in this regard, surely the gates of this art are closed to him. Lying in front of the entrance, he will dream the dreams of the deranged about the motion of the eighth sphere, and receive what he deserves for supposing that he should support his own hallucination by defaming the ancients.[46]

nian Library, Cracow; see Anna Lewicka-Kamińska, *Nieznane ekslibrisy polskie XVI wieku w Bibliotece Jagiellońskiej* (Cracow, 1974), p. 29; p. 36, n. 76. Burmeister (II, 40; repeated in Burmeister, *Achilles Pirmin Gasser* [Wiesbaden, 1970], I, 73) mistakenly reports Rheticus' presentation copy as consisting, not of Werner's entire 1522 collection of essays, but only of "De motu octavae sphaerae tractatus duo." It was the first of these "Two Treatises on the Motion of the Eighth Sphere" that was sent by Wapowski to Copernicus.

[43] As a private communication not intended for publication, Copernicus' letter to Wapowski of course bore no title. But that highly appropriate title Copernicus' *Letter against Werner* (*Epistola Copernici contra Wernerum*) was introduced by the manuscript which we shall call "B."

[44] PI², 221, 223.

[45] See Introduction at n. 18, above.

[46] *Letter against Werner*, at n. 90, below.

It has recently become somewhat fashionable to link Copernicus with Pythagoreanism, neo-Pythagoreanism, Neoplatonism, and hermeticism. The evidence adduced for such linkage would easily pass through the eye of a needle without noticeably deforming the needle. On the other hand, Copernicus' familiarity with the writings of Aristotle, that well-known critic of the Pythagoreans and Plato, is quite apparent in his *Letter against Werner*. Its opening paragraph quotes from the *Metaphysics*, mentioning Aristotle by name. Later on, it echoes a striking statement in Aristotle's *Physics*, without even alluding to that work or its author.[47] Such familiarity with Aristotle's treatises does not of course make Copernicus an Aristotelian in the sense that he regarded the Stagirite as infallible. On the contrary, where he detected a flaw in Aristotle, as in the Stagirite's division of simple motion into three mutually exclusive types, he did not hesitate to correct it.[48] But he did not undertake to overthrow Aristotelianism, as he did the Ptolemaic astronomy. On the other hand, what he believed was sound in both systems, he retained with gratitude and affection, an attitude which some of our contemporaries would do well to consider.

MANUSCRIPTS OF THE *LETTER AGAINST WERNER*

Although Wapowski did not try to have Copernicus' *Letter against Werner* printed, he felt that his contemporaries should not be deprived of the opportunity to read it. For that reason (perhaps without consulting the author) he permitted it to be copied by hand. One such copy was in turn "copied in Prague from Mr. Hájek's copy in the month of January 1531."[49] This date has been questioned[50] on the ground that Thaddeus Hájek (1525–1600), who gave Brahe a copy of *Commentariolus*,[51] was only six years old when the *Letter against Werner* was copied in Prague in 1531 "from Mr. Hájek's copy." But Thaddeus' father, Simon (c. 1489–1551), provided his son with a living model of the avid collector of valuable manuscripts. Simon received the baccalaureate degree from the University of Prague in the autumn of 1509.[52] He did not thereafter become a medical doctor, like his more illustrious son Thaddeus, who was later called "Dr. Hájek." By contrast, the 1531 notation refers, not to Dr. Hájek's copy, but to *D. Hagecii exemplari*, where *D.* is the abbreviation for *Dominus*, then the Latin equivalent of our simple Mr. We still do not know how a copy of the *Letter against Werner* was acquired by Mr. Simon Hájek in Prague in or before January 1531. But we recall that Wapowski in Cracow, acting as secretary of the king of Poland, was frequently in touch with other capitals, including Prague, the capital of the kingdom of Bohemia.

A sixteenth-century journal of Bohemian affairs records events that occurred day by day throughout the calendar year. Under 29 October, it reports the burial

> on this day in 1551, as was noted by Kollin, of Thaddeus Hájek's father, Simon Hájek, bachelor of liberal arts of the University of Prague, thereafter a citizen of Prague, a pious,

---

[47] *Letter against Werner*, at n. 199, below.

[48] NCCW, II, 17/24–27. For an excellent discussion of Copernicus' philosophical outlook, see Aleksander Birkenmajer, "Copernic philosophe," StC IV, 612–646.

[49] PII, 171/8. The Prague copy was made in January 1531, 6 $1/2$ years after Copernicus sent the autograph to Wapowski on 3 June 1524. This interval of 6 $1/2$ years was mistakenly reduced to $1/2$ year by Polkowski, III, 308.

[50] MK, pp. 503–504.

[51] See *Commentariolus*, Introduction, at n. 5.

[52] *Liber decanorum facultatis philosophicae universitatis pragensis*, II (Prague, 1832), 228. A xerox copy of this page was kindly sent to me by Zdeněk Horský of the Czechoslovak Academy of Sciences, Astronomical Section.

upright, and learned man. He was a very careful student of the character of our native language, as is shown by the Latin tablet which he wrote and had printed. With the addition of some examples of our tongue and observations about genders, numbers, and in particular the two [forms of the] participles etc., Kollin inserted it in his grammar because it most decidedly deserved to be known, especially by clergymen.[53]

In the Czech language, the active participles, present and past, each have two forms, one adjectival and the other adverbial, while the past passive participle also has two forms, one determinate and the other indeterminate. Simon Hájek's tablet was incorporated in Matthaeus Collinus (Matouš Kollin), *Elementarius libellus in lingua latina et boemica* (Prague, 1557), six years after Simon Hájek's death.

On 7 September 1504 Simon Hájek's mother had bought a house near the Bethlehem Church in Prague's Old Town for herself and her son Simon. After his death in 1551, his son Thaddeus inherited it, and in 1584 he made it available to the visiting English mystic John Dee (1527–1608), who described the start of his stay, together with his three companions, in Prague as follows:

Augusti 15. *Wednesday*, we began ... in the excellent little Stove, or Study of D. *Hageck* his house lent me, by *Bethlem* in old *Prage*. Which Study seemed in times past (Anno 1518.) to have been the Study of some Student, or A — skilfull of the holy Stone: a name was in divers places of the Study, noted in letters of Gold, and Silver, *Simon Baccalaureus Pragensis*.[54]

Dee's reference to "the holy Stone" reminds us that his presence in Prague was not connected with collecting manuscripts. But that had been the primary purpose of his fellow-countryman Henry Savile (1549–1622), who later founded the Savilian professorships of geometry and astronomy at Oxford, the first such chairs in any English university. "In 1578 he travelled on the continent, where he made the acquaintance of the most eminent scholars of his time, and collected a number of manuscripts."[55] When in Prague, Savile visited Thaddeus Hájek.[56] During this visit a copy of the *Letter against Werner* was made for him. Since the copyist who wrote the first folio produced a barely legible sheet, he was replaced by a scribe who wrote the remaining four folios quite clearly.[57] These five folios are now numbered 28–32 in the composite MS Savile 47 in the Bodleian Library in Oxford.[58] Savile's copy of the *Letter against Werner* will be cited hereafter as "Ox."

---

[53] Prokop Lupáč, *Rerum boemicarum ephemeris, sive Kalendarium historicum* (Prague, 1584), sig. k 1ʳ. Professor Edward Grant of Indiana University kindly extracted this passage from the copy of Lupáč in the Lilly Library.

[54] *A True and Faithful Relation of What passed for Many Yeers between Dr. John Dee ... and Some Spirits*, ed. Meric Casaubon (London, 1659; reprinted, London: Askin, 1974), p. 212. A gap in Casaubon's edition was filled by C. H. Josten, "An Unknown Chapter in the Life of John Dee," *Journal of the Warburg and Courtauld Institutes*, 1965, 28: 223–257. The purchase of the Hájek house for Simon, and his bequeathal of it to Thaddeus, are recorded in Josef Teige, *Základy starého místopisu Pražského*, II (Prague, 1915), p. 874, no. 7; p. 875, no. 16. In the Czech language *háj* means a grove, and the diminutive *hájek* means a "little grove"; the house was in a little grove.

[55] *Dictionary of National Biography*, XVII (reprinted, 1967–1968), 856.

[56] Vetter, p. 292.

[57] According to L. A. Birkenmajer, who first uncovered the Savile copy (MK, p. 493).

[58] *A Summary Catalogue of Western Manuscripts in the Bodleian Library at Oxford*, Vol. II, Part II (Oxford, 1937), p. 1112, no. 6593; Vol. III (Oxford, 1895), p. 472, no. 15698 (now MS Smith 93), p. 173, a copy made

Thaddeus Hájek must be given credit not only for Ox but also for a considerable dissemination of the *Letter against Werner* even before Savile's arrival in Prague. That dissemination was made possible by the reemergence of the autograph. For, the copy of the *Letter against Werner* which is now in the Austrian National Library in Vienna (and which will be cited hereafter as "V") describes itself as being "Among the first transcriptions after the autograph" (*Ex primis post* αὐτόγραφον *lituris*), and gives the date of its completion as 30 March 1575.[59] Some seven months later Brahe received a copy of Copernicus' *Commentariolus* from Thaddeus Hájek.[60] Unfortunately Brahe does not say who gave him a copy of Copernicus' *Letter against Werner* (Brahe's copy will be cited hereafter as "Br"). But he does say: "The copy in my possession was transmitted to me after a second or third transcription from Copernicus' autograph."[61] Br was copied either directly from V as a second transcription, or as a third transcription from a copy of V, because Br misdates the *Letter against Werner* in 1534,[62] a misdating found only in V (folio 9^v/3 up) and in no other extant manuscript of the *Letter against Werner*. Moreover, Br gives Wapowski's name in the distorted form "Vapoushy,"[63] which occurs very prominently in the middle of the very first line of V. Although many of Brahe's papers after his death passed into what is now the Austrian National Library in Vienna, Br has not been found there.

V and Br, as first, second, and/or third transcriptions from Copernicus' autograph, were made in or about 1575. Why were transcriptions suddenly being made at that time from Copernicus' autograph after it had dropped out of sight for four decades following Wapowski's death on 21 November 1535?[64] One possibility suggests itself. Rheticus' library, after his death on 4 December 1574,[65] was in part bequeathed to Thaddeus Hájek.[66] From 1554[67] to 1571,[68] Rheticus had resided in Cracow, where he was recognized as Copernicus' only disciple. During those seventeen years, was he given Copernicus' autograph of the *Letter against Werner*, preserved among Wapowski's papers in Cracow? If so, did the autograph pass to Thaddeus Hájek as part of Rheticus' bequest? If so, it was from the autograph in Thaddeus Hájek's possession that the first, second, and/or third transcriptions emanated in 1575, following Rheticus' death on 4 December 1574 and the arrival of his bequest in Prague. If this was the actual course of events, the autograph passed from Copernicus to Wapowski to an unknown intermediary or intermediaries to Rheticus to Thaddeus Hájek, and then disappeared from view with the rest of Hájek's papers.

While Rheticus was living in Cracow, he received a visit from Johann Praetorius (1537–1616), who had studied at the University of Wittenberg. The professor of theology, a close friend and active correspondent of Rheticus, had made his personal copy of *Revolutions*

---

in the 17th century; Vol. V (Oxford, 1905), p. 202, no. 26244, fol. 59, a copy of MS Savile 47 by Stephen Peter Rigaud (1774–1839).

[59] V, folio 9^v/last line; MCV, I, 21, where Maximilian Curtze reported his discovery of V.

[60] See *Commentariolus*, Introduction, at n. 5, above.

[61] TB IV, 292/19–20.

[62] Ibid., 292/6.

[63] Ibid., 292/5.

[64] This date was recorded on Wapowski's tombstone (PI², 221).

[65] Burmeister, I, 176.

[66] TB VII, 333/39–40.

[67] Burmeister, I, 131.

[68] Burmeister's statement (I, 134) that Rheticus remained in Cracow until 1574 must be revised in the light of Erna Hilfstein, *Starowolski's Biographies of Copernicus*, StC XXI, 27.

available to Praetorius.[69] While the latter was in Cracow in 1569, he was allowed to make a handwritten copy of Rheticus' *Table of Triangles*.[70] At the same time, a copy of Copernicus' *Letter against Werner* was made for Praetorius,[71] presumably from the autograph then in Rheticus' possession. From 1576 to 1616 Praetorius taught mathematics at the University of Altdorf. After his death, he was succeeded by his pupil Peter Saxonius (1591-1625), who acquired his library, and whose library in turn passed in part into the municipal library of Schweinfurt.[72] The Schweinfurt copy of the *Letter against Werner* is the oldest surviving manuscript of that work; we shall call it "Sch."

While Praetorius was in Cracow, in 1571 he was summoned by the University of Wittenberg to teach mathematics there. Having brought Sch with him, he permitted an (as yet unidentified) professor of astronomy to make a copy of it. That copy was later found in Berlin (and will therefore be designated hereafter as "B") when European libraries were being searched for missing Copernican documents in connection with the 1843 tercentenary celebration of the first printing of *Revolutions* (1543). B, by no means a slavish copy of Sch, was intended for use in lectures to advanced students of astronomy. Thus, on folio 8^v, B arranged lines 3–7 in such a way as to make room for the following exhortation:

> Read, read, these rules about uniformatization in Reinhold or Peucer. How the apparent and mean motions are equal in the case of the sun is demonstrated by Nunes and Regiomontanus. The prosthaphaereses are equal in the vicinity of the apogees and perigees, but not at all equal in the vicinity of the middle longitudes.

Of the four astronomers named in this exhortation, the latest is Conrad Peucer (1525-1602). In his *Astronomical Hypotheses or Theories of the Planets* (Wittenberg, 1571) Peucer defined the term "prosthaphaeresis" as

> the difference by which a mean motion diverges from the true, nonuniform, and apparent [motion], ... the arc commonly called "equation," in Greek *prosthaphaeresis*, an expression composed of *prosthesis* and *aphaeresis*, that is, its diverse use, since in determining the true motions it is sometimes added to the mean motion, as supplied by the tables, and sometimes it is subtracted in order to arrive at the true motion (p. 52/3 up–p. 53/8).

Peucer also explained where the prosthaphaereses are equal:

> When a planet is located at points (whether of an eccentric or a concentric or the ecliptic) which are equally distant to either side of the apogee or perigee in opposite semicircles, or at points of the eccentric which are opposite each other along a straight line passing through the concentric's center, it has its equations or prosthaphaereses equal (p. 97/12–19).

On the other hand, at the midpoints between the apogee and the perigee, the prosthaphaereses reach their maximum. Thus, in the theory of the sun,

> The prosthaphaeresis of the center is the difference between both apogees, the true and the mean. ... This difference is zero when the epicycle's center occupies the eccentric's

---

[69] Z, p. 454/17–19.
[70] Burmeister, II, 23, no. 6.
[71] MK, p. 619, D. The handwriting of the copy is neither Rheticus' nor Praetorius' (Z, p. 428/23–25).
[72] Z, p. 435, no. 160; Schweinfurt, Stadtarchiv, MS H 87, fol. 9^r–13^v, first uncovered by Ernst Zinner.

apogee or perigee, because at those times the apogees are not separated nor are the [apsidal] lines distinct, but coalesce at one point in the heavens. The greatest [prosthaphaeresis], however, occurs at the midpoints [between the apogee and the perigee; p. 260/10-19].

Unmistakably, therefore, B exhorted his students to read Peucer's recently published *Astronomical Hypotheses*.

B's descent from Sch is established by a numerical error in Sch which equated (fol. 9^v/6 up-5 up) a certain day in the Egyptian year CXL with the corresponding day in the year of the Romans CXXX. This incorrect numeral CXXX was copied from Sch by B, who added in parenthesis that something "is missing" (*de est*; fol. 8^r/3 up). While one other manuscript (V) of the *Letter against Werner* has the same mistake, it writes the number 130 in Arabic numerals (folio 2^v/8 up). All the other extant manuscripts of the *Letter against Werner* have the correct number 139 in Arabic numerals. Hence it was from Sch that B took the wrong number, while calling attention to its incompleteness (see n. 51 on the *Letter against Werner*).

B, descended from Sch, in turn affected Sch. In other words, the borrower not only returned Sch to Praetorius, but also made B available to him. This counterinfluence of B on Sch is indicated by B's uncertainty about the number 14. Above that number (folio 8^v/3) B wrote "perhaps 54." That uncertainty (*fortassis 54*) was transferred to Sch (fol. 10^r/1) above the line by a second hand (Praetorius'?; see n. 59 on the *Letter against Werner*). B's uncertainty indicates his unfamiliarity with the dating of two observations in accordance with the ancient Egyptian calendar.

By contrast with Brahe, who possessed a copy of the *Letter against Werner*, the earliest well-informed biographer of Copernicus did not. The biography of Copernicus by Starowolski went through two stages. In the first version (1625), which he wrote by himself, Starowolski made no mention of the *Letter against Werner*. Two years later, however, in his second version Starowolski talked about the "Letter Concerning the Motion of the Eighth Sphere."[73] In preparing this second version, Starowolski benefited from the help of Brożek, who knew about the *Letter against Werner*. From the extensive collection of Copernican correspondence available to him Brożek selected two valuable items for publication, but the *Letter against Werner* was not one of them, so that it remained in manuscript. Relevant manuscripts were unavailable to the next important biographer of Copernicus, Pierre Gassendi (1592-1655), who relied exclusively on printed materials.[74] Since the first printed edition of the *Letter against Werner* was still two centuries in the future, Gassendi lamented that he could not put his hands on the *Letter against Werner*.[75] The very wording of Gassendi's lament plainly indicates that he obtained his information about the *Letter against Werner* from the 1627 edition of Starowolski.

Another important writer without a copy of the *Letter against Werner* was Johann Gabriel

---

[73] Hilfstein, StC XXI, 15.

[74] Gassendi, *Life of Tycho Brahe* (Paris, 1654), followed by *Life of Nicholas Copernicus*:

> The little I knew about the man [Copernicus] was what was easily learned by any reader of the books by him or by others, since I too at this time, a century after his death, had no access to other information about him or any other source (pp. 3-4 of the separate pagination).

[75] Ibid., p. 39:

> Quite a large number [of Copernicus' letters] are said to have been recently in the hands of Jan Brożek, professor of astronomy in the University of Cracow. Whether Brożek published any [of them] is not clear to me. At the least, I should like [to have] that letter which Copernicus is said to have written about "The Motion of the Eighth Sphere" to Wapowski, Cantor of Cracow.

Doppelmayr (1671–1750), whose *Historical Information about Nuremberg Mathematicians and Artists* (Nuremberg, 1730) contains a report on Werner that is still valuable.[76]

Even more instructive is the case of Erasmus Oswald Schreckenfuchs (1511–1579), professor of astronomy at the University of Freiburg-im-Breisgau. In his *Commentary on Sacrobosco* (Basel, 1569), Schreckenfuchs expressed the highest regard for both Copernicus and Werner. He called the latter "a very great expert in mathematical matters and a most accomplished investigator of the truth."[77] But Schreckenfuchs was more cautious about

> Nicholas Copernicus, a man of incomparable genius, whom I could deservedly call a miracle of nature, were I not afraid that certain men who cling most tenaciously to the revered dogmas of the ancient philosophers would be offended, and not undeservedly so.[78]

Schreckenfuchs dealt with the discussions of the motion of the eighth sphere by both Copernicus and Werner.[79] He compared the theories of these two astronomers with great care, but his comparison shows that he never heard of Copernicus' *Letter against Werner*.

Both Schreckenfuchs and Doppelmayr were diligent and prolific writers. Yet Schreckenfuchs never heard of the *Letter against Werner*, and Doppelmayr had no copy. Neither did Gassendi. As these three examples indicate, the distribution of the *Letter against Werner* was severely limited, being restricted to rather narrow circles around Prague about 1575, the prime mover being Thaddeus Hájek. Even Brahe did not have his staff reproduce the *Letter against Werner* for dissemination among astronomers, as he did in the case of *Commentariolus*,[80] which he evidently prized more highly.

Wapowski, however, had permitted the autograph of the *Letter against Werner* to be copied; one such copy was acquired by Simon Hájek, whose copy, it will be recalled, was itself copied in Prague in 1531.[81] This second-stage copy by an unknown scribe passed into the municipal library of Strasbourg.[82] This in turn, as a third stage, was "copied in Strasbourg in the month of June 1839 by Antoni Makowski,"[83] a Pole who is otherwise unknown.

Makowski's copy (cited hereafter as "Mk") was a copy of the Strasbourg copy of Simon Hájek's copy. Mk "entirely agrees" with V, and "offers ... no essential deviations" from V, according to P's earlier statement.[84] Later, P's collation turned up only two disagreements between Mk and V. Hence, P concluded that V was "presumably therefore also copied from Hájek's Prague copy."[85] In making this inference, P ignored his own quotation[86] of V's

---

[76] J. G. Doppelmayr, *Historische Nachricht von den Nürnbergischen Mathematicis und Künstlern* (Nuremberg, 1730), p. 35, n. ll.

[77] E. O. Schreckenfuchs, *Commentaria in sphaeram Io de Sacrobosco* (Basel, 1569), p. 407/25 up–24 up.

[78] Ibid., p. 36/15 up–13 up.

[79] Ibid., pp. 388–390.

[80] See *Commentariolus*, Introduction, at n. 5. There is no documentary basis for Vetter's assertion (p. 292) that Brahe circulated the *Letter against Werner* as he did the *Commentariolus*.

[81] See above, at n. 49.

[82] This second-stage copy was destroyed when the Strasbourg library burned down in 1870 during the Franco-Prussian War.

[83] PII, 171. This obscure copyist's surname was distorted into "Malczewski" by Polkowski, *Kopernikijana*, III, 308. Polkowski's mistaken "Malczewski" was propagated by Henryk Baranowski, *Bibliografia kopernikowska 1509–1955* (Warsaw, 1958), p. 55, no. 53. Baranowski was also unaware that the Strasbourg copy was destroyed in 1870 (see n. 82, above).

[84] PI², 285/14 up–10 up, 4 up–2 up, misattributing a collation of Mk with V to Franciszek Karliński.

[85] PII, 171/last 2 lines.

[86] PII, 170/n.*.

self-description as "among the first transcriptions from the autograph." That transcription from the autograph was completed on 30 March 1575, whereas the Strasbourg copy had been made in Prague more than forty years earlier, in January 1531. P characterized his collation of Mk and V as "precise." Whereas P's "precise" collation of Mk and V yielded only two disagreements, the variants listed in MK cover more than three of its quarto pages (pp. 506–509). These numerous divergences between V and Mk as a descendant, once removed, of Hájek's Prague copy, when coupled with V's self-description as a transcription made directly from the autograph, utterly demolish P's conclusion that V is descended from Hájek's Prague copy. On the contrary, even more directly than Br, V is derived from the autograph. On the other hand, Mk belongs with Ox as a descendant of a Hájek Prague manuscript, as we shall see later on.[87] At first Mk passed into the Polish Library in Paris, where it was copied by Leonard Niedźwiedzki (1811–1892), an émigré man of letters.[88] Of this copy a Polish translation made by Niedźwiedzki (or under his supervision) was promptly published by Polkowski (III, 309–315).[89] Previously (I, 68–74) Polkowski had printed a somewhat different translation of the *Letter against Werner* into Polish.[90] These two translations in Polkowski are different not only because they were made by two different translators, but also because these two translators used different Latin texts of the *Letter against Werner*.

The earlier translation in Polkowski reproduced the Polish version of the *Letter against Werner* that had appeared in the fourth edition of *Revolutions* (Warsaw, 1854), pp. 575–582, in parallel columns alongside the first printing of the Latin text of the *Letter against Werner*. For in Berlin B,[91] having been found by Wacław Aleksander Maciejowski (1793–1883),[92] was copied by a professional scribe, whose transcription was then compared with B by a professor.[93] The resulting text was printed in the fourth edition of *Revolutions* as the first edition of the *Letter against Werner*.[94]

After V was found in 1877, it was collated with B. The result of this collation was printed in MCV, I, 23–33 (with corrections in MCV, IV, 9), and in PII, 172–183. In addition, V was collated with Mk. Only one textual discrepancy was reported: Mk reads *inaequalitatis*, which was condemned as an error by PII, 172, who mistakenly asserts that V has *aequalitatis*. Actually, V agrees with Mk (and with every other MS except B) in reading *inaequalitatis*, which is correct. P's conclusion that Mk is in "complete agreement" with V must be rejected.

For, Mk's affinity is not with V at all, but with Ox and with another manuscript of the

---

[87] See below, at n. 95.

[88] From Paris Mk was transferred in 1889 to the Akademia Umiejętności in Cracow, which in turn sent it back to Strasbourg, where Makowski had originally produced it.

[89] Niedźwiedzki's copy was sent by Polkowski to Karliński, the director of the Cracow Astronomical Observatory (MK, p. 504). It is now MS 482/73 in the Jagiellonian Library, Cracow.

[90] This translation was reprinted in part, with slight modifications, by Ludwik Antoni Birkenmajer, *Mikołaj Kopernik Wybór pism* (Cracow, 1920), pp. 118–119.

[91] B consists of folios 8ʳ–10ʳ in a composite manuscript, whose components are described in MCV, I, p. 20, n. 7. B may have been sent by Hewelke, a zealous collector of books and manuscripts, to a fellow-astronomer, J. G. Rabener. The Hewelke-Rabener correspondence is preserved in the National Library, Paris, Fonds Latin, 10347–10349 :IX, 90–92; XI, 390–391; XIV, 181–185, 201–206, 274–282, 291–302, 307–310, 324–329, 352–353; XV, 11–17, 21–24, 173–174, 216–219, 321–323, 350–352. An examination of this correspondence might reveal how and where Hewelke acquired a copy of the *Letter against Werner*, and why he sent it to Rabener.

[92] Maciejowski, who found the Berlin manuscript, was confused by Vetter (p. 292) with Makowski, who copied the Strasbourg manuscript.

[93] Ignacy Polkowski, *Żywot M. Kopernika* (Gniezno, 1873), p. 205, n. 1; MCV, I, 19, n. 5; MK, p. 492.

[94] Reprinted by Hipler, pp. 172–179, and by PII, 145–153.

Letter against Werner which was found in the library of the Astronomical Observatory of Uppsala, Sweden (Hjorter Collection, H, III, 34) by L. A. Birkenmajer.[95] We shall call this manuscript "Up."

Mk, Ox, Up agree not only with respect to *inaequalitatis*[96] but also in the closing member of the salutation. After greeting Wapowski as "Cantor and Canon of the Church of Cracow" as well as "Secretary to His Royal Majesty the king of Poland," Mk, Ox, Up add: "most highly esteemed patron, greetings" (*Domino et fautori suo plurimum observando S[alutem] D[icit]*). This third member of the salutation in Mk, Ox, Up is lacking in B, Sch, while V lacks *S. D.*[97] For B, Sch, V, astronomy belongs to a category (*ex numero*) of subjects; *ex numero* is missing in Mk, Ox, Up.[98] In Mk, Ox, Up we see that the planets (*planetae*) do not twinkle; *planetae* is not found in B, Sch, V.[99] "To be measured" is *metiri* in B, Sch, V, but *mensurari* in Mk, Ox, Up.[100] For observations "added since Ptolemy," B, Br, Sch, V have *post Ptolemaeum adauctis*, whereas Mk, Ox, Up read *Ptolemaei adiunctis*.[101] For three points not "located" on a straight line, B, Br, Sch, V have *data*, but Mk, Ox, Up read *posita*.[102] For "He derives," B, Sch, have *Assumit* (V: *Adsumit*), but Mk, Ox, Up read *Asseruit*.[103] Whereas B, Sch, V refer to the 200 years B. C. (*ante nativitatem*), Mk, Ox, Up have A. D. (*a nativitate*).[104] For "from the maximum slowness" B, Sch, V have *a summa ... tarditate*, where Mk, Ox, Up omit *a summa*,[105] and soon thereafter omit *ad tarditatem (summam)* altogether.[106] For "absolutely certain," B, Sch, V write *certissima*, but Mk, Ox, Up have *verissima*.[107]

The foregoing selected examples demonstrate the kinship of B, Br, Sch, V, on the one hand, and of Mk, Ox, Up, on the other hand. The external history of Ox is quite clear: it was copied for Savile when he visited Thaddeus Hájek in 1578.[108] The external history of Up, however, is less well known.[109] But its connection with Hájek seems indisputable.

Consider the passage where the *Letter against Werner* mentions that Ptolemy "drew up" his star catalog. For "drew up," B, Sch, V read *constituit*. Up also has *constituit*, but with

---

[95] MK, p. 497. [96] In the notes on the *Letter against Werner*, see n. 142. [97] Ibid. n. 2.
[98] Ibid., n. 79. [99] Ibid., n. 82. [100] Ibid., n. 101. [101] Ibid., n. 117. [102] Ibid., n. 118.
[103] Ibid., n. 130. [104] Ibid., n. 143. [105] Ibid., n. 154. [106] Ibid., n. 157. [107] Ibid., n. 180.

[108] See above, at n. 57. It has been suggested that Savile visited Cracow in the company of Dee (MK, p. 497, n. 2). But Savile was not in Cracow when Dee was there. Dee reports that on 13/23 March 1584 "I and my wife" Jane "came to Cracow," where they were joined two weeks later by Dee's medium Edward Kelley (1555–1595; *A True and Faithful Relation*, p. 72). Some four months later, on 1 August 1584, Dee notes, "we entred on our journey toward *Prage*. ... We came by Coach, I, E. K. and his brother [Thomas], and *Edmond Hilton*" (*Relation*, p. 212). Dee's party also included his daughter Katherine as well as his infant son Rowland and the infant's nurse. The section of the *Relation* recounting the party's return from Prague to Cracow is missing (pp. 257–352). But on 24 November 1584 Dee donated a book to the library of the University of Cracow (W. Wisłocki, *Catalogus codicum manuscriptorum bibliothecae universitatis Jagellonicae Cracoviensis* [Cracow, 1877–1881], p. 195, no. 620). Then on 10/20 December 1584 Dee and his party left Cracow to go back again to Prague (*Relation*, p. 353). Their return from Prague to Cracow began on 5 April 1585 (*Relation*, p. 397). During these swings back and forth between Cracow and Prague, Dee does not mention Savile as being with him. For at that very time Savile was in England busily promoting his own academic career. He was "elected Warden of Merton in 1585, upon the recommendation of Lord Burghley," whose letter of recommendation is dated "this last of Februarye, 1585"; see George C. Brodrick, *Memorials of Merton College* (Oxford, 1885; Oxford Historical Society, Publications, IV; pp. 60–61). Hence, while Dee was in Cracow in 1584 and 1585, Savile was not there at the same time, his trip to the continent having taken place some six or seven years earlier.

[109] Up is sewn to the endpapers of the copy of the 2nd edition of *Revolutions* (Basel, 1566), which was acquired by Olof Peter Hjorter (1696–1750), an observer at the Astronomical Observatory in Uppsala, Sweden. His collection now forms part of the Observatory's library (see MK, p. 501). How and where Hjorter acquired Up has not yet been clarified.

redegit above the line. On the other hand, Ox has *redegit* in the line, with *constituit* above the line and deleted. Evidently the autograph had *constituit*. Its first syllable, abbreviated by a sign resembling our capital C written backwards, was misinterpreted by Mk, an incompetent copyist, as *instituit*. How are the double readings *constituit/redegit* in Ox, Up to be explained? Thaddeus Hájek preferred *redegit* to Copernicus' *constituit*. In one of his two manuscripts, therefore, he placed *redegit* above *constituit*, which he did not disturb. This situation was duplicated by Up. Ox, however, thought that *redegit* was the corrected reading, which he therefore placed in the line. To show what had been removed, he put *constituit* above the line and struck it out.[110]

In another passage, where Copernicus is talking about mean motions, the correct reading (*mediantium*) is found in B, Sch, V. But *ambientium*, an erroneous reading, had somehow insinuated itself (in Simon Hájek's copy?) Above *ambientium*, *mediantium* was inserted interlinearly in Ox, Up (presumably as a correction made by Thaddeus Hájek). Mk, in his usual blundering way, wrote *mediantiam* (sic) *ambientium* in the line.[111]

In still another passage, dealing with the complete circuit of a revolution, Copernicus used the term *circuitionem*, which appears in B, Sch, V. But Thaddeus Hájek preferred *revolutionem*, which was written interlinearly above *circuitionem* in Ox, Up, while Mk put *revolutionem* in the line with *circuitionem* above it (after being deleted in the line).[112]

In discussing a recurring motion of varying speed, Copernicus calls its fastest phase *concitatissimum* (B, Sch, V). Thaddeus Hájek preferred *velocissimum*, which Mk, Ox, Up placed in the line with *concitatissimum* above it.[113]

These four sets of coupled readings (*constituit/redegit*; *mediantium/ambientium*; *circuitionem/revolutionem*; *concitatissimum/velocissimum*) appear only in Mk, Ox, Up, and are completely absent from B, Sch, V. They confirm the twofold descent of the manuscripts of the *Letter against Werner*: B, Br, Sch, V are closely affiliated with the autograph; Mk, Ox, Up, with a Hájek manuscript.

Mk is a copy of the (lost) Strasbourg copy of Simon Hájek's (lost) copy. Ox was copied from a manuscript belonging to Thaddeus Hájek. Was Up also copied from a Thaddeus Hájek manuscript? Or was it perhaps copied from Ox?

Where Ox reads *honestius et re vera*, Up omits *et*.[114] Where Ox expresses "to be praised" by *commendari*, Up has *laudari*.[115] Where Copernicus would have replied "in such a way" that ..., Ox reads *sic* correctly, but Up has *sit*,[116] which is wrong. Where Ox has *ipsum*, at first Up wrote *eum*, then *illum* (with its second *l* heavily drawn over the earlier *e*).[117] Where Ox has *dubitare*, Up has *dubitari*.[118] Where Ox has *supra*, Up has *ultra*.[119] Where Ox has *experimentum*, Up has *experientiam*.[120] Where Ox has *cum temporum intervallis*, Up has *tum temporum intervalla*.[121] Where Ox has *uniformior*, Up has *uniformiter*.[122] Where Ox has *insensibilis* for an "imperceptible" error, Up has the contrary, *sensibilis*.[123]

The foregoing selected examples of disagreement between Ox and Up are enough to show that Up was not copied from Ox. On the contrary, Up was copied directly from a Hájek manuscript. In other words, both Ox and Up are descended, independently of each other, from a manuscript of the *Letter against Werner* in the possession of Thaddeus Hájek.

The margins of Ox, which was copied for a man of great intellectual distinction, have numerous corrections of its readings. Nevertheless, blemishes remain. For instance, where

---

[110] In the notes on the *Letter against Werner*, see n. 27.
[111] Ibid., n. 106.   [112] Ibid., n. 109.   [113] Ibid., n. 126.   [114] Ibid., n. 9.   [115] Ibid., n. 8.   [116] Ibid., n. 16.   [117] Ibid., n. 24.   [118] Ibid., n. 35.   [119] Ibid., n. 64.   [120] Ibid., n. 65.   [121] Ibid., n. 87.   [122] Ibid., n. 138.   [123] Ibid., n. 174.

Up reads *Romanorum* for a year "of the Romans," Ox mistakenly has *Quoniam*.[124] Again, Ox has a non-existent form cognoscitum, where Up has *cognoscuntur* for things which "are known."

Both Ox and Up were descended from a common source. Ox was copied in 1578. Was Up written before or after that date? V was finished, it will be recalled, directly from the autograph on 30 March 1575. Indirectly from the autograph, Br followed soon thereafter. Somewhat earlier B was copied from Sch, which in turn had been copied from the autograph in 1569.

Mk was copied, as we saw above, by Niedźwiedzki, who started to correct Mk, but soon abandoned that effort. Between the lines of the Latin text, Niedźwiedzki inserted a Polish translation based, not on Mk's defective Latin text, but in the main on the Latin text in the Warsaw 1854 edition of the *Revolutions*. Niedźwiedzki's Polish translation was published by Polkowski, who thereafter sent Niedźwiedzki's manuscript to Karliński. The latter collated Niedźwiedzki's Latin text with the Warsaw edition, correcting the former in order to bring it into agreement with the latter. Karliński did not collate Mk with V, as is mistakenly asserted by P.

A collation of B and V formed the foundation of the Latin text in MCV and PII, on which was based the first translation into English of the *Letter against Werner*.[125] The present translation benefited from a collation of B, Br, Mk, Ox, Sch, Up, and V by Dr. Erna Hilfstein.[126]

---

[124] Ibid., n. 52.   [125] TCT, 1st ed., 1939, pp. 93–106.
[126] MK, pp. 494–509, collated Mk, Ox, Up with MCV, however, not with B and V; only Br's excerpt was collated with the manuscripts themselves.

MANUSCRIPTS OF COPERNICUS' *LETTER AGAINST WERNER*

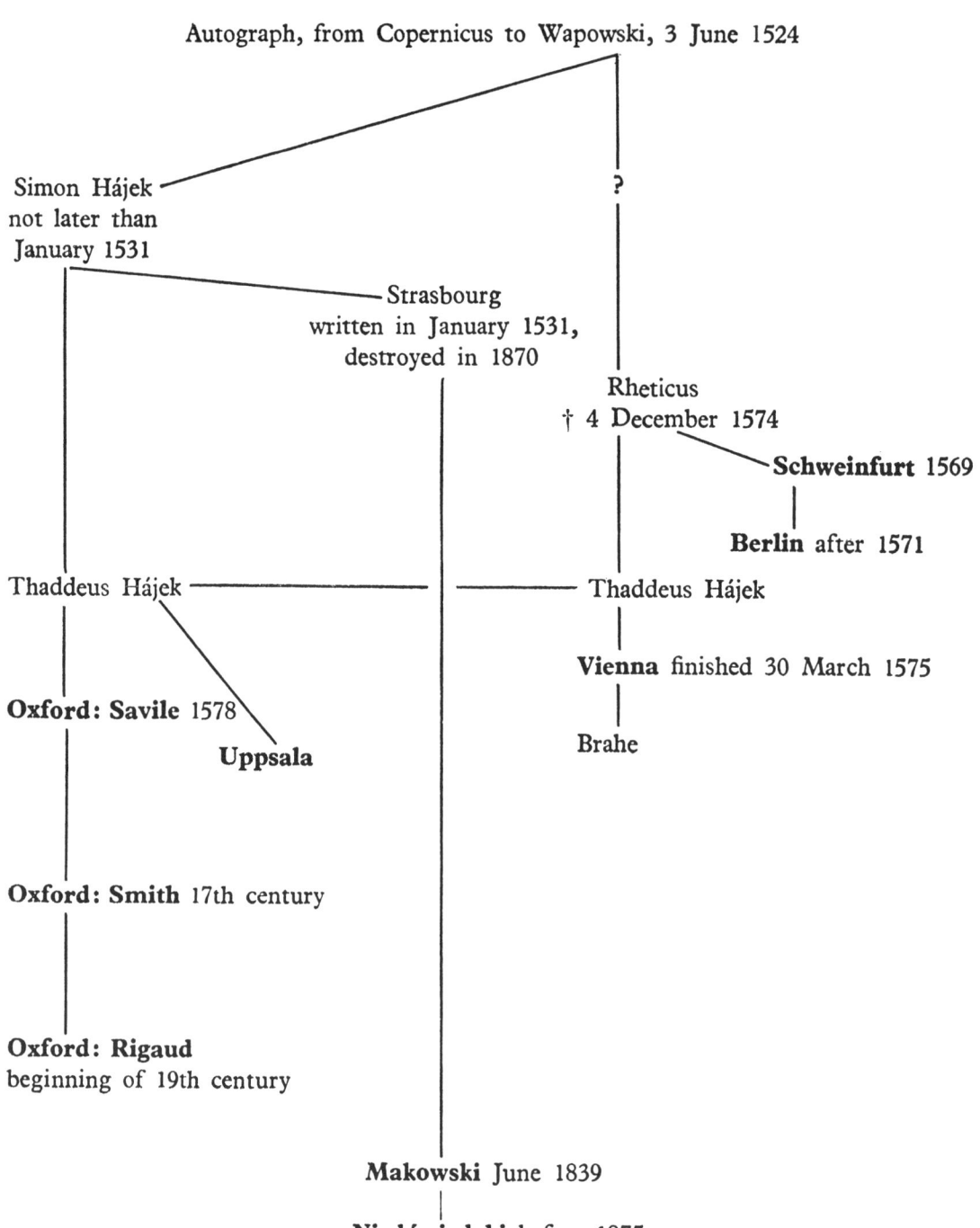

The extant manuscripts are printed in boldface

LETTER AGAINST WERNER

To the Reverend Bernard Wapowski,[1] Cantor and Canon of the Church of Cracow, Secretary to His Royal Majesty the king of Poland, and my most highly esteemed patron, greetings[2] from Nicholas Copernicus.[3]

Some time ago, my[4] dear Bernard, you sent me a little treatise on "The Motion of the Eighth Sphere," published by Johann Werner of Nuremberg.[5] Your Reverence stated[6] that the work was widely praised and asked me too to give you my[7] opinion of it. I would surely have done so gladly to the extent that I too could have really commended[8] it wholeheartedly.[9] Yet I may laud the fellow's zeal and effort. Moreover, it was Aristotle's advice[10]

> to be grateful not only to the philosophers who have spoken well but also to those who have spoken incorrectly[11]

because to those who wish to follow the right road, it is often no small advantage to have noted the blind alleys too.[12] Besides, faultfinding is of little[13] use and scant profit, for[14] it is the very mark of a shameless[15] mind to prefer the role of the censorious critic to that of the creative poet. Hence I even fear that I may arouse anger if I reprove another while I myself produce nothing better. Accordingly, I wanted to leave these matters, just as they are, to the attention of others, and I would have replied in such a way that your Reverence[16] would learn my attitude[17] expressed concisely. I am aware,[18] however, that it is one thing to snap at a man and assail him, but another thing to set him right and redirect him when he strays, just as it is one thing to praise, and another to flatter and play the fawner. Hence I see no reason why I should not comply with your request or[19] why I should appear to hamper the pursuit and cultivation of these studies, in which you have a conspicuous[20] place. And therefore, lest I even[21] seem to condemn the man gratuitously, I shall try to show as clearly[22] as possible in what respects he erred regarding the motion of the sphere of the fixed stars and maintains an unsound position. This may perhaps even contribute[23] not a little to the formation of a better understanding of this subject.

In the first place, then, he[24] went wrong in his calculation of time. For he thought that the emperor[25] Antoninus[26] Pius' second year, when Claudius Ptolemy drew up[27] the catalog of the fixed stars as observed by himself,[28] was A.D. 150,[29] when in fact it was A.D. 139.[30] For in the *Great Syntaxis*, Book III, Chapter 1, Ptolemy says that the autumnal equinox observed in the 463rd year after Alexander the Great's[31] death fell in Antoninus' third year.[32] But from Alexander's death to Christ's birth there are counted 323 uniform Egyptian[33] years,[34] 130 days, because from the beginning of Nabonassar's reign to Christ's birth 747 uniform years, 130 days, are reckoned. This is not questioned,[35] I observe, certainly not by our author, as is evident in his Proposition 22, except that he adds one day,[36] in accordance with the *Alfonsine Tables*.[37] The reason for this [discrepancy of one day] is that Ptolemy takes noon of the first day[38] of the first Egyptian month Thoth as the starting point of the years reckoned from Nabonassar and Alexander the Great,[39] while Alfonso begins with noon of the last day[40] of the preceding year, just as we compute the years of Christ from noon of the last day of the month December.[41] Now from Nabonassar to Alexander the Great's death 424 uniform[42] years are counted by Ptolemy,[43] Book III, Ch. 8,[44] with whom Censorinus, relying on Marcus Varro,[45] agrees in his *Natal Day*,[46] dedicated to Quintus[47] Caerellius.[48] [This interval of 424 years, when subtracted] from 747 years, 130 days, leaves a remainder of 323 years, 130 days, that is, from Alexander's death to Christ's birth. And from that time to the aforementioned observation of Ptolemy [there are] 139[49] uniform years, 303 days.[50] Therefore, the autumnal equinox observed by Ptolemy, it is clear, occurred on the ninth day of the month Athyr, 140 uniform years after the birth of our Lord, but 139[51] Roman[52] years, 25 September,[53] Antoninus' third year.

Again, in his *Great Syntaxis*, Book V,[54] Ch. 3, in his observation of the sun and moon in Antoninus' second year Ptolemy counts 885 years of Nabonassar, 203 days.[55] From Christ's birth, therefore, 138 uniform[56] years, 73[57] days, would have elapsed.[58] The fourteenth[59] day thereafter, that is, 9 Pharmuthi,[60] when Ptolemy observed Regulus [in the constellation] of the Lion,[61] was 22 February,[62] in the 139th Roman year after Christ's birth. And this was Antoninus' second[63] year, which our author thinks was 150 [A.D.]. Hence he went wrong by eleven years too much.[64]

If, however, anyone is still in doubt and, not satisfied by the foregoing [criticism], wants to make a further test[65] of this matter, he should remember that time is the number or measure of the motion in heaven considered as "before" and "after."[66] For by this motion we determine our years, months, days, and hours. But the measure and the measured, being related, are mutually interchangeable.[67] Besides, as far as Ptolemy's tables are concerned, since in addition[68] they were built up on the basis of his own recent observations,[69] it is unbelievable that they contain any deviation from the observations which is detectable by the senses, or any[70] discrepancy that would make them inconsistent with the foundations on which they rest. Since this is so, if [our skeptic] consults Ptolemy's tables and computes the positions of the sun and moon with reference to Regulus as found by Ptolemy using the astrolabe[71] in Antoninus' second year on the ninth day of the month Pharmuthi[72] at $5^1/_2$ hours after noon,[73] he will find these positions, not 149 years after Christ, but 138[74] years, 88 days, $5^1/_2$ hours, equal to 885 years after Nabonassar, 218 days, $5^1/_2$ hours.[75] In this way the error is now exposed which frequently vitiated our author's investigation of the motion of the eighth sphere when he mentions time.[76]

A second error, no less serious than the first, is involved in his hypothesis expressing his belief that in the 400 years before Ptolemy the fixed stars moved only with a uniform motion.[77] For the purpose of further explaining and clarifying[78] what will be said below, it should in my opinion be pointed out that the science of the stars belongs to the category[79] of those [subjects] which[80] we learn in an order contrary to nature. For example, first nature knows that the planets are nearer than the fixed stars to the earth,[81] then as a consequence that the planets appear less radiant. We, on the other hand, first see that the planets[82] do not twinkle, and then we know that they are nearer to the earth.[83] So, by the same token, first we perceive that the motions of the heavenly bodies seem nonuniform, then we conclude that there are[84] epicycles, eccentrics, or other circles by which the bodies are carried in this way.[85] And therefore I would like it to be said[86] that, with the aid of instruments, the ancient scientists first had to mark the positions of the heavenly bodies together with the intervals of time,[87] and with this [information] as a sort of guideline, they had to devise a precise theory of the heavenly motions, lest the investigation of these matters remain interminable. They appear to have found this theory when it matched, with a certain agreement, all[88] the observed and noted positions of the heavenly bodies. Such is also the situation regarding the eighth sphere's motion, which the ancient astronomers could not pass on to us in its entirety on account of its extreme slowness. Those who wish to examine it[89] must follow in their footsteps, however, and hold fast to their observations,[90] bequeathed like a legacy. But if anyone, holding fast to his own view,[91] thinks that they are untrustworthy[92] in this regard, surely the gates of this art are closed to him. Lying in front of the entrance, he will dream the dreams[93] of the deranged about the motion of the eighth sphere, and receive what he deserves for[94] supposing that he should support his own hallucination by defaming the ancients. It is well known, however, that those who handed down to us many famous and praiseworthy discoveries made all these observations with the utmost care and expert[95] skill. Consequently I cannot[96] possibly be persuaded that in noting the positions of the heavenly bodies they[97] erred by $1/4°$ or $1/5°$ or even $1/6°$, as this author believes. [I shall say] more about this [subject] later on.[98]

In addition, it must not be overlooked[99] that in every heavenly motion involving an irregularity, what we want above all is the entire period during which the apparent motion is recognized[100] as having passed through all its variations. For, an apparent irregularity in a motion is what makes it impossible for an entire revolution and uniformity of motion to be measured[101] by their parts.[102] But in their investigation of the moon's path Ptolemy, and before him Hipparchus of Rhodes, divined with keen insight that the revolution of a nonuniform [motion] must[103] have four diametrically opposite[104] points. These are[105] the maximum

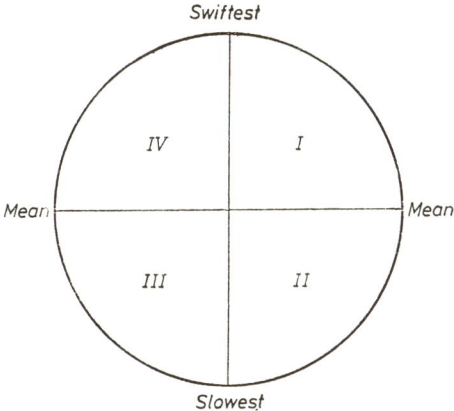

Fig. 13. Revolution of a nonuniform motion

swiftness and slowness, and the mean[106] and uniform [motion] at both [ends of the diameter] intersecting at right angles [the diameter connecting] both[107] maxima. The circle is [thereby] divided into four equal parts, with the result that in the first quadrant the swiftest motion diminishes;[108] in the second [quadrant] the mean motion diminishes; on the other hand, in the third quadrant the slowest motion increases, [as does] the mean [motion] in the fourth [quadrant]. By this device they could infer from the moon's observed and examined motions in what part of the circle it was at any specified time. Accordingly, when a similar motion had recurred, they understood that a revolution[109] of the nonuniformity had now been completed, as this was explained at considerable length by Ptolemy[110] in his *Great Syntaxis*, Book IV.[111] This [procedure] should have been adopted also in analyzing the eighth sphere's motion. As I said,[112] however, its extreme slowness, on account of which the nonuniform motion quite clearly has not yet[113] returned upon itself in[114] thousands of years,[115] does not permit an immediate solution of this [problem], because[116] it transcends many generations of men. Nevertheless it is possible by a reasonable conjecture to attain a solution even now with the aid of some observations added since Ptolemy,[117] observations which conform to the same pattern. For what is determinate cannot have innumerable explanations. For example, if a circumference is drawn through three points not located[118] on a straight line, superimposition of another circumference greater or smaller than the one drawn[119] previously is impossible.[120] But in view of my discussion of these matters elsewhere,[121] I may return to the point where I digressed.[122]

Hence we must now see whether our author[123] is correct in saying that in the 400 years before Ptolemy the fixed stars moved only with uniform motion. Besides, lest we be mistaken about the meaning of terms, by "uniform motion" I understand what we usually also call "mean" motion, which[124] is halfway[125] between the slowest and the swiftest.[126] Let him not mislead us by his statement in Proposition 7, Corollary 1,[127] that "the motion of the fixed[128] stars is slower" where according to his own hypothesis he puts the uniform motion, the rest of it being swifter, just as if it would never be slower. In these respects I do not know whether he is consistent with himself when later on he adduces a much slower [motion].[129] He derives[130] his measure of the mean motion, however, from the uniformity with which the fixed stars traversed equal distances from the earliest observers of the fixed stars, Aristyllus and[131] Timocharis, to Ptolemy, and in equal periods of time, to wit,[132] approximately 1° every one hundred years, as is quite clear in Ptolemy,[133] cited by our author in his Proposition 6.[134]

But being a great astronomer, he is not aware[135] that around the points of uniform motion, that is, the intersections[136] of the circles (the tenth sphere's ecliptic with [the circles] of trepidation, as he calls[137] them), the stars' motion cannot possibly appear more uniform[138] than elsewhere.[139] Indeed, the opposite conclusion must be drawn: at those times the motion appears to vary the most, but the least when the apparent motion is swiftest or slowest. He should have seen this even from his own hypothesis and system as well as from the tables based[140] on them, especially the last table,[141] which he drew up to exhibit the revolution of the entire nonuniformity[142] or trepidation.

In this regard, according to an earlier computation, for 200 years before the birth[143] of Christ the apparent motion is found to be only 49' of 1° in the first 100 years, and in the second century 57'. Then after Christ's birth in the first[144] 100 years the stars would have moved about[145] $1\,1/10$°, and in the second 100 years about $1\,1/4$°.[146] Thus in equal periods of time the successive motions increase by a little less than $1/6$°.[147] But if you combine the motion of the two centuries in either era, the first period's [total] will fall short of 2° by more than $1/5°$,[148] while the second [period's total] will exceed [2°] by about $1/4°$.[149] Thus again in equal times the later motion will exceed the earlier motion by about $1/2° + 1/15°$.[150] Yet previously in reliance on Ptolemy our author had reported that the fixed stars passed through 1° every 100 years. By the very law of the circles which he assumed, however, the opposite happens in the eighth sphere's swiftest motion, when a variation of scarcely 1' in the apparent motion is found in 400 years, as may be seen for the years A.D. 600–1000 in the same table,[151] and likewise in the slowest [motion] also, as for the 400 years after 2060.[152]

Now the reason for the nonuniformity is, as was said above,[153] that in one semicircle of the trepidation, namely, that which [extends] from the maximum[154] slowness to the maximum swiftness, there is always some increase in the apparent motion.[155] In the other semicircle, which [is reckoned][156] from the maximum swiftness to the maximum slowness,[157] the motion which had previously increased decreases steadily.[158] The greatest increase and decrease[159] occur at the diametrically[160] opposite points of uniform motion.[161] In the apparent motion, consequently, equal motions are not to be found in two continuous equal[162] periods of time. One [of the two motions][163] becomes[164] greater or smaller than the other,[165] except in the vicinity of the maximum swiftness or[166] slowness. Only there do [the motions] on either side traverse equal arcs[167] in equal times; starting or ceasing to increase or decrease, at those times they counterbalance each other.[168] Consequently it is by no means correct that the motion in the 400[169] years before Ptolemy was the mean motion. On the contrary, it was rather the slowest motion. Indeed I see no reason why we should speculate[170] about another slower motion concerning which we have heretofore been unable to obtain any hint. For, no observation of the fixed stars made before Timocharis has come down to us, nor did any come down to Ptolemy.[171] Besides, since the swiftest motion has already passed, we are now as a consequence in the second, post-Ptolemaic, semicircle, in which the motion decreases, and no small part of it too has passed.[172]

Accordingly it should not seem surprising that with these assumptions of his our author could not approach more closely to what was reported by the ancients, and that in his opinion they[173] erred by $1/4°$ or $1/5°$ or even $1/2°$ and more. Yet nowhere does Ptolemy seem to have exercised greater care than in striving to pass on to us the motion of the fixed stars free from error. For this [precision] would have been available to him only in that restricted portion of the motion from which he undertook to devise that entire revolution. An error, however imperceptible[174] when occurring[175] in the restricted portion, could undoubtedly[176] emerge as significant in that whole immense framework. Ptolemy seems to have linked Aristyllus[177] with his contemporary Timocharis of Alexandria, and Agrippa[178] of Bithynia with Menelaus[179] of Rome, in order in this way to have absolutely certain[180] and unchallenged evidence in the agreement of these [observers] from such widely separated places. Hence[181] it is unbelievable that such great errors were made by them or by Ptolemy, men who were able to understand many other even[182] more difficult matters down to the last detail, as the saying goes.

Finally, nowhere is our author more foolish than in Proposition 22 and especially its Corollary. Wishing to praise his own work, he censures Timocharis with regard to two stars, namely,[183] Arista[184] in the Virgin, and the most northerly of the three stars in the Scorpion's brow, by claiming that[185] in the first case Timocharis' computation falls short, and is excessive in the second case.[186] Here our author prattles in an exceedingly childish way. For with regard to both of the stars under consideration,[187] the displacement between Timocharis and Ptolemy is the same, namely, $4\ 1/3°$ in almost exactly the same time interval,[188] and the numerical result of that computation is therefore practically identical. Yet our author completely fails to notice that adding $4°7'$[189] to the place of the star found by Timocharis in $2°$ of Scorpion[190] could not properly fill out the $6°20'$ of Scorpion where it is found by Ptolemy.[191] Conversely, the subtraction[192] of the same number from $26°40'$ for Arista according to Ptolemy[193] could not restore $22\ 1/3°$,[194] as it should,[195] but remained at $22°32'$.[196] Thus our author thought that in the first case the computation was deficient by as much as it was excessive in the second case,[197] as though this disparity were inherent in[198] the observations, or as if the road from Athens to Thebes were not the same as the road from Thebes to Athens.[199] Besides,[200] had he either added or subtracted the number in both cases, as parity of reasoning required, he would have found that both computations proceed in the same way.

Moreover, between Timocharis and Ptolemy there were in fact not 443 years,[201] but only 432, as I indicated in the beginning.[202] Hence, the interval being shorter, the amount [of the precession] should be smaller, so that[203] our author will deviate[204] from the stars' observed motion not merely by $13'$[205] but by $1/3°$. This is how he imputed his own error to Timocharis, while Ptolemy barely escaped. But while he thinks that their reports[206] are untrustworthy,[207] what[208] else remains but to distrust his observations too?

So much for the eighth sphere's motion[209] in longitude. From the foregoing [remarks] it can also be easily inferred[210] what[211] we should think about the motion in declination[212] too. For our author complicates it[213] with two trepidations, as he calls[214] them, piling[215] this second one on top of the first.[216] But now that the underpinning itself has been destroyed, the superstructure must necessarily collapse, being weak and incohesive.

Lastly, what do I myself think about the motion of the sphere of the fixed stars? Since [my views] are to be stated elsewhere,[217] I deemed it superfluous and improper to extend this communication further here.[218] For it is enough if I satisfy your desire to have my opinion of this little work in compliance with your request. May your Reverence enjoy the best of health.[219]

Nicholas Copernicus[220]

Frombork, 3 June 1524[221]

BIBLIOGRAPHY

Album studiosorum universitatis cracoviensis, II (Cracow, 1892)

Bachmann, Siegfried, "Johannes Werner, kaiserlicher Hofkaplan, Mathematiker und Astronom zu Nürnberg als Chronist der Jahre 1506 bis 1521," *Historischer Verein für die Pflege der Geschichte des ehemaligen Fürstbistums Bamberg*, 102. Bericht (Bamberg, 1966)

Birkenmajer, Ludwik Antoni, *Stromata Copernicana* (Cracow, 1924)

Briefwechsel *Willibald Pirckheimers Briefwechsel*, ed. Emil Reicke (Munich, 1940–1956; Veröffentlichungen der Kommission zur Erforschung der Geschichte der Reformation und Gegenreformation, Humanisten Briefe, IV–V)

Burmeister, Karl Heinz, *Georg Joachim Rhetikus*, 3 vols. (Wiesbaden, 1967–1968)

Herold, Max, *Die St. Johanniskirche in Nürnberg* (Erlangen, 1917) = *Beiträge zur fränkischen Kunstgeschichte*, VIII (1917)

Hipler, Franz, *Spicilegium Copernicanum* (Braunsberg, 1873)

Kist, Johannes, *Die Matrikel der Geistlichkeit des Bistums Bamberg 1400–1556* (Würzburg, 1960), p. 431, no. 6570 (Veröffentlichungen der Gesellschaft für fränkische Geschichte, IV Reihe: Matrikeln fränkischer Schulen und Stände, Band 7)

Kressel, Hans, "Hans Werner," *Mitteilungen des Vereins für Geschichte der Stadt Nürnberg*, 1963–1964, 52:287–304

Polkowski, Ignacy, *Kopernikijana*, 3 vols. (Gniezno, 1873–1875)

Schottenloher, Karl, "Der Mathematiker und Astronom Johann Werner aus Nürnberg," *Festgabe zum 7. September 1910, Hermann Grauert zur Vollendung des 60. Lebensjahres* (Freiburg-im-Breisgau, 1910), pp. 147–155

Vetter, Quido, "Nicolas Kopernik et la Bohême," *Bulletin scientifique de l'Ecole polytechnique de Timisoara*, 1932, 4:292–294 (Kindly provided by Dr. Jerome Ravetz, University of Leeds)

NOTES

[1] This name is given in a nearly correct form only by Sch (which was written in Cracow) and by Mk (who was a Pole).

[2] Mk, Ox, Up, V: *Domino et fautori suo plurimum observando*; Mk, Ox, Up: *S. D.*; not in B, Sch. After *observando* V writes φ as the sign for the end of the salutation, a sign which V repeats at the end of a paragraph on fol. $1^v/6$ up, $3^v/9$, $4^v/8$ up, $5^v/10$, $7^v/3$ up, $9^v/3$ up.

[3] B, Sch: *Nicolaus Copernicus*; not in Mk, Ox, Up, V.

[4] Sch: *mi*; not in B, Mk, Ox, Up, V.

[5] Copernicus acknowledges receipt, not of Werner's entire second collection of essays (Nuremberg, 1522), but only of "De motu octavae sphaerae," which occupies sig. k 1^r–v 4^r in that collection. No convenient short title is available for Werner's 1522 collection because, like his first collection of essays (Nuremberg, 1514), it bears no distinctive title. In fact, the title page of both collections begins in the same way: *In hoc opere haec continentur* (In This Work the Following Are Contained). Then the individual titles of the component essays are listed.

[6] Mk, Ox, Up, V: *dicebas*; B, Sch: *ducebas*.

[7] Mk, Ox, Sch, Up, V: *meam*, consistent with *significarem*, with which B's *nostram* is inconsistent.

[8] B, Mk, Ox, Sch, V: *commendari*; Up: *laudari*.

[9] B, Mk, Ox, Sch, V: *honestius et re vera*; Up omits *et*.

[10] B, Mk, Sch, Up, V: *admonuit*; Ox: *admonet*.

[11] Aristotle, *Metaphysics*, I minor, 1; 993b11–13. Copernicus is not quoting directly from the Greek text, which refers to those who "agree" and those who "have spoken superficially," whereas he mentions those who "have spoken well" and those who "have spoken incorrectly." Hence he used an imperfect Latin translation, which has not yet been identified. It is neither the *vetustissima*, nor the *vetus*, nor the *anonyma*, nor Bessarion's. Since Copernicus' formal study of Aristotle took place at the University of Cracow, an unpublished Latin translation of Aristotle's *Metaphysics* in the Jagiellonian Library there may be the source from which he took his imperfect quotation.

[12] Aristotle advised gratitude to superficial thinkers because "they too made some contribution, since they did our warm-up exercises." This justification was inappropriate for Copernicus' thinkers who "have spoken incorrectly." Hence he had to replace Aristotle's justification by his own: "to those who wish to follow the right road, it is often no small advantage to have noted the blind alleys too." Those who "have spoken incorrectly" lead us down blind alleys, whereas warm-up exercises may be done for us by superficial thinkers. Copernicus' substitute justification was misattributed to Aristotle by being included in the quotation from his *Metaphysics* by MCV, I, 23/13–14, followed by PII, 173/7–8.

[13] B: *admodum*, corrected to *ad modicum*, the reading of all the other MSS.

[14] B, Sch, V: *quia et*; Mk, Up: *quin et*; Ox: *quoniam* (*est*, deleted).

[15] B, Mk, Ox, Sch, Up: *impudentis*; V: *prudentis*, a mistaken reading.

[16] B, Ox, Sch: *sic V[enerationi] tuae, ut*; Mk: *sicut V[eneratio] tua*; V: *sic V[eneratio] tua et*; Up: *sit ut V[eneratio] tua*.

[17] Ox, Sch, Up: *mentem meam acciperet*; V: *mentem* [*tuam*, deleted] *meam acciperet*; Mk: *mentem meam accipiet*. B mistakenly replaced *mentem* by *libenter*, and *meam* by *nostra* (like B's *nostram* in n. 7, above). The first printing of the Latin text of the *Letter against Werner* was based on B as the only manuscript then known. The editor was therefore unaware of the extent to which B deliberately diverged from his prototype (now identified as Sch).

[18] B, Mk, Ox, Up, V: *animadvertam*; Sch: *animadverterem*.

[19] B, Sch, V: *aut*, missing in Mk, Ox, Up.

[20] Mk, Ox, Sch, Up, V: *praecipue*; B: *praecipua*, under the influence of the preceding word *qua*. Copernicus' reference to Wapowski's eminence in the learned world is not empty flattery, for Wapowski is regarded as the founder of Polish scientific cartography. On 18 October 1526, some two years after Wapowski received Copernicus' *Letter against Werner*, Wapowski's publisher was granted by King Sigismund I of Poland the exclusive right to distribute two of Wapowski's maps, printed not long before, and a third, about to be printed; see Edward Rastawiecki, *Mappografia dawnej Polski* (Warsaw, 1846), pp. 11–12.

[21] B, Mk, Ox, Sch, V: *etiam*, missing in Up.

[22] Sch, V: *aptissime*, forgetting to draw a horizontal stroke across the lower vertical member of the *p* as the standard abbreviation for the syllable *-per-*. This was done by Ox, Up, and B (not by a second hand in B, as misreported in MCV, I, 24/on 5). Mk inadvertently omitted *apertissime ostendere in quibus ille*, perhaps because he skipped a line in his prototype. If so, a line in his prototype (Simon Hájek's copy) was about 35 letters long.

[23] B, Mk, Ox, Sch, Up: *conducet*; V: *conduceret*.

[24] B, Sch: *eum*; Mk: *illum*; Up: *eum*, replaced by *illum*; Ox, V: *ipsum*.

[25] Mk: *Augusti*; Ox, Up, V: *Aug[usti]*; omitted by B, Sch.

[26] B, Mk, Ox, Up, V: *Antonini*; Sch: *Antonii*, corrected interlinearly to *Antonini*.

[27] B, Sch, V: *constituit*; Up: *constituit*, with *redegit* above it; Ox: *redegit*, with *constituit* above it and deleted; Mk: *instituit*.

[28] Copernicus firmly believed that Ptolemy based his star catalog on his own observations. That belief was later called into question by a critical trend that peaked in J. B. J. Delambre's *Histoire de l'astronomie ancienne* (Paris, 1817; reprinted New York/London, 1965):

> Did Ptolemy observe? Would not the observations which he tells us he made be computations from tables? (p. xxv).

Coupling Ptolemy with a relatively unimportant astronomer, Delambre remarked that

> the two authors instead of observing the stars' positions as they were in their time, would have computed them as they imagined the position should be (p. xxviii).

> If he [Ptolemy] observed, he had to make these comparisons [between Hipparchus' observations and his own]; if he suppressed them in order not to discredit his catalog and his observations too much, he lacked good faith; he did not have that astronomical probity which is one of the most indispensable qualities for an observer (p. xxxi).

Ptolemy

> did not observe at all; he made calculations on the basis of Hipparchus' tables, and he gave us these calculations as observations. One is therefore compelled to discard his catalog. ... One subtracts the 2°40′ which one supposes he simply added to Hipparchus' longitudes. ... He has lost the

NOTES

right to be taken at his word, and one is disposed to deny the reality of all his observations (p. xxxii).

In our own time the pendulum continues to swing back and forth. On the one hand, Copernicus' attitude toward Ptolemy's catalog reappears, with some reservations, in Otto Neugebauer, *A History of Ancient Mathematical Astronomy* (Berlin/New York, 1975), pp. 280–288. On the other hand, a successor of Delambre exclaims:

> all of Ptolemy's own observations connected with the sun and moon are fabricated. All of his observations of the stars that he uses are also fabricated. ... all of his theories depend heavily upon fabricated data, and some of them seem to depend completely upon such data.
>
> Ptolemy says that he has measured the coordinates of all the stars that it is possible to observe, down to stars of the sixth magnitude. He identifies the instrument with which he made the measurements, he describes the procedure that was followed, and he presents the alleged results in his star catalog. However, ... the coordinates were not obtained by measurement at all. They were not obtained with the instrument that Ptolemy claims to have used, and they were not obtained by any other instrument or procedure of observation. They were fabricated, and Ptolemy lied about what he did.
>
> Ptolemy ... developed certain astronomical theories and discovered that they were not consistent with observation. Instead of abandoning the theories, he deliberately fabricated observations from the theories so that he could claim that the observations prove the validity of his theories.
>
> Ptolemy did not observe the stellar coordinates by the method he describes, nor by any method at all. Instead he took an older star catalog and simply added a fixed quantity to the longitudes (Robert R. Newton, *The Crime of Claudius Ptolemy* [Baltimore/London, 1977], pp. 344, 353, 354, 358–359).

[29] Werner, Prop. 4: "... in the second year of Antoninus [called Antonius by Werner] ... in the year 150 of our Lord's incarnation" (sig. k 2^v/4 up, k 3^r/9).

[30] Antoninus became the Roman emperor on 11 July 138, so that his second regnal year began on 11 July 139.

[31] B, Mk, Sch, Up: *Magni*; V: *M.*; missing in Ox.

[32] PS 1515, III, 1: "... the third year of the years of Attamenes, which was the 463rd year after the death of Alexander" (fol. 28^r/7–8). Recognizing "attamenis" as an Arabic distortion, Copernicus wrote *Antonini* in the margin of the copy belonging to his Chapter (MK, p. 254, no. 12). See P. Czartoryski, "The Library of Copernicus," StC XVI, 358; 372, no. 17.

[33] B: (*Aegyptii*); Sch: *Aegyptii* (interlinear); V: *Aegypt.* (interlinear); missing in Mk, Ox, Up.

[34] Copernicus, *Revolutions*, III, 6:

> In computing the heavenly motions ... I shall use Egyptian years everywhere. Among the civil [years], they alone are found to be uniform. ... The Egyptian year ... presents no ambiguity with its definite number of 365 days. [They comprise] 12 equal months [of 30 days each] ... and the 5 remaining days are termed intercalary (NCCW, II, 130/22–31).

[35] Mk, Sch, Up, V: *dubitari*; B, Ox: *dubitare*.

[36] Werner, Prop. 22:

> The interval between the years of Christ and Nebuchadnezzar is 747 uniform years, 131 days, according to the *Alfonsine Tables* (sig. q 2^v/14 up–12 up).

Here Copernicus does not call attention to Werner's confusion of Nabonassar with Nebuchadnezzar. But in *Revolutions*, III, 11, which explains that Ptolemy established as the earliest of his eras "the start of the reign of Nabonassar," Copernicus adds:

> Most [scholars], misled by the similarity of the name, have thought that he was Nebuchadnezzar, who lived much later (NCCW, II, 141/9–10).

[37] In the margin of B a later hand wrote:

> In the entire *Revolutions* Copernicus never thought that the *Alfonsine Tables* should be cited.

[38] B: (*Calendis non pridie Cal.*). The parenthesis ("first day, not the preceding day") was interpolated by

B in order to emphasize the difference between Ptolemy and the *Alfonsine Tables*. This parenthesis is not found in Mk, Ox, Sch, V, Up.

[39] PS 1515, fol. 34^r/13–16:

> ... from the beginning of the reign of Nebuchadnezzar [Nabonassar] to the death of Alexander ... and from the death of Alexander to the beginning of the reign of Augustus ... and from the beginning of the first year of the Egyptian years of the reign of Augustus, which [beginning] was on the first day of the month Thoth and at midday, because the reckoning of the heavenly bodies begins at noon.

PS 1515 places this passage in III, 8 (not III, 7, as in modern editions of PS). For, PS 1515 made a separate chapter (III, 6) out of the closing paragraphs of what is III, 5, in modern editions. Up to this point PS 1515 (Gerard of Cremona's translation of PS) was based on the 827–828 "translation made by al-Haggag ibn Yusuf ibn Matar and Sergius, son of Elias, ... from Greek into Arabic," according to Ibn as-Salah, as cited by Paul Kunitzsch, *Der Almagest: Die Syntaxis mathematica des Claudius Ptolemäus in arabisch-lateinischer Ueberlieferung* (Wiesbaden, 1974), p. 23/5–7.

[40] Mk jumped from *ultimi diei* to the next occurrence of that expression, omitting *anni praecedentis, quemadmodum nos a meridie*. By contrast, when Sch reached the second *ultimi diei*, he went back to *anni praecedentis*. Then, noticing his error, he deleted *quemadmodum nos a meridie ultimi diei anni praecedentis*, and proceeded with *quemadmodum nos a meridie ultimi diei mensis Decembris*.

[41] Werner, Prop. 16, Corollary 3:

> ... Roman years ... that is, years computed from the Savior's birth and from noon of the last day of December (sig. o 1^r/11 up–9 up).

[42] Ox, Up, V: *pariles*, missing in B, Sch; Mk: *pariter* (cf. n. 56, below).

[43] PS 1515, fol. 34^r/12–14:

> The total number of years which elapsed from the beginning of the reign of Nebuchadnezzar [Nabonassar] to the death of Alexander is 424 Egyptian years.

[44] Copernicus cites Chapter 8, instead of 7, for the reason explained in n. 39, above.

[45] Censorinus, *De die natali*, 21/1:

> Now, however, I shall discuss that period of time which Varro calls the "historical" [period, by contrast with the preceding "obscure" and "mythical" periods].

[46] Censorinus, 21/9:

> Certain years ... are called the years of Nabonassar ... of which the present year is the 986th; in like manner ... years are counted from Alexander the Great's death, and when extended to the present year amount to 562.

986 – 562 = 424. Copernicus may have learned from Censorinus that the astronomers' earliest era started with Nabonassar, not Nebuchadnezzar (see n. 36, above).

[47] Mk, Ox, Sch, Up, V: *Q.* (the abbreviation for Quintus, the correct praenomen); B: *C.* (the abbreviation for Gaius).

[48] Sch: *Cerelium*; V: *Cerilium*; B: *Cornelium*; Mk, Ox, Up: *C*. The edition of Censorinus which was published on 12 May 1497 in Bologna, while Copernicus was studying at the university in that city, on folio 2^r named the dedicatee as *Q. Cerelius*, a relatively rare name understandably confused with the much more common name *Cornelius*; see *Gesamtkatalog der Wiegendrucke*, VI (Leipzig, 1934), col. 370, no. 6471.

[49] Sch: CXXXLX, by scribal error for CXXXIX.

[50] The observation was made in the 463rd year after Alexander's death (see the *Letter against Werner* at n. 31) about an hour after sunrise on the 9th day of Athyr, the third Egyptian month, and is therefore assigned to the 68th astronomical day of the year. As Copernicus reckons,

from Alexander's death to the observation: $463^y 68^d = 462^y 433^d$
from Alexander's death to Christ's birth: $\quad\quad\quad\quad -323\ \ 130$
from Christ's birth to the observation: $\quad\quad\quad\quad\ \ 139^y 303^d$

But according to PS 1515 (fol. 28^r/18–19), the equinox occurred "in the 463rd year after Alexander's death" (*post mortem Alexandri in · 463 · anno*). Hence Copernicus should have reckoned with an interval of 462^{y}68^d (not 463^{y}68^d, which would put the equinox in the 464th year after Alexander's death).

[51] Mk, Ox, Up: *139*; V: *130*; Sch: *CXXX*; B: *CXXX (de est)*.

[52] B, Mk, Sch, Up, V: *Romanorum*; Ox: *Quoniam*.

[53] In moving from the uniform Egyptian year (= 365^d) to the Roman year (= 365$^{1/4d}$), Copernicus deducts 1^d for every 4^y:

$$1/4 \times 140^y = 35 \text{ leap years. Then,}$$
$$303^d - 35^d = 268^d = 25 \text{ September.}$$

[54] B, Sch: *V*; Mk, Up, V: *5*; Ox: *8*.

[55] PS V, 3 (ed. 1515, fol. 48^r/22 up–21 up):

> I observed the sun and moon in the 2nd year of the years of Antoninus.

Here PS 1515 prints *antonini* by contrast with "attamenis" at fol. 28^r/7 (see n. 32, above). PS 1515, fol. 48^r/15 up–13 up:

> The time which elapsed from the place of the sun and moon in the first of the years of Nebuchadnezzar [Nabonassar] to this observation was 885 years and 203 days.

[56] B, Ox, Sch, Up, V: *pariles*; Mk: *pariter* (cf. n. 42, above).

[57] B, Mk, Ox, Up, V: *73*; Sch: *730*.

[58] From Nabonassar to observation: 885^{y}203^d
From Nabonassar to Christ's birth: —747 130
From Christ's birth to observation: 138^y 73^d

[59] Mk, Ox, Up, V: *14*; B: *fortassis 54* / *14*; Sch: *14*. Sch^2: *fortassis 54*

[60] According to PS 1515, the observation was made "after the completion of 25 days of the month 'camenut,' which is one of the Egyptian months" (fol. 48^r/21 up–20 up). Copernicus recognized "camenut" as a distortion of Phamenoth, which he wrote in the margin (MK, p. 256, no. 50). Hence he reckoned from 25 Phamenoth to 9 Pharmuthi, the following month: 5+9 = 14^d.

[61] PS VII, 2 (ed. 1515, fol. 74^v/29): "the longitude of the star Heart of the Lion [Regulus] was observed."

[62] 138^{y}73^d + 14^d = 138^{y}87^d. In 138 Roman years there are 34 leap days, which must be subtracted from the Egyptian 87th day in order to obtain the corresponding Roman day:

$$87^d - 34^d = 53^d = 31 \text{ January} + 22 \text{ February.}$$

By a different computation *Revolutions*, II, 14, made this date 24 February (NCCW, II, 371–372).

[63] Referring to the end of the preceding paragraph, B interpolates parenthetically: (Previously he mentioned the third). B was evidently unaware that Antoninus' second regnal year ended on 10 July 139. Hence the observation of Regulus on 9 Pharmuthi, made in February, was executed while Antoninus' second regnal year was still in progress. On the other hand, the observation of the autumnal equinox on 9 Athyr was made in September, after Antoninus' third regnal year had begun. In other words, A.D. 139 was coextensive with the latter part of Antoninus' second year and the early part of his third year.

[64] B, Mk, Ox, Sch, V: *supra*; Up: *ultra*. Either reading means that Werner's error was "eleven years too much," not "more than eleven years," as in PI^2, 225/11, followed by Zinner's "about 12 years" ("Die fränkische Sternkunde," p. 26/6 up).

[65] B, Ox, Sch, V: *experimentum*; Mk, Up: *experientiam*.

[66] Aristotle, *Physics*, IV, 12 (220b32–221a1; 221b7): "time is the measure of motion," translated as *tempus mensura sit motus* by Joannes Argyropoulos (1410–1480), whose translation of Aristotle's *Physics* was first printed in Rome about 1480, and three times thereafter before 1524. Aristotle, *Physics*, IV, 11 (219b1–2; 220a24–25): "time is the number of motion according to before-and-after," translated by Argyropoulos as *tempus numerum esse motus per prius posteriusve*. Copernicus' *tempus esse numerum sive mensuram motus caeli secundum prius et posterius* conflates both of Argyropoulos' expressions and, in addition, specifies that the measured motion occurs in heaven (*caeli*), a specification not made by Aristotle and therefore not found in Argyropoulos' translation.

[67] Copernicus restates and generalizes the relation between time and motion formulated in Aristotle's *Physics*, IV, 12 (220b14–18):

> We measure not only motion by time, but also time by motion, because they are defined by each other. For time, being the number of motion, defines it, while motion defines time.

[68] Sch omitted *adhuc*. Noticing his error, he deleted *ex recenter*, and added *adhuc ex recenter* in the left margin.

[69] See n. 28, above.

[70] V: *aliquam* (not in B, Mk, Ox, Sch, Up), interpolated from *aliquem* in the preceding line. Like V (and Mk, Ox, Sch, Up), B too has *aliquem* (MCV, I, 25/on 25, to the contrary notwithstanding).

[71] Copernicus: *organis astrolabicis*; PS 1515, fol. 74^v/17: *Per unam quidem duarum armillarum ...*

[72] PS 1515, fol. 74^v/23–24: *transactis novem diebus mensis carmothi qui est ex mensibus aegyptiorum* (after the completion of nine days of the month "carmothi," which is one of the Egyptian months). Copernicus recognized "carmothi" as one of the Arabic distortions of the name of the Egyptian month Pharmuthi.

[73] Copernicus: *quinque horis et dimidia a meridie*; PS 1515, fol. 74^v/25: *post medietatem diei ... quinque horis et medietate horae*.

[74] Ox: *158*, corrected in the left margin to *138*.

[75] From Nabonassar to Christ's birth: $747^y 130^d 0^h$
From Christ's birth to observation: $+138 \ \ 88 \ \ 5\frac{1}{2}$
From Nabonassar to observation: $885^y 218^d 5\frac{1}{2}^h$

PS 1515's Table of the Sun's Mean Motion (fol. 29^r) gives for

810^y 163° 4′13″ (neglecting smaller fractions)
72   342 29 42
3   359 16 14
885^y 210^d 206 59 0
        8   7 53 6
       218^d 5^h 12 19
               $\frac{1}{2}$ 1 14
            $5\frac{1}{2}^h$
                  1079 55 48

initial position at the beginning of Nabonassar's reign (fol. 28^v/11 up–10 up)   +265 15 —
                           1345°10′48″

subtract 3 whole revolutions   —1080 — —
                           265 10

anomaly (fol. 33^v)   +2 23
                           267 33

place of solar apogee (Twins 5°30′)   +65 30
                           333°3′ = Fishes 3°3′

"The sun was ... in $3° + \frac{1}{2}$ ($\frac{1}{10}°$) of the Fishes, approximately" (PS 1515, fol. 74^v/31). A similar computation for the moon yields the same result.

[76] Sch: *Et tantum de primo errore* (And so much for the first error). Interpolated by Sch, this caption is not found in B, Mk, Ox, Up, V.

[77] Werner, Prop. 6:

> To prove that the fixed stars' motion in the zodiac for about 400 years before Ptolemy's era was nearly uniform and equal. In many passages of his *Great Syntaxis* Ptolemy shows with reference to the stars' motion that previous to himself and to his observation of the fixed stars, they moved for about 400 years only 1° in each century. Therefore if for 400 years the fixed stars' motion completed 1° in each century, the consequence is that the fixed stars' motion for 400 years before Ptolemy was nearly uniform and equal (sig. k 3^v).

Prop. 8:

> Ptolemy ... shows that prior to his analysis of the motions of the fixed stars, in their own motion the fixed stars traversed 1° in 100 years, and moved only 4° in 400 Egyptian years. Since in a continuous period of 400 years the motion of the fixed stars was always uniform and equal, in each 100 years it completed 1°. It is therefore clear that the fixed stars moved only with equal motion and lacked unequal motion, or if they had any unequal motion, it was very small and almost imperceptible (sig. 1 1^v/17 up–7 up).

Werner's "almost imperceptible" deviation from perfectly uniform motion is ignored here by Copernicus, who shows later on (see text below at n. 147) that the deviation is quite perceptible, amounting to almost $1/6$°.

[78] B, Mk, Ox, Sch, Up, V: *magisque*; MCV, I, 26/14: *magis*, omitting *-que*, and then reintroducing it through the unnecessary emendation [*appareant, utque*].

[79] B, Sch, V: *ex ... numero*, missing in Mk, Ox, Up.

[80] B, Sch, V: *quae*; Mk, Ox, Up: *quaecumque*.

[81] B, Mk, Ox, Sch, V: *terrae*; Up: *terra*, a faulty reading.

[82] Mk, Ox, Up: *planetae*, missing in B, Sch, V.

[83] Copernicus' example is based on Aristotle's argument that the stars twinkle while the planets do not, because the planets are closer to us:

> When our sight is extended a great distance, it wavers on account of its weakness. This may also be the reason why the fixed stars seem to twinkle, whereas the planets do not. For, the planets are near, so that our sight reaches them at full strength. But being extended too far all the way out to the fixed stars, it trembles on account of the distance. Its trembling, however, makes that motion appear to be performed by the star (*Heavens*, II, 8; 290a17–23).

Moreover, in Copernicus' epistemological analysis, our observation that the planets do not twinkle provides the empirical basis for the logical inference that they are nearer to us than the stars are. His reasoning is an extension of Aristotle's discussion:

> Let C stand for "planet"; B, for "non-twinkling"; A, for "being near." Then it is true to say B with regard to C, since the planets do not twinkle. But it is also [true to say] A with reference to B, because what does not twinkle is near, whether this was assumed because of a logical deduction or sense perception. Therefore A must be applicable to C. As a result, it has been proved that the planets are near. For, their nearness is not caused by their non-twinkling, whereas their non-twinkling is caused by their nearness (*Posterior Analytics*, I, 13; 78a31–38).

[84] B, Mk, Sch, V: *esse*, omitted by Up; Ox: *eorum*, a faulty reading.

[85] In Copernicus' thinking, the heavenly bodies are carried by epicycles, eccentrics, or other circles (*circulos*); see n. 326 on *Commentariolus*.

[86] B, Mk, Ox, Up, V: *dictum*; Sch: *datum*, an erroneous reading.

[87] B, Ox, Sch, V: *cum temporum intervallis*; Up: *tum temporum intervalla*; Mk: *tum temporum intervallis*, an erroneous reading.

[88] B, Mk, Ox, Up: *consideratis visisque stellarum locis adstipulatione quadam omnibus conveniret*; a redundant *omnibus* before *stellarum* (as in Sch, V) was printed by MCV, I, 27/7–8, followed by PII, 176/15–16.

[89] PII, 176/19's *cum* is a misprint for *eum*.

[90] B, Mk, Ox, Sch, Up: *considerationibus*; V: *considerationibus*, with *observationibus* inserted above, misreported by MCV, I, 27/on 11, as though V did not have *considerationibus* in the line. In this context *consideratio* is a standard technical term for "observation," and does not mean "contemplation" (*Betrachtung*), as was mistakenly supposed by MCV, I, 27, followed by PII, 176.

[91] Mk, Ox, Up: *quis suo sensu(i) inhaerens*; Sch, V: *secus aliquis*; B: *sensui inhaerens*. B's divergence from Sch indicates familiarity with the alternative reading presented by the group Mk, Ox, Up.

[92] Mk, Ox, Sch, Up: *non concedendum*; B, V: *non credendum*.

[93] Up at first wrote *more* (in the manner of), then deleted it.

[94] B, Sch, V: *utpote*; Mk, Ox, Up: *ut puta*, a mistaken reading.

[95] B, Mk, Sch, Up, V: *solerti*; Ox: *solertia*, a mistaken grouping with *diligentia et*.

[96] B, Mk, Ox, Sch, Up: *possum*; V: *possim*, a mistaken reading, printed by MCV, I, 27/19, followed by PII, 177/8.

[97] B, Mk, Ox, Sch, Up: *eos*, omitted by V and MCV, I, 27/20.

⁹⁸ In the *Letter against Werner*, see the paragraph following n. 172.

⁹⁹ Mk, Ox, Sch, Up, V: *praetereundum*; B: *praetermittendum*.

¹⁰⁰ B, Sch, V: *intelligatur omnes motus apparentis*;
  Ox: *intelligatur omnes motus apparentes*;
  Up: *intelligantur omnes motus apparentes*;
  Mk: *intelligemus omnes motus apparentis*.

¹⁰¹ B, Sch, V: *metiri*; Mk, Ox, Up: *mensurari*.

¹⁰² As a professor of astronomy, B demonstrates his familiarity with the literature of his subject by inserting within parentheses on the left side of the last four lines but one of fol. 8ᵛ a reference to fol. 89 of the 1543 edition of *Revolutions*:

> A mean and uniform motion is established with greater numerical accuracy, the better it is distinguished from its variations which are nonuniform (*inaequalitatis*, misreported as *aequalitatis* by MCV, I, 27, followed by PII, 177/n.*).

B has generalized Copernicus' specific statement in *Revolutions*, III, 18, by omitting *centri terrae* and changing *reddetur* to the present tense *redditur* (NCOO, II, 159/5–7).

¹⁰³ B, Mk, Ox, Up: *oportere*; Sch, V: *oportet*.

¹⁰⁴ Mk, Ox, Up, V: *invicem*, not in B, Sch.

¹⁰⁵ B, Mk, Sch, Up, V: *ut puta*; Ox: *utpote* (contrariwise, Ox has *ut puta*, where *utpote* is correct; see n. 94, above).

¹⁰⁶ B, Sch, V: *mediantium*; Ox, Up: *ambientium*, with *mediantium* above the line; Mk: *mediantiam ambientium*.

¹⁰⁷ B, Mk, Ox, V: *amborum*; Sch, V: *ambarum*.

¹⁰⁸ Mk, Ox, Sch, Up, V: *decrescat*; B: *decrescit*, corrected to *decrescat* in the line.

¹⁰⁹ B, Sch, V: *circuitionem*; Ox, Up: *circuitionem*, with *revolutionem* above the line; Mk: *revolutionem*, with *circuitionem* above the line after being deleted in the line.

¹¹⁰ B, Mk, Ox, Sch, Up, V: *Ptolemaeus*, misreported by MCV, I, 28/on 11–12, as missing in B.

¹¹¹ PS 1515, IV, 2, where Hipparchus' name is given as "Abrachis" (for example, fol. 36ʳ/4 up). PS 1515 prints a Latin translation of *Syntaxis* that was based, not on Ptolemy's Greek, but on a prior translation into Arabic. That language lacks the sound *p*, for which it substitutes *b* in non-Arabic names; see Kunitzsch, *Almagest*, p. 160, no. 1. PS 1515 did not include our Figure 13 (or its equivalent), so that the device of a quadrisected circle representing the places of a recurring nonuniform motion was inferred, rather than taken directly, from PS 1515, IV, 2.

¹¹² See the *Letter against Werner*, just before n. 89, above. The passage beginning here and ending at our n. 120 was quoted by Tycho Brahe in a work printed in 1588, 1603, and 1610. This quotation provided Brożek with his information about the *Letter against Werner*. For in his copy of the third edition of *Revolutions*, Brożek wrote the following note: "See Tycho's book on the comet of 1577; in it is a fragment of Copernicus' letter to Bernard Wapowski" (MK, p. 502, n. 1).

¹¹³ B, Mk, Ox, Sch, Up, V: *nondum*; Br (TB IV, 292/10): *non*, an inferior reading.

¹¹⁴ B, Mk, Ox, Sch, Up, V: *in annorum millibus*; Br (TB IV, 292/9) inserts *aliquot* (several).

¹¹⁵ Brahe alters the syntax by interpolating *est, ut* (TB IV, 292/10). In general, his quotation from Br is not meticulously precise.

¹¹⁶ Ox: *quia*; Up: *ǫa*, resolved incorrectly above the line as *qua*; Mk: *quia*, with *quae* above the line; Sch, V: *quae*, an incorrect reading, printed by MCV, I, 28/15, followed by PII, 178/6; B: q̃; Br (TB IV, 292/11): *quod*.

¹¹⁷ B, Sch, V: *post Ptolemaeum adauctis*, with which Br agreed (TB IV, 292/13); Mk, Ox, Up: *Ptolemaei adiunctis*.

¹¹⁸ B, Sch, V: *data*, with which Br agrees (TB IV, 292/16). Before *data*, B interpolated within parentheses: (like three lunar eclipses, three evening risings) of an outer planet, referring to *Revolutions*, V, 4 (NCCW, II, 244/2–6). Mk, Ox, Up: *posita*; in Sch, *data* was deleted, and *posita* written above it by a second hand.

¹¹⁹ B, Mk: *transmissa*, with which Br agrees (TB IV, 292/18); Ox, Sch, Up, V: *transmissae*, an incorrect reading.

¹²⁰ Mk, Ox, Sch, Up, V: *licet*, with which Br agrees (TB IV, 292/17); B: *licebit*, printed by MCV, I, 28/21, followed by PII, 178/11.

¹²¹ *Revolutions*, III, 6 (NCCW, II, 128); written before 3 June 1524.

¹²² In the *Letter against Werner*, at n. 98, above.

¹²³ B, Sch: *autor*, missing in Mk, Ox, Up, V.

[124] Mk, Ox, Sch, Up, V: *qui*; B: *quod* (because).

[125] After *medius*, B interpolated within parentheses: (arithmetical mean).

[126] B, Sch, V: *concitatissimum*; Mk, Ox, Up: *velocissimum*, with *concitatissimum* above the line.

[127] B, Ox, Sch, Up: *primo*; V: *primae*, a mistake due to the influence of the following word *septimae*; Mk: *primo*, changed to *primae* by a second hand.

Werner, Prop. 7:

> The motion of the fixed stars is known to have been in past centuries and times sometimes slower but sometimes faster. For the purpose of accounting for this inequality in the motion of the fixed stars, it is therefore necessary to place two small circles within the concavity of the tenth sphere. These are opposite each other along the diameter of the universe. Their vertices or poles are attached to the ecliptic of the tenth sphere. In these two circles two points, likewise separated from each other by the diameter [of the universe], revolve along the ecliptic of the ninth sphere. ...

Corollary 1:

> The motion of the fixed stars is slower when the two assumed revolving points of the ecliptic of the ninth sphere are near the intersections of the tenth sphere's ecliptic with the small circles, while when the same points are farthest removed [from those intersections] the same motion is faster (sig. l 1^r/3–11, 13 up–8 up).

[128] B, Mk, Ox, Sch, Up, V: *fixorum*; V's *fixorum* was misreported as *fixarum* by MCV, I, 28/on 30.

[129] Werner, Prop. 13:

> During the 400 Egyptian years before Ptolemy's observation, the fixed stars moved 1° in 100 years [but faster thereafter]. Therefore in Ptolemy's time the motion of the fixed stars was slower or the slowest (sig. m 3^r/7–13).

[130] B, Sch: *Assumit*; V: *Adsumit*; Mk, Ox, Up: *Asseruit*, a mistaken reading inconsistent with the present tense of *dicit* (twice), *ponit*, *animadvertit*, and *vocat* in this paragraph.

[131] B, Mk, Ox, Sch, V: *Aristarcho et*; Up: *Aristarcho Samio*, with *et* written over *Samio*. PS 1515, fol. 73^r/22 up–20 up: Hipparchus

> found that there were very few extant observations of the fixed stars before him, and he found that only the observations of Arsatilis and Timocharis were recorded.

Copernicus recognized that "Arsatilis" was an Arabic distortion of the name of a Greek astronomer. But at first he misidentified "Arsatilis" with Aristarchus, as he did here and also in a marginal note on fol. 75^v in his Chapter's copy of PS 1515 (Mk, p. 256, no. 64). At some time after sending this letter to Wapowski on 3 June 1524, however, he discovered the correct identification, Aristyllus, which he entered in a marginal note on fol. 73^r in his Chapter's copy of PS 1515 (MK, p. 256, no. 59). The same improvement may be seen in his autograph manuscript of *Revolutions*. There, at fol. 78^v, line 2, he originally wrote "Aristarchus," which he later corrected in the margin to "Aristyllus," after 3 June 1524. Hence it was before 3 June 1524 that he wrote folio 78^v as part of *Revolutions*, III, 6. This marginal correction on folio 78^v may be added to the other evidence Sw says he could not see when he contended (1974, p. 194) that

> in all probability the *Revolutions* was entirely a work of the 1530's, and I can see no evidence for giving any part of it an earlier date.

See "When Did Copernicus Write the *Revolutions*?" in *Sudhoffs Archiv*, 1977, 61:144–155, and n. 121, above.

When Copernicus at first misidentified "Arsatilis" with Aristarchus, at least he recognized that "Arsatilis" was wrong. By contrast, sixty or more years later Galileo repeated "Arsatilis" unchanged ("Arsatiris" in GG, I, 38/27, is a misreading, according to William A. Wallace, *Galileo's Early Notebooks* [Notre Dame, 1977], p. 263). The transition from Aristyllus to Arsatilis is explained by Kunitzsch, p. 161, no. 3. Arsatilis was misidentified with Aratus by Werner, Prop. 22 (sig. q 1^v/15 up, 13 up, 6 up, last line; q 2^r/4, 8).

[132] Mk, Ox, Sch, Up, V: *ut puta*; B: *utpote*, a mistaken reading; cf. n. 94, above.

[133] PS 1515, fol. 77^v/22 up: "1° in every 100 years of time."

[134] Copernicus mistakenly referred to Prop. 7, instead of Prop. 6, which was quoted in n. 77, above.

[135] B, Mk, Ox, Sch, Up: *advertit*; V: *animadvertit*.

[136] B, Ox, Sch, V: *sectiones*; Mk, Up: *sectionis*, under the influence of the preceding *aequalitatis*.

[137] Ox, Sch, Up, V: *vocat*; B: *notat*; Mk: *vere* (!).

Werner, Prop. 11:

> The apparent or nonuniform motion of the sphere of the fixed stars or of the eighth sphere is caused by the revolution of the first points of the Crab and the Goat of the ninth sphere's ecliptic on small circles. Thabit and the *Alfonsine Tables* call this revolution the forward and backward motion or trepidation of the eighth sphere. In addition, this trepidation proceeds sometimes in the order of the signs, sometimes in the opposite order. Hence the same motion of the fixed stars happens to be sometimes slow and sometimes fast. It is clear, moreover, that the same motion of the fixed stars is composed of the eighth sphere's uniform motion and the ninth sphere's trepidation or forward and backward motion on the small circles (sig. m 1ᵛ/5–15).

[138] B, Mk, Ox, Sch, V: *uniformior*; Up: *uniformiter*, a mistaken reading.

[139] Werner, Prop. 13:

> The first points of the Crab and the Goat of the ninth sphere's ecliptic were only a slight distance away from the intersections of the small circles with the tenth sphere's ecliptic in Ptolemy's time (sig. m 3ʳ/2 up–m 3ᵛ/2).

[140] B, Mk, Ox, Sch, Up: *contextis*; V: *contectis*, a mistaken reading, misreported as *confectis* and so printed by MCV, I, 29/16, followed by PII, 179/10.

[141] Werner's "Last Table of the Motions of the Eighth Sphere" is part of Prop. 30 (sig. s 4ᵛ–t 2ʳ).

[142] Mk, Ox, Sch, Up, V: *inaequalitatis*; B: *aequalitatis*, a mistaken reading, printed by MCV, I, 29/17, followed by PII, 179/11.

[143] B, Sch, V: *ante nativitatem*; Mk, Ox, Up: *a nativitate*, a mistaken reading.

[144] B, Sch, V: *primum*; Mk, Ox, Up: *priorum*, a mistaken reading.

[145] Mk, Ox, Sch, Up, V: *grad. 1 et decima fere parte unius, in secundo* (Mk: *eodem*, a mistake) *grad. 1 et*. B jumped from the first occurrence of *grad. 1 et* to the second occurrence, omitting *decima fere parte unius, in secundo*. The omitted words (except *unius*) were later written in the right margin by a second hand.

[146] In Werner's "Last Table" (sig. s 4ᵛ), the fifth column (True Motions of the Eighth Sphere or of the Fixed Stars) gives for the year 100 A.D. $1°18'26''$, from which $12'22''$ must be subtracted as the value at the beginning of the Christian era: $1°18'26'' - 12'22'' = 1°6'4'' \simeq 1\ 1/10°$. For the year 200 A.D., Werner's "Last Table" gives $2°32'40''$; subtraction of $1°18'26''$ leaves $1°14'14'' \simeq 1\ 1/4°$.

[147] $57' - 49' = 8'$; $1\ 1/10° - 57' = 9'$; $1\ 1/4° - 1\ 1/10° = 9'$ ($<10' = 1/6°$)

[148] $49' + 57' = 1°46' = 2° - 14'$; $14' > 12'$ ($= 1/5°$)

[149] $1\ 1/10° + 1\ 1/4° = 2°21'$. The excess over $2°$ is closer to $1/3° = 20'$ than to $1/4° = 15'$.

[150] $2°21' - 1°46' = 35' = 30' + 5' \simeq 1/2° + 1/15°$ ($= 4'$).

[151] B, Mk, Ox, Sch, V: *Canone*, omitted by Up inadvertently.

[152] Werner's "Last Table" (sig. t 1ʳ) gives the following values for

600	8°36′ 9″
700	10 17 0
800	11 58 35
900	13 39 29
1000	15 17 30

The centurial differences ($1°40'57''$, $1°41'35''$, $1°40'54''$, $1°38'1''$) vary by less than $1'$ in the first three instances, but the fourth varies by nearly $3'$. For the 400 years after 2060, Werner's "Last Table" (sig. t 1ᵛ–t 2ʳ) gives the following values:

2060	25°38′13″
2160	25 58 56
2260	26 17 38
2360	26 36 8
2460	26 56 7

Here the centurial differences ($20'43''$, $18'42''$, $18'30''$, $19'59''$) vary from $2'1''$ to $12''$. PII, 151/16's "2000" (repeating the misprint in the 1854 edition of *Revolutions*) is corrected to 2060 in PII, 179, n. *.

[153] See the *Letter against Werner*, near Figure 13.

[154] B, Sch, V: *a summa*, missing in Mk, Ox, Up.

[155] Mk, Ox, Sch, Up, V: *motui*; B: *motu*, a mistaken reading, somewhat smudged.

[156] B interpolates *computatus*, which is not found in Mk, Ox, Sch, Up, V.

[157] B: *ad tarditatem summam*; Sch, V: *ad tarditatem*; missing in Mk, Ox, Up.

[158] Mk, Ox, Sch, Up: *continuo*; V: *continue*; B: *contrario*.

[159] B interpolated parenthetically in the text: (προσθαφαίρεσις) as the Greek equivalent of "increase and decrease."

NOTES

[160] Sch: *diametro*; B, Mk, Ox, Up: *e diametro*; V: *in diametro*.

[161] B interpolated parenthetically in the text: (as in the case of the sun).

[162] B, Sch: *aequalibus*, missing in Mk, Ox, Up, V.

[163] Ox, Up, V: *qui*; Sch: *quae in est*; Mk: *quam*; B: *quorum*, an emendation which restores what Copernicus wrote, by contrast with the other manuscripts' various unsuccessful resolutions of an unfamiliar abbreviation of *quorum*.

[164] Mk, Ox, Sch, Up, V: *fiat*, unnecessarily emended by B to *sit*. MCV, I, 30/11's grammatically unsound emendation *sint* lacks manuscript support.

[165] B, Mk, Ox, Up: *altero*; Sch, V: *alteri*, the wrong case.

[166] Mk, Ox, Up: *aut*; B, Sch, V: *et*.

[167] Mk, Ox, Sch, Up, V: *circumferentiae*, an erroneous reading, corrected to *circumferentias* by B.

[168] In the text B inserted the passage quoted above, in the Introduction to the *Letter against Werner*, shortly after n. 72. From B, this exhortation was transferred to the left margin of Sch, folio 12^v (by Praetorius?).

[169] B originally wrote "4000," which he corrected by deleting the final zero.

[170] B, Sch, V: *divinemus*; Mk, Ox, Up: *divinemur*, a faulty reading.

[171] B remarked parenthetically in the text:

(It is a most wretched situation that the astronomical observations which we have begin with Timocharis, who lived 30 years after Alexander).

B's lament was transferred to the lower left margin of Sch's folio 12^v (by Praetorius?). *Revolutions*, III, 2, discussed an observation made by Timocharis in "the 30th year after the death of Alexander the Great" (NCCW, II, 120/31–35).

[172] Here Copernicus succinctly summarizes the preliminary version of *Revolutions*, III, 6. In the margin there (NCCW, I, folio 78^r) he drew a diagram that belongs with III, 7 (fol. 82^v). When he realized his error, he marked this diagram for deletion by passing three heavy curved strokes through it, but he failed to replace the deletion by an appropriate diagram. Hence we may turn to our own Figure 13. This complete revolution of a recurring nonuniform motion was conjecturally matched by Copernicus with the interval from Timocharis to his own age. He put Timocharis in what would be our quadrant II (in the text of *Revolutions*, III, 6, the quadrants are labeled differently, although the overall scheme is substantially the same). Ptolemy is placed in our quadrant III. Quadrant IV was completed several centuries before Copernicus' time, which he located toward the end of quadrant I. All the years since Timocharis were matched with the full cycle of all four quadrants. Here in the *Letter against Werner*, Copernicus put his own time late in the second semicircle: "no small part of it too has passed." By the same token, in the preliminary version of *Revolutions*, III, 6, "the cycle ... is nearly completed and is returning to where it began with Timocharis." Afterwards Copernicus revised these conjectural estimates, having found that "the motion ... had already completed its revolution, and exceeded it by 21°24'" (NCCW, II, 129/2–3, 9–10). Hence, when he wrote the *Letter against Werner*, he already had the preliminary version of *Revolutions*, III, 6, the later version of which must be dated after 3 June 1524.

[173] Mk, Ox, Up, V: *illos*; Sch: *eos*; B: *Eos eos*, at first with an erroneous upper-case *E*, then with a lower-case *e*, but forgetting to delete *Eos*.

[174] B, Ox, Sch, V: *insensibilis*; Mk, Up: *sensibilis*.

[175] B, Mk, Ox, Sch, V: *interveniens*; Up: *intervenire*, a mistaken reading under the influence of *evenire* eight words later.

[176] Mk, Ox, Sch, Up: *nimirum*, missing in V, and misread by B as *nimium*, which was printed in MCV, I, 31/16, followed by PII, 18/8.

[177] Here Copernicus repeats the error discussed in n. 131, above. Aware that Aristarchus came from Samos, which is some 500 miles northwest of Alexandria, Copernicus could link the Samian with Timocharis to form a pair of widely separated observers. Aristyllus, on the other hand, as an Alexandrian could not be paired with Timocharis for this purpose. Knowing that the astronomer Aristarchus was usually described as the "Samian" in order to distinguish him from the renowned literary critic Aristarchus, after *Aristarcho* Up wrote *Samio*, which he deleted when he realized that it was not present in his prototype.

[178] PS 1515, fol. 76^r/5 up: "Agrinus," correctly identified by Copernicus as Agrippa. In this instance the absence of *p* in Arabic led to its replacement, not by *b* (as in Hipparchus/Abrachis), but by *n*. The difference between *b* and *n* in the Arabic script is the location of the dot: above the curved stroke for *n*, and below it for *b*. Hence, a substitution of *n* for *b* is readily understandable. *Epitome*, VII, 5 (sig. h 7^r/5), called the Bithynian astronomer "Agrias," a form in which there is no equivalent for the original *p*. In earlier Arabic manuscripts the diacritical points were sometimes omitted. Seeing a curved stroke with no diacritical point above it or below it, and undecided as between *b* and *n*, a scribe may have omitted the character altogether, thereby emerging with

"Agrias"; this form is not discussed by Kunitzsch, p. 161, n. 4. *Epitome*'s "Agrias" and PS 1515's "Agrinus" could be discarded by Copernicus because he found the correct form Agrippa in Valla's *Seek and Avoid*, Book XVIII, 3 (sig. ff 6^v/19 up). By translating directly from the Greek text of Proclus' *Hypotyposes* (Chapter V, 7), Valla avoided the Arabic detour. Copernicus recorded the correct identification Agrippa in an annotation in his Chapter's copy of PS 1515 (MK, p. 257, no. 72). On the other hand, Galileo retained *Epitome*'s "Agrias" (which was changed to "Agrippa" by the editors of GG, I, 39/6). My colleague Professor Richard Lemay, formerly at the American University in Beirut, kindly helped me in questions concerning Arabic.

[179] PS 1515, fol. 76^v/5 up, 77^r/32: "Mileus"; the same distortion occurs also in *Epitome*, VII, 5 (sig. h 7^r/8 up), which relied on the same translation from Arabic; for the transformation of Menelaus into Mileus, see Kunitzsch, p. 162, no. 5. Copernicus found the full form of the name Menelaus in Valla's *Seek and Avoid* (fol. 6^v/13 up), and recorded it in an annotation in his Chapter's copy of PS 1515 (MK, p. 257, no. 75). On the other hand, Galileo wrote "Milaeus" (GG, I, 39/7). As late as 1817, in his *Histoire de l'astronomie ancienne* the great historian of astronomy Delambre asked whether Copernicus' Menelaus "would be Millaeus" (I, 186).

[180] B, Sch, V: *certissima*; Mk, Ox, Up: *verissima*.

[181] Mk, Ox, Sch, Up, V: *quo*; B: *quod* at first, with the *d* deleted later.

[182] Mk, Ox, Sch, Up, V: *etiam*; B: *et*, an incorrect reading.

[183] B, Sch, V: *ut puta*; Mk, Ox, Up: *utpote*, a mistaken reading (see notes 94, 105, above).

[184] Werner, Prop. 22: "... the fixed star called asichemech [aschimech, in PS 1515] inermis, but in Greek potrygetes [should be protrygetes], that is, grape-gatherer, in Latin, however, Arista" (sig. q 2^v/19–20). The more common Latin name *Spica* (Spike) was used by Copernicus in *Commentariolus*' section "Equal Motion Should Be Measured Not by the Equinoxes but by the Fixed Stars." See Paul Kunitzsch, *Arabische Sternnamen in Europa* (Wiesbaden, 1959), pp. 146–147, no. 66.

[185] Mk, Ox, Sch, Up, V: *quod*; B: *q* (a common abbreviation of *quod*) was reported by MCV, I, 32/on 3, as though it meant something other than *quod*.

[186] Werner, Prop. 22 (not 27, as in MCV, I, 31; repeated by PII, 181), Corollary:

> As regards Timocharis' observations, they clearly fall short of my computation for the fixed star called Arista, but exceed my calculation for the most northerly of the three bright stars in the Scorpion's brow. If, however, these observations made by Timocharis had both been correct, they should fall short of my calculation equally, or equally exceed it. My tables, therefore, are no less trustworthy than the observations and findings of the ancients (sig. q 3^r/5 up–q 3^v/3).

[187] B, Mk, Ox, Sch, Up: *consideratarum*; V: *consideraturum*, a faulty reading.

[188] Timocharis observed Scorpion 1

> in the 454th year of Nebuchadnezzar [Nabonassar], after the 16th complete day of the month "chaucha," in the morning of the 17th day (PS 1515, fol. 77^r/22–23).

In the Egyptian calendar the 4th month was Ch(o)iach, PS 1515's "chaucha." PS 1515's "aschimech inermis," identified with the Spike (PS 1515, fol. 83^r/7 up; MK, p. 256, no. 61; p. 257, no. 68), was observed by Timocharis

> in the 454th year of Nebuchadnezzar [Nabonassar], after the 5th complete day of the month "tobi," which is one of the Egyptian months, in the morning of the 6th day (PS 1515, fol. 76^v/25–27).

In the Egyptian calendar the 5th month was Tybi, PS 1515's "tobi." Hence, if PS 1515 was right, Timocharis' observations were made only 19 days apart, so that the time interval between them and Ptolemy's catalog of the fixed stars was "almost exactly the same." PS 1515's "chaucha," however, has been replaced in modern editions of PS by Phaophi, the 2nd Egyptian month. The fact that Timocharis' observations were made 79 days apart, rather than 19, does not seriously affect Copernicus' reasoning, since the interval between Timocharis and Ptolemy exceeds four centuries, as he points out in the next paragraph.

[189] Werner, Prop. 22, finds

> the fixed stars' true motion [in the interval] between Timocharis and Ptolemy [to be] 4° 7' 3" 28''' (sig. q 2^r/14 up–12 up).

The smaller fractions are disregarded by Copernicus.

[190] Werner, Prop. 22:

> Timocharis [whom he calls Timarchis] ... finds the most northerly of the three stars in the Scorpion's brow in 2° of the same sign Scorpion (sig. q 2^r/11–13).

NOTES

[191] Werner, Prop. 22:

> Ptolemy's tables place the same star in 6°20′ of Scorpion (sig. q 2^r/9 up–8 up).

Hence there would be a gap of 13′:

$$4°7'+2° = 6°7' = 6°20'-13'$$

[192] B, Ox, Up: *elevatus*; Sch, V: *elevatur*, a mistaken reading, shared by the (lost) Strasbourg copy, where Mk could recognize the ending *-tur* without being able to identify the rest of the word.

[193] Werner, Prop. 22:

> Subtraction of 4°7′57″ from Arista's true place as observed or computed by Ptolemy leaves 22°32′3″ of the Virgin (sig. q 3^r/5–6).

$$4°7'57''+22°32'3'' = 26°40'$$

[194] B, Ox, Sch, V: $22\ ^1/_3$; Mk, Up: $22\ \frac{s}{^1/_3}$. Mk copied the (lost) Strasbourg copy of Simon Hájek's copy. Since Up shares this peculiar fraction $\left(\frac{s}{^1/_3}\right)$ with Mk, Up too is descended from Simon Hájek's copy. This started to write the fraction as a number of minutes (*scrupula*, abbreviated *s*), but shifted to $^1/_3$ (of a degree) without deleting the *s*.

[195] PS 1515, fol. 76^v/28 up–26 up: Timocharis found "the distance of Aschimech inermis [Arista, Spike] from the summer solstitial point ... [to be] $82\ ^1/_3$°" = $22\ ^1/_3$° (+2 signs from the summer solstice).

[196] See n. 193, above.

[197] $2°+4°7' = 6°7'$, instead of 6°20′, a shortage of 13′; $26°40'-4°7'57'' = 22°32'3''$, instead of 22°20′, a surplus of 12′.

[198] Mk, Ox, Sch, Up, V: *in*, missing in B.

[199] Aristotle, *Physics*, III, 3: 202b13–14: "The road from Thebes to Athens [is the same as] the road from Athens to Thebes."

[200] Mk: *Alioquin*; Ox, Sch: *Alioqui*; B, Up, V: *Alioq̨* (= *Alioqui*). B's *Alioq̨* was misreported as *Alioq* by MCV, I, 32/on 17.

[201] Werner, Prop. 22:

> Between this observation of Timocharis and Ptolemy's investigation of the fixed stars, 443 Roman years and 64 days intervened (sig. q 2^r/13–15).

[202] In the *Letter against Werner* at n. 64 Copernicus declared that Werner dated Ptolemy's catalog 11 years too late.

[203] B, Mk, Up: *ut*; Ox, Sch, V: *et*.

[204] B, Sch, V: *dissidebit*; Mk, Ox, Up: *dissidebat*.

[205] Werner, Prop. 22:

> Therefore my tables would diminish the position of this star by 13′ (sig. q 2^r/8 up–7 up).

Again,

> However, my computation exceeds Timocharis' observation by 12′ (sig. q 3^r/8–9).

[206] B, Sch, V: *annotationibus*; Mk, Ox: *adnotationibus*; Up: *observationibus*.

[207] Werner, Prop. 22, Corollary:

> For this weakens not a little the reliability of the ancient observations of the fixed stars, since some of those observations exceed the computation based on the foregoing canons and tables, while certain of them fall short of this computation. For if all the results of the ancient observations of the fixed stars coincided exactly with the truth, they should, with perfect propriety, all together fall short of my calculation based on the aforesaid tables, or they should all equally exceed it. But it has been shown above that the ancient observations partly fall short of, and partly exceed, the calculation based on my tables (sig. q 3^r/15 up–5 up).

[208] Ox, Sch, Up, V: *quid*; B, Mk: *quod*, a mistaken reading.

²⁰⁹ Ox jumped from this occurrence of *motu* to its next occurrence, omitting *octavae sphaerae, e quibus etiam facile potest intelligi quid de*, these words being added in the right margin.

²¹⁰ Mk, Up, V: *e quibus etiam facile potest intelligi*, missing in B, Sch. This omission of about 35 letters in Sch (which was copied from the autograph) may indicate the length of a line in that lost document. By the same token, the omission of about 35 letters in Mk (see n. 22, above) suggests that Simon Hájek's copy was also arranged in lines of that length, like Mk itself.

²¹¹ B, Sch, Up, V: *quid*; Mk: *quod*, misattributed to B by MCV, I, 32/on 27–28, a misattribution corrected in MCV, IV, 9.

²¹² Here *motus declinationis* refers to the variation in the obliquity of the ecliptic, or the angle at which the ecliptic intersects the equator.

²¹³ B, Sch, V: *ipsum*; Mk, Ox, Up: *ipsam*.

²¹⁴ B, Mk, Ox, Sch, V: *ait*; Up: *vocat*.

²¹⁵ B, Sch, V: *instruendo*; Up: *inserendo*; Ox: a misreading, changed to *inserendo*; Mk: *in se redescendam*, a misreading of *inserendo secundam*.

²¹⁶ Werner, Prop. 18:

> The first trepidation or forward and backward motion is a property of the ninth sphere and its small circles. This trepidation of the ninth sphere is called the first trepidation because, by reason of the variation in the sun's maximum declination, a revolution or upward and downward movement on small circles must be assigned to the ecliptic of the tenth sphere also. This movement will, then, be named the second trepidation (sig. o 3ʳ/13–18).

²¹⁷ *Revolutions*, III, 1–12.

²¹⁸ B, Ox, Up, V: *hic*; Mk, Sch: *his*. In antiquity it had been discovered that the stars' celestial longitudes, as measured from the vernal equinoctial point, slowly increased over the centuries. This increase was attributed to an eastward motion of the sphere of the stars, or eighth sphere, while the earth remained motionless. Copernicus' earth, however, was a moving planet. A slight westward displacement of its axis would account for the increased longitudes of the stars, which remained motionless in Copernicus' universe. But to say something along these lines would have been inappropriate in a review of the treatise by Werner, who believed that the earth stood still and the stars moved. Remaining silent about these assumptions, Copernicus confined his review to showing that Werner's theory was grossly defective within the traditional framework.

²¹⁹ B, Sch, V: *faustissime*; Mk, Ox: *quam faustissime*; Up: *quam sanctissime*.

²²⁰ B, Sch: *Copernicus*; Mk: *Coppernicus*; Ox, Up, V: *Copphornicus*, a version of his surname never used by Copernicus himself. In his personal copy of his Greek-Latin dictionary, he indicated his ownership by writing in Greek that the book belonged to Κόπερνιχου (facsimile in PII, Plate Ve). In the Greek version of his surname he did not include the letter φ. The statement to the contrary (StC XVI, 356/20) improperly transfers to his dictionary an entry made by someone else in an entirely different work.

In 1512 Dantiscus published in Cracow a poem celebrating the marriage of King Sigismund I. In the manuscript of this *Epithalamium*, which Brożek examined in Lidzbark in 1618, he found a poem, five lines long, in praise of Dantiscus. Brożek transcribed this short poem on a sheet sewn into his copy of the third edition of *Revolutions*, which is now in the Jagiellonian Library in Cracow (shelf-mark 311204). The poem, which was not printed in the 1512 edition, survives only because it was transcribed by Brożek. His transcription attributes the authorship of the poem to Copernicus, whose surname is written in Greek with a φ. This Greek version of Copernicus' surname, written by someone else, differs from the Greek version written by Copernicus himself in his own dictionary, in that Copernicus does not include the letter φ, whereas the attributor of the poem does.

From Brożek's transcription, the poem and its attribution to Copernicus were taken by Martin Radymiński (1610–1664), a professor at the University of Cracow. His *Sketch of the Life and Writings of Nicholas Copernicus*, composed in 1658, remained in manuscript until it was printed (Cracow, 1873) in a celebration of the four-hundredth anniversary of Copernicus' birth (*Natalem N. Copernici ...*), with the poem at p. 21. There Radymiński incorrectly states that the poem preceded Dantiscus' *Epithalamium* in the edition of 1512. The poem, with its attribution to Copernicus, was published also by Dominik Szulc in his *Żywot M. Kopernika* (Warsaw, 1855), p. 74.

These two publications diffused the mistaken impression that Copernicus sometimes spelled his surname, in Greek, with a φ, and with the equivalent *ph* in Latin. No authentic example of Copernicus' signature shows φ in Greek or *ph* in Latin.

Yet his surname was unquestionably so written by others, including the attributor to him of the five-line poem in praise of Dantiscus. A medical book bequeathed by Copernicus to the Chapter's notary, Fabian Emmerich, bears the legatee's name "*D. Fabiani*," then the testator's name in the form "Copphernici," and the explanation "bequeathed to Fabian Emmerich in the will" (PI², 306; StC XVI, 370/1–2). Those who, like Emmerich, included *ph* in Copernicus' surname presumably connected it with the German word for copper, *Kupfer*.

Whereas Emmerich wrote *Copphernicus*, the *e* is changed to *o* in Ox, Up, V. In some handwritings of this period, the *e* is written in such a way that it is understandably mistaken for *o*.

[221] B, Sch: *1524*; Ox, Up: *MDXXIIII*; V: *MDXXXIIII*. In Mk, the place and date of writing appear, not here at the end of the *Letter against Werner*, but at the top of Mk's fol. 1. MCV, I, 33/11–13, repeats MCV, I, 23/2–3, without any basis in the manuscripts, while MCV, I, 33/14, belongs at MCV, I, 23/4. The same may be said about PII, 183/8–10, in relation to PII, 172.

COPERNICUS AND THE MONEY QUESTION

INTRODUCTION

On 3 November 1516 Copernicus was chosen by his Chapter to administer its holdings in the southern part of the diocese of Varmia. Eight days later, on 11 November, he entered upon his office in Olsztyn. There his duties brought him into close contact with the peasants. One of their most vexatious problems was the debasement of the coinage. Paper money had not yet been introduced. Ordinary business transactions were conducted with coins made of an alloy of silver and copper. Without fanfare, the percentage of silver to copper was being steadily reduced by the authorities. The resulting economic dislocation was carefully considered by Copernicus in a study which is the earliest entirely empirical discussion of the monetary question. It is equally remarkable for its secular spirit, since it is absolutely devoid of theological arguments and biblical quotations. It was completed on 15 August 1517, when Copernicus described it as his *Meditata* (private reflections). He wrote it in Latin, presumably with the intention of circulating it among his colleagues and friends. For the sake of convenience we shall call it "M."

Before long M came to the attention of the West Prussian Estates, which at that time conducted its normal business in the German language. At their request[1] Copernicus prepared a German version of M. Although at the end of M he had noted the day, month, and year when he finished it, he indicated only the year 1519 at the end of the German version. Moreover, he did not provide it with any characterization corresponding to *Meditata* for M. Since the German version lacks an authentic title, let us label it *Denkschrift*, abbreviated "D." D was not simply a translation of M. For in moving from M to D, Copernicus introduced suitable changes. These are not substantial enough to warrant complete English translations of both M and D separately. Instead, the translation of M will be presented in the center of the page, while alongside at the left D's modifications of M will appear in the appropriate places.

Despite the urgency of the monetary problem, 1519 (the year in which Copernicus wrote D) was most inauspicious for currency reform.[2] For on New Year's Day 1520 war broke out between the Teutonic Knights, on one side, and West Prussia (including the diocese of Varmia) and its overlord, the kingdom of Poland, on the other side. Fighting continued until a truce put an end to hostilities in 1521. On 5 April a ceasefire went into effect for four years. When peace was restored, the money question came to the fore again, this time in a wider framework.

At a meeting of the West Prussian Estates in March 1522, a spokesman for the Polish crown set forth the king's desire to introduce a uniform currency throughout his realm, including West Prussia.[3] It so happened that Copernicus was in attendance at this meeting as

---

[1] PII, 29/6 up: *ad petitionem Consiliariorum harum terrarum*. "The Prussians were used to hearing the communications of the king's emissaries in no other language than German or Latin. For, Polish was incomprehensible if not to all, then to most" [of them; Lengnich, I, 70-71, n. *].

[2] Copernicus "was requested to carry out several monetary reforms, e. g., the Prussian Monetary Reform in 1519," according to Hegeland, p. 14. But there was no Prussian Monetary Reform in 1519, nor was Copernicus ever requested to carry out any monetary reform. He was asked to present his views at a meeting of the West Prussian Estates in 1522, and in 1530 he was appointed a delegate of his Chapter to a commission charged with the evaluation of foreign gold coins, where his remarks went unheeded (see text at nn. 4, 156). Without any supporting argument, D is dated at the end of 1519 by Erich Sommerfeld, *Die Geldlehre des Nicolaus Copernicus* (Vaduz, 1978), p. 13/4.

[3] Schmauch, p. 13/1-5.

a representative of his bishop, who was ill. An official record of the proceedings, preserved in the archives of Gdańsk, reports

> the announcement that the honorable and worthy Nicholas Copernicus once occupied himself very diligently with these topics, and wrote out a discussion. The Councillors asked him kindly to communicate it to them, and for the good of the matter not to keep it hidden. He obliged, and it was read aloud in the presence of the Royal Councillors.[4]

A copy of what was read aloud was inserted at this point in the Gdańsk record, which also reports that on the same day, 21 March 1522, Copernicus "made a certain addition at the end" of D. This short Appendix (which will be called "A" hereafter) dealt with the new question how to obtain uniformity between the Prussian and Polish coinages:

> This might be done in the following way. Mint 60 new shillings to 1 mark, equal in intrinsic value and face value to 20 Polish groats; likewise, better pennies than those used at present, with 6 equal in value to 1 new shilling. According to this reckoning, 1 Polish groat would equal 3 Prussian shillings,[5] and $^1/_2$ Polish groat would equal 9 Prussian pence.[6] Thus the Prussian and Polish pennies would be equal in value.[7] Just as the Hungarian guilder is exchanged for 38 groats in Poland, so [it would be exchanged] likewise in Prussia for its groats, that is, for 2 marks minus 6 shillings.[8] Such a proposal might perhaps advance the equalization of the currencies[9] and also the land of Prussia.

A suggestion in A was repeated $2^1/_2$ months later, on 3 June 1522, at a meeting of the West Prussian Estates. In the new coinage, the representatives of Gdańsk recommended, "3 [Prussian] shillings should be worth 1 Polish groat, and be accepted throughout the kingdom" of Poland. This linkage between the Prussian and Polish currencies was adopted after $4^1/_2$ months, on 21 October 1522, and reported on the following day to the representatives of the crown.[10]

While such plans for equating the Polish currency with the Prussian were under consideration, the problem acquired an added dimension as a result of startling political developments. In 1525 the Order of Teutonic Knights was dissolved, its territory was converted into a secular state, and its former Grand Master became duke of (East) Prussia, which he ceded to the Polish king, from whom he received it back as a fief owing allegiance to the crown. An agreement between the king and the duke, dated 8 April 1525, in Article 28, provided that

> the duke of Prussia, as well as the people of Elbląg, Gdańsk, and Toruń, should refrain from minting any coins, while His Royal Majesty will, between now and Pentecost of

---

[4] PII, 29/22–29.

[5] 1 Prussian mark = 60 Prussian shillings = 20 Polish groats; 1 Polish groat = 3 Prussian shillings.

[6] 1 Polish groat = 3 Prussian shillings = 18 Prussian pence; $^1/_2$ Polish groat = 9 Prussian pence.

[7] 9 Prussian pence = $^1/_2$ Polish groat = 9 Polish pence; 1 Prussian penny = 1 Polish penny.

[8] 2 marks = 120 shillings; 2 marks — 6 shillings = 114 shillings; 3 Prussian shillings = 1 Prussian groat; 114 Prussian shillings = 38 Prussian groats; 1 Prussian groat = 1 Polish groat; 38 Prussian groats = 38 Polish groats = 1 Hungarian guilder.

[9] It would make 1 Prussian penny = 1 Polish penny, and 1 Prussian groat = 1 Polish groat.

[10] Schmauch, pp. 36–37.

next year [20 May 1526], summon an official session for the purpose of coordinating these coinages [of East Prussia and West Prussia.].[11]

If Copernicus expected to exert any influence on the decisions to be taken at such a session, he had to revise M, D, and A. For in some respects his previous writings on the coinage question had been made obsolete by the disappearance of the Teutonic Knights and the absorption of East Prussia into the kingdom of Poland. As an alert student of the Prussian monetary problem, Copernicus confronted a new political situation. Taking cognizance, therefore, of recent events, he composed the final version of his evolving reflections on money, the *Essay* which we shall call "E." In E he reverted to Latin for the convenience of those officials in the royal Polish administration who did not understand the language of their German-speaking subjects. But there is no historical justification for the assertion that Copernicus wrote E "on the invitation of the king of Poland, Sigismund I, and of his chancellor."[12]

Although Copernicus had dated M "15 August 1517" and D "1519," he did not indicate when he wrote E. Nevertheless, its composition can be placed between two official acts.

E (¶ 20, 21) referred to the former Grand Master of the Teutonic Knights as the duke of Prussia (*dux/princeps Prussiae*), a title granted him in April 1525. The office held by previous Grand Masters, on the other hand, was termed *magistratus* by E (¶ 9, 11). Hence E was not written before April 1525.[13]

On the other hand, it was written before 17 July 1526, when a royal decree concerning the new Prussian currency was issued by King Sigismund during his stay in Gdańsk from 17 April to 24 July 1526. Meeting with the West Prussian Estates, as he had promised in his agreement of 8 April 1525 with the duke of Prussia, the king ordained that

The new currency shall be minted in three denominations, namely, groats, shillings, and pennies.[14]

But E's 6th Conclusion had asked:

Should groats as well as shillings be struck?

Surely E asked that question before the king ruled that both groats (3-shilling pieces) and shillings should be minted. Moreover, E's 6th Conclusion proposed that "skoters or groats ... should be minted simultaneously." But the king's pronouncement made no provision for skoters. On both these counts, then (groats as well as shillings, but no skoters), we may reasonably conclude that E was written before the royal decision was announced on 17 July 1526. Indeed it was written in imminent anticipation of that decision, to which E ¶ 22 referred as the "regulation to be established now" (*ordinationem nunc instituendam*). Hence, whether or not Copernicus attended the Gdańsk meeting of the Estates (his presence there or elsewhere is not attested by any known document), he wrote E between April 1525 and July 1526.

---

[11] *Corpus iuris polonici*, IV, 157–158.

[12] So Wolowski, pp. 3, 5, 6; Monroe, p. 46; Le Branchu, I, xxix; Nomi, pp. 205, 241; R-H, p. 310, n. 64; Macleod, pp. 425–429.

[13] This conclusion was correctly drawn by Feliks Bentkowski (1781–1852), who was the first to publish E.

[14] *Corpus iuris polonici*, IV, 239/¶ 31. This paragraph was numbered 30 in Lengnich's German version (I, 15; also his Introduction, p. 51), followed by PI², 193/4 up. The paragraph is quoted more fully below at n. 132.

This interval of 15 months can be narrowed somewhat by considering how E survived. Just as the original manuscripts of M, D, and A have been lost, so too the original manuscript of E has disappeared. But it was copied by Copernicus' fellow-canon, Felix Reich. Like Copernicus, Reich was keenly interested in currency reform. But unlike Copernicus, Reich collected various relevant discussions. On the title page of his collection (*Collectanea*) Reich wrote:

> These collected writings about money are to be given after my death to Nicholas Copernicus, if perchance they might benefit his studies in any way.[15]

Below this testamentary disposition, Reich placed his signature and the date 1538, 18 August, renewed 18 October. He died on 1 March 1539.

Reich had arranged the seven documents in his collection in (almost precise) chronological order. The seventh and last item is a speech delivered by Reich himself on 23 July 1528. His speech is preceded by five items, four of which are dated 1526,[16] and one 1527. The first document is E. Since the second was written in the spring of 1526 (probably in April, or shortly before),[17] we may with considerable confidence restrict the time of composition of E to the year between April 1525 and April 1526.

In his collection Reich entitled E

Monete Cudende Ratio per Nicola[um Copernicum]
(*Essay on the Coinage of Money* by Nicholas Copernicus).

Having become a canon of the Varmia Chapter on 29 October 1526,[18] Reich undertook to qualify himself to be one of its spokesmen on the currency question, which was still far from being settled. It was for this purpose that he made his own copy of E, which had been written not long before by another Chapter spokesman, his fellow-canon Copernicus. While making an intensive study of his copy of E (which we shall call "ER") in Lidzbark, Reich imagined that he had detected an error in a computation. Dashing off a letter (which has not been preserved) to Copernicus in Frombork, Reich requested an explanation. This was promptly supplied by Copernicus, who addressed Reich as *Venerabilis domine* and *dominatio tua*, terms regularly applied by Copernicus to his fellow-canons. In his haste to reply, Copernicus not only omitted Reich's name as addressee, but also left the year out of the date (*octava pasce*, the eighth day after Easter Sunday). He signed this answer only with his initials "N. C.," as he had signed M. But after receiving Reich's collection in 1539, Copernicus wrote the rest of his surname (oppnic) after the initial C; a horizontal stroke through the lower vertical member of p was the standard abbreviation of the syllable *per*. Below his completed surname, Copernicus added the name of the addressee, which he had originally omitted in his haste to answer Reich.[19]

---

[15] PII, 31/18–21. Reich's *Collectanea* is now in Berlin-Dahlem. Secret State Archives. HBA chest 752. fols. 2ʳ–6ᵛ.

[16] Dmochowski (p. cxxxxviii/12) dated one of these documents 5 November 1529, which should be corrected to 1526.

[17] A quotation from the second document appears in the text below at n. 158.

[18] Hipler, p. 195; Sikorski, p. 146. It was Reich who gave E its title. Ordinarily Copernicus refrained from naming himself as the author of his writings not destined for the press, a personality trait overlooked by Sommerfeld, who attributes the title of E to Copernicus himself (p. 39).

[19] PII, 156/17. Copernicus' letter to Reich was dated by Sommerfeld (p. 42/14, p. 110/last 2 lines) 8 April 1526. But Copernicus addressed Reich as *Venerabilis domine*, and wished him as *dominationem tuam* the best

INTRODUCTION

After Reich's death on 1 March 1539, as we saw above, his collection of documents pertaining to the monetary problem passed to Copernicus. Then, after the latter's death on 24 May 1543, the Reich collection entered the Chapter's archives in Frombork. There it was seized by an invading army and in 1626 transported to Sweden, where it remained for a century and three-quarters. Then in 1801[20] it was returned at the request of the Prussian government, which stored it in the Secret Archives in Koenigsberg.[21] ER was noticed by a professor at the local university, who had a copy made for Samuel Bogumił Linde, an educator in Warsaw, who was also a zealous collector of material illustrating the history of Poland. The faithfulness of the copy was certified on 1 February 1816 by Karl Faber, the superintendent of the archives. Using the official seal, Faber testified:

> The foregoing copy agrees word for word with the original in the Secret Archives of the kingdom of Prussia.[22]

Faber used the term "original" to describe ER because at that time he was convinced that it was the "autograph product of Copernicus," and "as is proved by the numerous corrections by the hand of the author, it is the very original of the first draft." To this supposed "very original of the first draft," Faber assigned the date 1526, without adducing any reason for that choice.

After asserting that ER had been written by the hand of Copernicus, Faber found in his archives a letter that had been sent on 21 June 1541 to the duke of Prussia by Copernicus. The text of this letter was published by Faber, together with a lithographed facsimile.[23] Now that he had before him an unquestionable specimen of Copernicus' handwriting, Faber realized that ER had been copied by someone else. Accordingly, he now said that ER came "from the hand of the bishop's secretary, with corrections in Copernicus' hand."[24] Abandoning his previous claim, made in 1816, that the Koenigsberg manuscript was written by Copernicus' hand, three years later, in 1819, Faber still insisted that the corrections were in Copernicus' hand. This mistaken assertion was repeated, after Hipler, by Prowe, who maintained that only the author could have modified E so extensively.[25] But this reasoning overlooked Reich's even more zealous interest in the currency problem, and his penchant for altering the documents in his collection. By the time that reached Copernicus in 1539, the monetary

---

of health, terminology regularly applied by Copernicus to his fellow-canons. Reich, however, did not become a canon until 29 October 1526, half a year later than Sommerfeld's date, 8 April 1526. On that day Reich, not yet a canon, was the notary of the bishop, who resided in Lidzbark. Since Copernicus in Frombork closed his letter by asking Reich in Lidzbark to convey to the bishop his desire to serve the prelate, Copernicus did not have a personal conversation with Reich on 8 April 1526, as was suggested by Sommerfeld (p. 104).

[20] Not 1798 (as in PII, 32, 141) nor 1804 (as in Nomi, p. 207/4 up).

[21] Wolowski, p. 4/35.

[22] Ibid.

[23] Faber, p. 266; see Letter XV, below.

[24] Faber, p. 267. Here Faber changed his dating of E to 1528, again without giving any reason. Faber's "1528" was repeated by Hipler, p. 185. Hipler also uncritically endorsed Faber's gratuitous assertion that ER came from the hand of the bishop's secretary, whose name was even supplied by Hipler as Valentin Steinpik. Faber's revised and unsupported choice of 1528 as the year in which E was written affected not only Hipler but also Prowe ("1526–1528," PII, 31/35; "about 1527," PI², 195/3–4); Schmauch ("early 1528," p. 17, n. 44/3); and R–H ("1527 or 1528," p. 293). These scholars failed to notice that E preceded the royal decree of 17 July 1526. That E followed the decree was maintained by PI², 195/last 4 lines, "for various reasons," which were prudently left unspecified.

[25] PII, 31/28–30.

question had already been settled. Hence there was no occasion for Copernicus to introduce changes in E after 1539.

Friedrich Fischer, the chancellor of the duke of Prussia, compiled his own collection of documents dealing with the currency problem. In large part, Fischer's collection duplicated Reich's. That duplication took place before September 1529, the month of Fischer's death.[26] His copy of E (which hereafter will be called "EF") was formerly kept in Koenigsberg. But since the incorporation of that city under the name of Kaliningrad in the Soviet Union, the Fischer collection (like the Reich collection) has been housed in Göttingen, West Germany.

Just as Reich made changes in ER for reasons of his own (chiefly clarification of E's highly compact style), so too Fischer altered EF to suit his own purposes. One example will lay bare his motives. E ¶ 20 recommended "one common mint for all of Prussia" (*unam et communem esse in tota Prussia officinam monetariam*), since "to maintain standardized production in several mints is harder than in one." True, but the one mint for all of Prussia would inevitably be controlled by the king, and located somewhere in Royal Prussia. Ducal Prussia would therefore have no mint of its own, and that potential source of income would dry up for the hard-pressed duchy. This offending passage was accordingly deleted from EF (and later on also from ER, when it came into the possession of the Prussian government). As a result, the passage recommending a single mint was omitted from the first printed edition of E (and from three later editions which followed it). Hipler, however, printed the passage as a footnote (p. 191). Accepting Hipler's mistaken claim that the changes in E were due to Copernicus himself, Prowe assigned the deletion to "the second redaction"[27] by Copernicus. But E never underwent a second redaction at the hands of Copernicus. After he finished it, it was modified by Reich (ER) and Fischer (EF). E ¶ 16 laments that "it is shameful and painful to say what the condition of Prussian money is now." For that deplorable situation, Fischer, as chancellor of the duchy of Prussia, could not avoid partial responsibility. Hence he deleted the incriminating words *in quo statu nunc sit, pudet et dolet dicere*. In order to find passages in E more expeditiously, Fischer added in the margins of EF captions in Latin or German summarizing the essential content of every paragraph.

Whereas ER was copied out by Reich, and EF under Fischer's supervision, EC was copied out by a scribe on the staff of Maurice Ferber, who was elected bishop of Varmia on 14 April 1523 (with Reich serving as notary) and held that office until his death on 1 July 1537. As bishop of Varmia throughout the protracted currency discussions, which were bound to affect the daily lives of the people of his diocese, Ferber took a keen personal interest in the debates. His principal adviser in these matters was Copernicus, and Ferber had a copy of E made for the files in the episcopal palace in Lidzbark. Part of those files later passed, presumably through someone like S. B. Linde (who had obtained Faber's certified copy of ER)[28] into the possession of the Czartoryski family. Those wealthy protectors of the national cultural heritage established a museum which they later moved to Cracow. When the inventory of the manuscript holdings of the Czartoryski Museum in Cracow began to be published in 1887, it disclosed the existence of EC[29] (as we shall call the copy of E formerly in Lidzbark). EC was not known to the earlier editors and translators of E (Bentkowski, Baranowski, Wolowski, Polkowski, Hipler, and Prowe). But even after 1887, nearly half a century passed before its variants were taken into account by L. A. Birkenmajer and Dmochowski. EC does

---

[26] PII, 32/18–19.
[27] PII, 39–40, n.*.
[28] See the text above, after n. 21.
[29] Catalogus Czartoryski, I, 46.

not show ER's changes for the purpose of clarifying the text, nor EF's alterations in the interest of Ducal Prussia. EC limits its emendations to questions of Latin style.

All in all, then, none of the three surviving copies of E can be regarded as an absolutely faithful transmitter of E, Copernicus' own (lost) draft. In composing E, he modified M in three ways. First, he omitted M ¶ 7, 9–13, 15–18, and part of 8. Secondly, he added to M as new material E ¶ 1, 4, 9–16, 8 and 17 in part, and from 18 to the end. Thirdly, he combined M ¶ 4 and 6 in E ¶ 7. The relation between E and M may be seen in Table 1.

TABLE 1

E	M	E	M
1	—	—	7
2	1	8 (part)	8
3	2	—	9–13
4	—	9–16	—
5	3	17 (part)	14
6	5	—	15–18
7	4+6	18–end	—

COPERNICUS' WRITINGS ABOUT MONEY
(PARALLEL TEXTS OF *MEDITATA*, *DENKSCHRIFT* WITH APPENDIX, AND *ESSAY*)

Copernicus began E with an introductory paragraph not found in M.

D

M

E

E ¶ 1 Although there are countless scourges which in general debilitate kingdoms, principalities, and republics, the four most important (in my judgment) are dissension, [abnormal] mortality, barren soil, and debasement of the currency. The first three[1] are so obvious that nobody is unaware of[2] their existence. But the fourth, which concerns money, is taken into account by few persons and only the most perspicacious. For it undermines states, not by a single attack all at once, but gradually and in a certain covert manner.

E ¶ 2–3 repeat M ¶ 1–2, but not in every detail. In what follows, M occupies the center of the page. To the right, E's significant modifications of M are indicated by Roman numerals. To the left, D's departures from M are marked by letters. D's and E's additions to M are shown by a plus sign (+), while D's and E's deletions from M are noted by a minus sign (–).

E ¶ 2

M ¶ 1 Coinage is imprinted gold or silver, by which the prices of things bought and sold are reck-

D

M

oned according to the regulations of any State or its ruler. It is therefore a measure of values.[i] A measure,[ii] however, must always preserve a fixed and constant standard. Otherwise, public order is necessarily disturbed, with buyers and sellers being cheated in many ways,[3] just as if the yard, bushel, or pound did not maintain an invariable magnitude. Hence this measure is in my opinion the coin's face value. Although this is based on the metal's purity, nevertheless intrinsic value must be distinguished from face value. For, the denomination of a coin may exceed its metallic content, and the other way around.

M ¶ 2 Coinage was introduced for a necessary reason. Things could have been exchanged for gold and silver by weight alone, because mankind's common judgment prizes gold and silver.[a,iii] But to carry weights around all the time was very inconvenient. The purity of the gold and silver, moreover, was not instantly[b] recognizable.[iv] Accordingly people[c] ordained that a coin should be marked with a universally recognized symbol to indicate that it contained the proper proportion[v] of gold or silver, and to add legal authority to confidence [in it].[d,vi]

[a] + *everywhere*

[b] − *instantly*
[c] + *thought it best and*

[d] *to have the symbol's legal standing strengthen the confidence [in the coin]*

E

[i] Therefore money is, as it were, a common measure of values.
[ii] That which ought to be a measure

E ¶ 3

[iii] + everywhere

[iv] + by everybody

[v] amount
[vi] to instill confidence in its reliability

D

Here E inserts new material.

M

M¶3 The face value of a coin is just and proper when the coin contains very little[vii] less gold or silver than may be bought with it, since only the expenses[e] of the minters should be deducted. For, the symbol should add some value to the metal.

M¶4 This [face-value][viii] may be[ix] corrupted in three[x] ways. First, the metal alone may be defective, when for the same[f] weight of coin more than the right amount of copper is alloyed with the silver. Secondly, the weight may be defective, even though the proportion of copper and silver[g] is correct. Thirdly, and this is worse,[xi] both defects may be present at the same time.

[e] + *and wages*

[f] *proper*

[g] *the alloy or grain*

E

E¶4 In the next place, copper is usually mixed with coins, especially those made of silver. [This is done,] I believe, for two reasons. First, the coinage is less vulnerable to the schemes of crooks and those who would melt it down if it consisted of pure silver. Secondly, when silver bullion is broken down into little pieces and the smallest coins, it keeps a convenient size when it is alloyed with copper. A third reason may be added, namely, to stop it from disappearing sooner by being worn down through constant use, and to make it last longer by strengthening it with copper.

E¶5
[vii] slightly

E¶7
[viii] + *Valor*   [ix] is   [x] many

[xi] the worst

D

[h] + *more than usual*

[i] *it equals the value of the silver*

[j] − *Only for this reason*
The rest of this sentence is placed at the end of M ¶ 6

[k] + *To change this,*[5] (*the coinage should be melted down and replaced.*)

[l] *not according to the previous face value of the old coinage, but according to the value of the silver found in it*

M

M ¶ 5 Money can lose[xii] its value also[xiii] through excessive abundance, if so much silver is coined as to heighten people's desire for silver bullion.[h,xiv] For in this way the coinage's market value vanishes when with it I[xv] cannot[q] buy as much silver as the money itself contains, and then[xvi] I find[xvii] greater[xviii] advantage in destroying the coin by melting the silver. The solution is to mint no more coinage until it recovers its par value.[i,xix]

M ¶ 6 The value of a coin deteriorates also by itself as the coin is worn down through long use. Only for this reason[j] should it be renewed or[xx] replaced. This is indicated if somewhat[xxi] less silver is found in the coin than is bought with it.[k,xxii]

M ¶ 7 Hence, whenever a new coinage should be minted, the use of the old coinage must be completely prohibited. People bringing it to the mint must receive new coinage[l] wherein the amount of silver matches precisely what is in the old coinage. But if this is not done, the old coinage will taint the value of the new coinage, for two reasons. For, the mixture [of the old coinage with the new coinage]

E

E ¶ 6
[xii] loses   [xiii] most of all
[xiv] + more than for coined money

[xv] it is not possible to
[xvi] − then   [xvii] is found
[xviii] EC: − greater; EF: much greater

[xix] + and becomes more desirable than silver

E ¶ 7 (after M ¶ 4)

[xx] and
[xxi] ER: considerably
[xxii] + This is the condition in which depreciation of the coinage is properly perceived.

E omits M ¶ 7

M

will reduce the aggregate below the proper weight, and perhaps[m] make it excessively abundant, with the consequences mentioned above [M¶ 5]. However, the worst mistake, which is absolutely unbearable, is [committed] if the ruler, or whoever governs the State,[n] seeks a profit from the minting of the coinage, to wit, by adding to the money in existence[o] a new coinage which, although defective in metallic content or weight, pretends to have the value of the old coinage. For the ruler cheats not only his subjects but also himself by enjoying a merely temporary and quite modest profit, like a stingy farmer sowing bad seeds to save good seeds: he will reap[6] exactly what he sowed. This evil damages the coinage's worth, just as blight[p] [ruins] grain. After this disease has taken hold and been discovered too late, the ruler will not easily[q] get rid of it without burdening his subjects again and with his reputation unstained, since he himself caused the harm.

M¶ 8 I shall now[r] give as an example the[s] Prussian coinage, which has heretofore been subject to many defects.[xxiii] It circulates under the names mark, skoter, and the like, which are also names of weights. As a weight, a mark is 1/2 pound, and 3 skoters make 1 ounce.[xxiv] As a coin, however, a mark consists of 60 shillings.[t],[xxv] 7 But lest ambiguity[xxvi] give

D

[m] — perhaps

[n] + or the municipalities
[o] + and in circulation

[p] + or other parasites

[q] + atone for or

[r] + for the sake of better understanding
[s] a report on our

[t] — the rest of M¶ 8

E

E¶ 8

[xxiii] Having discussed money in a general way above, let me turn to Prussian money in particular by first showing how it became so debased.
[xxiv] — and 3 skoters make 1 ounce
[xxv] + all this is very well known [xxvi] + as be-

D

rise to misunderstanding, in what I shall say hereafter,[xxvii] wherever "mark" is used, I want the coin to be understood.[xxviii] On the other hand, where weight is involved, I shall speak of a pound, for instance, 2 marks make a pound.[xxix]

M¶ 9 Now the[u] money which we use at the present time[v] consists of shillings, groats, and pence. I find,[w] however, that the coins now called "groats"[8] were once shillings, and that[x] 8 marks contained 1 pound of pure silver,[y] as is learned also from their content. For they consist half of copper and half of silver, and 8 of these marks, at 60 shillings to the mark, weigh almost 2 pounds.[9] These were called "new shillings" and the corresponding marks "new marks" or "good marks."

M¶ 10 For there were also the other "old shillings" and the corresponding "old mark" or "light mark." These were equal to the new coins in weight, but[z] had $1/2$ their value. For only $1/4$ of their content was silver, and 16 marks contained 1 pound of silver,[aa] while weighing 4 times as much.

M¶ 11 Later, when the country's status changed, cities were granted the right to mint coins.[10] As they exercised their new privilege, curren-

---

[u] + Prussian
[v] − which we use at the present time
[w] It is found
[x] + the corresponding
[y] + which is 2 marks by weight

[z] differed by half in value, that is, were worth half as much
[aa] + that is 2 marks by weight

E

tween weight and coin
[xxvii] − in what I shall say hereafter
[xxviii] let it be understood as denoting the coin

[xxix] interpret the term "pound" as a weight of 2 marks, but a "mark" by weight as $1/2$ pound
E omits M¶ 9–13

E¶ 12

E¶ 12

E¶ 13

D

^{bb} + *the mark [of silver] by weight [cost] 10 marks and*
^{cc} + *alongside the new coins*
^{dd} + *former*
^{ee} + *of the ordinary newly minted marks, reckoned at 60 shillings* ^{ff} *is*
^{gg} — *at that time*

^{hh} — *new shillings, having become*
^{ii} + *also*
^{jj} + *at their face value*
^{kk} + *and circulated*
^{ll} + *or elevating*

^{mm} + *in Prussia*

M

cy increased in quantity, though not in quality. Four parts of copper began to be alloyed with a fifth part of silver, until^{bb} 1 pound of fine silver cost 20 marks. Since part of the old coinage still circulated,^{cc} however, those^{dd} new shillings became skoters, reckoned at 24 to 1 light mark.[11] For 1 [60-]shilling mark^{ee} was^{ff} at that time^{gg} not much better than those 24 [skoters].[12]

M ¶ 12 Afterwards, however, those new shillings, having become^{hh} skoters,^{ii} disappeared because^{jj} they were accepted^{kk} also throughout the Mark [of Brandenburg] and Pomerania. It was decided to recover them by evaluating^{ll} them at 1 groat, that is, 3 shillings. This was a very bad miscalculation, quite unworthy of so distinguished a body of notables [as the Estates of West Prussia. It was] as though they rejoiced at their own misfortunes, because forsooth Prussia could not get along without those [groats], even though they were worth no more than 15 pence and without them there was more than enough currency.^{mm}

M ¶ 13 The groat therefore differed from 3 shillings, being worth $1/5$ or $1/6$ less.[13] Yet because it

E

E ¶ 14

E ¶ 15

M

was mistakenly evaluated as their equal, it dragged down the value of the shillings[mm] while attaching their good quality to itself. It brought about a confusion of the mixed coinage's market value and intrinsic value. Consequently the money's market value fell lower and lower day after day. Nevertheless it was decided not to interrupt the coining of money at all.[oo] The costs did not cover the minting of coins equal in value to the previous coinage. What was minted, therefore, always fell somewhat short of the steadily declining market value. The later coinage, always inferior[pp] to the earlier coinage, as it was introduced[qq] depressed the market value of the previous coinage, and drove it out. [This continued] until the shilling's market value equaled the groat's intrinsic value, and[rr] 1 pound of silver cost 24 light marks. This has not stopped yet, up to the present time. Indeed, even after the approximate equalization [of the shilling's market value] with the groat, this is now followed by ever new groats, which are deficient at least in weight. For, 26 marks weighing 2 pounds contain 1 pound of silver. What is to be expected except that soon 1 pound of silver will cost 26 marks,[ss] unless help is forthcoming in the meantime?[15]

D

[mm] — the rest of this sentence

[oo] — at all

[pp] + in value
[qq] + constantly

[rr] + now 1 mark [of silver] by weight costs 12 light marks

[ss] + and the mark [of silver] by weight will cost 13[14] marks

D

tt + *and decline*
uu *who take the value of the money into their own hands*

vv *other coins with more silver*

ww + *and the mark [of silver] by weight to 10 marks*
xx *how this reform might be brought about*

yy + *Estates and*

M

M¶ 14 Such grave evils, then, beset Prussian money and, because of it, the whole country. Its calamities*tt* benefit only the goldsmiths, who know the purity of metal by experience.*uu* For from the mixed coinage they collect the old pieces, from which they melt down the silver and sell it. From the inexperienced public they constantly receive more silver with the coinage.*vv* After these old shillings disappeared completely, those following next [in intrinsic value] are selected, like wheat being winnowed from chaff. Would that these [distortions] were reformed, while there is time, before a greater disaster! At least let 1 pound of silver be brought back again to 20 marks,*ww* and held there for the future by the method described above.

M¶ 15 I shall accordingly add an example of*xx* this reform. First, only one place should be designated for the minting of money, not for a single city or under its emblem, but for the entire country. In the absence of a decision by the country's*yy* nobility and cities, no new money should be minted henceforth. There should be no exception, moreover, to the rule that not more than 20 marks should be struck from 1 pound of fine silver.

M¶ 16 The procedure should be as follows. For the shillings, take 3 pounds of copper and 1 pound

E

E¶ 17

E¶ 24

D

M

E

of fine silver, minus $^1/_2$ ounce or only as much as must be deducted for expenses.^{zz} From the molten mass strike 20^{aaa} marks, which in purchasing power will be worth 1 pound of silver. Furthermore, mint the skoters from 2 pounds of copper and 11$^3/_4$ ounces^{bbb} of silver, making 20 marks, with 24 [skoters to the mark]. But perhaps it would be better if instead of shillings, half-shillings were minted in accordance with the above reckoning. Five half-shillings would be exchanged for 1 skoter,[16] and 1 half-shilling for 4^{ccc}, 17 of the present pennies.

M ¶ 17 When this minting has begun, however, the use of the old coinage should be prohibited.^{ddd} In the mint, for 13 old marks, 10 new ones should be exchanged, either in shillings or groats.^{eee} For, this loss will have to be suffered once, in order that it may be followed by many benefits and a lasting advantage, and that a single currency reform in 25 or more years may be enough.

M ¶ 18 Let these remarks about money suffice.^{fff} I leave them to be analyzed^{ggg} by anybody who has better understanding, as new situations constantly present themselves in the course of time.

^{zz} + *of the minters*
^{aaa} + *[60-]shilling*

^{bbb} *a pound of silver minus $^1/_2$ ounce [= 15$^1/_2$ ounces]*

^{ccc} *3*

^{ddd} + *and stopped*
^{eee} *skoters*

^{fff} + *for a framework*
^{ggg} *censured or improved*

1519

N. C. *Meditata*. 15 August 1517

185

MONEY

In place of M ¶ 9–13, E ¶ 9–16 presents the earliest history of Prussian coinage.

D   M   E

E ¶ 9   Now in the old records of proceedings and official correspondence[18] we find that when Conrad of Jungingen was Grand Master,[19] that is, just before the battle of Tannenberg,[20] $1/2$ pound, that is, 1 mark [by weight] of pure silver cost 2 Prussian marks, 8 skoters. That was when 3 parts of pure silver were alloyed with a 4th part of copper, and from $1/2$ pound of this [pure] material[21] they made 112 shillings.[22] To this amount add $1/3$, which is $37\,1/3$ shillings, making a total of $149\,1/3$ shillings.[23] The total weighs $2/3$ pound, that is, 32 skoters,[24] containing of course 3 parts (which are $1/2$ pound) of pure silver. Its price, however, as was just said, was 140 shillings for $1/2$ pound.[25] But the remainder of $9\,1/3$ shillings, which was missing, was made up by the money's face value. Its face value was therefore suitably linked with its intrinsic value.[26]

E ¶ 10   Such coins of Winrich,[27] Ulrich,[28] and Conrad are still found now and then in strongboxes. Then after the defeat of Prussia and the aforementioned battle [of Grunwald/Tannenberg], the damage to the Order began to be more and more apparent in the coinage day after day. For although the shillings of Heinrich[29] look like those just mentioned, they are found[30] to have no more than $3/5$ silver.[31] This mistake grew worse until the proportion was reversed and three parts of copper began to be alloyed with a fourth part of silver, so that "silver coinage" would no longer be the proper designation, but rather "copper coinage." Yet it kept the weight of 112 shillings to $1/2$ pound.

186

E

E¶ 11   It is not in the least advisable to introduce a new, good coinage while an old, debased coinage remains in circulation. How much worse was this mistake, while an old, better coinage remained in circulation,[32] to introduce a new, debased coinage, which not only spoiled the old coinage but, so to say, swept it away! When Michael Rusdorf[33] was the Grand Master, they wanted to eliminate this mistake, and restore the coinage to its former,[34] better state. They minted new shillings, which we now call "groats." But since the old, debased coins apparently could not be withdrawn without loss, they continued to circulate alongside the new coins, by an extraordinary error. Two old shillings were exchanged for 1 new one.[35] It then came to pass that two kinds of marks were inflicted upon the people, namely, the new or good mark consisting of the new shillings, by contrast with[36] the old or light mark consisting of the old shillings, with 60 shillings to each of the marks. However, the pennies remained as they had been, with only 6 pence being exchanged for 1 old shilling, but 12 pence for 1 new shilling.[37] For it can easily be surmised that originally the shilling was equal to 12 pence. For just as we usually say *mandel* for the number 15, so in most parts of Germany the word *schilling* is still used for the number 12. The term "new shilling," on the other hand, lasted right down to a time within our memory.[38] How they finally became groats, I shall explain below.

E¶ 12   As regards the new shillings, then, at 60 shillings to the mark, 8 marks contained 1 pound of pure silver, as is quite evident from their composition.[39] For they consist $1/2$ of copper, and $1/2$ of silver. At 60 shillings to the mark, 8 of these marks weigh nearly[40]

M

D

M¶ 9

2 pounds. The old shillings, on the other hand, although equal to the new shillings in weight, as has been said,⁴¹ were worth half as much. For since the old shillings contained only ¹/₄ silver, 16 marks were produced from 1 pound of pure silver and weighed 4 times as much.⁴²

E¶ 13   Later, when the country's status changed, cities were granted the right to mint coins. As they exercised their new privilege, currency increased in quantity, though not in quality. Four parts of copper began to be alloyed with a fifth part of silver in the old shillings,⁴³ until 20 marks were exchanged for 1 pound of silver. And so those new shillings, since they were now worth more than twice as much as the recent⁴⁴ shillings, were made into skoters, reckoned at 24 to the light mark. Hence, ¹/₅ of the money's value in the mark perished.⁴⁵

E¶ 14   Afterwards, however, the new shillings, having become skoters, disappeared because they were accepted also throughout the Mark [of Brandenburg]. It was decided to recover⁴⁶ them by evaluating them at a groat, that is, 3 shillings. This was a very bad miscalculation, quite unworthy of so distinguished a body of notables, as though Prussia could not get along without those [groats], even though they were worth no more than 15 pence of the coinage then in circulation, while its value⁴⁷ was already depressed also by its abundance.

E¶ 15   The groat in fact differed from the shilling in that it was worth ¹/₅ or ¹/₆ less than the standard. Through its false and unfair evaluation it dragged down the value of the shilling. Maybe the wrong previously inflicted by the shillings on the groats by forcing them to become skoters had to be avenged in this way. But woe to you, O Prussia, you who pay the penalty for a maladministered state by your ruin, alas! Thus, although the money's market value and intrinsic value were gradually vanishing at the same time, nevertheless there was absolutely no interruption in the coining of money. The costs [of minting] did not cover the difference by which the later coinage would be equalized[48] with the older coinage. Hence, a coinage always worse than the previous coinage was superimposed on it. This depressed the worth of the earlier coinage and drove it out, until the shilling's face value coincided in due proportion with the groat's intrinsic value,[49] and 24 light marks were equivalent to 1 pound of silver.[50]

E¶ 16 But at least some small remnants of the money's worth must have persisted in the end,[51] since no consideration was given to restoring it. Yet this practice or abuse of counterfeiting, clipping, and tampering with money was ingrained so long that it could not stop, nor has it stopped to this very day. For what kind of money it will become hereafter and what its condition is now, it is shameful and painful to say. For it has fallen so low today that 30 marks contain hardly 1 pound of silver. Then what remains in the absence of help[52] except that hereafter Prussia, drained of gold and silver, will have an exclusively copper currency? Consequently imports of foreign merchandise and all foreign trade will soon end. For what foreign merchant will want to exchange his goods for copper coins? Lastly, which of our[53] merchants will be able to buy foreign merchandise

E

in foreign lands with the same money? Yet those in authority scornfully[54] disregard this immense misfortune of the Prussian state. To their very dear country they[55] owe, not to mention the deepest devotion, after piety to God, even their very lives. Yet by their thoughtless indifference they let their country slip wretchedly downhill further and further day after day and crash.

E ¶ 17 While, then, such grave evils beset[56] Prussian money and consequently the whole country, its calamities benefit* only the goldsmiths and those who know the purity of metal by experience. For from the mixed* coinage they collect the old pieces, from which they melt down the silver and sell it. From the inexperienced public they constantly receive more silver with the mixed[57] coinage. But* after those old shillings now disappear† completely, the next best are selected, *while the inferior mass of money remains behind. Hence [arises] that widespread and incessant complaint: gold, silver, food, household wages,[58] workmen's labor, and whatever is customary in human consumption soar in price. But, being inattentive, we do not realize that the dearness of everything is produced by the debasement[59] of the coinage. For in line with the quality of money everything,[60] *especially gold and silver, rises and falls, prices being based not on brass or copper, but on gold and silver. For we declare that gold and silver are, as it were, the foundation of money, on which its value rests.

M

M ¶ 14
* only the goldsmiths, who know the purity of metal
* promiscuous

— mixed
* And † have disappeared
* Here E veers off from M

* EF underlined "especially ... on gold and silver."

D

E

E ¶ 18   But maybe someone will argue that cheap money is more convenient for human needs, forsooth, by alleviating the poverty of people, lowering the price of food, and facilitating the supply[61] of all the other necessities of human life, whereas sound money makes everything dearer, while burdening tenants and payers of an annual rental[62] more heavily than usual. This point of view will be applauded* by those who were heretofore granted the right to coin money and would be deprived of the hope[63] of gain. Nor will it perhaps be rejected by merchants and artisans, who lose nothing on that account *since they sell their goods and products in terms of gold, and the cheaper the money is, the greater is the number of coins they receive in exchange.

E ¶ 19   But if they will have regard for the common good, they will surely be unable to deny that sound money benefits not only the state but also themselves and every class of people,[64] whereas debased coinage is harmful. Although this is quite clear for many reasons, we learn[65] that it is so also through experience, the teacher of objective truth. For we see that those countries flourish the most which have sound money, whereas those which use inferior coinage decline and fall. Certainly Prussia too prospered when 1 Prussian mark [as coin] was worth 2 Hungarian florins and when, as was said above [E ¶ 9], 2 Prussian marks, 8 skoters, were exchanged for 1/2 pound, that is, 1 mark [by weight] of pure silver. But in the meantime, as its coinage was debased more and more day after day, our fatherland too declined, and as a result of this plague and other misfortunes it was brought down almost to its final destruction. Moreover, those places which use sound money, as is well known, have flourishing trades, excellent craftsmen, and an abundance of commodities. On the other hand, where cheap money prevails, through listlessness, lethargy, and slothful idleness the development of the fine arts as well as of the intellect is neglected, and the plentifulness of all goods is also a thing of the past. The memory of man has not yet forgotten that grain and produce were bought in Prussia with a smaller number of coins[66] while sound money was still being used. Now, however, as it is being debased, we experience a rise in the price of everything related to food and human consumption. Hence it can be seen that cheap money fosters laziness more than it helps poor people. An improvement of the currency will not be able to impose a heavy

* ER + loudly; EC + perhaps

* EF underlined from "since" to the end of this sentence

burden on tenants. If they seem to pay more than usual for their land,[67] they are going to sell the products of their fields, their livestock, and that kind of output at an even higher price. For the adjusted evaluation of the money will balance the mutual exchange of giving and receiving.

E ¶ 20 Therefore, if it is decided to let Prussia at last recover at some future time from its previous depression by restoring its currency, the most urgent task will be to avoid the confusion arising from the differences between the various mints where the coinage is to be struck. For, multiplicity interferes with uniformity, and to maintain standardized production in several mints is harder than in one.*† It would therefore be advantageous to have for all of Prussia one common mint producing coinage of every denomination. On one side, the device will be the arms or insignia of the lands of Prussia, surmounted by the crown to signify the overlordship of the kingdom [of Poland]. But the other side will display the arms* of the duke of Prussia with the crown of the kingdom resting thereon.[68]

E ¶ 21 If, however, this could not be done because of the opposition of the duke of Prussia on the ground that he wants to have his own mint,* let two places be designated at the most, one in his Royal Majesty's territory and the other in the duke's domain. Let the first mint strike coins showing the royal insignia on one side, and on the other the arms of the lands of Prussia. Let the second mint, however, issue coins stamped with the royal insignia on one side, and on the other side the duke's. Let both coinages be subject to royal control, and by His Majesty's order be used and accepted throughout the entire kingdom. This arrangement will produce no small effect on the reconciliation of attitudes and participation in trade.

E ¶ 22 It will be essential, moreover, that these two coinages should be of a single standard, intrinsic value, and face value, and remain forever, under the watchful supervision of the leaders[69] of the State, in agreement with the regulation to be established now. It is also essential that in both places the rulers should expect no profit from the minting of the coinage. Only as much copper should be added as would make the face value exceed the intrinsic value, so that it would be possible to recover the loss of the expenses [of the minting operation] *and remove the opportunity of melting down the coinage.

* EF: fewer; ER: at the most
† EF — from "It" to "... own mint" in E ¶ 21
* EF: *insignia* (instead of the singular form *insigne*)

*ER, EC: *officinam*; EF: *monetariam officinam*

*EF — from "and" to the end of this sentence

M

E

E ¶ 23 Furthermore, let us hereafter avoid falling into our present age's confusion arising from the mixture of new coinage with old. It seems necessary, when the new coinage is issued, to abolish the old coinage, wipe it out completely, and exchange it at the mints for the new coinage in proportion to its intrinsic value. Otherwise the work of renewing the money will be in vain, and the subsequent confusion will perhaps be worse than the earlier. For again the old coinage will spoil the value of the new coinage. The mixture will of course make the aggregate's weight less than is right and its quantity excessive.[70] The result will be the dislocation described above [M ¶ 5, E ¶ 6]. In this regard, somebody may think that the solution is to assign to the remaining old coins a value as much beneath the new coinage's as their intrinsic value is inferior or lower. But this cannot be done without a great error. For not only the groats and shillings but also the pennies are now so different in their many kinds that individual coins can hardly be rated according to the condition of their intrinsic value and differentiated from one another. Consequently the resulting variety of money would produce inescapable confusion and aggravate the difficulties, problems, and other annoyances of those who engage in business and enter into contracts. It will therefore always be better to withdraw the old coinage completely from circulation when money is being renewed afresh. *For so small a loss will have to be borne calmly once, if that can be called a loss which gives rise to increased production and steadier serviceability as well as enhances the state.

E ¶ 24 To raise the Prussian coinage, however, to its original worth is very hard and perhaps impossible after so drastic a collapse. Although[71] any renewal of the coinage is a matter of no small difficulty, still under the conditions prevailing at the present time it seems possible to restore it satisfactorily, at least with 1 pound of silver returning to 20 marks. The program would be as follows. For the shillings, take 3 pounds of copper, but as regards pure silver take 1 pound minus ½ ounce or as much as has to be deducted to cover the expenses [of minting]. From the molten mass mint 20 marks, which will buy 1 pound, *that is, 2 marks [by weight] of silver. According to the same proportion, skoters or[72] groats and pennies may be struck as desired.

* EF underlined this whole sentence

M ¶ 16

* ER's margin + the rest of this sentence

E

COMPARISON OF SILVER WITH GOLD

Gold and silver, as was said above [E ¶ 17], are the basis of coinage, in which its worth resides. Most of what has been set forth about silver coinage can be transferred also to gold coinage. It remains to explain the ratio for the mutual exchange of gold and silver.

Accordingly, it is first necessary to examine the relative value of pure gold to pure or unalloyed silver, in order to proceed downward from the general to the particular and from the simple to the compound. In the next place, the ratio of gold to silver is the same in bullion form as in coinage of the same standard. Furthermore, the ratio of gold coin to gold bullion is the same as the ratio of silver coin to silver bullion of the same standard of alloy and weight.

Now the purest gold coins found among us are the Hungarian florins. For they have the least[73] alloy, and perhaps only as much as was necessary was deducted for expenses in the mints. Hence they are rightly exchanged for pure gold of the same weight, with the authority of the symbol making up for the deficiency of the florins. It therefore follows that the ratio of pure silver bullion to pure gold bullion is the same as the ratio of that silver to Hungarian florins, the weights being unchanged. But 110 Hungarian florins of proper and uniform weight, namely, 72 grains, make 1 pound (by a "pound" I always mean the sum of 2 marks by weight). By this reasoning we find that generally among all people 1 pound of pure gold is worth as much as 12^{74} pounds of pure silver. Yet we observe that 11 pounds [of silver] once equaled 1 pound of gold. For this reason, apparently, it was ordained of old that 10 Hungarian gold pieces should weigh $1/_{11}$th of a pound.[75] But if the same price continued today for that weight, we would have a convenient interchangeability of Polish and Prussian money on the basis of the aforementioned ratio [gold : silver = 1 : 11]. For if 20 [Prussian] marks were made from approximately[76] 1 pound of silver, 1 gold piece would equal exactly 2 marks[77] or 40 Polish groats.[78] But after it became customary to exchange 12 parts of silver for 1 part of gold, the weight and price disagree, so that 10 Hungarian gold pieces are worth 1 pound of silver plus $1/_{11}$th of a pound. Therefore if 20 [Prussian] marks are made from $1\ 1/_{11}$ pounds of silver, the Polish and Prussian coinages will be matched in the correct ratio, groat for groat, with 2 Prussian marks equal to 1 Hungarian gold piece.[79] The price of silver, however, will be 9^{80} marks, 10 shillings, or thereabouts, for $1/_2$ pound.

On the other hand, if a debased coinage and the ruin of the fatherland are definitely desired, and so slight a restoration and adjustment seem too hard, and it is decided that 15 Polish groats should remain equal to 1 [Prussian] mark,[81] and 2 marks, 16 skoters, to 1 Hungarian gold piece,[82] that too will be accomplished with no great trouble in the manner described above, if 24 [Prussian] marks are made from 1 pound of silver.[83] This was certainly the situation recently when the price of $1/_2$ pound of silver was still 12 marks,[84] the amount of money for which [6] Hungarian florins were exchanged.[85]

This is said by way of an example and guideline. For there are countless ways of establishing a currency, and it is not possible to explain all of them. But a general agreement after mature deliberation will be able to make this or that decision, which will seem most advantageous to the state. But if the currency is correctly related to the Hungarian florin, and no mistake is made, other florins will also be easily rated according to their content of gold and silver in comparison with the Hungarian florins.

Let these remarks about the restoration of the coinage be sufficient at least to make clear how its value has fallen and how it can be restored. This, I hope, is evident from what was said above.

CONCLUSION REGARDING THE RESTORATION OF THE COINAGE

With reference to the restoration and maintenance[86] of the currency, the following recommendations seem worthy of consideration.

First, the currency should not be renewed without the deliberate advice and unanimous consent of the Councillors.

Secondly, only one place, if possible, should be designated for a mint. There coins should be struck in the name, not of one city, but of the whole country with its insignia. The validity of this recommendation is proved by the Polish coinage, which for this reason alone maintains its value over so vast an extent of territory.

Thirdly, when new money is issued, the old coinage should be demonetized and abolished.

Fourthly, it should be a permanent rule, without change and without exception, to strike only 20 marks, and no more, from 1 pound of pure silver, minus what must be deducted for the expenses of the operation. The Prussian coinage will in this way be definitely adjusted to the Polish, with 20 Prussian groats as well as 20 Polish groats being worth 1 Prussian mark.[87]

Fifthly, an excessive multiplicity[88] of coinage should be avoided.

Sixthly, money should be issued in all denominations at the same time, that is, skoters or[89] groats, shillings, and pennies should be minted simultaneously.[90] How big should the proportion [of each denomination] be? Should groats as well as shillings be struck? Should silver pennies also [be minted], worth $1/4$ or $1/2$ or even 1 whole mark? [These questions] are to be decided by those concerned, except that whatever the distribution, the decision should be made in such a way as to last forever after. Attention must also be paid to the ordinary pennies, since they are now worth altogether so little that a whole mark [of pennies by weight] contains hardly more than the silver in 1 groat.

Lastly, a difficulty arises from contracts and obligations made before and after the renewal of the money. In these matters a way must be found not to burden the contracting parties too much. This was done in former times, as is clear from what is copied out on the other side of this sheet.

"What is copied out" is a proclamation promulgated by the Teutonic Knights in 1418. This proclamation was copied out "on the other side of this sheet," according to ER (*in altero latere huius folii*). But **EF** copied it out "below" (*infra*). Lastly, **EC** copied it out "on the sheet that follows next" (*in proximo sequenti folio*). These variants indicate the partial independence of our three manuscript copies of E.

The more significant changes made by Reich in E were listed above. A typical intervention on his part is his insertion of *duas tertias* above the line over *bessem* in E ¶ 9, for the sake of those readers who might be unfamiliar with *bes* as the Latin expression for "$2/3$." Where E ¶ 20 recommended one mint for all of Prussia, Reich interpolated *ad summam* (at the most). Such examples show that he did not feel himself bound to adhere in every detail to the text of E as he had received it. On the contrary, he regarded himself as fully authorized to introduce modifications. These, however, in no way altered the essential content of what he found in E. Rather, his intention was to clarify or emphasize, as the foregoing selected instances suggest.

EF is derived from ER, as part of the dependence of Fischer's collection on Reich's. An illustration of that dependence is the numerical error CXLVIII, which was unthinkingly copied by Fischer's scribe from ER (see n. 52). Had either the scribe or Fischer bothered to do the simple arithmetical computation in this passage, the necessity of correcting ER's CXLVIII to CXLVIIII would have been obvious at once. But that correction was not made in EF. Fischer was not at all concerned with such minor details. As his marginal captions in Latin or German reveal, he wanted to be able to find his way quickly in a treatise with which his official duties required him to be familiar. He was on guard ageinst expressions which might prove harmful to the financial interests of Ducal Prussia, as against Royal Prussia and the kingdom of Poland. As chancellor of the duchy, he imagined that he knew the history of the Teutonic Knights better than Copernicus and Reich did. Thus in E ¶ 10 he changed the name of the Grand Master Winrich to Heinrich, but in this respect he was mistaken. He (or his scribe) was equally mistaken in changing the perfectly correct form *rei publicae* in the next sentence to "respublici" (!). He altered ER ¶ 20's reference to the arms of the duke of Prussia from the singular number to the more impressive plural, *insignia*. In the same spirit he made quite specific ER ¶ 21's mention of the duke's desire to have his own minting office (*officinam*) by inserting *monetariam*.

Changes were likewise made in E by the copyist responsible for EC. For instance, after enumerating three ways in which a coin may lose some of its value, E ¶ 7 remarks that it may also (*etiam*) lose value through being worn down by long use. EC replaced *etiam* by *enim* (for). But being worn down does not assign a reason for the first three losses of value. On the contrary, it adds a fourth way of losing value. In ¶ 7, then, EC substituted a bad reading for a perfectly good one. In E ¶ 18, Copernicus anticipates that an advocate of cheap money will be applauded by the established mints, fearful of losing their profitable privilege. According to ER, that advocate will be applauded loudly (*fortiter*), inappropriately changed to "perhaps" (*forsitan*) by EC. On the other hand, where the 6th Conclusion urged that attention must also be paid to the pennies since (*quoniam*) they are now worth so little, EC expanded E's abbreviation correctly as *quoniam*, whereas ER and EF were both off target.

E's 6th and last Conclusion acknowledged that the proposed monetary reform would affect existing financial obligations. Suppose that somebody borrowed 10 marks before the reform. If, after the reform, he is required to repay 10 (better) marks, he will suffer an unfair loss, while the lender enjoys a corresponding unfair gain. A similar injustice threatens a tenant farmer who, before the reform, rented land for the duration of his life (*Leibrente*) or with the right of bequeathal (*Erbe*). E refrains from offering specific proposals to cope with such tough questions. Instead, E recalls that a similar situation occurred about a century earlier, and presents as a model the

PROCLAMATION OF 6 NOVEMBER 1418

The following proclamation[91] *governing the entire country, and issued from Malbork on Sunday* [6 November] *following All Saints Day* [1 November] *1418.*

1. First, any rental negotiated in good old money 3 years before the regulation proclaimed on St. Martin's Day [11 November] 1416 shall be paid in good new money or its equivalent.

2. Likewise, any debt contracted in good old money after the year of the aforementioned proclamation shall be repaid in good new money or its equivalent.

3. Likewise, any debt contracted in this devalued money shall be repaid in devalued coinage or its equivalent.

4. Likewise, anybody who, between St. Martin's Day 1413 and St. Martin's Day 1414, negotiated a lifelong tenancy shall be charged 5 quarters in devalued currency for 1 mark [of the original obligation].

5. Likewise, a similar provision shall apply to anybody who, from 1414 to 1415, from St. Martin's Day to the following St. Martin's Day, negotiated a lifelong tenancy; he shall be charged 1 mark 4 skoters in devalued currency for 1 mark in devalued currency.

6. Likewise, anybody who negotiated a lifelong tenancy after the aforementioned years shall be charged 1 devalued mark [for 1 mark of the original obligation].

7. Likewise, any lifelong tenancy negotiated in terms of good old money shall be paid in good new money.

8. The aforementioned regulation of lifelong tenancy is proclaimed[92] and so on, so that whatever the value and worth of the money whenever the lifelong tenancy was negotiated, the payments shall be reckoned at that value of the same money.

9. The same rule shall apply also to any rental negotiated in devalued currency from 1413 to 1414, with 1 mark being charged off at 5 quarters in devalued currency, and the settlement also being made on the same terms.

10. Likewise, any rental negotiated from 1414 to 1415 shall be paid by 1 mark 4 skoters of devalued currency for 1 mark [of the original obligation], and the settlement shall also be made on the same terms.

11. Likewise, any heritable rental negotiated in terms of good old money shall be paid with good new money; but whoever negotiated a heritable rental from 1413 to 1414 shall pay 5 quarters in devalued money for 1 devalued mark.

12. Whoever negotiated a heritable rental from 1414 to 1415, however, shall pay 1 mark 4 skoters in devalued money for 1 light mark, but with the payment days of the present transactions being retained.

13. Likewise, if anybody negotiated a heritable rental encumbered by a previous obligation, in the firm belief that he would make the rental free and clear of the obligation by a specified time, and is prevented by the modification of the coinage from doing so, he shall comply with the obligation or cancel the agreement, in which case whatever money he paid for the rental shall be returned to him.

14. Likewise, if any rental was negotiated in terms of devalued money, but had previously been paid with good old money, and if anybody has negotiated such a rental and wants to pay with devalued money, whoever negotiated in terms of devalued money shall pay with no other money than good new money or its equivalent.

15. In the aforementioned cases, agreements and documents of every kind shall remain in force, and what is now paid shall remain paid.

16. Likewise, whoever negotiated in accordance with such an agreement about what is usually done or what is accepted as a mark, or in accordance with such an agreement about what money is customary, whether this agreement be confirmed in court or with notarized documents or the transactions be witnessed by both parties or both legal advisers agree with each other, the payment shall be made in good new money or its equivalent, and what is now paid shall remain paid.

During the intense debate over the monetary question in Prussia, a (still unidentified) member of the Toruń secretariat thought that D was important enough to be summarized in a Latin translation and inserted in his city's archives. This anonymous translation-summary (which we shall label "T") introduced the touchstone (*lapis probatorius*) in D ¶ 2; by contrast, Copernicus never mentioned the touchstone. D ¶ 16's formula for the minting of skoters called for alloying 2 pounds of copper with $15^1/_2$ ounces of silver. By contrast, two years earlier, M ¶ 16 had prescribed only $11^3/_4$ ounces of silver. T's adherence to D ¶ 16's higher ratio of silver to copper suggests that whoever wrote T was unfamiliar with M. This unfamiliarity is reinforced by differences between T's Latin terminology and M's. For instance, T always refers to the mark by weight ($= 1/_2$ lb) as *marca uncialis*, in order to distinguish it from the mark as a coin. T's convenient term *marca uncialis* was not used by Copernicus in M (or E). Another terminological difference between T and M concerns the penny, which T calls *nummis*, whereas M ¶ 16 uses *denariis*. T ends with the following remark about its source: "This was written in the year 1519." D ends with the date "1519."[93] Clearly, then, T is a translation-summary of D, independent of M.

T was discovered in the Toruń archives by L. A. Birkenmajer, who published it for the first time.[94] In the same year, 1924, an improved text of T was printed by Dmochowski, who placed T after the last page numbered in Roman numerals (p. CLXIV) and before the first page numbered in Arabic numerals (p. 1) in his volume.[95] As his placement of T on 9 unnumbered pages between the Roman and Arabic numerations shows, T (together with Dmochowski's translation of T into Polish) was inserted in his volume at the last moment.

In the following year, 1925, Dmochowski published his conclusions regarding the chronological order of Copernicus' writings about money.[96] T was the earliest (Dmochowski concluded) and was later expanded into M (as published in *Acta Tomiciana*);[97] M, which (like T) was written in Latin, was then translated into German by Schütz, who published his translation in his *Historia rerum prussicarum*.[98]

It was not long before Dmochowski's arrangement was undermined by the discovery of a dated copy of M in the Gdańsk archives. Whereas M as published in *Acta Tomiciana* bore no date, the Gdańsk copy of M, first published in 1940,[99] is dated 15 August 1517.[100] Dmochowski in 1925, having no date for M, felt free to treat M as later than T, which ends with the statement that its source "was written in the year 1519." Now that M is known to have been written in 1517, Dmochowski's view must be reversed: M in 1517 preceded T, which cannot be earlier than 1519.

Dmochowski was therefore wrong on still another count. According to him, M was an expansion of T. But T associates itself, not with M (which was written in 1517), but with a document written in 1519. Since this is the date borne by D, T is related to D, not to M.

According to Dmochowski, T was expanded into M. Actually, T was a contraction (or summary) of D. D was read to the West Prussian Estates on 21 March 1522[101] at a meeting which was attended by delegates from Toruń. Having exercised the privilege of minting for a long time, Toruń was naturally interested in the contents of D. Whereas the delegates of

Gdańsk, which operated another Prussian mint, inserted in their archives a complete copy of D as Copernicus had written it in German, Toruń placed in its archives a summary of D in Latin. T, as we have called this Latin translation-summary of D, was made by a member of the Toruń municipal secretariat, not by Copernicus. Hence, T must be removed from the list of Copernicus' writings about money.

This is what was done to D by Dmochowski. He maintained that M "was translated into German by Caspar Schütz, and inserted by him" in his *Historia rerum prussicarum*. But if M was translated into German by Schütz, then D, which is the German version of M, was not translated by Copernicus and should be removed from the list of his writings about money.

According to Dmochowski, Schütz translated M into German in 1522. But Schütz published the first edition of his *Historia rerum prussicarum* in 1592, two years before he died. His date of birth is not known. Was he alive as early as 1522? If so, was he precocious enough to translate a technical document from Latin into German at so tender an age? Whether precocious or not, Schütz himself stated that in his *Historia rerum prussicarum* he was reproducing "word for word" (*von wort zu wort*) what Copernicus communicated to the West Prussian Estates in 1522. Dmochowski reprinted (1924, p. 33) this statement by Schütz, which ascribes the communication to Copernicus so plainly that there is no possible foundation for any claim that Schütz translated a Latin essay (*aufzsatz*) by Copernicus into German.

That *aufzsatz*, as Schütz called it, or D, as we have been calling it, was translated in a summary form into Latin by the Toruń delegation before they knew about E. E in turn was written for circulation before the following proclamation was issued on 17 July 1526 by King Sigismund in Gdańsk, where he was meeting with the West Prussian Estates:

In Prussia, the old money is demonetized and, under our insignia and the insignia of the lands of Prussia, a new currency shall be minted in three denominations, namely, groats, shillings, and obols or pennies. The scale shall be: 20 groats to 1 mark; 40 groats or 2 marks to 1 Hungarian florin; 3 shillings to 1 groat; and 6 pence to 1 shilling. In the intrinsic value of the metal and face value, this [new Prussian] currency shall be matched with the Polish, which likewise shall be struck anew, in such a way that the Prussian groat shall equal the Polish groat; the [Prussian] shilling, $^2/_3$ [of the Polish shilling]; and the [Prussian] penny shall equal the [Polish] penny. By royal decree this Prussian money shall circulate throughout the entire kingdom.[102]

After returning to Cracow, his capital, on 15 October 1526 King Sigismund issued a decree concerning Polish coinage.[103] The Prussian problem, however, was still not settled in detail. It was to be considered further at the meeting of the West Prussian Estates to be held at Elbląg on 16 March 1528. One week earlier, on 9 March, Bishop Ferber wrote to the Varmia Chapter as follows:

The monetary matter is mixed up with many difficulties. To deal with it properly requires, in my judgment, a number of men experienced in this business. Accordingly, I choose and designate the venerable Dr. Nicholas Copernicus, our brother. I also ask you to choose and designate him so that, together with those previously selected, Dean [Johannes Ferber] and Felix Reich, in Elbląg on 16 March he may take his place, prepared to give advice and, in keeping with his skill in such monetary matters, to put forward what seems necessary and advantageous.[104]

Presumably the Chapter complied with the bishop's wishes and designated Copernicus as its third spokesman. Nevertheless, for reasons still to be clarified, the official proceedings of the meeting in Elbląg do not show Copernicus as a participant.[105] Nothing was lost by his absence, however, since the duke's delegates were not fully empowered to act. In April, however, the Estates of Ducal Prussia agreed to go along with the planned reform of the currency in both Prussias, Ducal and Royal. Meanwhile, on 29 March, Bishop Ferber asked the Varmia Chapter

> to send our venerable brother, Dr. Nicholas Copernicus, as soon as possible to me in the palace at Lidzbark, so that he may be able to leave from here and return to the cathedral [in Frombork] in time for the Easter ceremonies [on 12 April]. I am planning to go over the coinage question with him to some extent and discuss it.[106]

The conversations pleased the bishop so much that when Copernicus left Lidzbark on 7 April, he carried with him a letter in which the bishop pointed out to the Chapter that

> at the meeting of the Estates scheduled to take place in Malbork on 8 [not 7] May the monetary problem will be taken up. To deal with it properly requires informed men, thoroughly experienced in matters of this kind. I therefore ask you to elect our venerable brother, Dr. Nicholas Copernicus, and as a Chapter to designate him to accompany me to that meeting, so that he may be by my side with his knowledge of these questions and with the support of your mandate, and together with me put forward what seems necessary and advantageous.[107]

Again the Chapter did as the bishop wished, and this time the proceedings of the meeting (which actually lasted from 9 May to 20 May) report Copernicus' active presence.[108]

This Malbork meeting faced a new problem: the older coins could be withdrawn from circulation only as fast as the new coins were being issued. During the transitional period both kinds of money, old as well as new, were being used. It was decided that "every denomination of silver and gold coins should be evaluated in terms of the new money."[109] For this purpose a commission was created on 14 May 1528, comprising members from both Ducal Prussia and Royal Prussia. The latter side chose seven representatives, including "Dr. Nicholas Copernicus." On the following day, 15 May, the commission met and had a lengthy discussion, which ended with a request that the masters of the Prussian mints should conduct a complete assay and report their findings.[110]

Three months earlier, the Polish authorities had decided to establish a royal mint in Toruń.[111] It could begin operations without delay, the Malbork meeting was informed by the king's secretary (writing on 28 April 1528),[112] if the Estates would agree on the issuance of the new Prussian coinage. But they postponed action, as we just saw, until the next meeting of the Estates, which was held in Toruń.

This time, instead of writing a letter from a distance, the royal secretary appeared in person for meanwhile, on 15 June 1528, he had been appointed master of the royal mint to be located in Toruń.[113] Justus Ludovicus Decius (Dietz) was a native of Wissembourg (Weissenberg), of which his father was mayor. But at an early age, like many other young men, he left on account of the unrestrained private feuds. After trying his luck elsewhere, Decius went to Cracow. Studying at the Jagiellonian University, however, was far from his mind.[114] Instead, he went to work for the king's banker. Through such employment for some 14 years, Decius gained the royal favor and rose to be the king's secretary. His banking experience qualified Decius to take a prominent part in the projected reform of the currency.

As designated master of the mint in Toruń, on 22 July 1528 he read a paper to the West Prussian Estates meeting in that city.[115] On the following day he explained his position at great length. In the absence of Copernicus, Felix Reich responded in writing on behalf of the Estates at the afternoon session of 23 July.[116] The agreement reached on that day served as a guideline for Decius, who proceeded to mint new Prussian pennies worth 6000 marks. These new coins were distributed to the members of the West Prussian Estates, who would pay for them with old coins at a specified time.

Although Copernicus did not attend the Toruń meeting in July 1528, he did participate in the Elbląg meeting of the West Prussian Estates, 14–17 February 1529.[117] On the very first day, 14 February, a decision was adopted providing for the demonetization of the old penny. Copernicus obtained from Decius for the Varmia Chapter 40 marks' worth of the new pennies, which were sent to the Chapter by the bishop on 24 February 1529.[118] On 27 April the bishop requested the Chapter to send Copernicus to the next meeting of the Estates, which was scheduled to start in Malbork on 8 May. This time, however, the bishop did not ask that Copernicus should be present at the beginning of the session. Instead, he wanted Copernicus to appear on 10 or 11 May with a report on the Chapter's decisions.[119] The Malbork meeting ended on 14 May 1529. Copernicus' presence is not attested.[120]

On Monday, 7 March 1530, a royal decree permitted Gdańsk and Elbląg to resume their minting operations,[121] which had been temporarily suspended since the agreement on 8 April 1525 between the king and the duke of Prussia. The latter signed a ten-year contract on 15 March 1530 with Decius, who took over the management of the mint in Koenigsberg.[122] As a result, Polish coinage became the standard for Ducal Prussia too. But some questions still remained unanswered, in particular the evaluation of foreign gold coins in terms of the new Prussian currency.

This topic was debated at a meeting of the West Prussian Estates in September 1530. But when it became clear that a large deliberative body such as the Estates was not the best forum in which to reconcile conflicting interests, a special small commission was created for this purpose.

On 15 October 1530 the bishop notified Reich that he had been chosen to represent Varmia, together with the "venerable Dr. Nicholas Copernicus, because of their experience in the questions to be considered."[123] Reich, however, begged to be excused on the ground that he soon had to present the annual financial report to the Chapter, which elected another canon in his stead.[124] "In the place of, and on behalf of, the Reverend Maurice [Ferber], bishop of Varmia, the respected and worthy Dr. Nicholas Copernicus and Alexander Sculteti, canons of Frombork,"[125] represented their diocese on the commission. It began its work in Elbląg on 28 October 1530. The next day Decius argued at some length that the Hungarian florin and one other foreign gold coin should be equated with the new Prussian groat in an amount below a specified ceiling. On the following day, 30 October 1530, Decius' recommendation was opposed by Copernicus. According to the report of the proceedings,

Dr. Nicholas Copernicus, canon of Varmia, spoke at length about the evaluation and fundamental indication of the value of gold. With some reference to Decius' long speech yesterday, Copernicus mainly pointed out that if the exact value of gold was to be ascertained, it was to be sought initially and basically not from coined or minted gold (since nobody knows whether it has much or little alloy) but from pure gold and silver. It is necessary to investigate and consider carefully how many coins may pay for 1 mark by weight [$^1/_2$ lb] of silver or gold. Thereafter the value of coined gold would be examined further.[126]

Copernicus' call for market research was not heeded by the other members of the commission, who were looking for instant answers. When they debated whether to raise the evaluation of a foreign gold coin or keep it at its former level, Copernicus spoke in favor of the latter alternative. That was the last time he took part in such proceedings, which continued to drag on.

THE LETTER TO DECIUS

Copernicus' authorship of M, A, and E has never been disputed. On the other hand, his authorship of D has been denied by the device of turning D into a translation of M by Schütz, a device which is chronologically preposterous and contrary to Schütz's own testimony. By an opposite device, the *Letter to Decius* has been ascribed to Copernicus as author (or co-author).

In a long communication about good and bad money to the West Prussian Estates meeting in Gdańsk in the summer of 1526, Decius submitted his ideas for the approval or disapproval of the Estates. He added that he "would wish others to offer different and better proposals of such merit that [his own] plans would deserve to be dropped."[127] He advised the utmost secrecy, however,

> for fear that if these recommendations were publicized, many greedy people would be found who would turn this plan of mine to their own benefit in such a way that the aforementioned good money would vanish, whereas through this disappearance the community would meanwhile be deprived of its advantage. In all these matters His Royal Majesty will issue a prudent edict.[128]

In response to this invitation and advice, the Estates quietly dispatched their *Letter to Decius*. Specifying no individual as author, their *Letter* was sent in the name of the Councillors of Prussia (*per Consiliarios Prussiae*) from their public meeting in Gdańsk in 1526 (*Ex publico conventu Gedanensi 1526*).[129]

Seven passages of this *Letter to Decius* resemble E to some extent. For instance, according to the *Letter*,

> merchants familiar with the value of money sell their goods in accordance with the value of gold and silver, not in accordance with the face value of coinage, for the defectiveness of which they make up by [increasing the] number [of coins]. For what usually happens in everything else needed for human consumption occurs especially in the case of silver, the price of which must rise as money depreciates.[130]

Here the *Letter* (*mercatores ... merces ... ad auri ... valorem vendunt*) repeats E ¶ 17 (*mercatores ... ad auri valorem merces ... vendunt*). On the other hand, the *Letter* goes on at once to point out something of importance that is not said explicitly in E:

> If silver were priced in terms of gold, we do not believe that silver would be much dearer than it was long ago.

Thus, a pound of silver once cost 5 Prussian marks and now it costs, say, 25 marks, a fivefold increase in the price of silver when bought with coined money. During that same period, however, a pound of silver cost $^1/_{11}$ or $^1/_{12}$ lb of gold, an almost stable price of silver when bought with gold bullion. In other words, the value of the precious metals, gold and silver, rose more or less as a unit against depreciating currency. This perceptive insight was put forward by the collective authors of the *Letter* with the expression "we do not believe" (*non putamus*). By contrast, Copernicus spoke of himself in the first person singular: "I can," "I find" (M ¶ 5: *mihi ... licet*; *sentiam*; D ¶ 5: *ich ... kann*; *ich ... smeltze*). Such differences in style and content indicate the relation of Copernicus to the *Letter*: its authors were familiar with E, but Copernicus was not the author of the *Letter*.

How well the collective authors knew E may be illustrated by the following passage of the *Letter*:

> But when we peruse the country's old documents, we learn that Prussian shillings contained 3 parts of silver and 1 part of copper while Prussia was in a very flourishing condition. Then, reversing the order, they alloyed only 1 part of silver with 3 parts of copper. Afterwards they added to 4 parts of copper a fifth part of silver. Nor is that the stopping place. Indeed, the situation becomes worse every day, with very great damage and loss to the community, so that it is to be feared that in the end the money is going to be all copper.[131]

The *Letter*'s historical summary and forecast are quite evidently dependent on E ¶ 9, 10, 13, 16, 19. Even certain expressions reappear in modified form:

E	LETTER
inverso ordine	*ordine praepostero*
tribus partibus aeris quartam argenti misceri	*tribus partibus aeris unam dumtaxat argenti miscebant*
mere cupream	*tota cuprea*
quatuor partibus aeris quinta argenti	*quatuor aeris partibus quintam argenti*
floruit ... Prussia	*Prussia in magno flore fuit*

In another passage the *Letter* compares bad minting practice with poor farming:

> For the ruler will reap what he sowed. If he sowed weeds, he will reap weeds with interest, which usually arrives even by itself. But if he broadcast wheat on the fields of his domain, he will also gather wheat more abundantly.[132]

This striking metaphor may be an elaboration of M ¶ 7 (which was omitted from E):

M	LETTER
metet idem quae seminaverit mala semina	*Id ipsum enim metet, quod seminaverit zizaniam*

The following passage in the *Letter* is unmistakably dependent on E ¶ 23, end:

> The currency cannot be changed and renewed without a loss to His Royal Majesty. It certainly is true that so great a change cannot be accomplished without some loss to the duke and the people. But the damage (if it deserves to be so called) will be amply repaid by the results of the future advantage.[133]

E	LETTER
si modo damnum dici possit ... fructus et utilitas	*damnum (si modo ita dici meretur) futurae utilitatis fructus*

The following passage in the *Letter* echoes E ¶ 3–4:

> For, a coin's face value should exceed its intrinsic value, with the stamp or symbol making up for the deficiency in the intrinsic value or silver, so that such a face value reimburses the mint's expenses and costs without loss to the ruler, and the coin cannot be melted down at a profit.[134]

E	LETTER
potest enim pluris aestimari moneta quam eius qua constat materia	*aestimari enim debet moneta pluris quam valeat*

Less close is the connection between the following passage in the *Letter* and E ¶ 2:

> When a denomination receives its proper proportional quantities in being alloyed with copper, coins (especially pennies) become distinguishable even in their lawful size.[135]

E	LETTER
massa argenti in minutas partes et scrupulos nummorum fracta retineat cum aere admixto convenientem magnitudinem	*moneta in suo quoque genere iustam quantitatis proportionem per aeris mixturam assequatur (praesertim oboli), ut etiam legitima magnitudine conspicua sit*

Our last passage in the *Letter* is almost a quotation from E ¶ 21:

> This arrangement will have a great effect on participation in trade and contracts as well as on generating friendliness.

E	LETTER
Quae res ad animorum conciliationem et negotiationum communionem non parum ponderis est habitura.	*Quae res ad negotiationum et contractuum ac amicitiae conciliationem magnum pondus habebit.*[136]

The foregoing comparison of the *Letter*'s seven passages with the corresponding material in E leaves no room for doubt that whoever wrote the *Letter to Decius* was quite familiar with E. On the other hand, there is no reason whatever to attribute the authorship of the *Letter* to Copernicus. Certainly Copernicus himself did not do so. When he acquired Reich's collection, in it he found the *Letter to Decius*, ascribed by Reich to the "Councillors of Prussia," whereas Reich had specified Copernicus as the author of E.[137] These ascriptions were left unchanged by Copernicus, although he completed his surname as the author of his letter to Reich, whose name he recorded as addressee.[138] Contemporaries handling these documents knew who wrote what: Copernicus was the author of E; the *Letter to Decius* emanated from the Councillors of Prussia, in other words, the West Prussian Estates. When Decius wrote to them on 28 April 1528, he referred to the letter he had received from them two years before as "Your Excellencies' letter, in which much was written about the method of coining money."[139] The *Letter to Decius* was attributed to Copernicus neither by Decius nor by Copernicus nor by Reich nor by the Estates nor by any contemporary.

Into this perfectly clear picture, a confusing shadow was introduced by the description of the Czartoryski Library's Manuscript 259.[140] Pages 32–43 of this manuscript were said to contain the "advice of the leaders of Prussia regarding the renewal of Prussian money" (*consilium primorum Prussiae de restitutione monetae Pruthenicae*), the "author being Nicholas Copernicus" (*adscr. Autore Nicolao Copernico*). Copernicus was indisputably the author of what appears on pages 32–43, namely, E. But he never pretended that E was the "advice of the leaders of Prussia." Nor was any such claim ever publicly advanced on behalf of E before the appearance of the Czartoryski *Catalog*.

That *Catalog* listed, as one of the items on pages 43–53, the *Letter of the Councillors of Prussia to Ludwig Decius of Cracow concerning the Method of Renewing Prussian Money (Epistola consiliariorum Prussiae ad Ludovicum Decium Cracoviensem de ratione restituendae monetae Pruthenicae)*. This *Letter to Decius* was correctly ascribed by the *Catalog* to the Councillors of Prussia, not to Copernicus. But the *Catalog* mistakenly described E as "advice of the leaders of Prussia" (*consilium primorum Prussiae*). Later on, this (mistakenly described) *consilium primorum Prussiae* somehow became confused with the *Epistola consiliariorum Prussiae*. As a result of this confusion, Copernicus, the actual author of E (erroneously called *consilium*) was transformed into the (alleged) author of the *Letter to Decius*.

This transformation first surfaced in 1924. In that year, in six passages of his discussion Dmochowski named Copernicus as the author of the *Letter to Decius*, without any qualification whatever.[141] Yet in still another passage he made Copernicus either the author or a collaborator,[142] while in his Preface Copernicus became "most probably" the author or collaborator.[143] In his Latin text and Polish translation of the *Letter*, Dmochowski listed Copernicus as its author, followed by a question mark.[144] In the same year as Dmochowski (1924) Bujak concluded that the "author of the *Letter to Decius* could be only Nicholas Copernicus" (p. 58). Then in 1957 Dunajewski in a special Appendix (pp. 403–405) argued strenuously for Copernicus' authorship. From his unqualified position, *Studia Copernicana* (VII, VIII) retreated in 1973 to Copernicus as probable author.[145]

Copernicus, however, was not the probable author, the author, the author (?), or a collaborator with the author or authors of the *Letter to Decius*. It is not even known whether Copernicus was present at the Gdańsk meeting of the West Prussian Estates,[146] in whose name and on which occasion the *Letter to Decius* was written. That *Letter* unquestionably incorporated (without acknowledgment) ideas, attitudes, and expressions found in E (and probably in M too), so that the *Letter to Decius* demonstrates the extent to which Copernicus' writings about the monetary question entered the mainstream of thought about that vexing problem.

Those writings include neither the *Letter to Decius* nor T, being confined to M, D, A, and E. These four discussions, regarded as an organic unit, constitute Copernicus' solid contribution to the understanding and solution of a nagging quandary that troubled his contemporaries. His contribution is substantial enough not to require any adventitious adornments. There is no need to gild his lily.

Nor should it be taken away from him, as was done by Wolowski's assessment:

The teachings concerning money, of which Nicholas Oresme made himself the skillful interpreter, found an illustrious defender at the beginning of the 16th century (p. 3).

> Copernicus' views on money are quite close to those of Nicholas Oresme. They are the same sound and vigorous insights, the same understanding of the importance attached to maintaining the means of exchange at the proper fineness and weight, the same judgment expressed on the nature of the ruler's power over the regulation of the value of money (p. vi).

This last judgment may be regarded as the principal thesis of the *De moneta* by Oresme (c. 1320–1382). He was mainly concerned about the monarch's power to enrich himself by debasing the currency at his subjects' expense. The French king, in whose service Oresme was employed, had such authority. But the Polish king sought to obtain the concurrence of the West Prussian Estates in any monetary reform proposed by him for them. As between Oresme and Copernicus, therefore, there is no identity of "judgment expressed on the nature of the ruler's power over the regulation of the value of money."

In fact, these two writers are poles apart. Copernicus confronted the contemporary currency crisis in West Prussia as part of the kingdom of Poland. Unlike Oresme, he did not begin with a quotation from the Bible. Nor did he say, as Oresme did: "In the present treatise I intend to write about what seems to me should be said, mainly according to the philosophy of Aristotle." Copernicus did not introduce a single quotation from Aristotle or from any of the other ancient writers who were constantly cited by Oresme. Instead, Copernicus looked back to the unpublished records of the Estates of West Prussia and of the Teutonic Knights. There is not the slightest indication that he ever consulted Oresme's *De moneta*,[147] whether in any of its numerous manuscripts in Latin or French, or in either of the two editions printed before Copernicus wrote M. It is by no means clear that Copernicus knew of the existence of Oresme's *De moneta* or, for that matter, of Oresme himself. Whether Copernicus ever heard of Oresme is a question of importance not only for the history of monetary theory but also for the development of astronomy. For, Oresme discussed the possibility that the earth moves in a way that has made some people think he was in some sense a pre-Copernican. But nobody has ever shown any dependence of Copernicus' astronomy on Oresme's. The same may be said about Copernicus' monetary writings. Whatever slight similarity may be found between Copernicus (M, D, A, E) and Oresme's *De moneta* arises from their treatment of the same topic, a topic that had been discussed since ancient times. Apart from their debt to these common sources, they diverge very widely, in monetary theory as well as in astronomy. Let us therefore gratefully restore to Copernicus the (ungilded) lily he earned through his own persistent efforts.

BIBLIOGRAPHY

Bentkowski, Feliks, "Rozprawa Mikołaja Kopernika o urządzeniu monety napisana 1526 roku," *Pamiętnik Warszawski*, 1816 (May-August), *5*:381-423; Latin text of E on even-numbered pages 386-422; Polish translation on odd-numbered pages 387-423

Bieda, Ken (Kazimierz), "Copernicus as an Economist," *Economic Record*, 1973, *49*:89-103

Birkenmajer, Ludwik Antoni, *Stromata Copernicana* (Cracow, 1924)

Bujak, Franciszek, *Traktat Kopernika o monecie* (Lvov/Warsaw, 1924)

Corpus iuris polonici, ed. Oswald Balzer, IV (Cracow, 1910)

Dmochowski, Jan, "Nicolas Copernic économiste," *Revue d'économie politique*, 1925, *39*: 101-126

Dunajewski, Henryk, *Mikołaj Kopernik, Studia nad myślą społeczno-ekonomiczną i działalnością gospodarczą* (Nicholas Copernicus, Studies of his Socioeconomic Thought and Administrative Activity; Warsaw, 1957)

Faber, Karl, "Ein Beitrag zur Lebens-Geschichte des Nicolaus Kopernikus," *Beiträge zur Kunde Preussens*, 1819, *2*:263-267

Hegeland, Hugo, *The Quantity Theory of Money* (Göteborg, 1951; reprint, New York, 1969)

Hirschberg, Aleksander, "O życiu i pismach Justa Ludwika Decyusza" (The Life and Writings of J. L. Decius), *Przewodnik naukowy i literacki*, 1874, *2*

Jamiołkowska, Danuta, "Memoriale Łukasza Watzenrodego — analiza paleograficzna," KMW, 1972, pp. 633-648

Johnson, Charles, *The De moneta of Nicholas Oresme and English Mint Documents* (London, 1956)

Kaussler, Ernst, "Jost Ludwig Decius gegen Copernicus," *Pfälzer Heimat*, 1974, *25*:84-87

Le Branchu, Jean-Yves, *Ecrits notables sur la monnaie (XVIe siècle) de Copernic à Davanzati*, 2 vols. (Paris, 1934)

Lengnich, Gottfried, *Geschichte der preussischen Lände königlich-polnischen Antheils, seit dem Jahr 1526 biss auf den Todt Koniges Sigismund I*, 9 vols. (Danzig, 1722-1755)

Macleod, Henry Dunning, *Theory of Credit*, 2nd ed. (New York/London, 1893-1897)

Meissner, Herbert, "Die ökonomischen Forschungen des Nicolaus Copernicus," *Wirtschaftswissenschaft*, 1973, *21*:229-234

Monroe, Arthur Eli, *Monetary Theory before Adam Smith* (Cambridge, Ma, 1923; reprint, New York, 1966)

Moore, George Albert, *Copernicus, Treatise on Coining Money* (Chevy Chase, Md: Country Dollar Press [1965])

Nomi, Federigo, *L'Oro assistito* (Perugia, 1972); part 3, pp. 205-243: Copernicus' Latin text and translation into Italian by Nomi; corrected version in *Opere di Nicola Copernico*, ed. Francesco Barone (Turin, 1979), unavailable for this edition

Oko, Jan, "Paweł Deusterwalt nieznany humanista XV wieku," *Ateneum Wileńskie*, 1930, *7*:786-798

R—H Reiss, Timothy J. and Hinderliter, Roger H., "Money and Value in the Sixteenth Century: The *Monete Cudende Ratio* of Nicholas Copernicus," *Journal of the History of Ideas*, 1979, *40*:293-313

Schmauch, Hans, "Nikolaus Coppernicus und die preussische Münzreform," *Staatliche Akademie, Personal- und Vorlesungsverzeichnis, 3. Trimester 1940* (Braunsberg, 1940), pp. 3-40; the MS on which Schmauch based his publication of M has since disappeared; for later MSS of M, see *Filomata*, 1972, p. 13; 1975, pp. 197-199

Schütz, Caspar, *Historia rerum prussicarum* (Zerbst, 1592)

Schwinkowski, Walter, *Das Geldwesen in Preussen unter Herzog Albrecht 1525-69* (Berlin, 1909)

Semrau, Arthur, "Jost Ludwig Dietz und die Münzreform unter Sigismund I," MCV, 1906, *14*:33-48

Sommerfeld, Erich, "Copernicus über das Geld, *Wirtschaftswissenschaft*, 1975, *23*:868-877

————, *Die Geldlehre des Nicolaus Copernicus* (Vaduz, 1978)

Szelągowski, Adam, *Pieniądz i przewrót cen XVI i XVII wieku w Polsce* (Money and the Price Revolution in the 16th and 17th Centuries in Poland; Lvov, 1902)

Taylor, Jack, "Copernicus on the Evils of Inflation," *Journal of the History of Ideas*, 1955, *16*:540-547

Vossberg, Friedrich August, *Geschichte der preussischen Münzen und Siegel von frühester Zeit bis zum Ende der Herrschaft des Deutschen Ordens* (Berlin, 1843)

Waschinski, Emil, *Die Münz- und Währungspolitik des Deutschen Ordens in Preussen* (Göttingen, 1952; Göttinger Arbeitskreis, No. 60)

———, "Nikolaus Kopernikus als Währungs- und Wirtschaftspolitiker 1519–1528," ZGAE, 1958, *88*:389–427

Wolowski, Louis, *Traictie de la première invention des monnoies de Nicole Oresme et Traité de la monnoie de Copernic* (Paris, 1864)

Żurawicki, Seweryn, "Copernicus on Money," *Journal of European Economic History*, 1974, *3*: 126–128

NOTES ON COPERNICUS' MONETARY TREATISES

¹ E ¶ 1's *Tria* was omitted from Nomi's translation (p. 213/5).

² E ¶ 1's *nesciat* is not matched by *in Abrede stellen wird* (will dispute; PI², 195/18).

³ M ¶ 1's (E ¶ 2's) *multipliciter* (in many ways) is not matched by "at every moment" (as in Wolowski's *à tout moment*, p. 23/1; p. 49/3 up).

⁴ Nomi's *si possa* (p. 217/8) omits *non*, and thereby turns Copernicus' denial (I cannot buy) into its opposite: it *is* possible to buy as much silver with the money as the money itself contains.

⁵ Reading *Das* (with Schütz, sig. h 5ᵛ/24) instead of *Dan*.

⁶ D ¶ 7's *mehen* (reap; modern spelling: *mähen*) was misread by Schütz (sig. h 5ᵛ/6 up) as *mehr* (more). Hence he interpolated *des bösen* (of the bad [seeds]), thereby emerging with an incomplete combination: he will [no verb] again more of the bad which he sowed. Schütz was followed by Prowe (PI², 147/last line: *mehr*; PII, 24/3 up: *meher*).

⁷ As a part of the Prussian monetary system, the shilling was originally not actually minted, but was used only for purposes of reckoning as money of account. The Prussian shilling was first issued by Winrich of Kniprode, Grand Master of the Teutonic Knights, whose order, dated 2 February 1380, was printed by Vossberg, pp. 94–95. The skoter was equal to $2^1/_2$ shillings.

⁸ The Prussian groat was first minted by order of Johann of Tiefen, who was Grand Master from 1 September 1489 until his death in battle on 25 August 1497. In an effort to overcome the inconveniences resulting from the disparity between the coinages of the Order and of Poland, he introduced the Prussian groat on a footing of equality with the previously established Polish groat (Vossberg, p. 190).

⁹ By putting specimen coins on the balance pan, Copernicus checked their actual weight against the weight they were expected to have. In his time coins were still made by the hammering process. Within a generation after his death, the mill-and-screw process was introduced as a safeguard against clipped edges.

¹⁰ On 4 February 1454 the towns and rural nobility of Prussia revolted against the oppressive rule of the Teutonic Order. Unable to defeat the Order without outside help, the towns and nobles sought and obtained an alliance with Poland. In this way "the country's status changed," with West Prussia ceasing to be part of the Order's domain and entering the kingdom of Poland. The Prussian coinage had previously been minted by the Order, which was now an enemy state. Hence on 9 March 1454 King Casimir IV issued the following decree:

> It is our wish that, only as long as the present war continues, ... money shall be minted in 4 places of the aforesaid lands [Prussia], namely, Toruń, Elbląg, Gdańsk, and Koenigsberg (*Volumina legum*, I: *Leges, statuta, constitutiones, et privilegia regni Poloniae* [Petersburg, 1859], 81/left column/26–29).

Since Koenigsberg sided with the Teutonic Order, it was dropped from the list of towns receiving the minting privilege. In 1457, on 15 May Gdańsk, and then on 26 August Toruń, were granted the minting privilege in perpetuity (Acten, IV, 560, 605–606). Schwinkowski's "1453" (p. 132/7 up) should be corrected.

¹¹ 3 skoters = 1 ounce; 24 skoters = 8 ounces = $^1/_2$ pound = 1 mark.

¹² "Skoters" was supplied by Schütz (sig. h 6ʳ/16 up).

¹³ The groat, officially equated with 3 shillings = 18 pence, was actually worth 15 pence; 18−15 = 3; $^3/_{18} = ^1/_6$. In 1511 the West Prussian Estates met in Gdańsk from 29 May to 18 June. On 12 June they decided to introduce the Prussian groat (ASPK, V, part 3, p. 26/22).

¹⁴ "13," as in PI², 148/8 up; not "8" (as in PII, 27/15).

¹⁵ D ¶ 13's *wo es nicht vorkommen wyrdt* was paraphrased by Waschinski (1958, p. 412/20) as *wofür es noch kaum zu haben sein wird* (for which reason it will in addition scarcely be available). But M ¶ 13's *nisi interim succursum fuerit* shows that D's *vorkommen* means "help," a sense it no longer has in modern German (J. and W. Grimm, *Deutsches Wörterbuch*, XII, 2 [reprint, Leipzig, 1960], column 1237, no. 5). Waschinski handled E ¶ 16 better (see n. 52, below).

[16] 1 mark = 60 shillings = 24 skoters; 5 shillings = 2 skoters; 5 half-shillings = 1 skoter. Copernicus' earliest prescription (in M) for the skoter called for $11^3/_4$ oz (slightly less than $^3/_4$ lb) silver with 2 lb copper. In D he raised the silver component to nearly 1 lb silver ($15^1/_2$ oz). But for the shilling, in both M and D, he prescribed nearly 1 lb silver with 3 lb copper. This higher ratio of silver for the skoter vanishes in E ¶ 24, where the skoter and the shilling are to be struck "according to the same proportion," nearly 1 lb silver to 3 lb copper.

[17] Traditionally, 1 Prussian shilling = 6 Prussian pence, and 1 half-shilling = 3 pence. M ¶ 16's new half-shilling, however, would have been worth more than 3 of the pennies then in circulation. Hence, M ¶ 16 proposed, 1 new half-shilling = 4 pence. D ¶ 16, however, withdrew this suggestion by substituting 3 for M's 4.

[18] As the prescriptive presiding officer of the Estates of West Prussia, the bishop of Varmia kept files of the pertinent documents in the episcopal palace in Lidzbark. Here Copernicus, as a canon of Varmia, had access to the "old records of proceedings and official correspondence" (*in antiquis recessibus ac litterarum munimentis*).

[19] Conrad of Jungingen was elected Grand Master of the Teutonic Order on 30 November 1393 and served until his death on 30 March 1407 (SRP, III, 190, 395).

[20] Or Grunwald (15 July 1410).

[21] The document used here by Copernicus (the monetary reform of 1415, referring back to the situation under Conrad) was printed in Acten, I, 266.

[22] The amount of silver contained in 112 shillings, weighing $^1/_2$ lb of the alloy $^3/_4$ silver $+ ^1/_4$ copper, is $^3/_4 \times ^1/_2 = ^3/_8$ lb; in other words, 112 shillings contain $^3/_8$ lb silver. Copernicus wants to know in how many shillings $^1/_2$ lb silver is contained. For, he wishes to compare that number with the number of shillings needed to buy $^1/_2$ lb silver bullion. Since 112 shillings contain $^3/_8$ lb silver, $^1/_3 \times 112$ contains $^1/_8$ lb silver, and the sum $[112 + (^1/_3 \times 112)]$ contains $^3/_8 + ^1/_8 = ^1/_2$ lb silver.

[23] $112 \times ^1/_3 = 37^1/_3$; $112 + 37^1/_3 = 149^1/_3$. ER read "cxlviii," an error repeated in EF. With 149 being written, not as cxlix, but as cxlviiii, the omission of the fourth i was inadvertent. This erroneous numeral was printed by Bentkowski, Baranowski (p. 565/13), and Wolowski (p. 54/17). Yet Baranowski's and Wolowski's translation (p. 55/16) read 149 correctly. In Nomi, however, both text and translation (p. 218/17; p. 219/15) are defective: *cxlviii, centoquarantotto* (148).

[24] $^1/_2$ lb $+ (^1/_3 \times ^1/_2 = ^1/_6) = ^3/_6 + ^1/_6 = ^4/_6 = ^2/_3$ lb. After "skoters" ER inserted above the line "silver, of course." In a (lost) letter to Copernicus, Reich questioned the equivalence of $^2/_3$ lb with 32 skoters, on the ground that $^2/_3$ lb = 8 ounces. Reich was thinking in terms of the assayer's 12-ounce pound, which continued the tradition of ancient Rome. Thus a document dated about 1350 says: "a pound of pure gold and a pound of silver, by weight, both consist of 12 ounces" (Johnson, p. 83/4 up–3 up). Copernicus' computation, however, was based on the 16-ounce pound. As used in business, the 16-ounce pound contained 48 skoters (3 skoters = 1 ounce; 1 skoter = $^1/_3$ ounce, not "about" $^1/_3$ ounce, as in Nomi, p. 217, n. 1). Hence Copernicus replied to Reich in part as follows:

> I matched $^2/_3$ [of a pound] with 32 skoters, as our whole pound contains 48 skoters. I should not have said "8 ounces." For, the pound, used especially by pharmacists, which is divided into ounces, is different, being lighter by $^1/_4$ (see below, Correspondence, Letter IV).

The pharmacist's or assayer's 12-ounce pound was $^1/_4$ lighter than the commercial 16-ounce pound: $16 - ^1/_4(16) = 12$.

[25] In E ¶ 9, 1st sentence, $^1/_2$ lb pure silver was priced at 2 marks, 8 skoters; 1 mark = 24 skoters = 60 shillings; 2 marks, 8 skoters = $120 + 20$ shillings = 140 shillings.

[26] Face value ($149^1/_3$) − intrinsic value (140) = $9^1/_3 = ^1/_{15}(140)$. In a sound metallic currency, Copernicus believed, face value should equal $^{16}/_{15}$ of intrinsic value (see Letter IV, n. 11).

[27] Winrich of Kniprode was elected Grand Master on 16 September 1351 and served until his death on 24 June 1382 (SRP, III, 394). EF's erroneous substitution of "Heinrich" for Winrich was followed by Wolowski (p. 54/4 up; p. 55/6 up) and Nomi (p. 218/4 up; p. 219/4 up). For Winrich's minting of the earliest Prussian shillings, see n. 7, above.

[28] Ulrich of Jungingen was elected Grand Master on 26 June 1407 and was killed in the battle of Grunwald/Tannenberg on 15 July 1410 (SRP, III, 395). Had Copernicus named these three Grand Masters in chronological order, Ulrich (1407–1410) would have followed Conrad (1393–1407).

[29] Heinrich of Plauen was elected Grand Master on 9 November 1410 (SRP, III, 335–338, 396) and was dismissed on 14 October 1413 (Wilhelm Nöbel, *Michael Küchmeister, Hochmeister des Deutschen Ordens 1414–1422* [Bad Godesberg, 1969], pp. 61–62). The dismissed Grand Master's cousins, Heinrich Reuss of Plauen and another Heinrich of Plauen, protested against the dismissal of their relative (Acten, I, 226–233).

[30] By omitting *reperiuntur* (they are found) from his translation (p. 57/1), Wolowski obscured Copernicus' handling of the coins themselves.

[31] $3/5$ silver, $2/5$ copper, instead of $3/4$ silver, $1/4$ copper.

[32] ER's margin added *vetere meliore remanente* (while an old, better coinage remained in circulation). Wolowski's translation (p. 57/10) omitted *meliore* (better).

[33] Michael Küchmeister of Sternberg was elected Grand Master on 9 January 1414 and resigned for reasons of health in March 1422 (SRP, III, 396; Nöbel, pp. 69, 128). He died on 15 December 1423 (Nöbel, p. 130). His successor, Paul of Rusdorf, was elected Grand Master on 10 March 1422 and served until his resignation on 2 January 1441, one week before he died (Carl August Lückerath, *Paul von Rusdorf, Hochmeister des Deutschen Ordens 1422 bis 1441* [Bad Godesberg, 1969], pp. 16, 204). Copernicus' mistake in assigning Küchmeister's given name Michael to his successor, Paul of Rusdorf, was followed by Wolowski (p. 56/12; p. 57/13) and Nomi (p. 220/12; p. 221/13).

[34] ER underlined "former," which was omitted by EC.

[35] In accordance with a decision adopted at a meeting of the Estates of the Teutonic Order on 7 July 1416 at Malbork, Grand Master Michael Küchmeister on 24 August 1416 proclaimed that "2 of the old shillings should be exchanged for 1 new shilling" (Acten, I, 281/8 up–7 up).

[36] Reading *vero* (with Wolowski, p. 56/8 up), not *vera* (as in PII, 36/14).

[37] Copernicus may have seen the regulation issued by the Grand Master on 24 June 1460:

> Of the small pennies which we now order to be minted, 12 shall be equal to 1 good shilling, and
> 6 to 1 old shilling of the currency minted formerly (Acten, V, 34/19–21).

[38] Reading *dicam*, with Wolowski (p. 58/3), not *dicavi* (as in PII, 36/21). Copernicus fulfills his promise in E ¶ 14. E ¶ 11's *quomodo demum grossi facti sunt [novi solidi]* (how the new shillings finally became groats) was not matched by Nomi's *come i grossi furono posteriormente considerati* (how the groats were considered later; p. 223/2–3).

[39] Wolowski's translation, *il est facile de le calculer* (p. 59/5–6; it is easy to compute) conceals Copernicus' actual handling of the coins themselves (*ex eorum compositione satis apparet*; see n. 9, above).

[40] Waschinski's *viel unter* (much below; 1958, p. 407/22, p. 408/1) is a mistranslation of *prope* (E ¶ 12, M ¶ 9), which is matched by D ¶ 9's *fyl na* (modern spelling: *viel nah*, very nearly). Finding that the surviving coins do not weigh "much below" 2 pounds, Waschinski thought that Copernicus made a mistake which, however, was made by himself. When Schütz modernized D's grammar and spelling according to the standards prevailing in his own time, almost three-quarters of a century after D, he confused D ¶ 9's *na* with *nach* (sig. h 6^r/11).

[41] It was in M ¶ 10 that Copernicus said: *antiqui ... pondere ... pares illis ... valore ex dimidio*. He evidently intended to repeat this statement in E, but did not do so.

[42] These old shillings, made from an alloy of $3/4$ copper + $1/4$ silver, toward the end of E ¶ 10 were called "copper coinage." 4 lbs of this alloy, containing 1 lb of silver, yielded 16 marks = 960 old shillings, as against 8 marks = 480 new shillings, for 1 lb silver.

[43] E ¶ 13 modified M ¶ 11 here by inserting "in the old shillings," which have just been described in E ¶ 12 as having $1/4$ silver.

[44] E ¶ 13's *recentibus* (recent), denoting the municipal coinage, does not mean "old" (R–H, p. 308/4; *anciens*, Wolowski, p. 59/19; *vecchi*, Nomi, p. 223/19). The old shillings, $1/4$ silver, "were worth half as much" as the new shillings, $1/2$ silver, which were "worth more than twice as much as the recent shillings," $1/5$ silver.

[45] During the minting of the old shillings, when the alloy consisted of $3/4$ copper + $1/4$ silver, 16 marks were produced from 1 lb of pure silver. But 20 marks were exchanged for 1 lb silver in the production of the new shillings. Copernicus reckons the reduction as $1/5$ of the new value: $1/5 \times 20 = 4 = 20 - 16$.

[46] E ¶ 14's *revocare* (recover) was omitted by Wolowski's translation (p. 59). But the purpose of equating the skoter with the groat (= 3 shillings) was to attract the skoter back into West Prussia from Brandenburg. That purpose was not conveyed by Nomi's *riattribuire* (p. 223/6 up).

[47] Reading *aestimationem* (with PII, 37/9), not *aestimatione* (as in Wolowski, p. 60/1, and Nomi, p. 224/1). Here *aestimationem* cannot mean "face value" (as in R–H, p. 308/13). E ¶ 14's *ubi iam multitudo etiam premebat aestimationem ipsius* (while the value of the coinage in circulation was already depressed also by its abundance) was grossly mistranslated by Nomi's *quando già la moltitudine faceva pressioni per la valutazione di questa* (when the crowd was already pressing for the appraisal of the currency; p. 223/last line). Nomi explained (p. 223, n. 1) that his "crowd resented the effects of the tyrannical monetary actions" (*la moltitudine mal sopportava le conseguenze dei soprusi monetari*). Copernicus' *multitudo*, far from denoting a crowd of people, referred to a currency's excessive abundance, which depressed its value. This statement in E ¶ 14, taken together with the corresponding remarks in E ¶ 6, M ¶ 5, and D ¶ 5, links Copernicus with a rudimentary stage of the quantity

theory of money. Before the advent of paper money, an excessive volume of metallic coins tended to arouse public fears that the authorities were surreptitiously introducing a debasement which could not be detected by the man in the street.

[48] Reading *cuderetur* with EC, as against Bentkowski's *redderetur*, followed by Wolowski (p. 60/11), Dmochowski (p. 9/6 up), and Nomi (p. 224/11), as well as Hipler's *videretur*, repeated by PII, 37/18. The translations by Wolowski (p. 61) and Nomi (p. 225) omitted *quibus aequivalens priori redderetur/videretur/cuderetur posterior* (the difference by which the later coinage would be equalized with the older coinage).

[49] E ¶ 15's distinction between *solidorum aestimatio* (the shilling's face value) and *valore grossorum* (the groat's intrinsic value) was ignored by Wolowski's *la valeur des sous et celle des gros* (the value of the shillings and that of the groats; p. 61/14–15). But in E ¶ 2 (M ¶ 1) Copernicus insisted that "intrinsic value must be distinguished from face value" (*oportet ... valorem ab aestimatione discerni/discernere*).

[50] According to E ¶ 12's last sentence, 1 lb silver then yielded 16 marks, as against 24 of these light marks to 1 lb silver.

[51] E ¶ 16's *Debuerant autem iam tandem saltem reliquiae tantillae dignitatis monetae permansisse* (But at least some small remnants of the money's worth must have persisted in the end) was not matched by Wolowski's *Tels devaient être les résultats de la détérioration de la monnaie* (The results of the deterioration of the money had to be of this nature; p. 61/18–19).

[52] E ¶ 16's *si non succurratur* was correctly paraphrased by Waschinski (1958, p. 419/10): *Wenn nicht bald Abhilfe geschaffen werde* (If help is not forthcoming soon), by contrast with his earlier blunder (see n. 15, above).

[53] E ¶ 16's *nostratium* (our) was not matched by Wolowski's *nous* (us; p. 61/last line).

[54] Reading *contemptim* (scornfully) with EC, rather than *contempti* (being despised; Wolowski, p. 62/5; PII, 38/2–3; Nomi, p. 226/5) and *contemplari* (contemplating; Dmochowski, p. 10/18–19).

[55] E ¶ 16's *debent* (they owe) was not matched by *wir ... verdanken* (we owe), a mistranslation by which PI[2], 197/9, sought to intensify Copernicus' emotional attachment to Prussia.

[56] Reading *laboret* (with EC, ER, Wolowski, p. 62/10, and Nomi, p. 226/10), not *labor et* (as in PII, 38/7).

[57] E ¶ 17's *cum moneta mixta* (with the mixed coinage) was not matched by Wolowski's *avec la même somme de monnaie* (with the same sum of money; p. 42/3 up, p. 63/13), nor by Nomi's *con la stessa somma di moneta* (p. 227/15), translating Wolowski rather than E ¶ 17. Copernicus was present at a meeting of the West Prussian Estates in Elbląg on 18 January 1504 when "goldsmiths were forbidden to melt down Prussian coinage" (ASPK, IV, part 1, p. 115).

[58] Waschinski's *Bedürfnisse* (1958, p. 414/23) translated E ¶ 17's *mercedem* (wages) as though it were *merces* (goods, necessities; used twice in E ¶ 16). Wolowski (p. 43/4) said *les salaires*, which he omitted from his translation (p. 63).

[59] Reading *vilitate* (with Wolowski, p. 62/22), not *utilitate* (as in PII, 38/17).

[60] Reading *omnia* (with EC, EF, ER, and PII, 38/17), not *etiam* (= also, as in Wolowski, p. 62/6 up, and Nomi, p. 226/6 up).

[61] E ¶ 18's *facilius suppeditantem* (facilitating the supply) was omitted from Nomi's translation (p. 229).

[62] Wolowski's translation added *et tous ceux qui ont à faire des payements* (and all those who have to make payments; p. 65/2–3), without any warrant in E ¶ 18.

[63] E ¶ 18's *spe* (hope) was omitted from Wolowski's translation (p. 65).

[64] E ¶ 19's *omni hominum ordini* (every class of people) or *omnium ordinum hominibus* (people of all classes) was not matched by PI[1], 197/25's *die Mehrzahl der Einwohner* (the majority of inhabitants).

[65] Reading *discimus* (with E ¶ 19 and PII, 39/2), not *dicimus* (as in Wolowski, p. 64/17, and Nomi, p. 228/17).

[66] Nomi's translation omits E ¶ 19's *minori pecuniarum numero* (with a smaller number of coins), and writes "Russia" instead of "Prussia" (p. 231/7).

[67] E ¶ 19's *dominio* (for their land) was matched neither by PI[1], 198/7's *ihren Herren* (to their masters) nor by Waschinski, 1958, p. 415/23's *ihrem Herrn* and Nomi, p. 231/14's *al loro signore* (to their master, as though *dominio* were *domino*). Wolowski's translation (p. 67) omitted E ¶ 19's *si plus solito suo dominio pendere videantur* (if they seem to pay more than usual for their land).

[68] Copernicus' recommendation repeats a resolution proposed by Grand Master Heinrich Reuss of Plauen on 15 February 1467, and adopted three days later, that both Prussias, Ducal and Royal, should share a single currency bearing the royal Polish insignia (Acten, V, 224–225).

[69] Nomi's text (p. 232/9), following Wolowski (p. 68/9), reads *prematum* instead of *primatum* (of the leaders). As a result, Nomi's translation (p. 233/10) interpolated "the primitive regulation" (*il primitivo regolamento*), for which there is no warrant in E ¶ 22. Treating *primatum* as the accusative singular instead of the genitive plural, R–H translate: "uphold the supremacy of the State" (p. 310/6 up); but this would require *perseverent*

to be a transitive verb. E ¶ 22's *ordinationem nunc instituendam* (the regulation to be established now) anticipates King Sigismund's proclamation of 17 July 1526 (see text after n. 14, above, and notes 72, 89, below).

[70] E ¶ 23's *et nimium multiplicatam* (and its quantity excessive) was left untranslated by Wolowski (p. 69) and Nomi (p. 233). Le Branchu's *les sommes ... seront trop compliquées* (the totals will be too complicated, p. 16/8 up) resembles *le problème se compliquera* (the problem will become complicated; Dmochowski, 1925, p. 120). When a recoinage was to be undertaken, the previous coinage must be completely demonetized (Copernicus never fails to recommend).

[71] Reading *cum* (with Dmochowski, p. 15/16), not *tum* (as in Wolowski, p. 70/16; PII, 41/12; Nomi, p. 234/16).

[72] Nomi's translation (p. 235/2 up) omitted *seu* (or), which helps determine when E was written. Copernicus' reform program envisaged the minting of skoters *or* groats. On 17 July 1526 King Sigismund ordered the minting of groats, but not skoters (see text above, at n. 14). Hence, E was written before 17 July 1526. See n. 89, below. The groat ($= 3$ shillings) was more convenient than the skoter ($= 2^{1}/_{2}$ shillings).

[73] Reading *minimum* (with Wolowski, p. 72/16), not *nimium* (as in PII, 42/2).

[74] Not "11" (*undici*, as in Nomi, p. 239/2). Copernicus emphasizes the change to the current ratio 1:12 from the former ratio 1:11 by the adversative conjunction *tamen* (yet), which Nomi mistranslates by *anche* (also). According to Nomi, the ratio remained unchanged at 1:11. But, according to Copernicus, it changed from 1:11 to 1:12.

[75] 10 florins weigh $^{1}/_{11}$ lb; 110 florins weigh 1 lb. Hence, R–H's "112" florins (p. 312/10) should be corrected.

[76] E: *factis enim XX marcis circiter ex libra una argenti*; Wolowski (p. 75/6): *une livre d'argent donnant environ 20 marcs*; Dmochowski (p. 67/22-23): *przy biciu bowiem około 20 grzywien z funta srebra*; Le Branchu (p. 18/5 up): *en frappant avec une livre d'argent 20 marcs environ*; Waschinski (1958, p. 416/23-24): *Falls man etwa 20 Mark aus 1 Pfund Silber machen wurde*; Sommerfeld (p. 63/262-263): *Wenn nämlich ungefähr 20 Mark aus einem Pfund Silber hergestellt würden*. All five of these translators mistakenly attached *circiter* (approximately) to "20 marks" instead of to "1 pound of silver." R–H omit *circiter* (p. 312/21).

[77] 11 lb silver = 1 lb gold; 1 lb silver = $^{1}/_{11}$ lb gold; from $^{1}/_{11}$ lb gold: 10 Hungarian florins; from $\simeq$ 1 lb silver: 20 Prussian marks; 10 Hungarian florins = 20 Prussian marks; 1 Hungarian florin = 2 Prussian marks.

[78] 1 Prussian mark = 20 Polish groats; 2 Prussian marks = 40 Polish groats. With 40 Polish groats, Nomi equated 2 marks of gold (*due marchi d'oro*, p. 239/10), as though the mark were a gold coin, instead of an alloy of silver and copper. As a result, Nomi omitted from Copernicus' equation *aureo* (1 gold piece). With 1 gold piece, Copernicus in 1522 (see text at n. 8, above) had equated 38 Polish groats. Now in 1526, he shifts to 40 groats, the rate soon proclaimed by the king (see text at n. 102, below).

[79] 10 Hungarian florins = $1^{1}/_{11}$ lb silver (not 12 florins, as in R–H, p. 312/24); from $1^{1}/_{11}$ lb silver > 20 Prussian marks; ∴ 20 Prussian marks = 10 Hungarian florins; 2 Prussian marks = 1 Hungarian florin.

[80] Copernicus supposes that 20 marks are made from $1^{1}/_{11}$ lbs of silver. Now 20 marks = 20×60 shillings = 1200 shillings, from $^{12}/_{11}$ lb silver. Then, 100 shillings from $^{1}/_{11}$ lb silver, and 550 shillings from $^{5.5}/_{11}$ lb = $^{1}/_{2}$ lb silver. But 550 shillings = 540+10 shillings = 9 marks 10 shillings. Hence ER's *marche VIIII et solidi X* is correct, whereas EC's wrong *marche VIII et solidi X* was followed by Wolowski, p. 74/11 up, p. 75/16; PII, 42/9 up; Dmochowski, p. 17/11, p. 67/2 up; Nomi, p. 238/11 up, p. 239/12 up; Sommerfeld, p. 63/269; and R–H, p. 312/29.

[81] Here Copernicus abandons what he had proposed in A: equating 20 Polish groats to 1 Prussian mark (see text near n. 5, above). Lowering the Prussian mark from 20 to 15 Polish groats devalues the Prussian mark by $^{1}/_{4}$.

[82] Copernicus has just proposed equating 1 Hungarian florin with 2 Prussian marks. Now he reluctantly agrees to equating 1 Hungarian florin with 2 Prussian marks plus 16 skoters. Since 24 skoters = 1 mark, and 16 skoters = $^{2}/_{3}$ mark, with this devaluation of the Prussian currency, 1 Hungarian florin = 2 marks, 16 skoters = $^{6}/_{3} + ^{2}/_{3} = ^{8}/_{3}$ marks, instead of $^{6}/_{3} = 2$ marks. This change from ($^{6}/_{3}$ marks = 1 florin) to ($^{8}/_{3}$ marks = 1 florin) entails a reduction of $^{2}/_{6} = ^{1}/_{3}$ in the value of the Prussian mark as against the Hungarian florin.

[83] In order to devalue the Prussian mark by $^{1}/_{3}$, Copernicus suggests minting 24 marks from 1 lb silver, or 1 mark from $^{1}/_{24}$ lb silver. He had previously proposed minting 20 marks from $1^{1}/_{11} = ^{12}/_{11}$ lb silver, or 1 mark from $^{12}/_{(20 \times 11)} = ^{12}/_{220} \simeq ^{1}/_{18}$ lb silver. A mark made from $^{1}/_{24}$ lb silver would be worth approximately $^{1}/_{3}$ less than a mark made from $^{1}/_{18}$ lb silver.

[84] A price of 12 marks for $^{1}/_{2}$ lb silver would devalue the mark by about $^{1}/_{3}$ of the price (9 marks, 10 shillings) explained in n. 80, above. Before Copernicus expounded his proposals for monetary reform to the tough-minded members of the West Prussian Estates, in M ¶ 17 he had recommended an upward revaluation of the

Prussian coinage in the ratio $^{13}/_{10}$, or nearly $^{1}/_{4}$. Now in E, after being exposed to the views of the business community, he is reconciled to a downward devaluation of about $^{1}/_{4}$ or $^{1}/_{3}$.

[85] PII, 42–43: *florenus ungaricus commutabatur* (a Hungarian florin was exchanged); Wolowski, p. 74/last line: *florenis ungaricis commutabantur* (Hungarian florins were exchanged). PII's text would put 1 Hungarian florin = 12 marks. Copernicus' preferred ratio, however, would be 1 Hungarian gold florin = 2 Prussian marks (*marchae duae pruthenicae pro aureo ungaricali*; see text above, at n. 79). Hence, our passage should read *VI floreni ungarici commutabantur*, making 12 marks = 6 florins (instead of an unspecified number of florins or 1 florin).

[86] Reading *conservationem* (with PII, 43/12) instead of *conversationem* (as in PI², 200/26).

[87] Copernicus' 4th Conclusion repeats the recommendation he had made in A (see the text near n. 5, above).

[88] E's *multitudine* was translated by Dmochowski (p. 69/17) as *ilości*; by Le Branchu (p. 20/10 up) as *quantité*; by Nomi (p. 243/15) as *troppa*; and by R–H as "excessive quantity of money" (p. 313). But the 5th Conclusion's warning is directed against, not an excessive quantity of coins of all denominations, but an excessive multiplication of denominations. In this respect the 5th Conclusion is closely related to the 6th Conclusion's recommendations concerning the various denominations.

[89] The 6th Conclusion's *sive* (or) was omitted by Wolowski's translation (p. 79/7). But Copernicus' indecision as between skoters or (*sive*) groats is decisive in dating the composition of E before 17 July 1526, when a royal edict ordered the minting of groats and said nothing about skoters (see text above, at n. 14, and n. 72). Whereas Copernicus was undecided as between the skoter or the groat, he felt sure that the shilling and (*et*) the penny were both needed; R–H's second "or" should be changed to "and" (p. 313/6/2).

[90] This 1st sentence of Copernicus' 6th Conclusion was omitted from Nomi's translation (p. 243).

[91] EF changed *verramung* (proclamation; *vorramunge*, in Acten, I, 320) to *vereinigung* (meeting). As this mistaken change shows, *verramen* (proclaim) had already become obsolete when EF was written; see Grimm, *Deutsches Wörterbuch*, XII, 1 (Leipzig, 1956), column 983. The proclamation selected by Copernicus was by no means unique; in fact, it followed closely in time and in spirit the proclamation of 12 November 1416 (Acten, I, 292–294). Somewhat closer to Copernicus' own period was the order issued on 24 June 1460:

> All debts contracted since 11 November 1459 shall be repaid between now and 11 November of the present year 1460 in terms of the money in which the debts were contracted (Acten, V, 34/10 up–7 up).

The proclamation of 6 November 1418 was printed in Acten, I, 320–321, and in CDW, III, 538–540, no. 436 (with only 15 sections, because our 11 and 12 are combined in one section).

[92] Here EF changed *verramet* (*vorromet*, in Acten, I, 320) to *vereinet*, as EF had substituted *vereinigung* in the preamble of the proclamation.

[93] See text at M ¶ 18, above. T "was specifically written for the Little (regional) *Sejm* meeting in Toruń in 1519," according to K. Bieda, "Copernicus as an Economist," *Economic Record*, 1973, 49:93. There was no such meeting in Toruń in 1519. With regard to the first of Copernicus' six conclusions, Bieda says (p. 98) that "the reform must be brought about by a unanimous agreement of the regional Little *Sejm* of Polish Pomerania." The "unanimous consent of the Councillors" of the Estates of West Prussia must be obtained, according to the recommendation of Copernicus, whose reform program was not intended for Pomerania. In Copernicus' second recommendation, Bieda similarly substitutes Pomerania for Prussia.

[94] L. A. Birkenmajer, *Stromata*, pp. 260–261.

[95] Although the title page of Dmochowski's *Mikołaja Kopernika rozprawy* does not indicate the year in which the book was published, internal evidence points to 1924.

[96] Dmochowski, "Nicolas Copernic économiste," *Revue d'économie politique*, 1925, 39:105. This article is cited herein as "Dmochowski 1925," in order to distinguish it from the full-length book he published in 1924.

[97] AcTom (V, 167–169) in 1855 published M for the first time, without any indication who its author might have been. The anonymous piece was first recognized as a work of Copernicus in 1902 by Szelągowski (p. 283, n. 5). More than two decades later, in 1924, and without being aware of Szelągowski's contribution, L. A. Birkenmajer (*Stromata*, p. 257, n. 2) correctly characterized the anonymous piece as Copernican, but mistakenly pointed to EC as the manuscript underlying AcTom's publication of M. The manuscript actually underlying AcTom's piece was properly identified later in the same year by Dmochowski (pp. clxii, 22) as No. 199 in the Ossoliński Library. Dmochowski cited pages 404–409 of MS No. 199, and located the Library in Lvov; the page numbers are now 264–269, and the MS is now in Wrocław (StC VIII, 80, no. 144). Because the Ossoliński MS copy of M lacked a date, the editor of AcTom V had to guess when M was written, and he placed M among his documents dated 1520. We now know that M was completed on 15 August 1517 (see M ¶ 18, above). The

Ossoliński copy also lacked a title. As a result of this gap, "De aestimatione monetae" was added in pencil. This later intrusion should be eliminated from the literature of the subject now that the true title *Meditata* has been recovered. The title was not *mediata considerationes*, as in Seweryn Żurawicki, "Copernicus on Money," *Journal of European Economic History*, 1974, 3:128/10.

[98] Sig. h 5^r–6^v; pp. 480–482 in the posthumous 2nd ed. ([Leipzig], 1599).

[99] Schmauch, pp. 27–34. According to StC XVIII, 181, this copy of M was written for the city council of Gdańsk by Reich, who had a copy of M in his *Collectanea*. Actually, Reich's *Collectanea* did not contain a copy of M, nor is the Gdańsk copy of M in Reich's handwriting.

[100] See M ¶ 18, above.

[101] See text near n. 4, above.

[102] *Corpus iuris polonici*, IV, 239/¶ 31; the sentence prescribing the three denominations was quoted in the Introduction at n. 14, above.

[103] Ibid., IV, 267–270.

[104] PI², 203/16 up–8 up.

[105] Schmauch, p. 16, n. 43.

[106] PI², 204/12 up–7 up.

[107] PI², 204/14–21.

[108] Schmauch, p. 17/2–8.

[109] Ibid., p. 17/18–19.

[110] Ibid., p. 19/21–24.

[111] Semrau, p. 36/5 up.

[112] Ibid., p. 42/30.

[113] Ibid., p. 37/8 up–7 up; p. 43/4–6.

[114] This motive was suggested, without any foundation whatever, by Dmochowski, p. lxxxiv, n. 1, line 5. Decius "attended no university," said Kazimierz Römer, *De Jodoci Ludovici Decii vita scriptisque* (Wrocław, 1874), p. 21. Römer also pointed out that "Germans moved there [to Cracow] for the sake of making money" (p. 5). See also Ernst Kaussler, "Jost Ludwig Decius gegen Copernicus," *Pfälzer Heimat*, 1974, 25: 84/left/24–27.

[115] Schmauch, p. 21/1–2.

[116] The choice of Reich to be the spokesman in response to Decius indicates that Reich's associates did not regard him as "obtuse" as he was thought to be by Dunajewski (p. 109, n. 1). Nor had Reich been regarded as obtuse by the authorities who entrusted him with the public proclamation on 18 August 1526 of the comprehensive settlement following the suppression of the Lutheran uprising in Braniewo (SRW, II, 444–463).

[117] Schmauch, p. 22/13/14.

[118] PI², 206/12 up–9 up.

[119] PI², 207, n. *, lines 9–13.

[120] Schmauch, p. 23, n. 63/3–4.

[121] Ibid., p. 23/4–8.

[122] Semrau, p. 40/1.

[123] PI², 208/last 4 lines.

[124] PI², 209/2–4.

[125] Schmauch, p. 24/3–6.

[126] Ibid., p. 25/9–20.

[127] Dmochowski, p. 139/3–5.

[128] Ibid., p. 136/2 up–p. 137/6.

[129] Ibid., p. 167/3, p. 173/last line. The copy in the Czartoryski Library in Cracow is dated 18 July 1526 (ibid., p. clxi/8 up; Dunajewski, p. 403/10–11). The Councillors dispatched their *Letter to Decius* the day after they heard the king's "prudent edict."

[130] Dmochowski, p. 167/3 up–p. 168/4.

[131] Ibid., p. 168/12 up–3 up.

[132] Ibid., p. 169/3–7.

[133] Ibid., p. 169/16–20.

[134] Ibid., p. 171/17–22.

[135] Ibid., p. 172/4–7.

[136] Ibid., p. 172/last 2 lines.

[137] PII, 31/14; the last syllable of the author's given name and his entire surname have been eaten away by bookworms (see text above, just before n. 18).

[138] PII, 156/17.

[139] Semrau, p. 41/7–8: *Dominationum vestrarum literae quibus de ratione cudendae monetae abunde scriptum fuerat*.

[140] Catalogus Czartoryski, I, 46.

[141] Dmochowski, pp. lii–liii, lxxxi, lxxxiii, lxxxvi, clviii, clxi; Dmochowski, 1925, p. 106.

[142] Dmochowski, p. cxxxxviii/14–15.

[143] Ibid., p. vii/3 up–2 up.

[144] Ibid., pp. 165/5, 166/8, 167/4, 175/5, 177/4. In *Przegląd Historyczny* (1913, *17*:195) Michał Grażyński had expressed a suspicion that Copernicus was the author of the *Letter to Decius*.

[145] Marian Biskup, ed., *Regesta Copernicana*, Document 274 (StC VII, 141; VIII, 129). Ignoring this retreat, Sommerfeld clung to the claim that Copernicus was the author of the letter to Decius (p. 14/10 up; p. 19/16–17; p. 38/15–16; p. 40/2–3; p. 111/4). That letter was dated in 1526, 18 July (not June, as in Sommerfeld, p. 14/11 up).

[146] In a medical treatise, a marginal note indicating that the writer was in Gdańsk on 20 August 1526 was formerly declared to be in the handwriting of Copernicus (MK, p. 574). But now that photofacsimiles of manuscript material are more readily available, the handwriting in question is recognized as not that of Copernicus. Moreover, a legal document discovered later proves that one of the witnesses present in Frombork on that very day, 20 August 1526, was Copernicus (StC VIII, 130). Whether he was in Gdańsk while King Sigismund was meeting with the West Prussian Estates there from 17 April to 24 July 1526 (Schmauch, p. 15/last line) is still an undecided question. But he certainly was not in Gdańsk on 20 August 1526, since on that day he was in Frombork.

[147] Macleod, pp. 428, 429, 432: "The treatise of Oresme was written a hundred years before printing became general, and was merely drawn up for the consideration of Charles the Wise, and consequently did not get into general circulation and become known. It was not known to Copernicus." "Copernicus had no knowledge of the treatise of Oresme, written 160 years before his own." "Copernicus wrote his treatise entirely without the knowledge of the preceding one of Oresme."

COPERNICUS' ADMINISTRATIVE DOCUMENTS

COPERNICUS' SCHOLASTRY

While Copernicus was studying medicine at the University of Padua, he was informed that he had been appointed scholaster of the church of the Holy Cross in Wrocław (Breslau, in German). Being unable under the circumstances to leave Padua and go to Wrocław in order to take possession in person of his new position, he resorted to the legal device of a proxy, nominating two Wrocław canons to act in his behalf. To make his proxy unassailable, he engaged the services of Stefano Venturato, a public notary in Padua. The Paduan Notarial Archive (*Archivio Notarile*) still preserves Venturato's files, where Copernicus' draft proxy was found in Volume I, folio 173, by the archivist Dr. Erice Rigoni, who published it in *Archivio Veneto*.[1]

The scholaster of the church of the Holy Cross was expected to supervise the instruction given there. Yet during the thirty-five years of Copernicus' scholastry, from 1503 until he resigned early in 1538, he is not known to have set foot in Wrocław. His duties there were performed for him by a vicar, while he continued to enjoy the income derived from this sinecure. When he gave it up, it was bestowed "on the recommendation of [King] Ferdinand [of Bohemia etc.] on Dr. Johannes Benedict [Solfa], physician of the king of Poland, ... after the resignation of Nicholas Copernick. Prague, 4 February 1538."[2]

Copernicus did not owe his appointment to a recommendation by the king of Bohemia. On the other hand, "the right of advowson in the collegiate church of the Holy Cross in Wrocław was held"[3] by the local bishop. Perhaps to ingratiate himself with the bishop of Varmia, he recommended the appointment of Lucas Watzenrode's nephew, Nicholas Copernicus. Eager to enhance his nephew's income, Watzenrode promptly sent the necessary information to Italy. Furnishing the names of two suitable Wrocław proxies, the bishop urged his nephew to attend to the legal formalities in Padua without any delay. As soon as Copernicus received the bishop's message, he hurried to a notary's office, where he wrote a draft of the required document. He wrote it in such haste that in its nine lines he committed no less than five errors. For in his highly excited state, he mistakenly described himself as a canon of Wrocław instead of Varmia, and a proxy, who was a canon of Wrocław, as a canon of Varmia. He was so impatient to take possession of the said (*dictae*) scholastry that he put *dictae* before *possessionem*, but had to delete it and place it after *possessionem*. By the same token, he started to write *substituendi* without a *b*. Besides these four mistakes which he noticed and corrected, he made a fifth, which he overlooked. In referring to the other (*alia*) perquisites of his scholastry, he misspelled *alia* with a double *l*.

In addition to its value as a source of information about Copernicus' ecclesiastical career, this Paduan document is important as the oldest surviving example of Copernicus' handwriting. It was twice published in facsimile by Hans Schmauch.[4]

---

[1] Year 81, 5th series, 1951 (issued in 1952), no. 48–49, at p. 149 (not p. 25, as in StC VII, 62, no. 42, and StC VIII, 44, no. 42, which mistakenly refers to this first printing of Copernicus' draft proxy as a "reprint").

[2] *Diplomataria et scriptores historiae germanicae medii aevi*, II, eds. C. Schöttgen and G. C. Kreysig (Altenburg, 1755), 27, no. XLVII, quoted in PI², 263. Solfa was raised to the ranks of the nobility by King Ferdinand on 14 May 1546 (*Diplomataria*, II, 46, no. 95).

[3] PI¹, 314/3 up–2 up; see Correspondence, Letter VII, n. 3.

[4] *Archiv für schlesische Kirchengeschichte*, 1955, *12:* opposite p. 154; *Studien zur Geschichte des Preussenlandes*, Festschrift für Erich Keyser, ed. Ernst Bahr (Marburg, 1963), opposite p. 425.

I

PROXY OF THE REVEREND NICHOLAS COPERNIK, BY THIS PRESENT[1] [DOCUMENT]

I, Nicholas Copernik,[2] canon of Varmia[3] and scholaster of the church of the Holy Cross in Wrocław, revoking etc.,[4] designate as proxies the honorable man Apicius Colo,[5] chancellor and canon of the cathedral of Wrocław,[6] and Michael Jode,[7] canon of the same cathedral of Wrocław, for the purpose of taking[8] possession of the said[9] scholastry recently conferred on me, and whatever other[10] etc.,[11] with the power of [naming their own] substitutes[12] in the presence of the venerable Leonard Redinger of the diocese of Passau[13] and Nicholas Monsterberg of the diocese of Włocławek[14] as witnesses.[15]

... taking, accepting, and receiving occupancy and physical, real, and actual possession of the scholastry recently conferred on him and obtaining ...

1503, Tuesday, 10 January, in the episcopal chancellery, in the official register, folio 37, 1502

II

PROXY OF THE REVEREND NICHOLAS COPERNIK, CANON OF VARMIA

In Christ's name, amen. In the 1503rd year of His birth, 6th indiction, Tuesday, 10 January, 11th year of the most sacred pontificate of Alexander VI, our father in Christ and ruler, by divine providence pope. In the presence of myself, a notary public, and of the witnesses named below, who were called and summoned especially for this purpose,[16] the reverend Nicholas Copernik, canon of Varmia and scholaster of the church of the Holy Cross in Wrocław, appearing in person as principal, and principally for himself, while revoking whatever proxies he has constituted heretofore in any way, in every most excellent manner, way, and form, with every most excellent law and reason for which he could and should [proceed] most excellently and most effectively, has made, constituted, created, named, and solemnly ordained as his true, established, lawful, and indubitable proxies, agents, factors, managers of his affairs mentioned below,[17] and special and general deputies, in such a way, however, that the specialty does not derogate from the generality and vice versa, to wit, the venerable and honorable man, Apicius Colo, chancellor and canon of the cathedral of Wrocław, and Michael Jode, canon of the same cathedral of Wrocław, in their absence as though they were present, solely and wholly, especially and expressly, for the purpose of receiving in the name of the designator and for him, accepting and retaining occupancy and bodily, real, and actual possession of the scholastry recently conferred on him, and for the purpose of obtaining whatever collations[18] and provisions have been made or are to be made for the same designator, from whatever ecclesiastical benefices, wherever constituted, by apostolic or episcopal or any other authority,[19] and for initiating and sending appropriate letters, and presenting them to executives, and seeking and obtaining appropriate procedures for decisions about those letters, and also presenting, publishing, publicizing, and exhibiting the letters and procedures to all and sundry as needed, and advising and requesting them and each of them to heed and obey the letters and procedures under the penalties and censures contained therein, and whatever benefits under whatever grants made and to be made to him etc.

Concerning all and each of these matters, the same Nicholas, acting for himself, asked me, the undersigned notary public, to make and draw up one or more public instrument or instruments.

These [instruments] were drawn up in Padua, in the chancellery of the bishop of Padua, in the year, indiction, month, day, and pontificate, as above, while there were present in the same place the venerable Leonard Rodinger[20] of the diocese of Passau, and Nicholas Monsterberg of the diocese of Włocławek, as accepted witnesses to the foregoing, especially called and summoned.

BIBLIOGRAPHY

Bronzino, Giovanni, *Notitia doctorum sive catalogus doctorum qui in collegiis philosophiae et medicinae Bononiae laureati fuerunt ab anno 1480 usque ad annum 1800* (Milan, 1962)

Diplomataria et scriptores historiae germanicae medii aevi, II (Altenburg, 1755)

Rigoni, Erice, "Un autografo di Niccolò Copernico," *Archivio veneto*, 1951, anno 81, vol. 48–49, 5th series, no. 83–84, pp. 147–150

Schmauch, Hans, "Des Kopernikus Beziehungen zu Schlesien," *Archiv für schlesische Kirchengeschichte*, 1955, 12:138–156

——————, "Um Nikolaus Copernicus," *Studien zur Geschichte des Preussenlandes*, Festschrift für Erich Keyser, ed. Ernst Bahr (Marburg, 1963), pp. 417–431

NOTES

[1] Reading *pr[a]esens*, with Schmauch, p. 154, rather than *pressen*, with Rigoni, p. 149. This heading was written by the notary over Copernicus' draft proxy.

[2] This is the only extant example of Copernicus' writing his surname with a final -*k*.

[3] At first Copernicus wrote that he was a canon of Wrocław, which he deleted. There is no documentary basis for Schmauch's assertion (p. 146/12–13) that Copernicus had a canonry in Wrocław.

[4] Copernicus set aside all previous proxies, as is made clear by the corresponding passage in the notary's official proxy.

[5] Apicius Colo entered the University of Leipzig in the summer semester of 1466 (*Die Matrikel der Universität Leipzig*, ed. Georg Erler, I [Leipzig, 1895], 258/10, left: Apatcz Kolo; *Codex diplomaticus Saxoniae regiae*, II Haupttheil, Bd. 16). By 18 March 1491 Colo was a canon as well as the chancellor of the Cathedral Chapter of Wrocław. In addition, from 14 March 1497 on he was a canon of the church of the Holy Cross in Wrocław, where Copernicus was to become the scholaster. Colo (who was born about 1450) died on 14 February 1517 (Gerhard Zimmermann, *Das Breslauer Domkapitel im Zeitalter der Reformation und Gegenreformation* [Weimar, 1938], p. 211).

[6] At first Copernicus wrote that Colo was a canon of Varmia, which he deleted.

[7] Michael Jode (or Jod) was born in Toruń, like Copernicus. He entered the University of Cracow in the summer semester of 1483 (*Album studiosorum universitatis cracoviensis*, I [Cracow, 1887], 255, left column/22: *Michael Libory de Thorun*). Again like Copernicus, Jod proceeded from Cracow to the University of Bologna, which granted him the doctoral degree in medicine on 19 December 1498 (Giovanni Bronzino, *Notitia doctorum sive catalogus doctorum qui in collegiis philosophiae et medicinae Bononiae laureati fuerunt ab anno 1480 usque ad annum 1800* [Milan, 1962], p. 7: Michael Polachus). Jod became the personal physician of two bishops of Wrocław, one of whom recommended him on 13 October 1495 for a vacant Varmia canonry (Schmauch, p. 142, n. 11/3–9). Although Bishop Lucas Watzenrode of Varmia did not obtain a Varmia canonry for Jod, it was he who secured the Wrocław scholastry for his nephew, Nicholas Copernicus. Jod was a more successful candidate in Wrocław, where he obtained a canonry in the Cathedral Chapter on 16 April 1501. But in less than three years, he gave up his canonry early in 1504 when he married. His death occurred after 1521 (Zimmermann, pp. 307, n. 5; 308/7 up–6 up, 318–319).

[8] Reading *accipiendam*, with Schmauch, p. 154, not *accipiendum*, with Rigoni, p. 149.

[9] Copernicus wrote *dictae* before *possessionem*, then deleted it and rewrote it after *possessionem*.

[10] Instead of *alia*, Copernicus wrote *allia* with a double *l*.

[11] The *etc.* refers to the other emoluments connected with the scholastry.

[12] Copernicus started to write *substituendi* as *sus-*, which he deleted.

[13] The *pataviensis* diocese (also called *passaviensis*) was misidentified with the Paduan (*paduanensis*, *patavinensis*) by Rigoni, p. 148/7 up, as was pointed out by Schmauch, p. 142/5 up–2 up; see Conrad Eubel, *Hierarchia catholica medii aevi*, 2nd ed., I (Münster, 1913), 385, 392.

[14] The *vladislaviensis* diocese was misidentified with that of Wrocław (*vratislaviensis*) by Rigoni, p. 148/6 up, as was pointed out by Schmauch, pp. 142–143; see Eubel, I, 533–535.

NOTES

[15] This concludes Copernicus' draft proxy. What follows was added a little lower on the page as a guide jotted down by the notary to help him in drafting the official proxy. His copy of the official proxy was found by Dr. Rigoni in Book I, folio 125, of Venturati's files, and published in *Archivio Veneto* as Document II (pp. 149–150). These two documents are translated here for the first time.

[16] Reading *hoc*, with Schmauch, p. 154/2 up, rather than *h[a]ec*, with Rigoni, p. 149, II/4.

[17] Reading *infrascriptorum*, with Schmauch, p. 155/8, not *instrumentorum*, with Rigoni, p. 149/4 up.

[18] Reading *collationes*, with Schmauch, p. 155/18, not collactiones, with Rigoni, p. 150/5.

[19] Schmauch omits the rest of this paragraph.

[20] Redinger, according to Copernicus' draft proxy.

COPERNICUS' LEASES OF ABANDONED FARMSTEADS

INTRODUCTION

Copernicus, having been duly elected to that office, became the "Administrator of the Property Held in Common by the Venerable Chapter of Varmia" for a term of three years, beginning 11 November 1516.[1] The Chapter's statutes provided that the "elected administrator shall swear that he will present to the Chapter or its deputies in its entirety the whole income from the districts of Melsac and Olsztyn, when he has collected it."[2] That income was collected from the peasants (*subditi*) bound to the soil owned by the Chapter and divided into parcels (*mansus*). A peasant (*colonus, villanus*)[3] who took possession of a parcel owed an annual money rent (*census*) as well as several unpaid labor servitudes ("*rusticalia servitia*, which in the vernacular are called *Scharwerk*").[4] The administrator was required to maintain the "ledger (*registrum*) in which the villages, rentals, and property of the Chapter are recorded," and "to render an accounting of the same and present it to the Chapter" "when he is relieved of the office of administrator."[5]

The combined burden of rent and servitudes was so heavy that many a peasant ran away, either to look for a job in a town or for a less oppressive situation in another village. The frequent wars made conditions even worse, so that in 1500 nearly one-fourth of the Olsztyn parcels were unoccupied.[6] In order to attract a settler, the administrator could exempt him for one or more years from the rent or servitudes or both. The new settler would usually receive the buildings (house, barn, stable), livestock (horse, ox, cow, pig, goat), equipment (wagon, plough, tools), and seed. The transaction was recorded by the administrator in a special ledger called *Locationes*. In the law of ancient Rome, *locatio* was the term for a lease. In these Varmia *Locationes* the lessor was the Chapter, which retained ownership (*dominium*) of the land. The lessee[7] was often required to furnish one or more guarantors, who faithfully

---

[1] According to L (p. 30/12 up), Copernicus held the office only "until 8th November 1519." L's mistaken notion that Copernicus' term as administrator ended on 8 November is a result of Sikorski's misinterpretation of *tunc administratorem* in the minutes of the Chapter meeting of 9 November 1519 (Hipler, p. 276, no. 43/2; PI², 98, n. **/3-4). For, according to Sikorski (p. 10/17-21; p. 54, no. 206), because Copernicus was called the "administrator at that time" when money for his office was being collected in a certain village, therefore he was no longer the administrator on 9 November 1519, when that money formed part of the annual distribution to the canons. Copernicus was the administrator, however, not only when the money was collected, but also when it was distributed at the Chapter meeting on 9 November 1519, the meeting which elected his successor to take office on 11 November 1519.

[2] Hipler, p. 257, no. 34/1-4; PII, 511, no. 34/1-3. Copernicus wrote *Melsac*, not Melsack (as others did, assimilating the place name to the German word for "flour sack"). When the town was founded in 1312, the place was "called in the Old Prussian language *Malcekuke*, which is pronounced *Melzak* in German" (CDW, I, 283, 11-13; CDP, II, 81/7 up-6 up).

[3] L, p. 86/6, 11 up.

[4] Hipler, p. 257, no. 37/2-3; PII, 512, no. 37/2. The German word *Scharwerk* was taken into Polish as *szarwark*.

[5] Hipler, p. 257, no. 35/1-3, 7-8; PII, 512, no. 35/1-3, 7.

[6] For the Olsztyn district in the year 1500, the administrator reported that $400^{1}/_{2}$ parcels out of a total of 1688, or more than 23%, were abandoned (Schmauch, 1929, p. 542, n. 3).

[7] In French he is still called *locataire*, and in Polish *lokator*.

224

promised (*fideiusserunt*)[8] that he would not run away within a specified time. If he stayed beyond that time, presumably he was satisfied and would remain until the end of his working life. In that case his holdings would be inherited by his heirs.[9] But if more desirable parcels became available during his lifetime, with the Chapter's permission he could sell his holdings[10] and buy others, either in his own village or in another village[11] within the Chapter's domain. He could not move into the bishop's domain in the diocese of Varmia or leave the diocese.

When a peasant ran away (*fugit*), if he left anything of value on his parcels, it was taken into temporary custody by the village's overseer (*scultetus*). Whatever the administrator later awarded to the new occupant was given to him by the overseer, particularly the animals. Sheep do not appear in these *Leases*, a *pastor* being a herdsman rather than a shepherd. The parcels contained pasture land (*prata*) as well as arable land (*agri*).[12] The chief cereal was rye (*siligo*), with oats (*avena*) and barley (*ordeum*) less prominent. Buckwheat (*gricka*),[13] millet (*frumentum aestivum*),[14] and flaxseed (*lini semen*)[15] were also grown. Among vegetables, peas (*pisae*)[16] are mentioned. Farm equipment (*suppellex*) comprised a wagon (*plaustrum*), plough (*aratrum*), iron plowshare (*ferramentum pro aratro*),[17] ax (*securis*), scythe (*falcastrum*),[18] sickle (*falx*), kettle (*caldar*),[19] and water barrel (*tona aquaria*).[20] Conspicuous features of the landscape were the mill (*molendinum*),[21] producing flour (*farina*),[22] and the tavern (*taberna*),[23] producing good cheer, if not inebriety.

It was the overseer's duty to find a runaway peasant and bring him back.[24] Pending his return, the parcels he abandoned might be leased to another peasant. This new lease was conditional, however, being valid only until the fugitive reappeared on the scene.[25] If he stayed away a long time and was not replaced, the abandoned parcels would naturally revert to their original wooded state (*insilvatus*).[26] The longer they remained overgrown, the longer the exemption from rent and servitudes required to attract a cultivator.

If a peasant ran away after having sown the winter fields, and he was replaced before harvest time, the new lessee usually received one-quarter of what had been sown (*quarta pars satorum*).[27] Presumably this fraction was his reward for bringing in the crop, while the remaining three-quarters escheated to the Chapter as overlord. Exceptionally, the new lessee received the entire winter crop (*sata hiemalia*).[28]

The precise provisions of a lease were decided by the administrator, who used his own discretion in granting or withholding an exemption (*libertas*) from the rent and/or servitudes. If the new lessee was fairly prosperous, and the parcels in good condition, the lease carried no exemption (*sine libertate, absque libertate*).[29] But if the situation warranted, the administrator could grant relief from the next annual payment of rent (*libertas census proximi*).[30] In more serious cases the exemption could be extended to three years (*libertas triennis, libertas iii censuum*),[31] four years (*libertas iiii annorum a censu et servitio*),[32] or six years (*libertas annorum vi*).[33]

Exemption from the servitudes usually included all of them, but exceptionally a peasant was required to assist in the hunt (*venatio*) while being excused from all the other servitudes.[34] In one case, the exemption granted by Copernicus to an elderly couple without sons was permanent (*concessi libertatem*).[35] In the same spirit when Peter, an aged peasant, sold his 1½ par-

---

[8] L, p. 77/9. [9] L, p. 92/8 up. [10] L, p. 77/10. [11] L, pp. 82/last line–83/2. [12] L, p. 82/8 up.
[13] L, p. 86/1. [14] L, p. 83/7. [15] L, p. 81/5 up. [16] L, p. 82/12. [17] L, p. 82/14. [18] Ibid.
[19] L, p. 82/13. [20] L, p. 86/1–2. [21] L, p. 91/1. [22] L, p. 82/12. [23] L, p. 83/18. [24] L, p. 83/3–4.
[25] L, p. 94/last 2 lines. [26] L, p. 86/5. [27] L, p. 77/17. [28] L, p. 84/5. [29] L, pp. 82/1, 78/8 up.
[30] L, p. 77/17–18. [31] L, pp. 82/6–7, 90/1. [32] L, p. 90/18–19. [33] L, p. 89/16. [34] L, p. 90/7 up–6 up.
[35] L, p. 81/13.

cels, the annual rent ($^1/_2$ mark) due from the $^1/_2$ parcel was kept by the Chapter, "which graciously donated the 1 mark [due from the remaining parcel] to the aforesaid Peter for life."[36] In helping a new settler to establish himself, Copernicus sometimes added to the inventory left by the previous occupant: "In addition, I promised him 2 horses."[37] Where an outright gift seemed excessively generous, a loan took its place: "I lent him 3 sacks of oats until St. Michael's" day, 29 September, when the harvest would be gathered in.[38] While such help in the form of gifts and loans was frequently extended to those who worked hard and obeyed the law, harsh punishment was meted out to criminals. A thief was hanged,[39] a traitor was beheaded.[40]

The inhabitants of a village constituted a community (*communitas villae*),[41] which acted as a unit in matters affecting them all. Thus, a parcel associated with a certain abandoned mill was leased by a community, not by an individual. But finding the rent too burdensome because it entailed maintaining the bridge, the community gave the parcel back. However, when another community promptly asked for the parcel, the first community changed its mind and petitioned to have the parcel restored to it, even at a higher money rental.

The transactions recorded in these *Leases* were entered (for the most part) in chronological order under the appropriate year. The Chapter's administrative year, however, did not coincide exactly with the calendar year, for the former began on St. Martin's day, 11 November. Hence, a transaction occurring on 10 December 1516 appears in the *Leases* under 1517.[42] Copernicus used both the ecclesiastical and secular calendars. Thus, he referred to one and the same day as *xii Martii* (12 March) and *Sabbatum ante Invocavit*.[43]

As witnesses of some of the transactions which he authorized, Copernicus names either or both of his servants. He calls the older one, Albert Szebulsky, his *famulus*,[44] while referring to the younger one, Jerome (whose surname is unknown), as his errand boy (*puer*).[45]

In recording the names of the numerous Polish peasants involved in these *Leases*, Copernicus "registered the phonetic characteristics of the Polish language correctly," according to L (p. 33/28–29). Yet in the same entry Copernicus wrote both *Czepan* and *Zcepan*.[46] This transposition of the first two letters indicates that he was trying to reproduce phonetically the sounds he heard, and provides no basis for the conclusion that he "not only knew the Polish language but spoke it fluently enough to be able to use it in his dealings with Polish peasants."[47] He wrote the Polish names as well as could be done by an intelligent and conscientious administrator essentially unfamiliar with that language. How did he and his German-speaking predecessors and successors as administrators communicate with their Polish subjects? In this region and period of rapid demographic change, Poles were migrating northward in increasing numbers into a land previously populated predominantly by German-speaking peasants. In these commingled groups, persons familiar with the limited bilingual vocabulary required for these transactions may not have been hard to find.

The earliest Copernican scholar to pay any attention to these *Leases* was Hipler, who incorporated them to a restricted extent in the Regesta Copernicana section of Hipler (pp. 272–277). Then PI², 90–93, printed some extracts. The first complete publication of the *Leases*, with photofacsimiles of the manuscript and translation into Polish (Plates I–XV, pp. 97–104), was the contribution of L, which was made possible by a notable example of inter-

---

[36] L, p. 78/6 up–3 up.   [37] L, p. 77/8.   [38] L, p. 82/4.   [39] L, p. 77/5.   [40] L, p. 92/12–13.
[41] L, p. 77/3, 8.   [42] L, p. 91/2.   [43] L, pp. 90/2 up, 91/last line.   [44] L, p. 78/8.   [45] L, p. 77/5 up–4 up.
[46] L, Plate 2/7, 10; p. 78/12, 14.
[47] L, p. 33/12 up–10 up. Although the English version says "suppose," the French and German say *conclure*, *schliessen* (L, pp. 42/14 up, 51/12), closer to L's *wyciągnięcie wniosku* (p. 26/11 up).

INTRODUCTION

national scholarly cooperation. For, the manuscript had suffered severe damage during World War II, only its last four[48] pages being preserved intact. Fortunately, these and all the other pages had been photocopied for Schmauch, whose untimely death in 1966 prevented him from publishing the entire manuscript. After his death, his material was made available to L by the management of ZGAE. Such heartwarming collaboration between Germans and Poles bodes well for the future relations of these two neighboring great nations.

All but the first of Schmauch's photocopied pages survive. Most of the missing first page had been printed, however, in Hipler, pp. 272–273, no. 31–32, while the substance of the rest had been summarized by Schmauch, 1942, pp. 488/5 up–489/1, 507, no. 48/6 up–4 up.

What follows is the first English translation and analysis of the leases supervised by Copernicus in his capacity of Chapter administrator in the years 1516–1519 and 1521.

---

[48] The last page contains the entries for 1521, while those for 1517–1519 in the Melsac district cover the three preceding pages. Instead of three, L says "two" (pp. 7/6 up, 27/12 up, 35/11 up, 43/9 up).

LEASING OF FARMSTEADS BY ME, NICHOLAS COPPERNIC, A.D. 1517

Jonikendorf [10 December 1516]

Merten Caseler took possession of 3 parcels,[1] [vacant] because Joachim was hanged on account of theft. They were not sown last year. I canceled the payment[2] for this year, and he will pay next year and thereafter. ... He got 1 cow, 1 heifer, an ax and a sickle and, as for grains, a sack[3] of oats and barley for the sowing omitted by his predecessor. Done on weekday 4,[4] 10 December 1516.[5] In addition, I promised him 2 horses.[6] The overseer[7] was his guarantor[8] for 4 years.

Voytsdorf [11 December 1516]

Hans Bodner took possession of $2^1/_2$[9] parcels, which Andreas Daumschen sold[10] to him. Done on 11 December.

Voytsdorf [undated; perhaps 11 December 1516][11]

Gregor Knobel adds to his 2 parcels 1 more parcel that belonged to Peter Glande, who died in a fire.[12] Gregor is the guardian of his brother Peter's sons, who are minors, and promises to satisfy them when they are grown up.

A.D. 1517 LEASING OF ABANDONED FARMSTEADS BY ME, NICHOLAS COPPERNIC, CANON AND ADMINISTRATOR[13]

Spigelberg [undated; after 7 January 1517, and probably on or just before 29 January 1517][14]

Valentin,[15] [from] Passenhaim,[16] takes possession of 1 parcel, previously held by Augustin, who is very old and incapacitated. The overseer is Valentin's guarantor.

Greseling [29 January 1517]

Jan took possession of 3 parcels, from which Asman ran away.[17] Jan will have one cow and a fourth[18] part of the sown[19] [grain]. He will have[20] exemption from the next annual rental, and he will pay for the first time in the year 1519. Brusien, Andres,[21] and Hensel of the same place were his guarantors for three years. Done on the antepenultimate[22] [day] of January, in the presence of the chaplain and my attendant, Jerome.[23]

Godkendorf [30 January 1517]

Jan of Vindica took possession of three parcels,[24] from which Niclis Cleban[25] withdrew, his right hand being lame or crippled. Done on the penultimate [day] of January 1517, in the presence of Andres the overseer and Jerome.

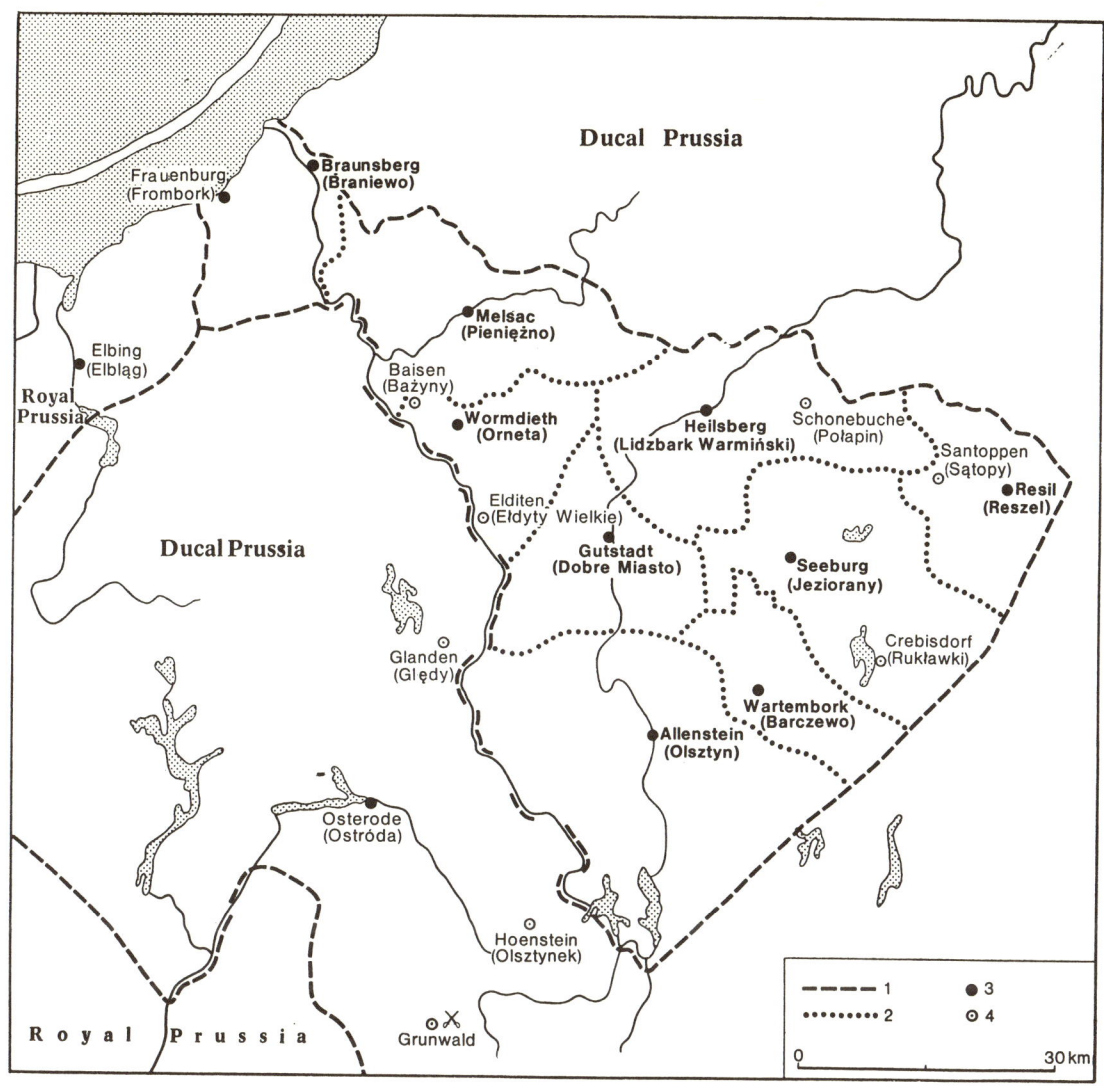

2. Varmia
1 — Boundaries of Lands, 2 — Boundaries of Districts, 3 — Cities, 4 — Villages

Berting Teutonica [26 February 1517]

Lorencz of Marquardshoffen took possession of four parcels, sold[26] by the heirs of Aldejorge, deceased.[27] For these parcels Lorencz will perform all[28] the customary farm servitudes. Done[29] on 26 February.

Plauczk [4 March 1517]

Andrhe,[30] the overseer, having sold the overseership[31] to Bartosch, took possession of 2 parcels, from which Matz ran away. I gave back to Andrhe one horse, one cow, 3 she-goats, one sow,[32] and he will be exempt for 2 years, and will pay rent for the first time in the year 1520. Done on 4 March, in the presence of Nicolaus the chaplain and Albert,[33] my servant.

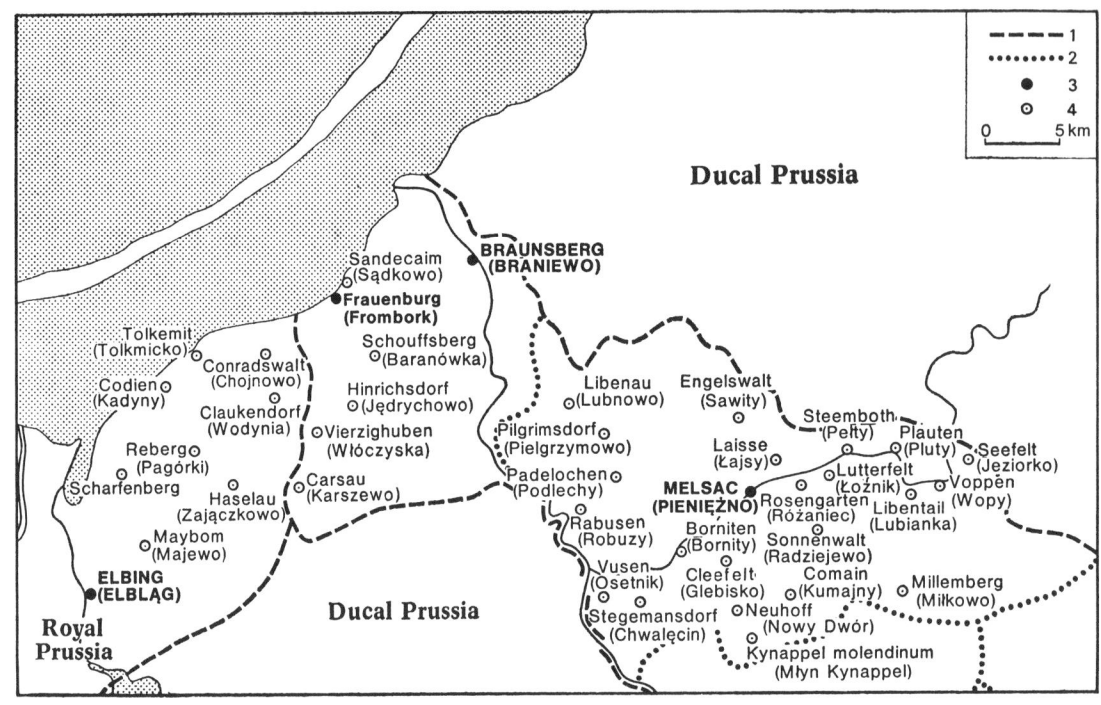

3. Districts of Braunsberg and Melsac
1 — Boundaries of Lands, 2 — Boundaries of Districts, 3 — Cities, 4 — Villages

Scaiboth [5 February 1517]

Nickel[34] Pippelk took possession of two parcels, which Jan Roman sold to him. Done on 5 February,[35] in the presence of the local overseer and Martzyn Baytz.[36] Nickel's brother, Bartolomeus in Petrica, was his guarantor for 2 years.[37]

In the same place [5 February 1517]

Gregorhs[38] Czepan took possession of 1½ parcels, from which Jacob Wayner[39] ran away.[40] Gregorhs will pay the next rental. Done on 5 February. He received 2 cows. Zcepan Wayner was his guarantor in perpetuity.

Miken[41] [5 February 1517]

Borchart Crix took possession of 2 parcels, which Mertin sold to him. Done on 5 February. Mertin was his guarantor for five years. Done on 5 February,[42] in the presence of Albert, my servant, and Jorge Nimsgar.[43] Frederich in Vadang was guarantor for the same [period of time].

Naglanden [23 March 1517]

Martzyn Voyteg took possession of 4 parcels, which were ceded to him by Jorch Voteg, without exemption. Martzyn's brother, Jan, was his guarantor for 4 years. Done on the second weekday after Laetare.[44]

Leynau [23 March 1517]

Bartolt Faber[45] of [46] Schonewalt took possession of $1^1/_2$ parcels, sold[47] by Peter Preus, who is very old. As regards these parcels, Bartolt will give the overlord $^1/_2$ mark as rent for the half-parcel. But as regards the other parcel, the Chapter graciously donated 1 mark to the aforesaid Peter for life.[48] After his death the entire rent will revert to the overlord. Done on the second weekday[49] after Laetare, 1517, in the presence of Albert, my servant, and Jerome,[50] etc.

Plutzk [23 March 1517]

Brosien Trokelle took possession of 3 parcels, made vacant by the death of Peter. In addition to these parcels, Brosien got 2 horses, 1 cow, 3 she-goats, 2 piglets, 1 sack of rye, and 3 sacks[51] of oats. He will pay rent for the first time

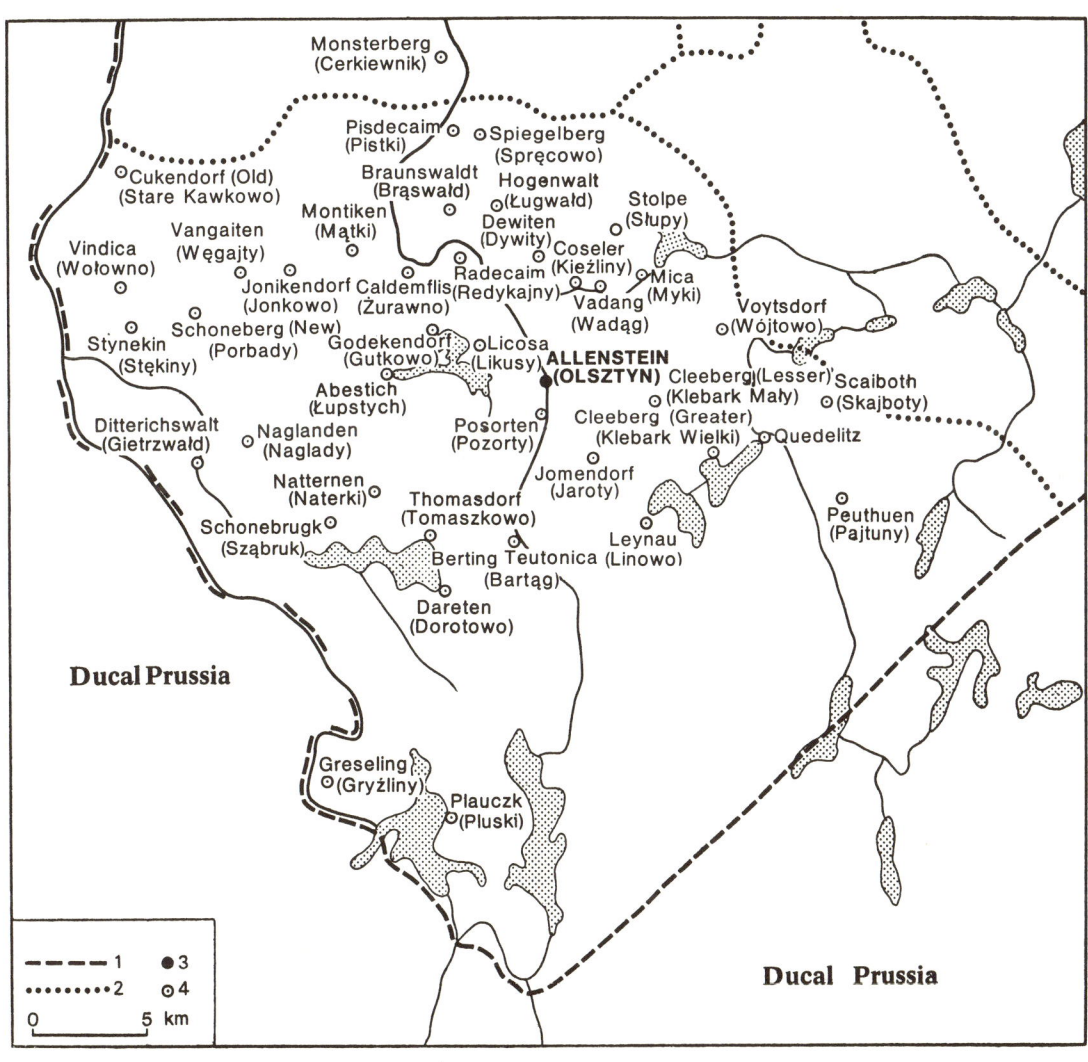

4. District of Allenstein
1 — Boundaries of Lands, 2 — Boundaries of Districts, 3 — Cities, 4 — Villages

next year. His brother, Augustyn, was his guarantor for 4 years. Done on the second weekday after Laetare.[52]

[This transaction] did not take effect[53] on account of the man's excessive[54] dishonesty,[55] and the aforementioned [properties] were given back.

Voytsdorf [26 March 1517]

Jorge Woyteck took possession of 2½ parcels, from which death removed the late Peter Glande, who was burned.[56] Jorge will pay the next rental. Done on 26 March.

In the same place [29 March 1517]

Martzin, who has[57] two parcels, took possession of one in addition, his third,[58] which was sold to him by Alde Urben, who is aged in fact[59] and in name. Done on the antepenultimate[60] [day] of March. To this[61] Urban and his wife, who are very old[62] and have no sons, I granted exemption.[63]

Lesser Cleberg

Hans[64]
[After starting this entry, Copernicus struck it out.]

Brunswaldt [19 April 1517]

Hans Woppe, who has 1½[65] parcels in this place, in addition took possession of one parcel, released by Greger Gadel, who kept 3 parcels[66] elsewhere as enough for himself. Done on the eighth [day after] Easter.[67]

Hogenwalt [23 April 1517]

Stenzel the herdsman took possession of 3 parcels,[68] from which Hans Calau ran away.[69] Stenzel got one ox, 1 cow, 1 piglet, 2 sacks of rye seed, nothing else. And I promised to add 1 horse. Hans and Lorencz Hinzke[70] were his guarantors for four years that he would not run away. In addition, he will be exempt from the annual payment, and he will pay for the year[71] 1519. Done on the day [23 April] of St. Adalbert, father of [our] country[72] and apostle [of Prussia]. I made available to Stenzel 4 sacks of oats, which he will return on the feast of [St.] Michael [29 September].

Stenzel also took the horse I promised.[73]

New Schoneberg [8 May 1517]

Greger Noske took possession of 1½ parcels, from which Matz Leze ran away because he was suspected of thievery. I gave Greger 2 horses, 1 heifer,[74] 2 pigs, 1 sack of flaxseed, 3 sacks of barley.[75] He will pay the next rental. His father, Maz Noske, was his guarantor in perpetuity. Done on 8 May in the presence of the chaplain and J[erome], my errand boy.

Schonebrugk [14 May 1517]

Martzyn took possession of 2 parcels, which Cosman left when he ran away.[76] I allotted to Martzyn 3 horses, 2 cows, 1 heifer, 3[77] pigs, and a wagon, without exemption. Andreas of Dareten and[78] Jorge[79] the overseer in the same place were his guarantors that he would not run away for two years.[80] Done on 14 May.

I lent him 3 sacks of oats until [St.] Michael's [day, 29 September].[81]

Stolpe [17 May 1517]

Stenzel took possession of 3 3/4 parcels, from which death removed Pavel. I gave Stenzel 6 sacks of rye, nothing besides. He will be exempt for three years, and will make his first[82] payment in the year 1521.[83] Stenzel the overseer[84] and Andres of Micken were his guarantors in perpetuity. Done on Rogation Sunday,[85] in the presence of the chaplain and Albert, my servant.

Vindica [22 May 1517]

Jan[86] took possession of 4 parcels, from which death removed Czepan Copetz, his maternal uncle, without exemption.[87] He got 4 horses, 1 colt, 4 cows, 6 pigs, 1 leg of pork, 1 sack of rye, 1 sack of flour, 1/2 sack of peas, 4 sacks of barley, 5 sacks of oats, 1 large kettle, 1 wagon, iron plowshares, 1 ax,[88] 1 scythe. His son-in-law Stentzel was his guarantor in perpetuity. Done on the sixth weekday of the Lord's Ascension,[89] in the presence of Albert the forester,[90] and the overseer of that place.[91]

Lycosen [25 May 1517]

Jacob of Jomendorf[92] took possession of 2 parcels, which were sold to him with my permission by Marcus Kycol, who is very old. Done on the second day of the week following Ascension.[93]

New Cleberg [4 June 1517]

Petrus, a herdsman in Thomasdorf, took possession of 2 parcels, which are vacant because Hans ran away. Petrus got 2 cows, 5 pigs, one horse.[94] But this summer he will be satisfied [to have] only the meadows, while the sown fields[95] remain with tenants, and he will pay his first annual rental in the year 1519. The overseer in that place, Salomon, and Matz were his guarantors for 6 years. Done on the fifth weekday after Pentecost.[96]

In the same place [29 June 1517]

Salmon,[97] who has 1 1/2 parcels, took possession of an additional parcel. With my permission it was ceded to him by Jacob Cusche[98] as unnecessary and burdensome to himself, since he kept[99] two parcels. Done on the day[100] of Peter and Paul.[101]

Voytsdorf [12 July 1517]

Jacob of Lesser Scaibot bought unoccupied parcels in that place last year with the permission of the venerable[102] provost.[103] Here[104] [in Voytsdorf] Jacob has 2 parcels and sold them with my permission to Lorenz, the overseer's brother. Done on 12 July.

Scaibot [2 August 1517]

Jacob Wayner, who with his wife ran away last year,[105] has now been brought back by the overseer. Jacob took possession of one parcel, from which death removed Caspar Casche.[106] The building[107] is in ruins, and the parcel is of little value,[108] and for that reason was abandoned by Caspar's heirs and guardians. When Jacob took possession, I gave him one horse, a quarter[109] of the previously planted millet,[110] and exemption from the next annual payment. His brother Michel Wayner was his guarantor in perpetuity. Done on 2 August.

LEASING OF FARMSTEADS BY ME, NICHOLAS COPPERNIC, A.D. 1518

Berting Teutonica [14 March 1518]

Voytek, who has 2 parcels in the same place, took possession of two additional parcels, which have been abandoned for a long time on account of the flight long ago of Stenzel Rase. Voytek will pay the next annual rental. And later I gave him 2 sacks[111] of rye, nothing more. However, with regard to the 2 aforesaid [additional] parcels,[112] I granted him exemption from the peasant servitudes for four[113] years so that he will begin to serve from the feast of St. Martin [11 November] in the year 1522. Done on Laetare Sunday,[114] in the presence of the chaplain and the overseer.

Old Cukendorf [22–25 March 1518]

Lurenz, having bought the tavern[115] in Branswalt,[116] with my consent sold[117] 4 parcels, of which Merten took possession and for which he will perform the customary [servitudes]. His brother Peter was his guarantor for 3 years. Done on a weekday[118] after Judica [Sunday].[119]

Jonikendorf [26 March 1518]

Urban Hillebrant, having leased[120] his 2 parcels in Montikendorf, took possession here [in Jonikendorf] of 3 parcels, from which Borkart Crix ran away. Urban got 1 horse, 1 cow, 1 calf, 1 sow[121] with 2 piglets, and a fourth part of the sown[122] [fields] in the same place. He will be[123] exempt for three years, and in the 4th year, 1522, he will pay for the first time, a payment and one-half. The overseer was his guarantor in perpetuity that he would not run away. Done on 26 March.

Montiken [26 March 1518]

Matz Santke took possession of the 2 aforesaid parcels, which the aforementioned Urban ceded to him without exemption. Tewes in Jonikendorf was his guarantor, and Steffan Gerber[124] in Monsterberg also promised in perpetuity that Matz would not run away. Done on the same day, 26 March.[125]

Vindica[126] [27 March 1518]

Jan, having withdrawn[127] from 3 parcels and leased[128] them in[129] Godekendorf, without exemption took possession here [in Vindica] of 3 parcels, abandoned because Paul ran away.[130] I gave back to Jan 3 horses, 1 cow, 2 piglets, and the sown winter [fields]. The overseer was his guarantor for 6 years. Done on the Sabbath before Palm [Sunday].[131]

Godekendorf [27 March 1518]

Merten took possession of the 3 parcels mentioned just above. Having leased[132] his parcels in Abestich, he will make[133] the next payment and those following it. Done, as above.

Abestich [27 March 1518]

Stenczel, the overseer in Naglanden, took possession, under the usual conditions, of the 3 parcels released by Merten, mentioned above.[134] Stenczel promises[135] that the overseership will be filled[136] within a year. Done, as above.[137]

Glandemansdorf [3 May 1518]

Matz Wanczke[138] ran away from $1\frac{1}{2}$[139] parcels, leaving behind 3 goats, nothing else. Hans Cucuc and Jorge Poppe took possession of these parcels, dividing them in half. In this way each will have, together with his previous holdings,[140] $2\frac{1}{4}$[141] parcels, for which they will perform all the customary [servitudes]. Done on 3 May.

Thomesdorf [4 May 1518]

Hans[142] Clauke has 2 parcels, for which[143] he was bound by hereditary payments to the church in Berting. As a man incapacitated for a long time, he sold those parcels to Simon Stoke with my permission. Done on 4 May.

Pisdecaim[144]

Greseling [12 July 1518]

Stanislaus took possession of 3 parcels, from which Cranzel ran away five years ago. I gave Stanislaus 1 horse, 1 cow, 1 bullock, and a quarter of the fields sown[145] this year, together with an exemption from the next payment and from the servitudes of this year and the following year,[146] if he will have built well.[147] He will therefore make his first annual payment in[148] the year 1520 A.D. The overseer and Simon in that place were his guarantors. Done on the 2nd weekday before [St.] Margaret's.[149]

Naglanden [18 October 1518]

Bernt took possession of 4 parcels, which were ceded to him with my permission by Peter the overseer, after he bought the overseership in that place.¹⁵⁰ The same overseer was his guarantor in perpetuity. Done on 18 October 1518.

[Since there were no transactions to be recorded during the 3½ weeks between 18 October 1518 and the beginning of the next fiscal year on 11 November 1518, Copernicus left the rest of this page 9 blank, and at the top of page 10 he wrote:]

A.D. 1519 LEASING OF FARMSTEADS BY ME, NICHOLAS COPPERNICUS, ADMINISTRATOR

*Brunswalt*¹⁵¹ *[22 November 1518]*

Jacob the herdsman took possession of one parcel, which was ceded to him by Cristof, the overseer in that place, who was his guarantor¹⁵² for five years, with a promise to help him build, etc. Done on St. Cecilia's [22 November], in¹⁵³ the presence of inhabitants in that place.

*Pisdecaim*¹⁵⁴ *[3 January 1519]*

Peter of Caldeborn took possession of 2 parcels, which were ceded to him by the taverner, without exemption. The same taverner, Stenczel, was his guarantor for 5 years. Done on 3 January.

Natternen [10 January 1519]

Voitec took possession of 3 parcels,¹⁵⁵ which Jan sold to him. Martin, the overseer, and Martzyn Wayner, and Jan, the seller, were Voitec's guarantors. Done on the day of Paul, the first hermit.¹⁵⁶

Mica [28 February 1519]

Brosche, who has 2½ parcels here, took possession of one more,¹⁵⁷ which was ceded to him with my permission by Simon, the overseer of that place. Done on the last [day] of February.¹⁵⁸

*Old*¹⁵⁹ *Scaibot [28 February 1519]*

Matz Slander took possession of 2 parcels, from which Marczyn Baicz¹⁶⁰ ran away. Matz will make the next payment. Done on the last [day] of February. He received nothing on the parcels.

*Old*¹⁶¹ *Cleeberg [28 February 1519]*

Stenzel Zupky took possession of 2 parcels, which Matz Slander with my permission sold¹⁶² to him for 33 marks.¹⁶³ Done on the last [day] of February.

*Dewiten*¹⁶⁴

Ditterichswalt [6 April 1519]

Urban Gunter took possession of 4 parcels, from which Jacob Rape ran away. He got 4 horses, 4 pigs, 2 cows, 10 sacks of oats, 2 sacks of barley, 1$^1/_2$ sacks of flaxseed,[165] one kettle, wagon, and plough. Pavel Gunter[166] was his guarantor in perpetuity. Done on 6 April.

Greseling[167] [6 April 1519]

Pawel took possession of 3 parcels, made vacant[168] by the death of Broschius Broch, without exemption.[169] He got 1 horse, 1 cow, 1 pig,[170] 5 sacks[171] of rye seed, 2 sacks of buckwheat, a wagon, pitchfork,[172] sickle, and 1 water barrel.[173] Pavel, the overseer,[174] and[175] Simon,[176] a peasant, were his guarantors.[177] Done on 6 April.

Dewyten [10 April 1519]

Since 13[178] parcels in this place were abandoned long ago and are now covered with woods, Augustinus,[179] the priest, took 4 of them under his supervision to provide within 8[180] years a cultivator on them who will satisfy the overlord with regard to the annual payment and servitudes. Done on Judica Sunday.[181]

In the same place [11 April 1519]

Hans, the overseer, took possession of 2 of the above-mentioned parcels, for which he will have an exemption for five years so that he will begin to perform all the customary [obligations] in the year 1525. Done on the 2nd weekday before Palm [Sunday].[182]

Montikendorff [14 April 1519]

Bartolmis took possession of 3 parcels, which with my consent were ceded to him by Jorge Wolf, who is an incapacitated little old man.[183] The taverner in Brunswalt[184] was the guarantor in perpetuity that Bartolmis would not run away. Done on the fifth weekday[185] before Palm [Sunday].

Schonebrugk [14 April 1519]

Benedict took possession of 2 parcels, which were abandoned by Martzyn, who ran away.[186] Matz in Degten and Hans in Schonebrugk were his guarantors for 6 years. Done on the fifth weekday [Thursday, 14 April] before Palm [Sunday].[187]

Coseler[188] [31 May 1519]

Alex[189] took possession of 2 parcels, released by the overseer in Vadang, without exemption. The same overseer, Jacob Walgast, as well as Michel Han[190] and Matz, peasants in the same place, were guarantors for 6 years that Alex would not run away. Done on the last [day] of May.

New Cleeberg[191] *[undated]*

Michel took possession of 2 parcels, which were sold[192] to him by Andres Gnik[193] with my permission.[194]

Hogenwalt [14 August 1519]

Stenczel the herdsman with my consent relinquished 3 parcels,[195] of which Greger took possession without exemption. Niclas made a promise for Greger for 6 years. Done on 14 August.

Vindica [undated]

Brosien[196] took possession of 4 parcels, abandoned because Simon ran away. Brosien got 4 horses, 1 cow, 2 piglets, and 5 sacks of rye for planting. He will make his first payment in the[197] year 1521. Augustin in Plauczk[198] was his guarantor for 7 years.[199]

MELSAC

A.D. 1517 LEASING OF ABANDONED FARMSTEADS BY ME, NICHOLAS COPPERNIC, CANON AND ADMINISTRATOR

Vusen [7 January 1517]

Pavel Ebert took possession of 3 parcels, which Andres Hoveman[200] sold to him. Done on 7 January.[201]

Comain [10 February 1517]

Hans Molner took possession of 2 parcels, which Jorge Hausberg[202] sold to him. Done on 10 February.[203]

Steemboth [11 February 1517]

Melcher Tolkesdorf took possession of 2 parcels, which Urban Tile sold to him.[204] Done on 11 February.

Schonebrucke[205] *[2 March 1517]*

Hans Smith took possession, without exemption, of 3 parcels, from which Cosman withdrew. Done on 2 March, in the presence of Albert and Jerome.

Libentail [undated]

George Strewbyr took possession of one parcel that had been abandoned for a long time, with an exemption for 6 years. He will therefore make his first payment in the year 1524. I gave him 3 sacks of rye.[206]

Millemberg [21 October 1517]

Theus Messing took possession of 3 parcels which had been abandoned for a long time by Stenzel Hoveman.[207] Done on the day of the 11,000 Virgins,[208] in the presence of the overseer and burgrave in Melsac.[209]

A.D. 1518 LEASING OF ABANDONED FARMSTEADS BY ME, THE AFORESAID NICHOLAS COPPERNIC

Sonnenwalt [22 October 1518]

Michel Hun[210] took possession of 3 parcels, from which death removed Ryman two years[211] ago. Michel got 3 horses, 2 cows, 3 oxen, and certain other farm equipment. He will be exempt from 3 annual payments, and will make his first [payment] in the year 1521. Done on 22 October.

Laisse [24 October 1518]

Peter Brun[212] took possession of one parcel.[213] In the same place he holds 3 adjoining parcels, to which the parcel he has taken possession of once belonged. From that parcel the aforementioned[214] Michel Hun withdrew with my permission.[215] Done on 24 October in the year 1518.

Libenau [25 October 1518]

As Jacob Treter[216] is moving[217] to Zager with my permission, Andris Radau took responsibility for the 3 released parcels with all their obligations, promising that together with his sons he would lease them within 2[218] years. Done on 25 October.

A.D. 1519

Lutterfelt [14 November 1518]

Merten[219] Scholze took possession of 4 parcels, which Andres Eglof, being incapacitated,[220] ceded to him. He will perform all[221] the customary [obligations]. Done on the day after [St.] Brixius'[222] [day].

Stegemansdorf [undated]

M[223]

Voppen [11 March 1519]

Frantzke Gilmeister, the miller in that place, took possession of $3\frac{1}{2}$[224] parcels, from which Merten Haneman ran away a year ago. In addition to[225] these parcels, Frantzke got 2 horses, 1 cow, 2 oxen,[226] 2 pigs. He will be exempt for 4 years from the annual rent and servitudes, except for his hunting [duties], and he will make his first payment in the year 1523. Done on 11 March.

In the same place[227] *[12 March 1519]*

Bendict Eler[228] in Seefelt, who has 4 parcels in that place, took possession of $3^1/_2$ [229] parcels here [in Voppen], on the understanding that in the course of time he would provide a cultivator for them.[230] I granted him exemption from payment and servitudes up to the beginning of the year 1521. Done on 12 March.

Kynappel Mill

When in recent days certain persons from Neuhof[231] gave back one mill parcel, previously leased to the village of Ne[u]hof by Tiedemann,[232] and the village of Clefelt wanted to take possession of that parcel, the people from Neuhof denying that they had authorized...

[Copernicus started to write the entry in this way, and then with four diagonal strokes he deleted the passage, and instead wrote the following entry.]

Kynappel Mill [12 March 1519]

The community of the village of Neuhof had formerly taken possession[233] of one parcel, associated with this abandoned[234] mill, for an annual payment of $^1/_2$[235] mark. Abandoning the parcel in recent days, they turned it back to the overlord because it was burdensome to them at such a rental, and[236] mainly because they did not want[237] to maintain the bridge[238] located[239] there. Then the community in Cleefelt asked that the parcel should be assigned to them. On the other hand, the Neuhof community, motivated by regret, in numerous petitions asked to have the parcel back. Therefore I, Nicholas Coppernic as administrator,[240] having obtained the advice as well as the consent of the Venerable Chapter about this matter, for certain cogent[241] reasons have restored the parcel to the aforesaid Neuhof community, and have assigned it to them again, on the understanding that hereafter they are to pay 15 skoters[242] for it every [fiscal] year ending on [St.] Martin's [day].[243] They[244] are also obligated[245] to maintain the bridge as long as the mill is abandoned. On the other hand, if at any time it happens to be reactivated, the parcel will have to be ceded to the mill. Done on the Sabbath day before Invocavit.[246]

A.D. 1521, AFTER THE CEASEFIRE WENT INTO EFFECT ON 10 APRIL, AND ALL THE DETACHMENTS OF ARMED MEN IN THIS DISTRICT WERE DISBANDED, THE FOLLOWING LEASES OF ABANDONED FARMSTEADS WERE MADE, AT FIRST,[247] BY THE VENERABLE NICHOLAS COPPERNIC, ADMINISTRATOR[248]

Licosa [6 May 1521]

Stanislaus Czichotzinsky took possession of 3 parcels, made vacant by the death of Michel the one-eyed, without exemption. Stanislaus got nothing on the parcels except the building. Done on 6 May.

Jomendorf [6 May 1521]

Jacob, moving back from Licosa[249] with permission, took possession of 2 parcels, made vacant by the beheading of Peter in Hoensteyn for plotting treason, and obtained nothing on the parcels except the sown[250] rye. Done on 6 May, without exemption.[251]

Lycosa [6 May 1521]

Mattheus took possession of 2 parcels, released by the aforesaid Jacob. Mattheus got nothing on the parcels except the building[252] and the sown rye,[253] without exemption. Done on 6 May. He did not go to the parcels.[254]

In the same place [Lycosa] and in Radecaim [6 May 1521]

Peter, having given up 2 parcels in Radecaim with permission, bought 4 parcels here [in Lycosa] from the heirs of Nickel Rabe, who died at an early age. Peter pledged that this summer he would provide a cultivator in Radecaim. In addition, the seller's heirs, Cristof, overseer in Braunswalt,[255] and Peter Ludike, overseer in Licosa, were his guarantors. Done on 6 May.[256]

Lesser[257] Cleberg [20 May 1521]

Petrus the overseer, complaining about the small number of his parcels in the overseer's portion, asked to have added to his holdings the $1^1/_2$ parcels listed[258] as abandoned by Thomas Polen, who ran away. Petrus will manage them like the other cultivators.[259] He got them conditionally, as long as Thomas does not return. About[260] five sacks of rye were found there, nothing else, not even[261] the buildings, because they were destroyed[262] by fire. He promised to build.[263] Done on the second weekday of Pentecost.[264]

In the same place [Lesser Cleberg; 23 May 1521]

Merten, father of five sons and holder of $1^1/_2$ parcels, complained about the small extent of his land. Therefore,[265] with permission, he bought $1^1/_2$ additional parcels from Niclis Ruche. Niclis took possession of 2 other parcels, that were ceded to him by Merten Micher, who is very old and incapacitated, having lost his sons and wife. Done on the 5th weekday of Pentecost.[266]

Jomendorf[267] [31 May 1521]

Steffen, having sold with permission $1^1/_2$ parcels in Glandemansdorf, bought 2 parcels here [in Jomendorf] from Jeschki's heirs. Done on the last [day] of May.

BIBLIOGRAPHY

Dmochowski, Jan, *Mikołaja Kopernika rozprawy o monecie i inne pisma ekonomiczne* (Warsaw [1924])
Hipler, Franz, *Spicilegium Copernicanum* (Braunsberg, 1873)
L *Mikołaja Kopernika lokacje łanów opuszczonych (Nicolai Copernici locationes mansorum desertorum)*, ed. Marian Biskup (Olsztyn, 1970)
Schmauch, 1929 Schmauch, Hans, "Die Wiederbesiedlung des Ermlandes in 16. Jahrhundert," ZGAE, 1929, *23*:537–732
Schmauch, 1942 ——————— "Nicolaus Coppernicus und die Wiederbesiedlungsversuche des ermländischen Domkapitels um 1500," ZGAE, 1942, *27*:473–541
Sikorski, Jerzy, *Mikołaj Kopernik na Warmii* (Olsztyn, 1968)

NOTES

[1] The term *mansus* used here (and throughout these *Leases*) by Copernicus denoted the basic unit of cultivated farmland.

[2] The term *census* used here (and throughout these *Leases*) by Copernicus denoted the annual money payment collected from the peasant by the Chapter for his use of the arable land cultivated by him and owned by him in a restricted sense, with the Chapter retaining the right of the overlord (*dominium*). The Latin word *census* gave rise to the German word *Zins* and the Polish word *czynsz*, both meaning "rent." The annual rent per parcel or *mansus* was 1 mark (Leynau transaction, 23 March 1517).

[3] Hipler, p. 273/1's misreading *modicum* was corrected to *modium* by PI², 93/11.

[4] Weekday 4 (*feria iiii*) was always Wednesday, as the days of the week were counted in the ecclesiastical calendar, with Sunday as the first day of the week.

[5] This transaction, dated 10 December 1516, is listed under 1517 because the Chapter's administrative year ran from 11 November through 10 November. The caption above this transaction was written by Copernicus soon after he took office as administrator on 11 November 1516 (Hipler, p. 272/12–17).

[6] Unlike the animals mentioned previously, which had been left behind on the farm when Joachim was hanged, these 2 horses were taken from the Chapter's stables and given to Merten by the overseer.

[7] The term *scultetus* used here (and throughout these *Leases*) by Copernicus was taken into medieval Latin from *Schultheiss*, a German word (meaning "obligation-enforcer"), compounded from *Schuld* (obligation) and a form of *heissen*. *Schultheiss* was later contracted to *Schulze* and *Schulz*.

[8] A peasant who took possession of a farmstead with its financial and labor obligations was not free to leave if he later found his situation not to his liking. Lacking the legal right to move away without the Chapter's permission, he might run away surreptitiously. Hence, before the Chapter gave him possession of the farmstead, it might require him to furnish a guarantor that he would not run away within a specified time. In this transaction the village's overseer was the guarantor, who had to assume the obligations abandoned by the fugitive.

[9] Hipler, p. 273/4, and Schmauch, 1942, p. 488/4 up, read $2^1/_2$, changed to *3*, without any justification for doing so, by StC VIII, 70, no. 111, and L, p. 77/10.

[10] A peasant had the right to sell his holdings to another peasant, with the buyer acquiring not only the land but also the financial and labor obligations associated with the land so transferred. The village Voytsdorf was bought by the Varmia Chapter after 1470, with $^3/_4$ of the rental income therefrom going to the mortuary office, and $^1/_4$ to the canonical hours of Our Lady (*Regestrum horarum dominae nostrae*; CDW, II, 289, n. 1).

[11] The Latin text of this transaction, which was deleted by Copernicus, is no longer available. It was summarized in German by Schmauch, 1942, p. 488/last 3 lines.

[12] His death in a fire is mentioned again in the transaction of 26 March 1517.

[13] This second caption for 1517 was written by Copernicus on or shortly after 1 January 1517.

[14] The Latin text of this transaction is no longer available. It was summarized in German by Schmauch, 1942, p. 507/14–16.

[15] In Spiegelberg (Spigelberg) on 9 December 1536 Valentin took possession of 3 parcels, abandoned from

time immemorial, for which he had to pay rent from 1542 on, while being granted 4 marks and 6 sacks of rye (Schmauch, 1929, p. 701, no. 40/4 up–2 up).

[16] Passenhaim, as printed in Schmauch, 1942, p. 507/16, looks like a surname, and is so treated by L, pp. 77/14, 109. Actually, it is the name of a Prussian town, situated about $3^{1}/_{2}$ miles east of the southernmost part of Varmia. A settler coming from Passenheim (*ex Passenheim veniens*) appears in Schmauch, 1929, p. 703, n. 45/1.

[17] Copernicus started to write *reliq-*, which he deleted in favor of *fugit*. In 1514 Asman took possession of these 3 parcels from his predecessor, who also had run away. Asman was exempt from rent until 1516 (Schmauch, 1942, p. 491/15–17). When his exemption expired, he ran away.

[18] Copernicus forgot to put a stroke over the final -a, to give his *quarta* the required ending. Asman having sown the grain before he ran away, Jan will have one-fourth of the sown grain, with the remaining three-fourths escheating to the Chapter as overlord.

[19] Copernicus' *satorum* does not mean "sowing," as in StC VIII, 71, no. 114.

[20] Copernicus inadvertently wrote habuturus instead of *habiturus* (see n. 123, below).

[21] This guarantor may be Andres Preusse, who had previously taken possession of 3 parcels in this village (Schmauch, 1942, p. 490/3 up).

[22] L, p. 77/6 up's *ante penultimam* was corrected to *antepenultima* in StC VIII, 71, no. 114. The date of this transaction (29 January 1517) is given incorrectly as 30 January 1717 by Schmauch, 1942, p. 491/18.

[23] Copernicus regularly refers to Jerome as *puer meus*, without giving him any surname. Copernicus' use of *puer* implies that Jerome was younger than Copernicus' other servant, Albert.

[24] These three parcels in Godkendorf (Godekendorf) were given up by Jan of Vindica on 27 March 1518, when he took possession of three parcels in his own village, Vindica, after having obtained four parcels there on 22 May 1517.

[25] Niclis Cleban took possession of parcels in 1506, when his right hand was in good physical condition (Schmauch, 1942, p. 490/10–11).

[26] Copernicus inserted *venditos* in the left margin. A peasant's heirs had the right to sell the land they inherited if they did not wish to work it themselves.

[27] Aldejorge ("Old George") married a widow in 1506, and settled Niclis Cleban on her former husband's parcels (Schmauch, 1942, p. 490/7–11; see n. 25, above).

[28] Copernicus' *omnia* was misread as *iam* by L, p. 78/2. When the identical abbreviation recurred (p. 84/16), L read it correctly, but did not return to correct p. 78/2.

[29] Copernicus wrote *Actum* prematurely, before *super*. He deleted it there, and wrote it afresh here.

[30] Misread as Andrze by Schmauch, 1942, p. 501/5.

[31] Such a sale transferred both the office and the land associated with the office.

[32] Copernicus' *porcam* (feminine) was not matched by *Schwein* (instead of *Sau*) in Schmauch, 1942, p. 501/7.

[33] Copernicus does not use Albert's surname in these *Leases*. But on 15 March 1518, when Copernicus as administrator drew up a legal document in the Chapter's castle in Olsztyn, one of the witnesses present was Albert Szebulsky (Hipler, p. 164, no. 1/2 up). There Copernicus calls Albert his *familiaris*, while in these *Leases* he uses the synonym *famulus*.

[34] Copernicus started to write what may be the name of Nickel's brother, and then deleted this false start.

[35] This entry is the first of three out of chronological order. All three concern transactions that occurred in the Olsztyn district on 5 February, but were not recorded in the official ledger on that day. Several days later Copernicus was in the Melsac district, recording transactions that occurred there on 10–11 February. Later on in that month he was back again in the Olsztyn district, recording transactions that occurred there on 26 February and 4 March. Only on (or after) the latter date did he record the three transactions of 5 February. In other words, his field notes for that day were not transferred to the official ledger right away, perhaps because he was busy preparing for his trip to Melsac. After returning to Olsztyn, however, and recording the Plauczk transaction of 4 March, he came across his field notes of 5 February and recorded them, out of chronological order.

[36] This witness is later recorded as a fugitive on 28 February 1519. On that occasion Copernicus wrote his name Marczyn Baicz (L, p. 85/19), whereas here he appears as Martzyn Baytz. This difference in spelling, in an age when the spelling of personal names had not yet been standardized, prevented L from recognizing the identity of the witness and the fugitive, who are indexed (L, p. 107) as though they were two different men.

[37] Copernicus evidently squeezed this last sentence in as an afterthought.

[38] Copernicus inserted this given name above the line.

[39] Copernicus mistakenly wrote this surname as "Wayneson" here. But when this fugitive was brought

back on 2 August 1517, Copernicus wrote his surname correctly, in agreement with the surname of the guarantor on this occasion, who was undoubtedly a relative.

[40] Copernicus wrote *a quib-* a second time, but deleted these five redundant letters.

[41] This entry was cancelled by a horizontal stroke through each line as well as by two diagonal strokes.

[42] Copernicus' repetition of this date may indicate that his field notes for this Miken transaction were written on two separate slips of paper, with the second containing the names of the witnesses and the additional guarantor. Earlier that same day, 5 February, Copernicus had supervised two transactions in Scaiboth, some seven miles southwest of Miken.

[43] Read as Trimsgar by L, p. 78/17.

[44] In 1517 Laetare Sunday fell on 22 March. The second weekday (*feria secunda,* always a Monday) thereafter was 23 March.

[45] This surname was read as an occupational name and so translated (*kowal*) by L, pp. 78/6 up, 98/14, and by StC VIII, 74, no. 126 ("smith"). There is no instance in these *Leases* of a smith becoming a peasant. Copernicus did not capitalize the initial letter of a surname, although he did capitalize the initial letter of the corresponding given name, which he evidently regarded as more important than the surname.

[46] Copernicus at first wrote *des*, leaving no blank space between the preposition *de* and the place name.

[47] Copernicus wrote *venditos* in the left margin, as a correction of *dimissos* in the line. When he was in Leynau, his understanding was that these parcels were released (*dimissos*). After his return to Olsztyn, however, he learned that the transaction was a sale (*venditos*). He was quite busy on 23 March 1517, if he was in Naglanden, some 11 miles northwest of Leynau earlier that day, and then later that same day in Plutzk, about 10 miles southwest of Leynau.

[48] The Chapter's charitable act is not properly expressed by the statement that "Copernicus left to the latter [Preus] 1 mark of rent from 1 mansus of his annuity" (StC VIII, 74, no. 126). As this transaction indicates, the annual rental per parcel was 1 mark.

[49] Copernicus wrote *L* as the beginning of *Laetare*, but deleted the letter when he remembered that he needed *post* before *Laetare*. The second weekday (Monday) after Laetare Sunday in 1517 was 23 March (see n. 44, above).

[50] This is the second transaction witnessed by both Albert and Jerome. The first had occurred three weeks earlier, on 2 March 1517, also in the Olsztyn district, although recorded for some reason in the Melsac ledger.

[51] Copernicus wrote *modus*, where *modios* is required.

[52] This date (= 23 March) was overlooked by PI², 93/18–19's remark that this entry is "without indication of the day and month."

[53] This entry was cancelled by a curved diagonal stroke drawn through it.

[54] Copernicus' *nimiam* (as in PI², 93/21) was misread as *inimicam* ("hostile") by Schmauch, 1942, p. 501/11, and by L, p. 81/5.

[55] Nevertheless, in the last Olsztyn transaction of 1519 Brosien took possession of four parcels in another village, with his brother again being his guarantor. This Brosien is indexed by L, p. 107/10, as though he were not Brosien Trokelle.

[56] L, p. 81/8's misreading *conflagrati sunt* led to StC VIII, 75, no. 128's reference to "the destroyed farm of Peter Glande." But referring only to Peter Glande, Copernicus wrote *conflagratus*, followed by *fuit*, which he deleted as ungrammatical. Peter Glande's death in a fire was mentioned in the undated Voytsdorf transaction that occurred after 11 December 1516. He had taken possession of $^1/_2$ parcel in 1502 (Schmauch, 1942, p. 488/9 up–7 up).

[57] Copernicus' $\overline{hns}$ (*habens*) was misread as a surname $h\bar{u}s$ by PI², 93/15, followed by Dmochowski, pp. 195/16, 210/11 up, and Sikorski, p. 43, no. 137.

[58] Copernicus' *adhuc unum et tertium* was mistranslated by Dmochowski, p. 210/11 up, as "$1^1/_3$."

[59] Copernicus' remark that Urben looked old ("Alde," *alt* in modern German) indicates that he was in Voytsdorf on 29 March 1517. If he was there also for the 26 March transaction, he traveled five miles northeast from Olsztyn to Voytsdorf twice in three days.

[60] L, p. 81/11's *ante ante penultimam* prints *ante* twice, and misinterprets the stroke over *antepenultia* as final *-m*, whereas it indicates the omission of a medial *-m*. The antepenultimate day of March is 29 March, not 30 March (as in Hipler, p. 273/15; Dmochowski, p. 210/12 up: *przedostatniego*; and Sikorski, p. 43, no. 137).

[61] Copernicus at first wrote *hic*, which he deleted in favor of *huic*.

[62] Copernicus added *-tis* above the line to *decrepi-* in the line.

[63] StC VIII, 75, no. 129, equates *concessi libertatem* with "emancipated," "permitted to leave the village even in spite of their overdue payments." But Copernicus does not indicate that this old couple, who had just sold a parcel, were behind in their payments. Nor does he mention their leaving the village. If, after selling 1

parcel, they remained in order to work their reduced holdings without any sons to help them, their exemption from rent and compulsory unpaid labor was an act of humane and sensible administration on the part of Copernicus.

[64] This Hans may be identical with the Hans who is reported on 4 June 1517 as having run away from New (*nova*) Cleberg, called here "Lesser (*minor*) Cleberg." The deletion of this entry may indicate that Copernicus went to Lesser Cleberg only to find that Hans had decided not to consummate the projected transaction which drew Copernicus to Lesser Cleberg.

[65] Not "1," as in StC VIII, 77, no. 133.

[66] These may be the 3 parcels of which Gregor Geddel took possession on 25 September 1498 (Schmauch, 1942, p. 484/7–12).

[67] In 1517 Easter Sunday fell on 12 April, so that the eighth day thereafter (as then counted) was Sunday, 19 April.

[68] Copernicus' *Stenzel pastor* (Stenzel the herdsman) becomes a "settler by the name of Hirt Stenzel" (*osadnik imieniem Hirt Stenzel*) in Sikorski, p. 44, no. 141, because Schmauch, 1942, p. 493/11, referred to *der Hirt* (the herdsman) Stenzel. He later relinquished these 3 parcels on 14 August 1519.

[69] When he took possession of these 3 parcels in 1514, he received an exemption for 3 years (Schmauch, 1942, p. 493/9–10). When the 3 exempt years elapsed, he ran away.

[70] Lorentz Hintzke took possession of 1 parcel in 1499 (Schmauch, 1942, p. 492/last 2 lines).

[71] Copernicus started to write *an-* a second time, then deleted the unnecessary repetition.

[72] Copernicus' reference to St. Adalbert as *patriae patris* recalls *nostrae patriae pater* in the dedication of his translation of Theophylactus to the bishop of Varmia. As a part of Prussia, Varmia was felt by Copernicus to be his fatherland, of which the kingdom of Poland was the suzerain.

[73] The last two sentences were added later with a different pen and in a darker ink. This change of writing materials, and the fulfillment of the previous promise of a horse, suggest that this transaction was consummated in two separate stages.

[74] Copernicus' *iuvencam* (feminine) was not matched by Schmauch, 1942, p. 499/3's *Zugochsen* (masculine; draft ox).

[75] Copernicus' *ordei* does not mean *Hafer* (oats, as in Schmauch, p. 499/4).

[76] After *profugus*, Copernicus started to write *acce-*, which he abandoned in favor of *assignavi*. On 2 March 1517 Cosman had withdrawn from 3 parcels (recorded in the Melsac ledger).

[77] Not "4," as in Schmauch, p. 504/18 up.

[78] Copernicus' *Dareten et* was misread as Daretenach by L, p. 82/1.

[79] When Copernicus learned that this was the overseer's name, he inserted it above the line.

[80] In the event Martzyn did run away before the end of two years, for on 14 April 1519 Copernicus records him as a fugitive.

[81] Copernicus added this last sentence later with a different pen and in a darker ink, so that this transaction took place in two stages.

[82] Copernicus at first wrote *primus*, under the influence of the preceding word *daturus*. When he noticed his error, he deleted the *-s* and drew a horizontal stroke over the *-u*.

[83] Copernicus originally wrote *1520*, then instead of the zero he wrote the numeral *1*, with a dot above it to call attention to the change.

[84] Stenzel the overseer has the same given name as Stenzel, the protagonist in this transaction. The popularity of this name is indicated by the appearance in these *Leases* of six different men named Stenzel.

[85] In 1517 Rogation Sunday fell on 17 May.

[86] Jan of Vindica took possession of 3 parcels in Godkendorf on 30 January 1517, but gave them up on 27 March 1518 when he acquired 3 parcels in his own village, Vindica.

[87] Before *sine libertate* Copernicus wrote *de quibus ob-*, which he deleted, and then wrote *de quibus obiit* after *sine libertate*.

[88] Copernicus' *securim* was misread by L, p. 82/14, as *servientia*, a plural form followed by *i* (= one).

[89] In 1517 the Lord's Ascension fell on Thursday, 21 May. The sixth weekday (always a Friday) of the Lord's Ascension in 1517 was 22 May.

[90] After *Alberto*, Copernicus wrote *fa* as the beginning of *famulo meo*. But after the first two letters he stopped when he realized that he needed *famulo silvarum* to identify the second witness. The latter was misidentified with Albert, Copernicus' *famulus*, in ZGAE, 1970, 34:57/8–9.

[91] To this transaction there were three witnesses: (1) Albert; (2) the forester; (3) the overseer.

[92] Jacob of Jomendorf later left Lycosen (Licosa) to return to Jomendorf, where he took possession of 2 parcels on 6 May 1521.

⁹³ The second weekday (always a Monday) of the week following (*octava*) Ascension in 1517 was 25 May.

⁹⁴ Here Copernicus wrote *habebitque hanc*, which he deleted when he decided to use a different expression. Copernicus' *Petrus pastor* (Petrus the herdsman) becomes a "settler by the name of Hirt Petrus" (*osadnik imieniem Hirt Petrus*) in Sikorski, p. 45, no. 147, because Schmauch, 1942, p. 494, no. 20/4, referred to *der Hirt Petrus* (the herdsman Petrus).

⁹⁵ StC VIII, 79, no. 140, equated *agris* with "soil," which does not convey the contrast with *pratis* (meadows), with which the herdsman is satisfied during the summer of 1517.

⁹⁶ In 1517 Pentecost fell on Sunday, 31 May. The fifth weekday (always a Thursday) thereafter was 4 June.

⁹⁷ This Salmon is not identical with Salomon, the overseer, of the preceding transaction. That Salomon was the guarantor again in a transaction of 28 December 1523 (Schmauch, 1929, p. 692, n. 17/10 up–9 up).

⁹⁸ He died when all his buildings burned down, and his 2 parcels were taken over by his stepson on 16 June 1521 (Schmauch, 1929, p. 692, n. 17/1–2).

⁹⁹ Copernicus wrote rent-, which he deleted because he needed *retentis*.

¹⁰⁰ Copernicus' abbreviation *d* (for *die*) was expanded to *d[ivorum]* by L, p. 82/2 up, and by StC VIII, 79, no. 141.

¹⁰¹ The day of Peter and Paul is 29 June. This second New Cleberg transaction was misdated 12 July by Hipler, p. 273/19, by confusion with the following transaction in Voytsdorf, a confusion repeated by Sikorski, p. 45, no. 149.

¹⁰² Copernicus at first wrote *d* as the abbreviation of *d[ominus]*. Then, realizing that he had forgotten to give his fellow-canon the customary title "venerable," he wrote a capital *V* over the lower-case *d*, which he repeated.

¹⁰³ Christopher of Suchten, provost of the Varmia Chapter, was Copernicus' predecessor as administrator in 1516.

¹⁰⁴ Copernicus' *hic* was misread as *his* by L, p. 83/1.

¹⁰⁵ He was recorded as a fugitive on 5 February 1517, when Copernicus did not know the correct form of his surname (see n. 39, above).

¹⁰⁶ He had taken possession of 1 parcel in 1513 (Schmauch, 1942, p. 505/6 up).

¹⁰⁷ Copernicus' *aedificium*, being singular in form, provides no basis for "buildings," as in StC VIII, 79, no. 143.

¹⁰⁸ Copernicus at first wrote *parvi valoris* (of little value) as a genitive of quality. But since he wanted to add that the parcel was also abandoned (*destitutum*), he inserted the participle *habentem* in order to obtain a construction parallel with *destitutum*. In so doing, he forgot that *habentem* governs the accusative, not the genitive. In any case, the intrusion of *ruinosum aedificium* ruins the grammatical structure of the sentence.

¹⁰⁹ Copernicus' *quartam* (with *partem* understood; see the transactions of 29 January 1517 and 26 March 1518) was misread as *quartum* by L, p. 83/6.

¹¹⁰ Copernicus' *quartam satorum quondam frumenti aestivi* does not mean *Saatgetreide* (as in Schmauch, 1942, p. 506/6) nor "sowing grain" (as in StC VIII, 79, no. 143; see nn. 122, 145, below). Among the *frumenta aestiva* (cereals planted in the summer), millet was listed first by Pliny (*Natural History*, XVIII, 49), who recommended that it should be planted late in April (XVIII, 250).

¹¹¹ Copernicus at first wrote *lastos* (lasts), which he deleted in favor of *modios* (sacks).

¹¹² The words *super ii mansos praedictos* were added by Copernicus in the right margin, in order to make clear that the exemption did not apply to Voytek's two previous parcels.

¹¹³ Copernicus originally wrote *triennium* (three years), which he deleted when he placed *quadrennium* above it.

¹¹⁴ In 1518 Laetare Sunday fell on 14 March.

¹¹⁵ Presumably he is the taverner mentioned in the transaction of 14 April 1519. On 6 July 1523 he took possession of 1 parcel, for which his first rent was not due until 1528, while in 1523 he was authorized to make use of the meadows of 2 abandoned parcels (Schmauch, 1929, p. 687, n. 3/1–3).

¹¹⁶ L, p. 83/18's Brauswalt does not appear among the various forms of this name in L's index, p. 111.

¹¹⁷ Copernicus at first wrote *reliquit* (abandoned), which he deleted, and above which he wrote *vendidit* (sold).

¹¹⁸ Copernicus forgot to specify which weekday (*feria*).

¹¹⁹ In 1518 Judica Sunday fell on 21 March. Hence this transaction took place on Monday, Tuesday, Wednesday or Thursday, 22–25 March, since two transactions of 26 March were recorded, the second "being done on the same day."

¹²⁰ Copernicus' *locatis* does not mean "giving," as in StC VIII, 83, n. 152.

NOTES

[121] Copernicus' *porcam* is better matched with *Sau* than with *Schwein* (as in Schmauch, 1942, p. 494/19 up).

[122] Copernicus' *satorum* does not mean *Saatgetreide* (as in Schmauch, 1942, p. 494/19 up) nor "sowing grain" (as in StC VIII, 83, no. 152); see n. 110, above.

[123] Copernicus inadvertently wrote habuturus instead of the required *habiturus* (see n. 20, above).

[124] Read as Gerke by Schmauch, 1942, p. 496/2 up.

[125] Since Montiken was only two miles northeast of Jonikendorf, Copernicus could easily have been present in both these villages on the same day.

[126] Copernicus at first started to write the peasant's name *Ja-*. Then, realizing that the place name should come first, he deleted the partly written personal name.

[127] Copernicus' *dimissis* does not mean "succeeded to," as in StC VIII, 83, no. 154. Jan had acquired these parcels on 30 January 1517.

[128] Jan leased these 3 parcels to Merten, as the following transaction shows.

[129] Copernicus started to write the name of a village beginning with *N-* (Naglanden or Nattern). Then, realizing that the village in question was Godekendorf, he deleted the *N-*.

[130] After *profugio Pauli*, Copernicus wrote *cui ab* (not turiab, as in L, p. 84, n. ee), at first intending to express the denial of exemption by *absque* (as in L, p. 78/8 up). But he deleted *cui ab*, and switched to *sine*.

[131] In 1518 Palm Sunday fell on 28 March, so that the preceding Saturday was 27 March.

[132] Copernicus' *locatis* does not mean "succeeded to," as in StC VIII, 84, no. 155, in conflict with StC VII, 98, no. 155.

[133] Copernicus wrote *dabit* a second time, forgetting that he had already written it in the preceding line.

[134] In taking possession of 3 parcels in Abestich, Stenczel relinquished his overseership in Naglanden, about $4\ 3/4$ miles southeast of Abestich.

[135] Before recording this promise, Copernicus wrote *Actum* (Done), which he deleted in order to include the promise before terminating this entry.

[136] Copernicus' entirely correct *habitatorem* was incorrectly branded by L, p. 84/12, as wrong. Stenczel kept his promise that his overseership would have a new occupant, for this office was bought before 18 October 1518, as noted by Copernicus on that day.

[137] In order to be present personally in each of the villages involved in these three transactions of 27 March 1518, Copernicus had to go from Vindica about nine miles southeast to Godekendorf, and then from there about two miles southwest to Abestich.

[138] Copernicus started to write this surname as *Wans-*, which he deleted.

[139] Copernicus' *ij* ($= 1 1/2$) was misread as *II* by PI², 93/24, followed by Dmochowski, pp. 195/20, 210/6 up.

[140] These amounted to $1 1/2$ parcels, like Matz Wanczke's.

[141] Copernicus' *et $\overline{qr}$ i* (and $1/4$) was omitted by PI², 93/26, followed by Dmochowski, pp. 195/12 up, 210/3 up. Jorge's $2 1/4$ parcels, having become unoccupied on account of his death when his buildings burned down, were taken over by someone else on 18 June 1521 (Schmauch, 1929, p. 698, n. 33/3–4).

[142] Before *Hans*, Copernicus started to write *S-*, which he deleted. He may originally have intended to record this transaction by beginning with the name of the buyer, Simon Stoke.

[143] Copernicus at first wrote *quos*, which he deleted and replaced by *de quibus*, not a particularly felicitous construction.

[144] Copernicus wrote the name of this village, but made no entry thereunder. A transaction in this village was consummated later, on 3 January 1519.

[145] Copernicus' *quartam satorum* does not mean "the fourth part of sowing grain," as in StC VIII, 86, no. 161; see n. 110, above.

[146] Copernicus' *sequentis anni* does not mean "coming years," as in StC VIII, 86, no. 161 (in conflict with StC VII, 100, no. 161: *rok*).

[147] Copernicus' *bene* was misread as *tamen* by L, p. 84/3 up. L's punctuation was also faulty, since the sentence ends with *aedificaverit*, as is clearly shown by the period and the capital D of *Dabit*.

[148] Copernicus at first wrote *ad* (misread as *sed* by L, p. 84/3 up), but deleted it when he decided to shift to the dative case, *anno*.

[149] In eastern Europe, St. Margaret's is 13 July, which in 1518 fell on a Tuesday. The 2nd weekday (*fe.ª 2*) before St. Margaret's was therefore Monday, 12 July. Hipler, p. 273's mistaken "23 August" was repeated by Sikorski, p. 48, no. 167. In Schmauch, 1942, p. 491/19, "1818" is a typographical error for 1518.

[150] Peter bought the overseership from Stenczel after the transaction of 27 March 1518.

[151] Copernicus at first forgot to record the name of this village, which he evidently squeezed in as an afterthought.

[152] He acted as a guarantor again on 6 May 1521.

[153] Copernicus wrote the letter *s*, perhaps to indicate the presence of the overseer (*scultetus*), but he deleted the *s* when he realized that the overseer was not present.

[154] This may be the transaction which Copernicus started to record in 1518, between 4 May and 12 July.

[155] One of these 3 parcels, with the administrator's consent, was given up by Voitec on 24 January 1522, and instead he took possession of 2 abandoned parcels bordering on his other 2 parcels (Schmauch, 1929, p. 694, n. 24/1–4).

[156] Paul the first hermit was honored on 10 January as well as 15 January.

[157] Brosche's $3^1/_2$ parcels, burned down and abandoned on account of his death, were taken over on 8 July 1521 by somebody else (Schmauch, 1929, p. 693, n. 21/1–2).

[158] This is the first of three transactions dated 28 February 1519. Mica is about 8 miles northwest of Old Scaibot, the scene of the next transaction on that day. The third and last transaction took place in Old Cleeberg, about $2^1/_2$ miles west of Scaibot, on the road back to Olsztyn.

[159] Copernicus wrote *a* (for *antiqua*, old) after Scaibot, in order to distinguish it from New or Lesser (*minor*) Scaibot, mentioned in the transaction of 12 July 1517.

[160] He had been a witness of the transaction of 5 February 1517.

[161] Copernicus wrote the initial letter *a* (for *antiqua*, old) as in the preceding transaction.

[162] Matz Slander sold 2 parcels in Old Cleeberg, after taking possession of 2 other parcels in Old Scaibot, earlier the same day.

[163] This is the only price mentioned in these *Leases*. If it is an average price, a parcel sold for $16^1/_2$ marks.

[164] Copernicus wrote only the name of this village, but made no entry thereunder. If he was there when he wrote this name, he had to return on 10 and 11 April 1519, when he recorded two transactions.

[165] L, p. 85/4 up, printed selini as a single word, and branded it as an error. But Copernicus wrote *se* separated from *lini*. Copernicus' *se* is an abbreviation of *seminis* (as written out in full at L, p. 81/5 up).

[166] Copernicus did not indicate the family relationship of Pavel and Urban Gunter. The latter may be the Urban (surname not given) who, on 20 August 1527, took possession of 4 abandoned parcels without buildings, got 1 cow, 2 sheep, 8 sacks of rye, and had to pay rent from 1530 on (Schmauch, 1929, p. 687, n. 5/3–6).

[167] After the name of this village, Copernicus wrote *a*, for *antiqua*, the usual way of distinguishing an older village from a more recent village of like name. But he deleted the *a* when he remembered that there was only one Greseling, where he recorded three transactions (29 January 1517, 19 July 1518, and 6 April 1519). It was 10 miles southeast of Ditterichswalt, where he had recorded the previous transaction earlier that same day.

[168] Copernicus at first wrote *dimissos per* (released by), but he deleted these two words when he learned the real reason for the vacancy.

[169] After *sine libertate*, Copernicus wrote, and then deleted, *providebat* (misread as *procedebat* by L, p. 85, n. mm).

[170] Copernicus wrote *porcos* in the plural, because at first he thought that the number of pigs was five (*v*). Then, when he changed *v* to *i*, he forgot to change *porcos* to the singular form. The *v* apparently belonged with the next item, the sacks of rye seed.

[171] Instead of the required *modios*, Copernicus wrote *modiorum* in the genitive case, under the influence of the following word *siliginis* in the genitive case.

[172] Copernicus' *vorge* (read as norge (?) in L, p. 86/1) was an old spelling of modern German *Forke*, which is derived, like English "fork," from Latin *furca*.

[173] Copernicus forgot to put *tona aquaria* in the proper case by placing a stroke over the final -*a* in both these words.

[174] This time Copernicus recorded the overseer's name, which he had omitted from the transaction of 12 July 1518. This overseer was a guarantor also on 18 June 1521 (Schmauch, 1929, p. 689/last 2 lines).

[175] Copernicus started to write *Sim*, but deleted these letters in order to insert *et* between the two guarantors.

[176] This Simon may be the holder of 2 parcels who on 28 August 1524 took possession of 2 other parcels that had been abandoned. For the latter, he was to pay his first rent in 1527, but begin to perform the servitudes in 1526, while his rent for 1524 was cancelled (Schmauch, 1929, pp. 689/last line–690/3).

[177] These two had acted as guarantors in the transaction of 12 July 1518.

[178] Copernicus at first wrote *xiiii*, which he deleted.

[179] Copernicus started to write this name, but after the first two letters (*Au*), he realized that he had omitted the appropriate designation *D* (for *Dominus*). He therefore deleted *Au*, wrote *D*, and then proceeded with the full name. This priest Augustinus was a guarantor in 1531 for the former sexton of the local church until he paid the annual rent three times (Schmauch, 1929, p. 688/7–8).

NOTES

[180] After *annos* Copernicus started to write some other number, which he deleted so effectively that it can no longer be read.

[181] In 1519 Judica Sunday fell on 10 April.

[182] In 1519 Palm Sunday fell on 17 April. The preceding 2nd weekday (always a Monday) was therefore 11 April. The preceding transaction is dated 10 April in the same village. Did Copernicus spend the night of 10-11 April 1519 there (and if so, where?), or did he ride back and forth the 5 miles between Dewyten and his headquarters in Olsztyn, due south of Dewyten?

[183] After *inutilis*, Copernicus wrote *percepit ille in ma-* (Bartolmis received on the parcels), and then deleted this material, presumably because Bartolmis received nothing on the parcels.

[184] Presumably Lurenz, whose purchase of the tavern in Brunswalt was mentioned in the transaction of 22-25 March 1518.

[185] The fifth weekday (always a Thursday) before Palm Sunday, 17 April 1519, was 14 April.

[186] On 14 May 1517 Martzyn's guarantors had pledged that he would not run away within two years.

[187] Schonebrugk is about 9 miles due south of Montikendorff, the scene of the earlier transaction on this day.

[188] Before the name of this village, Copernicus wrote *Vuriten*, which he deleted. It appears nowhere else in these *Leases*, although transactions in it from 1481 to 1513 are noted by Schmauch, 1942, pp. 510-511, no. 58. Did Copernicus go to Vuriten, about 9 miles southeast of Olsztyn, only to find that an impending transaction was not consummated?

[189] Alex may be identical with the Alexius who in 1520 took possession of 1 long-abandoned parcel, with an exemption of 3 years. He did not live to enjoy his exemption since he was killed in the war, and his 3 parcels were taken over by somebody else on 12 February 1524 (Schmauch, 1929, p. 692, n. 18/1-6).

[190] After *Han*, Copernicus wrote *in*, which he deleted to make room for the third guarantor. He then wrote the *in*, after describing the last two guarantors as *villani*.

[191] *Cleeberg b*, called *Cleberg nova* in the transaction of 4 June 1517. StC VII, 110, no. 193, and StC VIII, 96, no. 193, brand as erroneous L's entirely correct identification (p. 86/n. 73; p. 102/7 up) of *Cleberg b* with *Cleberg nova*.

[192] Copernicus at first wrote *emit* (he bought), which he deleted.

[193] After selling these 2 parcels, Gnik (Gnix) still had 2 other parcels, to which on 27 October 1523 he added 2 further parcels, made vacant by his predecessor's death. He was given $1/2$ mark, required to reconstruct the buildings, pay rent from 1526 on, and perform the servitudes from 1525 on, with all the grain taken from the parcels to the castle being restored (Schmauch, 1929, p. 692, n. 17/21-25).

[194] Copernicus forgot to date this transaction. It was assigned to 14 August by Hipler, p. 275/12, followed by Sikorski, p. 54, no. 203. But 14 August is the date of the following transaction. The New Cleeberg transaction may have taken place on that day or on any day from 31 May to 14 August 1519.

[195] He had taken possession of these 3 parcels on 23 April 1517.

[196] This is the Brosien Trokelle, who was to take possession of 3 parcels in Plutzk on 23 March 1517, but did not do so "on account of the man's excessive dishonesty."

[197] Copernicus first wrote *futuro* (next), which he deleted.

[198] Augustin, the brother of Brosien Trokelle, was his guarantor on 23 March 1517 for 4 years. On 26 April 1526 Augustin (Augstin) added to his 2 parcels a third, previously unoccupied, for which he did not have to pay rent until 1529 (Schmauch, 1929, p. 696, n. 28/3-4).

[199] Copernicus forgot to record the date of this transaction. It was assigned to 14 August, the date of the preceding transaction, by Hipler, p. 275/12, followed by Sikorski, p. 54, no. 205. It may have occurred on that day in 1519 or on some later day in that year.

[200] He is recorded as having 2 parcels in 1498, and the third in 1503 (Schmauch, 1942, p. 537/8-9, 11 up).

[201] Copernicus left the Olsztyn district at some time after 11 December 1516, when he was active in Voytsdorf.

[202] For Jorge Hausberg, Urban Tile (who is mentioned in the following entry) was substituted by PI², 91/19 up, followed by Dmochowski, pp. 193/15 up, 208/6 up. According to Schmauch (1942, p. 524/14-15), this sale was consummated with the approval of the administrator (Copernicus). Yet he did not explicitly record his approval here, as he did in a number of other transactions.

[203] After 7 January 1517 Copernicus went back to his main headquarters in Olsztyn, where he recorded three transactions in January: Spigelberg (undated), Greseling (29 January), Godkendorf (30 January). By 10 February he was back again in the Melsac district.

[204] According to Schmauch (1942, p. 534/1-2), this sale was consummated with the approval of the administrator (Copernicus). Yet he did not explicitly record his approval here, as he did in a number of other transactions (see n. 202, above).

[205] This village is in the Olsztyn district, where Copernicus recorded transactions on 26 February and 4 March. Why was this Olsztyn transaction of 2 March recorded in the Melsac ledger? If Copernicus was in Schonebrucke on 2 March, did he take the Melsac ledger along by mistake? Or if both ledgers always remained in the headquarters, did he simply transcribe his Schonebrucke field notes in the wrong ledger?

[206] Copernicus did not date this transaction. His presence in the Melsac district on 10–11 February 1517 is known from the transactions there in Comain and Steemboth. His return to the Melsac district later in 1517 may be dated by the following transaction in Millemberg on 21 October. Hence this transaction in Libentail, also in the Melsac district, took place after 2 August, when Copernicus was in Scaibot in the Olsztyn district, and before 21 October, when he was in the Melsac district. The most likely date is just before 21 October.

[207] Was Stenzel Hoveman of Millemberg related to Andres Hoveman of Vusen, about 15 miles due west of Millemberg?

[208] *Acta sanctorum Octobris 9* (Paris/Rome, 1869), 73.

[209] The presence of the Melsac burgrave in Millemberg may indicate that he accompanied Copernicus to this village, which was situated about 9 miles southeast of Melsac.

[210] Over the *u* in *Hun*, Copernicus wrote a short stroke (somewhat longer, when repeated in the following entry), which is characteristic of some German handwritings in this period.

[211] After *biennium*, Copernicus began to write *accep-*, which he deleted.

[212] On 3 February 1527 Peter Braun, adding to his 2 parcels, took possession of 2 abandoned parcels, for which he pays rent from 1529 on, but the servitudes from 1530 on (Schmauch, 1929, p. 715, n. 12/5–8).

[213] Copernicus wrote *a quo*, which he deleted because he needed these two words later on.

[214] Copernicus' *supradictus* (aforementioned) identifies this Michel Hun with the Michel Hun named in the transaction of 22 October 1518. Yet there he appears as *Michel han*, and here as *michil hun*, in PI², 91/5 up, 92/4.

[215] He left Laisse and moved to Sonnenwalt, about 4 miles to the southeast.

[216] Misread as *tector* by PI², 92/5, followed by Dmochowski, pp. 194/10, 209/19.

[217] Misread as *transmigravit* by PI², 92/5, followed by Dmochowski, p. 194/10. Copernicus at first wrote *transmigrans* (as in L, pp. 76/12, 92/11). Then, realizing that he needed the ablative absolute instead of a participial construction, he wrote *-te* over the final *-s*.

[218] Misread as *3* by PI², 92/7, followed by Dmochowski, pp. 194/12, 209/22.

[219] Copernicus wrote the first five letters of this name, and then deleted them when he realized that he had forgotten to indicate the place of the transaction.

[220] Copernicus' *inutilis* does not mean "without profit" (*bez zysku*, as in Dmochowski, p. 209/16 up). Thus, *dextera manu inutilis* (Olsztyn, 30 January 1517) describes the peasant as "lame in his right hand," and *vetulus inutilis* (Olsztyn, 14 April 1519) characterizes the "little old man" as "incapacitated."

[221] Copernicus' contraction *oīa (omnia)* was misread as *onera* by PI², 92/10, followed by Dmochowski, p. 194/16, who translated it as *ciężary* (burdens).

[222] St. Brixius was commemorated on 13 November. This transaction, which occurred in 1518 (not 1519, as in Schmauch, 1942, p. 528/21), was recorded under 1519 because the Chapter's administrative year ran from 11 November 1518 through 10 November 1519.

[223] Copernicus started to record an entry, which he did not complete. Did he go nearly 5 miles southwest of Melsac to Stegemansdorf only to find that his presence was not required?

[224] Misread as *IIII* by PI², 92/11, followed by Dmochowski, pp. 194/19, 209/13 up. After the numeral $3^1/_2$, Copernicus wrote something which he deleted so thoroughly that it cannot be read.

[225] Before *super*, Copernicus wrote *a*, which he deleted.

[226] PI², 92/13, reads incorrectly *bovem*, repeated by Dmochowski, p. 194/20.

[227] Copernicus wrote *Ibidem*, deleted it, and then wrote it again. Did he spend the night of 11–12 March 1519 in Voppen, or did he return to Melsac on 11 March, and then go east nearly 9 miles to Voppen a second time on the following day?

[228] Misread as *clez* by PI², 92/16, followed by Dmochowski, pp. 194/24, 209/8 up (Klez).

[229] Misread as *IIII* by PI², 92/17, followed by Dmochowski, pp. 194/25, 209/7 up.

[230] PI², 92/17, misread *ipsi*, instead of *ipsis*.

[231] At first Copernicus wrote *Neudorf*, which he deleted.

[232] In 1515 Tiedemann Giese, who was then the Chapter's administrator, declared with regard to Kynappel Mill:

> I, Tiedemann, with the consent of the venerable Chapter, leased in perpetuity one parcel connected with this mill to the community of the village of Neuhoff on condition that it pay $^1/_2$ mark in lieu of all rent and servitudes every year ending on St. Martin's day. But if at any time the mill

is restored, at that time the said parcel should revert to the mill in accordance with the terms of the grant (Schmauch, 1929, p. 715, n. 9/1–7).

[233] Copernicus started to write *acceptasset*, but after the first four letters he stopped, deleted them, and wrote the whole word at the beginning of the next line (in order to avoid dividing the word between two lines?).

[234] Copernicus' *desertum* was detached from the preceding word *molendinum* and instead attached to *mansum*, four words earlier, by Dmochowski, p. 209/3 up, and L, p. 104/13 up, thereby making the one parcel abandoned (*pusty, opuszczony*), instead of the mill. But later on in this entry Copernicus says: "as long as the mill is abandoned" (*dopóki młyn będzie niezajęty*, Dmochowski, p. 210/10).

[235] Misread as I by PI², 92/2, followed by Dmochowski, pp. 194/12 up, 209/2 up.

[236] Copernicus inserted this second reason vertically in the left margin.

[237] Copernicus' *nollent* was misread as *nullus* by PI², 92/23, followed by Dmochowski, p. 194/9 up, and as nullent by L, p. 91/5.

[238] Copernicus' *po[n]tem* (lacking the required stroke over the *o*) was misread as *potest* by PI², 92/23, followed by Dmochowski, p. 194/9 up.

[239] Copernicus' *existeñ* was misread as *existens* by PI², 92/23, followed by Dmochowski, p. 194/9 up. The three misreadings (*nullus, potest, existens*) prevented Dmochowski, p. 210/1–2, from producing a satisfactory translation.

[240] Copernicus' *administrator* was omitted by PI², 92/25, followed by Dmochowski, pp. 194/5 up, 210/5.

[241] Copernicus' *moventibus* was misread as *monentibus* by L, p. 91/9. Although Dmochowski, p. 194/4 up, followed PI², 92/27's *moventibus*, it was left untranslated by Dmochowski, p. 210/6–7.

[242] Hence the rent went up from $1/2$ mark (= 12 skoters) to 15 skoters (= $5/8$ mark).

[243] On St. Martin's day, 11 November, the Chapter's fiscal year ended, and began.

[244] Copernicus added this maintenance obligation at the bottom of the entry.

[245] Copernicus wrote *oblig-*, which he deleted because *obligati* belongs at the end of the phrase.

[246] In 1519, Invocavit, the first Sunday in Lent, fell on 13 March. This was Copernicus' last entry during the year 1519 in the Melsac district, while in the Olsztyn district his last 1519 entry, although undated, may have been written during the month of August. Thereafter his activities as administrator are not recorded in these *Leases*. As his three-year term of office approached its end, he proceeded to Frombork, where at the meeting of the Chapter on 9 November 1519 his financial accounts had to be approved by his fellow-canons.

Two letters written in the German language in the name of the Chapter to the City Council of Gdańsk and to the Grand Master of the Teutonic Knights on 18 November 1519 and 18 December 1519 are both said to be in Copernicus' handwriting. On this basis (and only on this basis) he is said to have been the Chapter's chancellor in the closing months of 1519. But the handwriting of these two letters is not identical, and both differ markedly from Copernicus' German handwriting (see NCCW IV, Plates XXII and XXIII). Hence the claim that as soon as Copernicus' term as administrator expired, he was made the Chapter's chancellor must be regarded as devoid of documentary support. This claim was first put forward by NM, p. 21 (with facsimiles of the two German letters in Plates XIII, XIV), and repeated in StC VIII, 98, no. 198–199; StC XVIII, 90; and ZGAE, 1972, *36*:182.

Copernicus' successor as administrator of the Chapter did not serve a full three-year term. For, war broke out on 1 January 1520, and the Chapter may have wanted to have a more experienced administrator in Olsztyn while the fighting was going on. Hence Copernicus was given a second term as administrator, perhaps beginning on 11 November 1520.

[247] Copernicus presided over the first seven transactions recorded in 1521. All seven concern the Olsztyn district. Whether Copernicus recorded any transactions in the Melsac district in 1521 is not known, because that ledger has not been preserved.

[248] By contrast with the entries for 1517, 1518, and 1519, when Copernicus wrote that the entries were made "by me," and when he did not give himself the title "Venerable," here in 1521 "by me" is absent, and "Venerable" is present. Hence, although the first seven transactions were supervised by Copernicus, the corresponding official entries were not written by him, but from his field notes by his successor, identified by his handwriting as Giese. The title Administrator is attached to Giese's name in the minutes of the Chapter's meeting on 20 August 1521 (Hipler, p. 277, no. 51/6). Immediately preceding Giese in this list of canons present at the meeting is Copernicus, *Warmiae commissarius*. His second term as administrator was interrupted on 31 May 1521, when he was replaced as administrator by Giese, while he became *Warmiae commissarius*, Commissioner for Frombork, which suffered exceptionally severe war damage and sorely needed a special Commissioner of reconstruction. As *Warmiae commissarius,* Copernicus was concerned only with Frombork (Varmia, the munic-

ipality), not with Varmia, the entire diocese, as was mistakenly thought by L, p. 30/3 up; p. 39/16: *Commissaire de la Warmie*; p. 47/11 up: *Kommissar für Ermland*. See Werner Thimm, "Nicolaus Copernicus Warmiae Commissarius," ZGAE, 1971, 35:171–179. When the municipality was founded, it was called in Latin *Castrum Dominae Nostrae* (Our Lady's Fortress), "which in German is known as *unservrowenburk*" (CDW, I, 266, no. 154/9–10; 269–270; dated 8 July 1310; CDP, II, 104/2 up, 107/14–15, misdated). When *unser* was later omitted from this long name, what remained was written as *Frauenburg*. The last syllable should not be confused with *-berg*, the German word for "mountain."

[249] Jacob of Jomendorf had moved to Licosa (Lycosen) in a transaction which Copernicus recorded on 25 May 1517. Hipler, p. 277, no. 49/5, 7, misread Jomendorf as Joncendorf.

[250] Copernicus' *sata* was branded as an error by L, p. 92/13, because L misread *siliginis* as *siliginem*. Copernicus' *sata* means "sown," not "sowing," as in StC VIII, 110, no. 225, in conflict with StC VII, 123, no. 225.

[251] Copernicus' *sine libertate* was tacked on after the date, as an afterthought. The date is 6 May, not 20 May (as in Hipler, p. 277, no. 49/6). Although this entry is listed under Jomendorf, it was written in Licosa, like the preceding entry and the two following entries, all four being dated 6 May 1521.

[252] Copernicus' *aedificium*, being singular, does not mean "all his buildings," as in StC VIII, 110, no. 226.

[253] Copernicus *sata siliginis* was misread as *satam siliginem* by L, p. 92/16, and mistranslated as "sowing rye" by StC VIII, 110, no. 226, in conflict with StC VII, 123, no. 226.

[254] This last remark was not present in Copernicus' preliminary notes, but was squeezed in after 6 May by Giese. The date is 6 May, not 20 May (as in Hipler, p. 277, no. 49/6).

[255] Cristof was the guarantor also for Jacob the herdsman on 22 November 1518.

[256] Not 23 May (as in Hipler, p. 277, no. 49/7).

[257] Hipler, p. 277, no. 49/6, and L, p. 92/4 up, misread *minor* as *maior*. The village called *Cleberg minor* by Giese was called *Cleberg nova* and *Cleeberg b* by Copernicus (4 June 1517; after 31 May 1519).

[258] Giese's *censitos* was misread as censuos by Schmauch, 1929, p. 690, n. 13/5–6.

[259] The principal perquisite of an overseer's office was his permanent exemption from the annual rental payment and the peasant servitudes. In this instance, however, the overseer agreed that his $1^1/_2$ additional parcels would be subject to the burdens that had induced the fugitive to run away.

[260] Giese at first wrote *ad*, which he deleted because he wanted *siliginis* before *ad*.

[261] Giese's *neque* was omitted by Schmauch, 1929, p. 690, n. 13/3 up, and misread as *atque* by L, p. 92/last line.

[262] Giese at first wrote *periit* (misread by L, p. 93, n. b, as *percepit*), which he deleted because he needed, not the singular form, but the plural *periere* in agreement with *aedificia*.

[263] Giese squeezed these words in after the date.

[264] In 1521 Pentecost fell on Sunday, 19 May. The second weekday (*feria secunda*, always a Monday) was therefore 20 May, not 23 May (as in Hipler, p. 277, no. 49/7). The following transaction is dated 23 May.

[265] L, p. 93/4, misread *quapropter* as *quippe*.

[266] The 5th weekday (*feria v*, always a Thursday) after Pentecost in 1521 (19 May) was 23 May, not 20 May (as in Hipler, p. 277, no. 49/7). The previous transaction is dated 20 May.

[267] Misread as Joncendorf by Hipler, p. 277, no. 49/7.

COPERNICUS' APPROVAL OF FOUR FINANCIAL TRANSACTIONS

When Copernicus was serving his Chapter as its administrator, one of his functions was to approve financial transactions of a certain kind. Four such transactions are known, all of them executed in the form of a sale (*venditio*). What is sold in these transactions, however, is not a material object, such as a house or a horse. Instead, what is sold is the rent (or part of the rent) of a farm cultivated by a tenant who pays an annual rent to the owner of the farm. In the four transactions approved by Copernicus as administrator, the owner sells the annual rent payable in future years (rent futures, so to speak) for an equal sum of money paid to him in full on the day of the transaction. At the same time he obligates himself (and his heirs or successors) to repay the purchase price in annual installments until the buyer is fully reimbursed. In Document I, for example, the seller receives 6 marks on 27 March 1518. In exchange, he promises to pay the buyer $1/2$ mark a year until the rent is repurchased. If all goes well, the buyer's claim on the rent expires at the end of twelve years ($12 \times 1/2^m = 6^m$).

What the seller gains in this transaction is the use of the buyer's 6 marks over a period of 12 years. What the buyer gains is less obvious. For this reason it has been suggested that the buyer may somehow in the end receive more than the 6^m he paid at the outset. But no such form of interest, overt or concealed, is explicit or implicit in any of these four documents. In fact, Documents III and IV plainly state that the sellers have the right to buy back the rent by repaying the purchase price to the buyer, a right "which has remained in their power" (*quod in eorum potestate remansit*). Hence, the buyer recovers only what he originally paid, and nothing more.

But if the buyer gains nothing, why does he enter into such a transaction? He surrenders the use of his capital for as long as twelve years, and in the end recovers only his original investment. Instead of having a large sum of cash on hand, which must be safeguarded from thieves, he is promised the equivalent in small annual payments by a landed proprietor, whose farm is security against a default. The buyer in Documents I and II is a wealthy elderly canon seeking a safe investment payable to the heir to whom he intends to bequeath some of his property. In Documents III and IV the buyer is a pious vicar who wishes to receive the annuity during his life; but if he dies before his purchase price is repaid in full, the balance is to be converted into an endowment payable to whoever holds the office of preacher in Frombork Cathedral. A transaction which, on the face of it, is a sale turns out to be a sort of short-term annuity. If the seller is a married man, his wife's consent is required to make the transaction valid (Documents I, II, IV). Does the absence of this provision from Document III imply that the seller had no wife?

Documents I and II contain an expression (*legitimo venditionis titulo in his partibus consueto*) which is not found in Documents III and IV. The reference in the first two documents to a "legal act of sale customary in these parts" may conceal an evasion of the rules forbidding the taking of interest on a loan. In both cases Stockfisch pays a lump sum of money for rent to be collected over a period of years. If the "legal act of sale customary in these parts" implied a claim on the land underlying the rent, then after the loan was fully repaid in 12 years, in both cases, the borrower would still have to eliminate the lender's claim. In that case he would be paying $8\frac{1}{2}\%$ interest a year.

Documents I–IV were discovered by Carl Peter Wölky in the cathedral archives of the

Frombork Chapter, and shortly thereafter were published for the first time in Hipler, pp. 163–165, 274–276. Then PI², 95–96, reprinted extracts from Documents I–III, and Document IV almost completely. All four documents were later reprinted in their entirety by Dmochowski, pp. 189–193, with a translation into Polish at pp. 205–208. What follows is the first translation of Documents I–IV into English as well as the first analysis of their contents. Only Document I survived the last war (Olsztyn, ODA Z. 2/1); it is reproduced in NCCW IV, Plate XLIII.

I

In God's name, Amen. To all and each who will see the present document, I, Nicholas Coppernig[1], canon of the church of Varmia, doctor of canon law, and administrator of the common property of the venerable Chapter of Varmia, etc., declare by the present [document] that Urban Scultetus, overseer in Ditterichswalt, holder of four exempt[2] parcels in that place together with the office of overseer, having sought and obtained my permission for this [transaction], by a legal act of sale customary in these parts, affecting and concerning the four exempt parcels which he owns in the same place, with the consent of his wife and heirs, whose approval he promised in good faith, has sold to the honorable Nicholas Vicke, vicar of Varmia and chaplain of the castle in Olsztyn, buying for the fourth allod in Zcauwer owned[3] by the venerable Baltasar Stokfisch, canon of Varmia, half a mark in good money of the annual rent, payable on the feast of St. Michael[4] every year, in exchange for six marks of the same good money paid[5] in full to him in cash. [Urban Scultetus] promises that he, his heirs, and the owners of the said parcels will pay at the next feast of St. Michael, in proportion to the [elapsed] time, half of the rent, that is, one-quarter [of a mark],[6] and every year thereafter the entire rent of half a mark, to the said Baltasar or whoever will be the owner of the said fourth allod, until [Urban] himself, his heirs, or the owners of the parcels, will have accomplished the repurchase of this rent in whole or in part[7] for himself or themselves, with money like that for which it was bought, rent in arrears, however, being previously paid in full in accordance with the time [involved]. In witness and confirmation of these [matters] the present document has been sealed with the seal of the office of administration. Done in the castle in Olsztyn on 15 March 1518 in the presence of Baltasar Lossau[8] and Albert Szebulski, servitors, both called and summoned as witnesses to the foregoing [proceedings].

II

In God's name, Amen. To all and each who will see the present document, I, Nicholas Coppernig, canon of Varmia, doctor of canon law, and administrator of the common property of the venerable Chapter of Varmia etc., declare by the present [document] that Thomas Moldyth, vassal, a subject of the said Chapter in Old Trynckus, having sought and obtained my permission for this [transaction], by a legal act of sale customary in these parts, affecting and concerning the two exempt[9] parcels which he owns in the said place, with the consent of his wife and heirs, whose approval he promised in good faith, has sold to the honorable Nicholas Vicke, vicar of Varmia and chaplain of the castle in Olsztyn, buying for the fourth allod owned in Zcauwer by the venerable Baltasar Stokfisch, canon of Varmia, 15 skoters in good money of the annual rent,[10] payable on the feast of St. Michael every year, in exchange for[11] seven and one-half marks of the said good money paid in full to him in cash. [Thomas Moldyth] promises that he, his heirs, and the owners of the said parcels will pay at the next feast of St. Michael, in proportion to the [elapsed] time, half of the rent, that is, $7^1/_2$ skoters, and every year thereafter the entire rent of fifteen skoters, to the said Baltasar or whoever will be the owner of the said fourth allod, until [Thomas] himself, his heirs, or the owners of the parcels, will have accomplished the repurchase of this rent in whole or in part for himself or themselves with money like that for which it was bought, rent in arrears, however, being previously paid in full in accordance with the time [involved]. In witness and confirmation of these [matters] the present document has been sealed with the seal of the office of administration. Done in the castle in Olsztyn on 27 March 1518 in the presence of Christopher Drawschwitcz, burgrave of the said castle, and Andrew, overseer in Gödekendorpf,[12] both called and summoned as witnesses to the foregoing [proceedings].

III

In God's name, Amen. To all and each, reached by the present document, I, Nicholas Coppernig, canon of Varmia, doctor of canon law, and administrator of the common property of the venerable Chapter of Varmia, declare that the honorable George Schonsze,[13] permanent vicar of Frombork Cathedral, has bought one-fourth [of a mark][14] of the monetary rent of three exempt parcels of the overseership in Stynekyn from the overseer Palm, which the seller owns in the said village, in exchange for three marks of the good coinage, which the seller received in cash and counted for himself. As a result, Palm himself and his successors in the said properties are effectively obligated every year on the feast of St. Michael the Archangel to pay the said rent to the said George Schonshe, the buyer, as long as he lives, and after he dies, to the preacher at that time in Frombork Cathedral, to whom the said George wished to donate in perpetuity a rent of this kind. [This obligation will remain in force] until the successors, or any of them, choose to repurchase the rent with three similar marks repaid in whole or in part, a choice which has remained in their power, any rent in arrears, however, having been previously paid in full. In witness and confirmation of this transaction, I have had the present document drawn up and furnished with the seal of the office of administration. Done in the castle in Olsztyn, A.D. 1518, 29[15] May, in the presence of Christopher Drawschwicz, burgrave of the said castle in Olsztyn, and Baltazar of Lossaw, both called and summoned as witnesses to the foregoing [proceedings].

IV

In God's name. Amen. To all and each, reached by the present document, I, Nicholas Coppernig, canon of Varmia, doctor of canon law, and administrator of the common property of the venerable Chapter of Varmia, declare that the distinguished George Frederici in Stygeyn, with the consent of his wife, has lawfully sold to the honorable George Schonszee, permanent vicar in the said church, half a mark [of the rent] from five exempt parcels which he owns in the said property, in exchange for six marks of the said good money, which the seller himself counted and received in cash. As a result, George Frederici and his successors in these properties are effectively obligated every year on the feast of St. Michael to pay the said rent to the said George Schonse as long as he lives, and after his death to the preacher at that time in Frombork Cathedral, to whom the said George wished to donate in perpetuity a rent of this kind. [This obligation will bind George Frederici and his successors] until they or any of them choose to repurchase the rent with six similar marks[16] paid in whole or in part, a choice which has remained in their power, any rent in arrears, however, having been previously paid in full. Done in the lodgings of the venerable Baltazar Stokfisch, canon of Varmia,[17] on St. Dorothy's day [6 February] 1519, in the presence of George Plastewigk and Jacob the overseer, both called and summoned as witnesses to the foregoing [proceedings].

---

[1] Although this document read "I, Nicholas Coppernig," it was not written by Copernicus himself, according to Hipler, who inspected the document when it was in the Frombork Cathedral archives, Z 2/i. Hipler judged the handwriting to be that of someone other than Copernicus, who never wrote his surname with a final -g, although he made no objection to it in this document (PI², 97/17 up–14 up).

[2] Exempt from the customary labor servitudes required of the holders of parcels outside the overseership.

[3] Stokfisch owned (*possidet*) the fourth allod; he is not its usufructuary (*Nutzniesser*, as in PI², 95, n. **/ 13–14), since a usufructuary enjoys the income from a property owned by another.

[4] 29 September.

[5] The scribe wrote *persolutas*, where *persolutis* is required to agree with *marcis*, ten words before. Through Vicke in Olsztyn, on 15 March 1518 Stokfisch in Frombork paid 6^m to Urban Scultetus, who undertook to pay $1/2^m$ a year to Stokfisch until the 6^m were repaid (or the rent was repurchased, to use the formulation of these documents).

[6] The elapsed time from 15 March 1518, when Document I was executed, to 29 September 1518, would be a little more than half a year, so that Urban Scultetus' payment for 1518 was set at $1/4^m$ = half of the regular annual payment of $1/2^m$.

[7] The whole rent would be repurchased by the Ditterichswalt party on 15 March 1530 after paying $1/2^m$ a year for 12 years to the Zcauwer party, which would thereby recover the 6^m lent to the Ditterichswalt party on 15 March 1518. The Ditterichswalt party could, however, repurchase the rent in part on any 29 September before 15 March 1530 by paying to the Zcauwer party the remaining unpaid balance. This transaction, therefore, although nominally a sale, was essentially a loan of 6 marks, to be repaid at the rate of $1/2^m$ per year in not more than 12 years. Should the borrower fail to repay the loan, his 4 parcels of arable land were available to the lender. If the borrower paid in full, the lender recovered the principal sum he lent, but no interest in addition thereto. This transaction was executed by Sto(c)kfisch not long before his death. As a wealthy man, he was thereby providing a twelve-year annual income or annuity for the heir whom he chose to inherit the price paid for his Zcauwer property.

[8] "Lossan" in Hipler, p. 164, no. 1/2 up, was corrected to "Lossau" by PI², 96/8.

[9] A vassal's parcels, like an overseer's, were exempt from the labor servitudes required of the peasants.

[10] According to StC VIII, 84, no. 157/5, Moldyth was permitted "to sell his annual rent from the fourth

256

farm in Zcauwer." But that fourth farm or allod belonged to Stokfisch, not to Moldyth. The annual rent in question flowed from Moldyth's two parcels to the owner of the fourth allod.

[11] Fifteen skoters ($= 37\frac{1}{2}$ shillings; see Copernicus and the Money Question, n. 7), if paid annually for 12 years, would be a delayed equivalent of $7\frac{1}{2}^m$ ($= 450^s = 12 \times 37\frac{1}{2}^s$). Yet, according to StC VIII, 84, no. 157/8–9, "The sold annual rent amounts to 15 skoters ... instead of 7.5 marks." Far from being "the sold annual rent," the 7.5 marks were the cash purchase price of the sold annual rent.

[12] He was a witness of a lease transaction on 30 January 1517.

[13] George Schonense was one of the witnesses to the oath of allegiance sworn by Copernicus and the other Varmia canons to the new bishop of Varmia on 5 April 1512 (StC VIII, 57, no. 72/last 4 lines). On 30 March 1519 George Scheneza (*sic*) was one of the proxies designated by Alexander Sculteti in Rome to help him obtain possession of the "canonry and prebend held by the late Andrew Copernicus [the astronomer's older brother] while he was alive" (*Coelum*, 1951, *19*:41).

[14] Document III's $fj = \frac{1}{4}^m$. The customary 12-year maturity would require $12 \times \frac{1}{4}^m = 3^m$, the cash purchase price.

[15] Not "19," as in Hipler, p. 274, no. 37/1, followed by Sikorski, p. 47, no. 165.

[16] The rent is repurchasable with 6 good marks, the amount paid for it on 6 February 1519 by Schonszee. The seller's repurchase of the rent for 6^m is obscured by StC VIII, 91, no. 176/8–10: "The rent will be paid as long as it is repurchased completely or partially ... together with 6 marks."

[17] Unlike Documents I–III, which were executed in the castle in Olsztyn, Document IV was drawn up in the Frombork lodgings of Stockfisch. He was not personally involved in this transaction, as he was in Documents I and II, where he was a distant buyer acting through a local agent. Why did Copernicus conduct this transaction in Stockfisch's lodgings (*curia*)? A Varmia canon's *curia* was located in Frombork, so that Document IV was not executed in Zcauwer (as was tentatively suggested by StC VII, 105, no. 176, and StC VIII, 91, no. 176).

COPERNICUS' INVENTORY OF 1520

INTRODUCTION

In compliance with papal policy, the Varmia Chapter preserved important documents affecting its status and rights. The earliest inventory, consisting of more than thirty documents, was compiled about the middle of the fifteenth century. Then, a decade or so later, the second inventory listed 38 documents, which in the main were different from those in the first inventory. The latter had identified its documents by assigning an Arabic numeral to each of them, but that method was not followed in the second inventory, whose documents were stored in a special wooden box kept in Frombork. The storage plan was changed at the time of the third inventory, which consisted of two sections. The first, comprising 32 documents, was assembled in Frombork on 7 March 1502. Then,

> by order of the venerable Varmia Chapter on 1 October 1502 A.D., I, Balthasar Stockfisch, canon and administrator, collected all the aforementioned documents in the Frombork Cathedral, and took them to Olsztyn, and deposited them in the treasury of the castle. Among them were the following twenty[1]

documents, making a total of 52. Frombork was dangerously near the border with the Chapter's hostile neighbor, the Order of Teutonic Knights. The wisdom of the transfer from Frombork to the Chapter's most heavily fortified castle in Olsztyn was later confirmed in the war of 1520–1521, when Frombork was overrun whereas Olsztyn remained intact.

The fourth inventory begins with a preamble explaining that it was drawn up

> ... in the course of the year 1508 by me, George of Delau, cantor and canon of Varmia, and administrator ... in the sixth and last year of my office.[2]

Delau arranged the "rights and other documents in alphabetic order." The number of entries having risen to 144, the Chapter acquired for its treasury in Olsztyn Castle a chest of drawers, each of which was designated by a letter of the alphabet. The drawers were protected by two doors which were hinged at the sides and opened outward from the center. The Olsztyn chest has not been preserved. But a similar (albeit somewhat larger) chest survives to this very day in Wrocław. Instead of assigning an individual designation to each docu-

---

[1] PI², 82, n. **/5–9, mistakenly calling Stockfisch's the first inventory. Obłąk (pp. 13/15, 70/17, 76/17) misunderstands Stockfisch's M^vC *secundo prima Octobris* as "21 October 1502." In Latin, "21" would be *vice(n)sima prima*. How can *secundo* (masculine) be combined with *prima* (feminine)? If it could, the year would be 1500 instead of 1502. Stockfisch's administratorship was erroneously terminated in 1500 by PI², 78, n/4; 89, n/4–5.

[2] Obłąk, p. 76/19 up–16 up. The English summary of Obłąk (p. 76/14 up: "During the year 1503 he [Delau] was the administrator") contains a typographical error, as may be seen by comparing it with the Polish text (p. 15/1–3). PI² (78, n/5; 89, n/6–7) misdated Delau's administratorship as 1500–1509, and mistakenly extended it to eight years instead of six. The administrator's term of office sometimes lasted three years, and was renewable. Delau served from November 1502 to November 1508. The 1622 inventory, fol. 28/6 up–5 up, lists a "Copy of the purchase, made by the cantor G. of Delau, of 31 parcels in the village Rosenau."

INTRODUCTION

ment, as had been done previously, Delau grouped related documents in a drawer, and listed them in his inventory under the letter of that drawer.

The next administrator served only one year, and did not compile a new inventory. Neither did his successor, Tiedemann Giese (1509–1515), who did, however, add some entries to the 1508 inventory, and also some notes. His successor, like his predecessor, served only one year, and was followed by Copernicus. But in his first term as administrator (1516–1519) Copernicus was occupied with more pressing problems than the inventory. His successor, who served only one year, likewise treated the inventory with benign neglect. After Copernicus had begun his second term as administrator on 11 November 1520,[3] however, he compiled the Chapter's fifth inventory in the next few weeks, since he dated the document 1520. The war against the Teutonic Knights was then in full swing, preventing Copernicus from leaving Olsztyn, as he had frequently in pursuance of his duties during his first term as administrator. But in November–December 1520 an up-to-date inventory was urgently needed in case the Knights penetrated Olsztyn's defenses. In April 1521 an armistice went into effect, and Giese replaced Copernicus as administrator after 31 May 1521, when Copernicus was reassigned. He had not attached his name to his inventory, which used to be misattributed to Giese.[4]

The credit for correcting this misattribution is due to Bishop Jan Obłąk, director of the Olsztyn diocesan archives (cited hereafter as "ODA"). In connection with the worldwide quincentennial celebration in 1973 of Copernicus' birth in 1473, Obłąk reexamined ODA in the hope of finding previously unutilized source material. In the course of his investigation he recognized the handwriting of the 1520 inventory[5] as Copernicus'. By promptly publishing it, he made a considerable contribution to Copernican studies. His long article[6] contains the inventory's Latin text; a photofacsimile; a translation into Polish; an extensive paleographical, bibliological, and historical discussion of the 1520 inventory in its place among the Chapter's inventories; and summaries in English, German, Italian, and Russian.

In the 1520 inventory Copernicus followed Delau's procedure in the 1508 inventory by refraining from assigning an individual number to each document. Instead, he placed related documents in the same drawer. Each drawer was designated by a letter of the alphabet, beginning with A and ending with R. He used 17 drawers in all, since in those days J was not recognized as a letter independent of I but rather as its consonantal variant.

In order to facilitate references to the entries in Copernicus' inventory, under each letter the entries will be numbered consecutively in square brackets. These numbers range from 4 (under A) to 18 (under E). The total number of entries under all the letters from A to R is 161. This total indicates the number of entries in the inventory, not the number of documents in the drawers. Thus, entry C 3 mentions "Seven bulls of indulgences," and G 14 refers to "Various wills." Hence, the number of documents inventoried exceeds 161. Although Copernicus did not indicate the general subject of the documents placed in any one drawer, he seems to have followed a plan, which will be indicated below in square brackets following the letter of the drawer. In three cases, at first glance Copernicus classified a particular docu-

---

[3] Leases, n. 246/last ¶. On an unspecified day in 1520 (presumably in early November) the acting custodian of the Chapter received payments from the administrator then in office as well as from Copernicus, administrator for the following year (*administratori anni sequentis*; *Studia i materiały z dziejów nauki polskiej*, Series C, 1963, no. 7, p. 72/7–12; ZGAE, 1972, *36*:183/6–11; StC VIII, 103, no. 212).

[4] PI², 82, n. **/last 5 lines.

[5] Now ODA Y 9.

[6] *Studia Warmińskie*, 1972, *9*:7–85.

ment as belonging in a certain drawer, but on second thought he changed its classification. For instance, he originally placed a letter from the king of France after no. 9 under B, with communications from other potentates concerning the boundaries of the diocese of Varmia. But then he realized that the king's letter dealt with a gift, and therefore more properly belonged under C, where it is listed as no. 6. By the same token he moved the entry following no. 7 under P, whose contents are mainly concerned with vicariates, to Q 6, where its subject matter (an annual rent) links it with Q 4. Less clear is his reason for shifting the entry originally listed after no. 14 under E to F 1.

After writing the first eight entries under F, Copernicus deleted all this material by drawing three parallel vertical lines through it. Leaving the rest of this column blank, he rewrote the deleted entries, with the addition of a ninth entry, under F on a second sheet. He shifted to another sheet at this point in order to preserve the alphabetical order of his inventory. He wrote it on paper that was approximately three times as long as it was wide. Folding a sheet in half vertically, and then in half again, he obtained long, narrow columns. On the recto of the first column he placed only the title:

Inventory of the Documents and Legal Papers in the Treasury of the Castle in Olsztyn 1520 A.D.

On the verso of the first column he put sections A and B, with sections C and D on the recto of the second column, and E on its verso. Below E, his longest section, he wrote the capital letter F, which he deleted when he realized that there was not enough space to accommodate the whole of section F on the verso of column 2. He therefore turned to the recto of column 3, and under F he wrote the first eight entries. But then he became aware that when he inserted the needed second sheet, section F would be hopelessly out of alphabetic order. Hence, folding the second sheet in the same way as the first, but cutting it once down the middle, he placed the second sheet inside the first and sewed them together. As a result, he could now write sections F and G in the proper alphabetic order on the recto of the first column of the inside sheet. Proceeding in this way, he wrote the rest of his inventory on the inside sheet, with his last two sections, Q and R, appearing on the recto of its fourth column. Thus he had no need for the verso of the fourth column of the inside sheet, and for the last two columns of the outside sheet.

To the left of most of the 161 entries, there is a dot. This may signify that the presence of the document (or documents) in the drawer was verified. In seven cases a horizontal stroke stands to the left of the entry.[7] In two of these cases[8] there is also a curlicue, which appears by itself in 16 other cases.[9] What these signs mean separately and together has not yet been clarified.

Although Copernicus' inventory is not yet completely understood in every detail, it has already thrown welcome added light on his second term as administrator of his Chapter as well as on the Chapter's earlier history.

---

[7] H 7, 8; I 6; N 6, 7; O 5; R 2.
[8] H 8, I 6.
[9] D 12; G 8, 13, 14; H 6; K 3, 9, 10; L 11; M 4, 5, 6; N 4, 5, 8; O 10.

INVENTORY OF THE DOCUMENTS AND LEGAL PAPERS IN THE TREASURY OF THE CASTLE IN OLSZTYN 1520 A.D.

A [ESTABLISHMENT OF THE DIOCESE OF VARMIA]

[1] 2 Bulls of Inno[cent] VI concerning the boundaries of the regions of Prussia
[2] 2 Golden Bulls of Charles IV concerning the delimitation of the regions of Prussia and the confirmation of the privileges of the diocese of Varmia
[3] Document of the said Charles concerning the renewal of the privileges of the diocese of Varmia
[4] Document of Anselm, the first bishop of Varmia, [concerning] the construction of a cathedral and the division [of the Prussian lands] into dioceses

B [DELIMITATION OF THE BOUNDARIES OF THE DIOCESE OF VARMIA]

[1] Delimitation of the Varmia and Samland dioceses, with 4 seals
[2] Agreement between the diocese of Varmia and the Teutonic Knights concerning the boundaries, in Latin, with 8 seals
[3] Another agreement, in German, with 6 seals
[4] Bull of Gregory XI committing the case of the boundaries to the archbishop of Prague
[5] Letter of the archbishop of Prague to the priest in Elbląg concerning the acceptance of witnesses
[6] Decision of the arbiters
[7] Copy of the delimitation of the boundaries, in the form of a book
[8] Letter transmitting the Grand Master's [agreement] concerning the boundaries [of the Varmia diocese] with the Osterode district [of the Teutonic Knights]
[9] Two letters, extracts, concerning the same [subject] [Document of the king of France concerning a gift of wood from the Holy Cross]

C [DOCUMENTS CONCERNING FROMBORK CATHEDRAL]

[1] Document concerning the gift of the village Santoppen for the workshop of Frombork Cathedral
[2] Legal documents of the said workshop
[3] Seven bulls of indulgences for the said cathedral
[4] Document concerning the transfer of the head of St. George from Lidzbark to Frombork Cathedral
[5] Letter concerning the conduct of the choir in the said cathedral
[6] Document of the king of France concerning a gift of wood from the Holy Cross

D [BUSINESS TRANSACTIONS CONCERNING PROPERTIES IN NORTHERN VARMIA]

[1] Grant of Tolkemit, together with its district and with its fisheries, to the Varmia diocese
[2] Grant of Tolkemit to the Chapter by Bishop Fabian
[3] Individual's property rights in Codyn, Scharfenstein, Reberg, and the Haselau mill
[4] Acknowledgment by G[eorge] of Baysen concerning the payment for the said properties
[5] King Sigis[mund's] document approving the purchase of the village Claukendorf
[6] Approval of the purchase of the Tolkemit district
[7] Document approving the assignment, with papal authorization, of the villages Crebisdorf and Carsau in perpetuity to the monastery of St. Brigit in Gdańsk
[8] Registration of Conradswalt etc.
[9] Mortgaging of Schonebuche
[10] Approval of the purchase of Conradswalt by Simon Rabenwalt
[11] Approval of the purchase of Conradswalt by Bishop Lucas
[12] Approval of the purchase of the village Maybom

E [BUSINESS TRANSACTIONS CONCERNING PROPERTIES NEAR OLSZTYN AND BRANIEWO]

[1] King Sigismund of Poland's document approving the purchase of villages near Stum
[2] Document concerning certain villages improperly named in the papal confirmation
[3] Document concerning the purchase of the land in front of the castle in Olsztyn and of two granaries
[4] Document concerning the purchase of the rent of the tavern from the overseer in New Cukendorf
[5] [Document concerning the] purchase of Quedeliz's estate
[6] Document of the Chapter concerning part of the pastureland of the community of Rosengarten
[7] Individual's rights in the village Glanden
[8] Individual's rights in the village Gabelen
[9] Individual's rights in the village Pilgrimsdorf
[10] Individual's rights in the estate of Bebir
[11] Individual's rights in the Borniten mill
[12] Decision concerning fishing in the mill pond, in the same place
[13] Decision concerning the Birckpusch woods, against the people of Frombork
[14] [Document concerning the] purchase of the estate of Caleberg
Document concerning the purchase of 12 marks, light money, [as rent] in the village Baisen
[15] Legal documents concerning Posorten, and a copy

[16] Claim concerning the arrears in the rent of the Caldemflis mill
[17] Document of acknowledgment in Olsztyn concerning the same [subject]
[18] Document concerning the transfer of the said mill to the Chapter

F [ADDITIONAL BUSINESS TRANSACTIONS CONCERNING PROPERTIES NEAR OLSZTYN AND BRANIEWO]

[1] Document concerning the purchase of 12 marks, light [money, as rent] in the village Baisen for the office of [the hours of] Our Lady
[2] Donation of the village Padelochen
[3] Confirmation of the said grant by the Council of Basel
[4] Three copies of the decision concerning the village Padelochen
[5] Document concerning the purchase of 8 parcels in the property of Peuthuen in the Olsztyn district
[6] Individual's rights in the village Voitsdorf
[7] Document concerning the purchase of the said [village], and other similar [documents]
[8] Copy of an individual's rights in Sandecaim
 [F 1–F 8 were deleted, and then rewritten on the second sheet, fol. 1^r]
[9] Document concerning the sale of parcels in Vusen for the [office of the] hours of Our Lady

G [FURTHER BUSINESS TRANSACTIONS CONCERNING PROPERTIES NEAR OLSZTYN AND BRANIEWO]

[1] Individual's rights in the village Wuszen
[2] Copy of the purchase of 8 parcels in the said place
[3] Decision against Fabian in the matter of the rent in the same place
[4] Implementation of the decision against the said individual concerning the mill
[5] Mortgage of 4 parcels in the said place by Fa[bian] Tolke
[6] Individual's rights in the village Scaibot
[7] Document concerning the payment for the said [rights]
[8] Document concerning the purchase of the Scaibot estate
[9] Document concerning the payment for the said [estate]
[10] Copy of the sale of parcels in Engelswalt for the late Christian Tapiau
[11] Contract concerning the properties in Dareten
[12] Document concerning the sale of parcels and the mill in Schouffsberg
[13] Register of the anniversaries [of benefactors' deaths]
[14] Various wills

H [ADMINISTRATION OF THE CHAPTER]

[1] Confirmation by Pope Boniface of certain statutes [governing the Chapter]
[2] Papal permission to the bishop of Varmia concerning disbursements by the custodian and other disbursements in addition to the main stipends [of the Varmia canons]

[3] Also, approval of the custodian's disbursements [only to canons in residence]
[4] Deposition of witnesses concerning disbursements given to [canons] absent on a mission of the diocese or Chapter
[5] Letter of Bishop Fab[ian] concerning the distribution of the income from Tolkemit
[6] Damaged bull concerning the canons' [obligation] to study for three years
[7] Agreement between the bishop and the Chapter concerning the canonical or episcopal lodging near Frombork Cathedral
[8] Agreement between the bishop and the Chapter concerning the election of canons and other matters
[9] Copy of the agreement [in H 8] between the bishop and the Chapter
[10] Concerning the provost's [right to] speak first

I [ELECTION OF BISHOPS]

[1] Creation of the archdeaconate
[2] Confirmation by Innocent VIII of the election of the late Bishop Lucas [Watzenrode]
[3] Copy of the resignation from the bishopric in favor of the same [Lucas Watzenrode]
[4] King Sigismund's document concerning the agreement about the election of bishops
[5] Articles of the oath [of allegiance] in the election of Bishop Fab[ian]
[6] Proceedings concerning the election of the said [bishop]
[7] Letter of the Chapter to the pope and the college of cardinals concerning the election of the late Arnold Venrade
[8] Letters of the bishops of Riga, Courland, Samland, and Pomesania concerning the same [subject]
[J was not yet recognized as a letter separate from I]

K [VARMIA AND SECULAR RULERS]

[1] Testimonial letter of the king of Poland on behalf of the bishop [of Varmia]
[2] Testimonial letter of the grand duke of Lithuania
[3] Letter of the king of Poland in which he took the diocese of Varmia under his protection
[4] Copy of the peace [treaty] between King Ladislaus of Poland and the Teutonic Knights
[5] Copy of the agreement between the Grand Master of the Teutonic Knights and Bishop N[icholas] of Varmia
[6] Document concerning the acceptance of the "perpetual peace" by the Chapter of Varmia
[7] Bundle of letters of King Matthias of Hungary and Bishop N[icholas] of Varmia
[8] Copy of the document naming Vincent Kelbas as administrator of the diocese of Varmia

[9] Draft of a petition of the bishop to the king of Poland
[10] Copy of an order to submit the diocese [of Varmia] to the protection of King Casimir of Poland

L [VARMIA'S SILVER AND OTHER POSSESSIONS]

[1] Document concerning the silver bequeathed by the late Bishop Henry
[2] Document concerning the cathedral's silver, [weighing] 306 marks, for the use of the bishop, by [permission of] the Chapter
[3] Copy of an Elbląg letter concerning a loan made by the Chapter
[4] Concerning the silver pledged in Livonia
[5] Letter concerning the loan made by Bishop Paul of Courland
[6] Rescript of Sixtus IV against the withholders of the diocese's lawful rights and possessions
[7] Also, a rescript of Inno[cent] VIII in a similar case
[8] Rescript against Sander of Vuszen
[9] Document of Benedict [of Macra], commissioner of Emperor Sigismund, against the Teutonic Knights concerning the restitution of what they had seized
[10] Document concerning the bishop's goods seized by the Teutonic Knights in Elbląg
[11] Copy of the arbitral decision by Emperor Sigismund in favor of Bishop Henry of Varmia
[12] Order of the said bishop concerning the recovery of the property of his diocese and the booty, in accordance with the preceding decision
[13] Account of what was received by B[artholomew] Libenwalt from the bishop of Courland

M [VARMIA'S LOST PROPERTIES]

[1] Trial of George of Schliven and his accomplices
[2] Agreement with Gutco and those who held the castle in Seeburg
[3] Royal letter to the people of Brau[n]sberg concerning the mortgaging of the villages
[4] Order of Bishop Paul [Legendorf of Varmia] for a subsidy to repurchase the castle in Seeburg
[5] Letter of the said [bishop] to certain priests concerning the same [subject]
[6] Copy of a letter to the commander of Brande[n]burg concerning the property left behind by the administrative officers
[7] Letter of the Gutstadt provost acknowledging receipt of certain goods left by him with the Varmia Chapter

N [ABSOLUTION, TAXES, AND RELATED MATTERS]

[1] Bull of Pius II concerning absolution from the censures of the [Prussian] League

[2] Concerning compliance with the League's instructions
[3] Circular letter in the matter of the League
[4] Letter of absolution of certain canons excommunicated by Bishop Francis [Kuhschmalz]
[5] Order of absolution in behalf of Bal[tasar] Scaibot
[6] Copy of a dispatch of Bishop Francis to the Grand Master concerning the levy
[7] Copy of [the letter of] the late Paul Rosdorf, [Grand Master of the] Teutonic Knights, to the bishop concerning the levy
[8] Document concerning Oporow[ski's] provision for the parish church in Resil
[9] Decision of the arbiters between the diocese [of Varmia] and the Teutonic Knights

O [AGREEMENTS AND LAWSUITS]

[1] Document of agreement between the Chapter and the overseers
[2] Another document of agreement in a similar matter
[3] Also, another document in a similar case
[4] Letter in the Greussing case
[5] Receipt for expenses paid for the Chapter
[6] Document concerning the arbiters' decision between Peter Polen and Andrew Melczer, citizens in Olsztyn
[7] Interrogation of the witnesses [in the case concerning the] house of C[hristopher] of Delen
[8] In the case of Philip Greussing
[9] In the case of Michael Bogener
[10] In the case of the 3 brothers from Plauten, who were condemned
[11] In the case of Tynappel

P [THE LESSER CANONRIES AND VICARIATES]

[1] Suit against the canons endowed with medium prebends
[2] Pope Martin V's suppression of the medium and minor prebends of the diocese of Varmia
[3] The bishop's and Chapter's consent to the vicariate [established] by the late Otto of Russen
[4] Copy of the establishment of 2 Frombork Cathedral vicariates, based on Degeten and Vangaiten
[5] Establishment of the vicariate of the 11,000 virgins
[6] Document of the Brethren of St. Anthony in response to the donation
[7] Document concerning the establishment of 3 vicariates in Frombork Cathedral
[Document concerning the sale of the annual rent in Elditen]
[8] Mortgage of 4 parcels in Vusen for the vicariate of the late Martin Achtesnicht

[9] Approval of the will of Provost Henry concerning the vicariate of the provostship in Frombork Cathedral
[10] Document acknowledging the receipt of 9 florins for the community of vicars in Vierzighuben and Hinrichsdorf

Q [ALMSHOUSE AND RENTALS]

[1] Establishment of an almshouse in Frombork Cathedral
[2] Document concerning a gift to the poor
[3] Concerning the expansion of $1^1/_2$ parcels in Rabusen
[4] Acknowledgment of 1 mark as an[nual] rent for a vicariate
[5] In the case of the land for the vicariate of St. Wenceslaus
[6] Document concerning the sale of the rent of 11 marks in Elditen

R [VICARIATES]

[1] Establishment of a vicariate in Melsac
[2] Also, another establishment of a vicariate in the same place
[3] Establishment of a vicariate of St. George in the same place
[4] Inventory of the gift of Olsztyn
[5] Inventory of the gift of Santoppen

BIBLIOGRAPHY

        Bańkowski, Piotr, "Rewelacyjne odkrycie. Mikołaj Kopernik, 'Felix Notarius,' u kolebki dziejów polskich archiwów," *Archeion*, 1974, *9*:61–80

        Biskup, Marian, "Articuli iurati biskupa warmińskiego Fabiana Luzjańskiego z r. 1512," *Rocznik Olsztyński*, 1972, *10*:297–303

        Codex epistolaris saeculi decimi quinti, ed. Anatol Lewicki, II (Cracow, 1891; Monumenta medii aevi historica res gestas Poloniae illustrantia, XII)

        Jahrbücher des Vereins für meklenburgische Geschichte und Alterthumskunde, 1868, *33*:27–31

        Lites ac res gestae inter Polonos Ordinemque Cruciferorum, III (Warsaw, 1935)

MKDW Mikołaj Kopernik, *Dzieła wszystkie:* I (Warsaw/Cracow, 1972); II (Warsaw/Cracow, 1976); III (Warsaw/Cracow, in press)

Obłąk, Jan, "M. Kopernika inwentarz dokumentów w skarbcu na zamku w Olsztynie," *Studia Warmińskie*, 1972, *9*:7–85

ODA Olsztyn Diocesan Archives

        Rosen, Edward, "Czy Kopernik był 'szczęśliwym notariuszem'?," KHNT, 1980, *25*:601–605

        ——— "Copernicus Was Not a 'Happy Notary'," *The Sixteenth Century Journal*, 1981, *12*:13–17

        Schmauch, Hans, "Der Streit um die Wahl des ermländischen Bischofs Lukas Watzenrode," *Altpreussische Forschungen*, 1933, *10*:65–101

Theiner, Augustin, *Vetera monumenta Poloniae et Lithuaniae gentiumque finitimarum historiam illustrantia* (Rome; I-II, 1860–1861)

Thunert, Franz, *Acten der Ständetage Preussens, königlichen Antheils*, I (Danzig, 1896)

        Voigt, Johannes, *Namen-Codex der Deutschen Ordens-Beamten* (Koenigsberg, 1843)

Weise, Erich, *Die Staatsverträge des Deutschen Ordens in Preussen im 15. Jahrhundert* (Koenigsberg/Marburg, 1939–1966)

NOTES

A 1   The boundaries of the regions of Prussia were defined by Pope Innocent IV in two bulls, dated 30 July and 8 October 1243. The first was addressed to the bishop of Prussia (CDW, I, 8–9, no. 6, and Theiner, I, 36–37, no. 76) and the second to the Grand Master of the Teutonic Knights (CDW, I, 10, no. 7). On 7 June 1333 the first bull was inspected in Frombork Cathedral by the bishops of Varmia and two neighboring dioceses (CDW, I, 437–438, no. 263). But Bishop Johannes Streifrock of Varmia later informed the papacy that this bull had been lost (the original was in the Koenigsberg secret archives as late as 1860). Pope Innocent VI responded by having his predecessor's bull copied and incorporated in a bull of his own, which was sent to Varmia on 12 December 1355 (Theiner, I, 563–565, no. 751, from the original in Pope Innocent VI's archives, and CDW, II, 232, no. 229). ODA J 7 is a copy, dated 8 July 1426, of both bulls.

A 2   The division of Prussia by Pope Innocent IV, as confirmed by Pope Innocent VI, was repeated by the Holy Roman Emperor Charles IV in his Golden Bull of 20 (not 25, as in PI^1, 187, n. */9) August 1357, printed in CDW, II, 256–257, no. 257. Charles IV's other Golden Bull of the same date, 20 August 1357, confirming the privileges of the diocese of Varmia, was printed in CDW, II, 254–255, no. 256. The Varmia Chapter possessed parchment originals of both Golden Bulls and a parchment copy of each (ODA L 32, J 7).

A 3   The Varmia Chapter's copy (ODA L 32) of Charles IV's pronouncement regarding the diocese's privileges lacked the closing reference to the golden seal (*bulla aurea*), the seal itself, and the imperial chancellor's authentication. For these reasons Copernicus failed to recognize ODA L 32 as a copy of the bull, and listed his own A 3 as a document (*Litterae*), by contrast with A 1 and A 2, which he called "bulls."

A 4   Obłąk's identification of Copernicus' A 4 (p. 41, n. 4) with ODA C 15, fol. 17, printed in CDW, I, 47–49, no. 26, is now withdrawn in favor of ODA L 14, printed in CDW, I, 85–86, no. 48 (dated 27 January 1264). ODA L 14's references to the construction of a cathedral and the division of the Prussian lands into dioceses (*divisione ... diocesum*) resemble Copernicus' A 4's *fundatione ecclesiae et diocesum divisione*. Although the parchment original of ODA L 14 is largely illegible, the missing readings are available in ODA L 14 (2), a copy.

NOTES

B 1   ODA L 52a, the original parchment, dated 20 October 1340, and printed in CDW, I, 500–501, no. 311. Of the four seals mentioned by Copernicus, the third was missing by 1860. Five seals were attached to ODA L 52, from which CDP, III, 33–34, no. XX, was printed.

B 2   Copernicus' B 2 was identified by Obłąk (p. 41, n. 6) with ODA II 37, a copy of a document dated 4 September 1288. On 2 September 1288 a commission of four arbiters settled a boundary dispute. The parchment original of this agreement having become barely legible, in the sixteenth century a copy was made on paper (ODA A 19, printed in CDW, I, 133–136, no. 78). The copyist reported that the seals, which were five in number, had been removed (*avulsis sigillis*). By contrast, Copernicus' B 2 had eight seals, still attached. They sealed an agreement dated 28 July 1374, and settling an external dispute between the diocese of Varmia and the Teutonic Knights (printed in CDW, II, 518–533, no. 497). The 1288 agreement, on the other hand, was internal, between the bishop of Varmia and the Varmia Chapter.

B 3   The German-language version of Copernicus' B 2 (CDW, II, 518–533, no. 497, printed in parallel columns alongside the Latin version); the German version was printed also in CDP, III, 158–163, no. CXIX, from the original in the Koenigsberg archives, which had eight seals, as against Copernicus' six.

B 4   A boundary dispute between Varmia and the Teutonic Knights was committed to the archbishop of Prague by Pope Gregory XI in a letter dated 24 September 1371 (printed in Theiner, I, 667, no. 898, and CDW, II, 447–449, no. 451). Then in a letter dated 10 October 1373 (ODA L 5) the pope instructed the archbishop to hurry the appointed arbiters along. Their decision is listed by Copernicus as his own N 9. ODA L 5 is an original parchment, from which CDW, II, 494–496, no. 484, was printed. CDP, III, 151–152, no. CXIV, was printed from a slightly later copy preserved in the Koenigsberg archives of the Teutonic Knights. Theiner, I, 694, no. 933, was printed from the original in Pope Gregory XI's archives.

B 5   The archbishop of Prague, having been commissioned by Pope Gregory XI to settle the boundary dispute between Varmia and the Teutonic Knights, had a panel of arbiters named. On 16 April 1372 the archbishop wrote to the priest in Elbląg to have the arbiters solemnly swear to be impartial. This is made clear in the 1622 inventory, fol. 2ʳ/6 up–3 up, "Letter of the archbishop of Prague as apostolic delegate, subdelegating the priest of Elbląg to receive oaths from the arbiters in the matter of delimiting the territory of the diocese of Varmia and the Teutonic Knights, to be decided in 1372." But Copernicus' cursory glance at the archbishop's letter (ODA L 23) gave him the false impression that the priest was to receive witnesses (*ad recipiendum testes*) instead of swearing in arbiters (*arbitros ... ad sancta dei evangelia iurare in vestris manibus tamquam executore a nobis deputato*). The arbiters were duly sworn on 3 October 1372 by the Elbląg priest (CDW, II, 465–467, no. 462). The archbishop's letter was printed in CDW, II, 461–462, no. 460 (not 459, as in Obłąk, p. 41, n. 8). Obłąk also misidentified Copernicus' "Letter of the archbishop of Prague to the priest in Elbląg," dated 16 April, with a letter from the bishop of Olmütz, dated 15 April, to "all and each to whom the present [letter] arrives" (CDP, III, 136, no. CIII/7). CDP, III, was published in 1848, not 1836 (as in Obłąk). CDP, I, was published in 1836. After *elbing* in B 5, Copernicus wrote the letter *b*, which he deleted.

B 6   ODA Q 4, dated 28 July 1374, parchment with seals, the decision of the arbiters in the boundary dispute between Varmia and the Teutonic Knights. As announced on the following day, the decision was printed in CDP, III, 158–163, no. CXIX, and CDW, II, 518–533, no. 497. Obłąk's misidentification (p. 41, n. 9) of the arbiters' decision with ODA L 23, dated 16 April 1372, and printed in CDW, II, no. 460, is now withdrawn. Copernicus' B 6's *Pronuntiatio arbitrorum* echoes *pronuntiatio arbitrorum* in ODA Q 4. The arbiters' decision was approved by Pope Gregory XI on 16 February 1375. Toward the end of the 15th century this approval was copied on parchment (ODA G 10). From this copy, CDW, II, 540–542, no. 503, was printed, while Pope Gregory XI's archives were the source of Theiner, I, 714–718, no. 965 (a defective version).

B 7   ODA D 14, a copy (made on a Friday in April 1482) of Pope Gregory XI's bull dealing with the boundary question, in the form of a small book consisting of six folios sewn together. Gregory XI confirmed the arbiters' decision on 16 February 1375 (CDW, II, 540–542, no. 503; not in the form of a book).

B 8   Obłąk (p. 41, n. 10) identifies Copernicus' B 8 with CDW, IV, no. 258, a document in the German language which emanated from a subordinate of the Grand Master.

[Because this document emanated from a French king, at first Copernicus put it in Drawer B with papers issued by eminent personages. On second thought, however, because the French king's document dealt with a gift to Frombork Cathedral, Copernicus moved it to Drawer C, with other documents concerned with the cathedral].

C 1   ODA L 51, the original parchment, dated 30 October 1343, a joint document of the bishop and Chapter of Varmia donating in perpetuity the village later called Santoppen, with its sixty parcels and income of every sort, to the workshop engaged in the construction of Frombork Cathedral (CDW, II, 27–28, no. 29; CDP, V, 6, no. VII, from a copy). See R 5.

C 3 (a) Pope John XXII, 12 November 1329; ODA J 5, the original parchment (CDW, I, 408–409, no. 244).

(b) Pope Clement VI, 21 May 1350; ODA J 11, the original parchment (CDW, II, 159–160, no. 160).

(c) Pope Innocent VI, 12 January 1356; ODA J 17, the original parchment (Theiner, I, 568, no. 756; CDW, II, 233–234, no. 232).

(d) Pope Urban V, 23 January (not February, as in CDW, II, 422, and Obłąk, p. 42, n. 12) 1367 (not 1357, as in Obłąk); ODA J 8 (Theiner, I, 642, no. 865; CDW, II, 422, no. 411).

(e) Pope Boniface IX, 17 December 1392; ODA J 31, the original parchment (CDW, III, 233–234, no. 263).

(f) Pope Boniface IX, 12 November 1393; ODA J 13, the original parchment (CDW, III, 250–251, no. 279).

(g) Pope Boniface IX, 1 November 1394; ODA J 24, the original parchment (CDW, III, 271, no. 293).

C 4 ODA L 60, dated 11 (rather than 19, as in Obłąk, p. 42, n. 13) January 1510. St. George's head had been preserved as a relic in the chapel of the episcopal palace in Lidzbark earlier than 1432 (CDW, IV, no. 396). On 11 January 1510, however, Bishop Lucas Watzenrode of Varmia personally brought the relic in an official procession from Lidzbark to Frombork (ZGAE, 24:53, n. 1).

C 5 ODA L 11, dated 16 (rather than 15, as in Obłąk, p. 42, n. 14) May 1515, the original letter sent by Bishop Fabian to the Chapter regulating the choir's activities.

C 6 Copernicus' C 6 referred to the cross on which Christ was crucified as *lignum vitae* (the wood of life, as in CDP, IV, 28, no. XXIV/11–12, dated 30 April 1384). But in the 1622 inventory, fol. 10/4–5, the gift from King Charles V of France in 1377 is said to be "wood of the Holy Cross."

D 1 On 10 February 1508 King Sigismund confirmed the Varmia diocese's rights to the town Tolkemit and its district, and also made a grant of forty boats for catching eels (Summaria, IV, 1, p. 15, no. 245; AcTom, II, 187/12; ODA T 18/1). ODA T 2/20, dated 2 June 1503, was also cited here by Obłąk (p. 42, n. 15), although it concerns King Alexander's consent to the purchase of the town Tolkemit with its villages by the bishop of Varmia from George of Baysen (Summaria, III, 51, no.792). Obłąk cited ODA T 18/2 as well, and misdated it 26 (instead of 25) February 1519. But ODA T 18/2 concerns King Sigismund's transfer of the Tolkemit property to the Chapter (Summaria, IV, 1, p. 168, no. 2900).

D 2 ODA T 2/9, a parchment document with a seal, dated 16 April 1513, when Varmia Bishop Fabian donated Tolkemit to the Chapter. See H 5, below.

D 3 In recording this entry as a note in the 1508 inventory, Giese wrote *Scharfenberg*, which Copernicus miscopied as *Scharfenstein* (ZGAE, 1974, 37:195/11–16). George of Baysen was later given permission on 27 February 1534 by King Sigismund to mortgage or sell his properties in Codyn and Reberg, but afterwards the permission was withdrawn (Summaria, IV, 3, p. 7, no. 17465).

D 4 Baysen, an Old Prussian place name, was adopted as their own name by the family of a brother of Bishop Henry Fleming of Varmia. On 10 July 1289 the bishop granted his brother 110 parcels *in campis Baysen* (CDW, I, 141/last line). The tombstone of Bishop Henry Fleming, who died on 15 July 1300, was formerly mistaken for Copernicus' (CDW, II, 53–54, n. 1).

D 5 *Claukendorf* was misread as Claudiendorf by Obłąk, p. 42 (ZGAE, 1974, 37:195/17). The village owed its name to the landowner Clauko Hoenberg (Hohenberg; CDW, II, 176, n. 1), Clauko being a nickname for Nicholas. On 7 December (not 1 July, as in Obłąk, p. 42, n. 19) 1518 King Sigismund consented to the purchase of Claukendorf by the bishop of Varmia from the municipal authorities of Elbląg (Summaria, IV, 2, p. 195, no. 11980; ODA C 44).

D 6 Copernicus' D 6's *Consensus exemptionis districtus Tolkemit* agrees fairly closely with *consensus ... eximendi oppidum Tolkmith cum villis*, a summary of King Alexander's action on 2 June 1503 "approving the purchase of the town Tolkmith with its villages" (Summaria, III, 51, no. 792). At that time Bishop Lucas Watzenrode of Varmia was permitted to *buy* the Tolkemit district "from the hands of George Baysen." Copernicus' *exemptionis* was mistranslated as *egzempcja* (exemption; Obłąk, p. 64/1). Obłąk (p. 42, n. 20) also mismatched Copernicus' D 6 with ODA T 18/2, which concerns, not the purchase of the Tolkemit district, but the transfer of Tolkemit to the Chapter (see D 1, above).

D 7 ODA J 6, dated 23 or 26 October 1518. Crebisdorf and Carsau were two of the villages granted by King Casimir IV for the construction of a monastery of St. Brigit in Elbląg (Summaria, I, 48, no. 928, dated 23 December 1472). But the project came to an end when the villages were abandoned. In this condition they were bestowed on the bishop and Chapter of Varmia by King Sigismund on 10 February 1508 (for the bishop's own account, see SRW, II, 162–163). The Order of St. Brigit, however, filed a claim in Rome for the property, which Bishop Fabian of Varmia then had to cede to the monastery of St. Brigit in Gdańsk (to the left of Copernicus'

D 7, a marginal note states that the villages "were given to the brethren of St. Brigit"). On 7 December 1518 this assignment was approved by King Sigismund. Yet on 25 February 1519 he transferred the villages to the Chapter (an action which was later overruled; Summaria, IV, 1, p. 15, no. 245, where Barsczaw is a mistake for Carsau; p. 168, no. 2900; IV, 2, p. 195, no. 11979; SRW, I, 247–248, n. 146). See AcTom, VIII, 112/16 up–9 up, not dated (1526, probably July), and the 1622 inventory, fol. 2ᵛ/8 up–5 up.

D 8   Obłąk's identification (p. 43, n. 22) of Copernicus' D 8 with ODA T 2/20, dated 2 June 1503, is now withdrawn. Conradswalt was usually associated with Reichnaw (Summaria, IV, 2, no. 10257, 10627, 10628).

D 9   Obłąk's identification (p. 43, n. 23) of Copernicus' D 9 with ODA C 1, fol. 132, dated 28 July 1349 (CDW, II, 136, no. 133) is now withdrawn. Copernicus' D 9 speaks of a *Schonebuche* mortgage, whereas CDW, II, 136, no. 133, says nothing about a mortgage in *Schonemburch*.

D 10 King John Albert on 15 January 1495 consented to the purchase of the royal villages Conradswalt (and Reichnaw) by Simon Rabenwalt, an official of Elbląg, from an Elbląg citizen (Summaria, II, 29, no. 481; ODA K 5).

D 11   The 1622 inventory, fol. 6ʳ/8–10, lists "King Alexander of Poland's consent to the purchase of Tolkemit, Conradswalt, Reichenau, and other villages for Bishop Lucas [Watzenrode] of Varmia, 1503." Watzenrode bought Conradswalt before he died on 29 March 1512. Several months later, on 9 July 1512 and again on 18 September 1515, King Sigismund permitted Conradswalt to be bought from the Varmia Chapter by Elbląg (Summaria, IV, 2, no. 10257, 10627, 10628).

D 12   ODA M 3/1, a parchment with seal, expressing King Sigismund's approval on 31 March 1510 of the purchase of the village Maybom for Bishop Lucas Watzenrode. Copernicus' D 12 appears in the 1622 inventory, fol. 6ᵛ/4 up.

E 1   Copernicus wrote *ad* above the line before *Stum*, and after it he wrote *villas*. On 28 August 1508 King Sigismund consented to the purchase by Varmia Bishop Lucas Watzenrode of "all the villages surrounding the castle in Stum from the hands of whoever holds them" (Summaria, IV, 1, p. 29, no. 452; AcTom, I, appendix, p. 29, no. 16/10–11).

E 3   ODA J 28, dated 15 June 1404, and recording the purchase by the Chapter administrator from an Olsztyn citizen of a plot of land, together with one garden (*cum uno ortu*), near the Chapter's castle in Olsztyn (CDW, III, 388–389, no. 398). The administrator also bought a granary outside the town (*horrii foris dictum oppidum*). Copernicus mistakenly says "two granaries" (*duorum horreorum*). When Gutstadt was founded on 26 December 1329, for every "inhabitant known to possess parcels," one parcel was exempted from taxation "for gardens and granaries" (*pro Ortis et Horreis*; CDW, I, 410/8 up). For a vegetable garden (*ortum olerum*) and a brick granary (*orreum laterum*), see CDW, I, 1/6 up, 425/6 up.

E 4   Obłąk's identification (p. 43, n. 28) of Copernicus' E 4 with ODA 11–13, dated 23 March 1440, is now withdrawn.

E 5   Copernicus' *quedeliz* (ZGAE, 1974, 37:196/2) was misread as Quedelig by Obłąk (p. 43, n. 29), and also misinterpreted as the name of a place (Quedelitz). But on 20 (not 19: *fer. 4 post dom. Invocavit*) February 1483 the Chapter bought ten parcels of a *Curia* from Andreas Quedlitz (CDW, II, 547, n. 2, from ODA F, fol. 138). The 1622 inventory, fol. 9/5–6, lists an "Extract from the bill of sale of the estate of Lesser Scaibot by Quedlitz, with details about the payments made by the offices."

E 7   ODA P 30, dated 21 January 1386 (CDW, III, 153–154, no. 190; not no. 191, as in Obłąk, p. 43, n. 31). The Varmia Chapter improved the conditions under which an individual held land in the village Glanden. Glande was the name of an Old Prussian (as contrasted with a German), who owned land in 1326 and became an overseer in 1337 (CDW, I, Regesta, p. 135, no. 360; Diplomata, p. 473).

E 8   ODA P 61, the original parchment, dated 25 July 1363 (CDW, II, 357, no. 348). The text of ODA P 61 refers to the "village called Kabe," which is also the name of the Old Prussian who received this grant. The village did not survive. Is this the reason why Copernicus placed a row of dots after *Gabelen*?

E 9   ODA P 26, dated 6 October 1301 (CDW, I, 215–217, no. 121). Copernicus' E 9's *Privilegium villae Pilgrimsdorf* repeats the opening words of the inscription on the verso of the parchment original. A copy of the original is in ODA F, fol. 24.

E 10   ODA P 55, dated 13 March 1287 (CDW, I, 127–129, no. 75). This is the original, which bears an inscription pointing out that the "Varmia Chapter bought the estate of Beber" (*Curiam Beber*), these last two words being echoed by Copernicus. CDP, IV, 53–54, no. XLIV, was printed from a defective copy, and was misdated 1387. On 16 February 1397 the Chapter bought property in Beber, with the purchase money being returned for divine services. Then on 21 January 1410 the Chapter sold the property (CDW, III, 338–339, no. 37; 456–457, no. 454).

E 11   ODA P 59, the original parchment, dated 6 May 1304 (CDW, I, 227–228, no. 129). Three-eighths of the mill were bought for the Chapter in 1450, and the remaining five-eighths by Stockfisch, who wrote a note to this effect on the verso of ODA P 59, in 1500, when he was the administrator of the Chapter. His will was listed in the 1622 inventory, fol. 24^v/2.

E 12   ODA J 22, dated 6 (not 12, as in Obłąk, p. 44, n. 36) April 1449, original "decision ... concerning the right to fish in the mill pond of Borniten." The 1622 inventory, fol. 19^r/6–7, makes it clear that the decision went "in favor of the Chapter."

E 13   ODA S 31, dated 29 (not 4, as in Obłąk, p. 44, n. 37) August 1442. The 1622 inventory, fol. 19^r/8 up–6 up, makes it clear that this decision affected the "ownership of the mountain" as well as the woods of Birckpusch.

E 14   ODA L 73, dated 20 August 1498 (not 18 August 1488, as in Obłąk, p. 44, n. 38).

[After E 14, Copernicus wrote a two-line entry, which he deleted because he transferred it to F 1. His deletion consisted of a long horizontal stroke through the first line, but in the second line he left *Baisen* undeleted.]

E 15   The estate of Posorten was sold in Olsztyn on 13 July 1449 by Johannes Plastwich, administrator of the Varmia Chapter and author of the *Chronicle of the Lives of the Varmia Bishops* (SRW, I, 13/9–11; ODA P 6). ODA P 31 has seven documents concerning Posorten and ranging in date from 1448 to 1507.

E 16   ODA L 83a/2, original document with seal, dated 1442. Its phraseology (... *impetitionis ... super censibus retardatis ... de molendino in Kaltenflÿss*) is echoed in Copernicus' E 16 (*Impetitio census retardati in molendino Caldemflis*). Obłąk's date, 26 July 1447, and his identification (p. 44, n. 40) of Copernicus' E 16 with ODA L 83 are now withdrawn. Caldemflis took its name from the mill stream's low temperature (... *fluvio, qui frigidus Rivulus appelatur*; CDW, I, 424/3 up–2 up, dated 1331). See E 17 and O 6, below. The 1622 inventory, fol. 8^v/6–7, lists a "Document concerning the decision and adjudication of the Kaldeflüss mill in favor of the Chapter."

E 17   ODA L 83a/1, undated, in German, entitled "in the matter of Peter Polen." The 1622 inventory, fol. 22^v/1–2, lists "Documents in the suit of Peter Polen against the Chapter and the citizens of Olsztyn."

E 18   ODA L 83, dated 23 July 1447, and entitled "Document concerning the claim and award of the mill in Kaltflüs to the Chapter." Copernicus' E 18's *Litterae devolutionis eiusdem molendini ad Capitulum* echoes ODA L 83's *Litterae ... molendini ... pro Capitulo*.

F 1   ODA H 1/8, dated 9 March 1439. Funds for the hours of Our Lady were provided by the Chapter after 1470 (CDW, II, 289, n. 1) and also by Bishop Lucas Watzenrode (SRW, II, 153, ¶ 2).

F 2   ODA P 43/19 (CDW, I, 299–300, no. 173) records the Varmia Chapter's confirmation on 21 May 1315 of a sale by Johannes Padluche of 22$^1/_2$ parcels. Copernicus' F 2's "Donation of the village Padelochen" is not among the documents listed by Obłąk (p. 44, n. 42) as present in ODA P 43. The 1622 inventory, fol. 9/5 up–4 up, refers to "documents of the arbitration decision confirming the village Padeluchen for the Varmia Custodian."

F 4   The three copies are now identified with ODA P 43/18, dated 30 April 1449. The decision was rendered in Malbork by Archbishop Sylvester of Riga, Abbot Nicholas of Oliva, and Eberhard of Weşenthau, commander in Balga (Johannes Voigt, *Namen-Codex der Deutschen Ordens-Beamten*, p. 21/8).

F 5   Copernicus' *Peuthuen* (ZGAE, 1974, *37*:196/2) was misread as Peuthnen by Obłąk (p. 44/14, n. 45). His identification of Copernicus' F 5 with CDW, II, no. 289, is now withdrawn, since that document belongs with Copernicus' F 6.

F 6   ODA V 2, dated 18 August 1360 (not 1359, as in Obłąk, p. 44, n. 46, and CDW, II, 289–290, no. 289).

F 7   ODA V 2a, documents ranging in date from 1447 to 1484. Obłąk's date, 30 January 1378 (p. 44, n. 47), is now withdrawn.

F 8   ODA C 1, fol. 7 (CDW, I, 92–95, no. 54), a copy of a land grant to a brother of Bishop Henry Fleming of Varmia. The grant is dated 1278, with no indication of the month or day. The copy is labeled *Privilegium super Sandekow*, agreeing with *Copia privilegii Sandecaim*, Copernicus' label, which recognized the document as a copy.

F 9   ODA L 30, now dated 1 October (rather than 24 September, as in Obłąk, p. 44, n. 49) 1497.

G 1   The original document, dated 27 July 1289 (CDW, I, 149, no. 83), was damaged by bookworms. Hence it was renewed on 19 (not 25, as in Obłąk, p. 45, n. 50) August 1404 (CDW, III, 391, no. 401 [1]; ODA W 1b); on 10 September 1487 (ODA W 1a); and on 9 April 1488 (ODA W 1d).

G 2   ODA W 1b, 3 August 1513, or 12 January (not 11 January, as in Obłąk, p. 45, n. 51), 1501.

G 3   ODA P 41, dated 24 December 1443, is entitled: "On behalf of the Varmia Chapter, judgment rendered against Fabian of Wusen and others for the payment of the rent of six marks." Copernicus' G 3 (*Sententia contra Fabianum occasione census ibidem* [= Wuszen]) echoes ODA P 41's ... *sententia ... contra Fabianum ... census*. Obłąk's identification (p. 45, n. 52) of Copernicus' G 3 with ODA W 1c is now withdrawn.

G 4   ODA W 1e, dated 24 December 1443. Its title, "Implementation of the decision regarding the mill in Wusen, against Fabian" (*Instrumentum sententiae super molendino ... contra Fabianum*) is echoed by Copernicus' G 4's *Instrumentum sententiae contra eumdem super molendino*.

G 5   The numerous properties bought by Fabian Tolke from his stepfather on 17 June 1490 included four parcels in Wuszen (SRW, II, 18/5 up–4 up). Those four parcels were mortgaged by Fabian Tolk(e) on 10 March 1507 (ODA, W 1g, 1h, captioned *Impignoratio mansorum quatuor in Wuszen per Fabianum Tolk*). These captions are echoed by Copernicus' G 5's *Impignoratio mansorum iiij ibidem* [= Wuszen] *per Fabianum Tolke*. *Tolk* is an obsolete German word meaning "interpreter."

G 6   ODA S 25, the original parchment, dated 11 November 1362 (CDW, II, 344–345, no. 333). A note on the verso reports that the parchment was given to the Chapter on 27 May 1487, when it made a first payment of 400 marks of good money toward the purchase of the village Schayboth (Scaibot).

G 7   ODA L 67. Its caption, *Litera solutionis villae Scaiboth*, was echoed by Copernicus' G 7's *Litterae solutionis eiusdem* [= *villae Scaibot*]. ODA L 67, which was dated 1430 by Obłąk (p. 45, n. 55), is now dated 1485.

G 8   ODA L 81. Obłąk's date, 19 March 1433 (p. 45, n. 56), is now withdrawn, and replaced by "Thursday, the third [day] before Laetare [Sunday] 1483" (6 March 1483). ODA L 81's caption, *Literae emptionis curiae Scayboth*, is matched by Copernicus' G 8's *Litterae emptionis curiae Scaibot*. A deleted note (read as *est supra Quedelig* by Obłąk, p. 45, n. m) in the left margin alongside Copernicus' G 8 refers to *superior Quedeliz*, Upper Quedlitz. This designation resembles *Scaybot inferior*, with which Quedlitz was identified in the caption of ODA F, fol. 148 (CDW, II, 547, n. 2). Scaibot is not far to the northeast of Quedlitz.

G 10   ODA C 101, 20 May 1482. Its caption, *Copia venditionis duorum mansorum in Engelswald pro domino Christiano*, is closely echoed by Copernicus' G 10's *Copia venditionis mansorum in Engelswalt pro q Christianno Tapiau*. Christian (not Christopher, as in Sikorski, pp. 16, 147, 148, 157) Tapiau was the chaplain of the castle in Olsztyn in June 1449. On 16 and 18 March 1456, when he held the title of vicar, he was one of the witnesses summoned to testify against George of Schlieben (SRW, I, 164/17, 165/14). By 19 November 1460 he was a Varmia canon. He served as the Chapter's custodian from 20 January 1467 to 8 June 1475, when he became the Chapter's dean. He held that post until his death on 16 May 1498 (SRW, I, 239, n. 112; II, 27, n. 1: 18 May in II, 47, n. 1, is wrong). PI[1], 266, n. */3–4, dated his death in 1497, thereby overextending the interval before his property in Zagern was opted by another canon on 7 February 1499; see ZGAE, *3*:356/5 up–3 up.

G 11   ODA L 64, dated 26 May 1486.

G 12   ODA P 17, the original parchment, dated 3 November 1366 (CDW, II, 417–418, no. 405). The Chapter, having previously bought twenty parcels and the mill in Schaphisberg, sold them "for a certain sum of money counted out for us by him [the purchaser] and paid in full." Copernicus' *Schouffsberg* (ZGAE, 1974, *37*:196/2–3) was misread as Schouffburg by Obłąk, p. 45/14, n. 59.

G 13   ODA L 15 (SRW, I, 213–220). This register was completed on 12 May 1393.

G 14   Identified with Y 4/1, fol. 24^{r-v}, in the 1622 inventory.

H 1   On 23 January 1384 the bishop and Chapter of Varmia jointly formulated a set of statutes (CDW, III, 119–127, no. 165, with later modifications at pp. 323–344, no. 358). A compact version (CDP, V, 27–28, no. XXVI) was submitted to the papacy under the original date, and was approved on 13 November 1393 by Boniface IX (CDW, III, 251–253, no. 280; ODA B 22, the original parchment).

H 2   The bishop of Varmia petitioned the papacy for permission to disburse certain emoluments in the form of bread, beer, wine, and cash (*denariis ... custodialibus*), not derived from the canons' main stipends (*corporibus prebendarum*). On 25 February 1373 Pope Gregory XI approved these distributions (*distributionibus*; Theiner, I, 682–683, no. 922, from a papal source; CDW, II, 480, no. 473, from ODA B 3). The expressions quoted above from ODA B 3 are echoed in Copernicus' H 2.

H 3   ODA H 2, the original parchment, dated 2 December 1343, the bishop of Varmia's approval (CDW, II, 33–34, no. 31) of the Varmia Chapter's unanimous resolution on 12 November 1343 that the custodian's fund *(pecunia custodialis)* should be distributed only to canons in residence. CDP, V, 7, no. VIII, was printed from a copy.

H 4   This testimony was apparently given in connection with the rule adopted in H 3, above.

H 5   ODA T 2/1, dated 5 September 1513, when Bishop Fabian discussed the distribution of the income derived from the grant he made in Copernicus' D 2.

H 6 Above H 6, in smaller letters Copernicus wrote: This "is an instruction to appeal against Jo[hannes] Rex," who was appointed a Varmia canon by the papacy on 25 August 1409 (SRW, I, 221, n. 25; CDW, III, 450, no. 448). An appeal may have been lodged against Rex, previously a canon of Włocławek, in connection with the obligation of a Varmia canon to complete three years of study at a recognized institution of higher learning. Rex died in 1447. Although his tombstone states that he died on 25 October (SRW, I, 221, n. 25), he was commemorated on 13 October (SRW, I, 242/9 up). See Q 5, below.

H 7 ODA D 33, dated 30 June 1429 (CDW, IV, no. 32), and labeled *Concordia inter episcopum et capitulum super curiam episcopalem*, in close agreement with Copernicus' H 7's *De curia canonicali sive episcopali ... concordia episcopi et Capituli*.

H 8 The parchment original, dated 2 September 1288, having become barely legible, a copy was made on paper in the 16th century (ODA A 19). The "other matters" dealt with in this agreement, in addition to the arrangements for the election of the canons, included the division of the diocese of Varmia into two parts, with $^2/_3$ going to the bishop and $^1/_3$ to the canons (CDW, I, 133–136, no. 78).

H 9 ODA A 19, a copy of Copernicus' H 8. CDW, I, 133–136, no. 78, was printed from the copy.

I 1 The cantor of the Varmia Chapter in 1498–1499 was Johannes Sculteti. He was deprived of his prelacy by George of Delau, who later wrote the Chapter's fourth inventory (see Inventory, Introduction, n. 2). To compensate Canon Sculteti for his loss, Bishop Lucas Watzenrode created the archdeaconate for him. After papal permission had been received, the Chapter approved on 26 December 1502, and three days later the bishop named Sculteti archdeacon (ZGAE, 3:594–595). In the 1622 inventory, Y 4/1, fol. 43^v, deals with this subject, and Y 4, fol. 15^v/3 up–2 up, lists the "rights of the vicariate of St. Jerome founded in 1510 by the late Johannes Sculteti, archdeacon."

I 2 ODA C 93, Pope Innocent VIII's confirmation, directed to the Varmia Chapter in 1489, of the election of Lucas Watzenrode as bishop of Varmia. Writing in 1520, Copernicus refers to his uncle as the late bishop, because Watzenrode died on 29 March 1512. He had been elevated to the bishopric on 18 May 1489 (Eubel, II, 288).

I 3 On 31 January 1489 Varmia Bishop Nicholas Tungen resigned for reasons of ill health in favor of Lucas Watzenrode; see Hans Schmauch, "Der Streit um die Wahl des ermländischen Bischofs Lukas Watzenrode," *Altpreussische Forschungen*, 1933, 10:69/22–24. ODA C 48 is a contemporary copy of the resignation.

I 4 ODA P 1, dated 7 December 1512. To the left of I 4, in the margin, an undeciphered word connects I 4 with I 5. Were the articles mentioned in I 5 referred to in I 4?

I 5 As soon as the Varmia Chapter heard the news that Bishop Lucas Watzenrode had died far from Frombork on 29 March 1512, it hastily convened as many canons as could be reached to choose his successor from among themselves, in order to prevent the Polish king from intervening. Before proceeding to the election on 5 April 1512, all the canons, including Copernicus, swore to abide by the newly adopted articles governing the new bishop, whoever he might be. Among those who signed was Canon Fabian. After the election he signed again as Bishop Fabian. Hence the oath preceded the election on the same day (not vice versa, as in PI^2, 34–35). The original parchment, dated 5 April 1512, is preserved in the Czartoryski Library, 716, II, no. 32. ODA A 4/1 is a sixteenth-century copy. The Latin text was published by Marian Biskup, *Rocznik Olsztyński*, 1972, 10:297–303, with a translation into Polish at pp. 303–310. Plates I–II (pp. 294–295) show the beginning of the document and Copernicus' signature (Plate II/4 up).

I 6 ODA C 16. On 7 April 1513 King Sigismund of Poland instructed his papal emissary to obtain confirmation of his agreement with the bishop and Chapter of Varmia regarding the future election of bishops in that diocese (AcTom, II, 194, no. 236).

I 7 Copernicus started to write this surname as Vend-, which he deleted. Venrade, an emigrant from Flanders to Varmia, was born in 1398, as is known from a document stating that he was 40 years old on 18 August 1438. By 1421 he was a secretary of the bishop of Varmia, and in 1425 his notary. By 30 July 1437 he was a canon of Varmia, and by 3 November 1448 the Chapter's cantor. In 1457 he was an unsuccessful candidate for bishop of Varmia, his candidacy being the occasion for this letter. The 1622 inventory, fol. 21^r/last 2 lines, makes it clear that there were others, in addition to the Chapter, who wrote on behalf of (not merely about) Venrade, in the year 1458. His will is listed in the 1622 inventory, fol. 24^r/7. It was dated 15 May 1461 (1561 is a slip in ZGAE, 3:587/17; see also ZGAE, 1:130–131, and SRW, I, 273, n. 224). Obłąk's identification (p. 46, n. 72) of Copernicus' I 7 with ODA A 18, dated 27 January 1458, is now withdrawn, since ODA A 18 concerns a document in Copernicus' I 8, not I 7.

I 8 ODA A 18, a letter written on 27 January 1458 by Paul, the newly elected bishop of Courland, to the College of Cardinals about the election of the new bishop of Varmia (ZGAE, 1:136). The other three letters are missing.

NOTES

K 1   In 1410 King Ladislaus of Poland wrote a testimonial letter to the effect that Henry Heilsberg of Vogelsang, bishop of Varmia, had not engaged in any plot against the Teutonic Knights. This letter was registered as L 22 in the list compiled in 1789 of the documents in the Chapter's archives in Frombork, but was missing when the research work was done three-quarters of a century later for CDW, III, 510, no. 498 (1). The 1789 entry reads: *Litterae domini Wladislai regis Poloniae testimoniales pro Episcopo Varm. Henrico, quod nihil sit machinatus contra Ordinem 1410*, with Copernicus omitting the second and third words as well as what follows *episcopo*.

K 2   By the same token, Copernicus' K 2 (*Litterae testimoniales magni ducis Lituaniae*) is an abbreviated form of L 9 in the 1789 list (CDW, III, 510, no. 498 [2]). It may have been removed from Frombork by Tadeusz Czacki when he was there in 1802. For whereas the editor of CDW, III, reported it as missing, it turned up in the Czartoryski Library in Cracow, vol. 2956, no. 287, fol. 6. It was printed in *Monumenta medii aevi historica res gestas Poloniae illustrantia*, XII = *Codex epistolaris saeculi decimi quinti*, ed. Anatol Lewicki, II (Cracow, 1891), 50–51, no. 45. Like his cousin the king of Poland, the grand duke of Lithuania wrote a testimonial letter on 17 January 1412 defending the bishop of Varmia against the accusation that he had plotted against the Teutonic Knights.

K 3   On 5 May 1464 King Casimir of Poland confirmed his agreement with the bishop of Varmia that he was taking this diocese under his protection. The king's confirmation was summarized in Erich Weise, *Die Staatsverträge des Deutschen Ordens in Preussen im 15. Jahrhundert*, II (Marburg, 1955), 258, no. 395. The document handled by Copernicus is no longer in ODA, which instead possesses a copy made early in the 17th century (ODA T 1, fol. 42–43). See K 10.

K 4   At the battle of Tannenberg/Grunwald in 1410, King Ladislaus of Poland defeated the Order of Teutonic Knights, and in 1411 he signed a treaty of peace with the Knights. A copy of this treaty was acquired by the Varmia Chapter, whose interests were vitally affected. This copy may have been among the documents in the Chapter's earliest inventory (about 1450). In the 1508 inventory it appears as the following entry:

> One copy on parchment of the [treaty of] peace between King Ladislaus of Poland etc. and the Order [of Teutonic Knights], which was made in the year 1411
> *Copia una in pergameno de pace inter regem Ladislaum Poloniae etc. et ordinem facta anno MCCCCXI*

In the space left blank between this entry and the next, a later hand inserted the following note:

> This copy was sent to the bishop by Nic[holas] Copp[er]nic [and] Felix [Reich], the notary, with the consent of the inspectors in the year [15]11
> *Haec copia missa fuit d[omi]no e[pisco]po p[er] d[ominum] Nic[olaum] Coppernic [et] felicem notarium de voluntate dominorum visitatorum anno [MD]XI*

According to Obłąk (pp. 15–16), this note was written by Copernicus. Had he written it, he would have followed his usual practice in official documents by saying that the copy was sent to the bishop "by me" (*per me*), Nicholas Coppernic (cf. Leases, Olsztyn, 1517, 1518, 1519; Melsac, 1517, 1518). Besides, the note's handwriting differs from Copernicus' in his autograph manuscript of the *Revolutions* (facsimile in NCCW, I) and agrees with Giese's (facsimiles in *Locationes*, Plate XV, and NM, Plate XXIVe). While Giese was the Chapter's administrator from 1510 to 1515, he did not make an inventory of his own, but chose instead to add some notes, including this one, to the 1508 inventory. It was Giese, not Copernicus, who wrote our note.

Obłąk also maintained (p. 15) that "Copernicus copied the peace treaty with his own hand." But the Chapter's copy was listed in the inventory of 1508, whereas the earliest document attesting Copernicus' presence in Olsztyn is dated 1 January 1511 (PI¹, 381; PI², 256). Hence the Chapter's copy was not written by Copernicus' own hand. Obłąk, however, imagined that Copernicus made a duplicate of the Chapter's copy, and that the duplicate was sent to the bishop. StC VIII (219, no. 64a) uncritically repeated that "Copernicus ... delivers a duplicate of a copy of the peace treaty ... to the bishop." But Giese's note in the 1508 inventory states explicitly that "This copy (*Haec copia*, the Chapter's copy) was sent to the bishop." If the Chapter's copy was duplicated, that was done in the bishop's residence in Lidzbark, not by Copernicus in Olsztyn. To this very day, ODA has no duplicate. It has only one copy (V 8), which was probably acquired soon after the treaty was signed, registered in the 1508 inventory, and again in Copernicus' inventory of 1520.

In addition, Obłąk mistranslated *de voluntate dominorum visitatorum* in the note as "with the Chapter's consent" (*za zgodą Kapituły*). The Chapter's consent was not requested with regard to this transfer of the document. In Obłąk's interpretation the part was confused with the whole, since the Chapter consisted of 16 canons, only two of whom were appointed as inspectors (*visitatores*) every year (CDW, III, 123, no. 25/1–3; 126, no. 41/1–3; 332, no. 23; Hipler, p. 258, no. 38; PII, 512–513, no. 38). It was "with the consent of the inspectors in the year 1511" that the Chapter's copy of the peace treaty was sent to the bishop. It so happens that Copernicus was one of the two inspectors in 1511, the year mentioned in our note:

> A.D. 1511, by order of the venerable Chapter, we, Fabian of Lussigein [Lossainen, Lutzingheim] and Nicholas Copernicus, appointed as inspectors [*visitatores*] by the venerable Chapter, in Olsztyn ... received ... money ... (PI^1, 381; PI^2, 256).

It was with the consent of these two inspectors of 1511, Fabian and Copernicus, that the Chapter's copy was sent to the bishop.

By contrast with this unexciting account of what actually happened, Obłąk imagined a dramatic scenario. He eliminated Reich from the entry by misinterpreting his given name as the adjective *felix* (happy). That elimination left a nameless notary in the note. Ignoring Felix Reich's self-description as a "public notary by the sacred authority of the pope and the emperor" (NM, p. 34), Obłąk made a notary out of Copernicus, who was never so licensed. In becoming a notary, Copernicus must also be happy, since in Obłąk's version the sequence was *Coppernic felicem notarium*. Copernicus the notary was happy because he was handling a document which put him in emotional rapport with King Ladislaus, the conqueror of the Teutonic Knights. But Copernicus deleted the note because a notary must not show his feelings in official documents. This combination of misconceptions was hailed as a "sensational discovery" by Piotr Bańkowski, *Archeion*, 1974, *9*:61–80.

By contrast with the case of Copernicus, his fellow-canon Stockfisch provides a useful example. Before becoming a canon on 18 May 1489 Stockfisch signed a document on 1 September 1484 as a notary (CDW, I, 150–151, n. 2).

By the same token, before becoming a Varmia canon on 29 October 1526, Felix Reich functioned as a notary. In this capacity he served Bishop Lucas Watzenrode of Varmia on 16 January 1510 (ZGAE, 1974, *37*:197/ 18–21). Through him, as the bishop's notary, and through Copernicus, the bishop's nephew, the Chapter's copy of the document was sent from Olsztyn, with the consent of the second inspector, to Lidzbark, the bishop's residence. When the Chapter's copy was later returned to Olsztyn, the note reporting its transmission was deleted. In the right margin above the deletion and alongside the entry proper, *est* (it is here) was written to indicate that the document was in place. It was still there in 1520, when Copernicus listed it as K 4 in his own inventory.

K 5   Before World War II, ODA C 97, but missing since the war; printed in Weise, *Staatsverträge*, III (Marburg, 1966), 60–63, no. 454.

K 6   On 20 January 1467 in Olsztyn the Varmia Chapter adhered to the "perpetual peace" following the Thirteen Years' War (1454–1466; Weise, *Staatsverträge*, II, 296, no. 409).

K 7   Two letters (ODA S 30, A 11) from King Matthias Corvinus of Hungary to Bishop Nicholas Tungen of Varmia. One, in code, is dated 1476. The other, not in code, is dated 12 March 1477. Since they were described by Copernicus as a bundle (*fasciculus*), were they separated from each other later on? The 1622 inventory, fol. 21^r/4 up–3 up, lists the "Documents of the alliance between Bishop Nicholas Tüngen and King Matthias of Hungary."

K 8   Vincent Kelbas (Kiełbasa), a secretary of the king of Poland, on 1 April 1467 was confirmed as bishop of Chełmno by Pope Pius II (Eubel, II, 156: Vincent Goslawski; see ZGAE, *6*:409–410). A few months later, when the bishopric of Varmia became vacant with the death of Paul Legendorf on 23 July 1467 (ZGAE, *1*:148), the appointment of Kiełbasa as the administrator of the diocese pending the election of a new bishop was unanimously requested by the Chapter (ODA L 85; Thunert, pp. 62–65), with the West Prussian Estates granting the request on 2 December 1467. Shortly thereafter the Chapter formally acquiesced (ODA K 3; Thunert, pp. 562–563). ODA K 3 is not dated. Obłąk's date, 1471 (p. 47, n. 76), is wrong, December 1467 being more likely, since ODA K 3 refers to ODA L 85, which the Chapter promises to observe faithfully. ODA has a copy, dated 4 November 1468, of Kiełbasa's acceptance of the administratorship.

K 10   In 1454 the Prussian League revolted against the Teutonic Knights and formed an alliance with King Casimir IV of Poland, whose overlordship was accepted by the League. On 11 June 1454 the diocese of Varmia, together with the other Prussians, swore allegiance to the Polish crown (Summaria, I, 13, no. 217; Weise, II, 146–147, no. 301). See K 3, above.

L 1   ODA J 19, the original parchment, dated 3 May 1415 (CDW, III, 508–510, no. 497). Bishop Henry Vogelsang wanted a notary to execute one or more public instruments (*instrumenta*) concerning his bequest. Copernicus describes L 1 as an *instrumentum*. But the 1622 inventory, fol. 10^v/1, refers instead to an "Inventory of Bishop Henry's silver, bequeathed to the Chapter in 1415."

L 2   Bishop Henry Sorbom of Varmia on 15 September 1398 bequeathed "306 $^3/_4$ marks of silver" objects to the Chapter, which could alienate them only by renting them to Bishop Henry's successors. The original parchment of the bequest (ODA, Wills, I, F, no. 19; CDW, III, 305–307, no. 333) describes itself as an *instru-*

mentum, the term used by Copernicus for L 2. In the 1622 inventory, fol. 10^r/5 up, the weight of the silver was given as 317 marks.

L 4   Copernicus' unnecessary stroke over the *a* changes the correct form *impignorato* to an incorrect form *impignoranto*. Bartholomew Libenwald, a Varmia canon, wrote a *Memorandum (Memoriale)*, in which he reported that in 1465

> I was sent by the Chapter to the prelates of Livonia ... in order to borrow money for the defense of the castle in Olsztyn, and to pawn jewels (SRW, I, 309/6–10).

From the list in ODA S 1, fol. 68, it is clear that the jewels in question were religious objects made of silver.

L 5   ODA C 20 has three copies of this document, dated 15 and 17 July 1461. See L 13, below. Obłąk's reference (p. 47, n. 79) to ZGAE, *1*:136, may be ignored.

L 6   ODA B 21, dated 28 (rather than 12) February 1483.

L 8   ODA C 98, a rescript of Pope Innocent VIII, dated 9 (rather than 5) April 1488.

L 9   After defeating the Teutonic Knights at the battle of Tannenberg/Grunwald in 1410, the victorious armies of Poland and Lithuania withdrew from Prussia. Taking advantage of this opportunity, the Knights seized Varmia. The resulting dispute was referred by the Knights and Poland-Lithuania to Sigismund, the Holy Roman Emperor, for settlement. The part of his decision, rendered on 24 August 1412, that concerned Varmia was printed in CDW, III, 486–487, no. 475. Emperor Sigismund commissioned Benedict of Macra to execute his decision. On 13 November 1412 the bishop of Varmia, still in exile, named deputies to ask Benedict to carry out the emperor's decision insofar as it affected Varmia (CDW, III, 487–489, no. 478). On 20 March 1413 Benedict issued the desired document (Czartoryski Library, no. 295; *Lites ac res gestae inter Polonos Ordinemque Cruciferorum*, III [Warsaw, 1935], p. 217/30–38). Obłąk's identification (p. 47, n. 82) of Copernicus' L 9 is now withdrawn.

L 10   ODA J 10, dated 5 November 1411 (CDW, III, 468–470, no. 468). The bishop of Varmia left certain valuables for safekeeping in the custody of a municipal official of Elbląg, who turned this property over to the Teutonic Knights. It is itemized in the public instrument (*instrumentum*) which was executed by a notary at their request. Copernicus' L 10 repeats the designation *Instrumentum*. In the 1622 inventory, fol. 10^r/last 2 lines, the instrument specifies "jewels, golden bulls, and other objects as well as legal documents of the bishop seized by the Order in Elbląg in 1411."

L 11   The part of the emperor's decree that affected Varmia was printed in CDW, III, 486–487, no. 475, from documents not under the Chapter's control.

L 13   During the Thirteen Years' War (1454–1466) the Varmia Chapter took the precaution of sending its ecclesiastical jewels away for safekeeping. Later, when hard pressed for cash, the Chapter decided to pledge its jewels as security for a loan and, if necessary, to sell some of them. On 2 April 1465 it commissioned Canon Bartholomew Libenwalt (Libenwald) "to borrow money for the defense of the castle in Olsztyn and to mortgage the jewels," as he reported in his *Memoriale* (SRW, I, 309/8–10). For these purposes the Chapter gave him power of attorney, on the strength of which he obtained a loan from Bishop Paul Einwald of Courland on 10 May 1465. Libenwald's extensive pursuit of money for the Chapter was traced in detail in SRW, I, 309–310, n. 24. His career was sketched in ZGAE, 1866, *3*:587–592. See L 5, above.

M 1   During the Thirteen Years' War between Poland and the Teutonic Knights, a contingent of the Knights was admitted to the Varmia Chapter's castle in Olsztyn, pursuant to an agreement made on 6 January 1455 between the Grand Master of the Knights and the Chapter. But the Varmia canons complained to the Grand Master about the misconduct of his subordinates, who for their part accused the canons of wishing to turn the castle over to the Poles (SRW, II, 310–312). The canons were imprisoned for a year, and then sent bound to Koenigsberg, where they were treated as traitors. The Chapter drew up a report, and the Order's infantry commander, George of Schlieben, composed a defense of his conduct. The testimony of ten witnesses was recorded on 18–22 March 1456. Copies of these proceedings in ODA S 6 and A 8 were printed in SRW, I, 138–207, where *complices* (p. 161/3) is echoed by Copernicus' M 1's *complices*. His reference to *Schliven* (ZGAE, 1974, *37*:196/4) was misread as Schlinen by Obłąk, p. 48, M/1, n. 64. For a detailed account of the Schlieben affair, see ZGAE, 1894–1897, *11*:401–434.

M 2   ODA L 38, an agreement made on 30 September 1461 by the armed men besieged in the castle in Seeburg to surrender it peacefully to the forces of Bishop Paul Legendorf of Varmia, who was struggling to obtain control of his diocese during the Thirteen Years' War (SRW, I, 122/6 up–last line; ZGAE, 1894–1897, *11*:435).

M 3   Copernicus wrote impignatione, where *impignoratione* is required.

M 4   While Bishop Paul Legendorf was fighting to gain control of Varmia, he found the castle in Seeburg held by his opponents. On 5 July and 23 July 1461 he authorized the collection of money from the Varmia churches and clergy to buy out his opponents (ODA L 79, M 14). He issued these authorizations during his siege of Seeburg, to ransom the castle in Lidzbark (SRW, I, 122, 6 up–last line). Copernicus (or his predecessor) confused Seeburg, where the authorizations were issued, with Lidzbark, the castle to be redeemed. Copernicus' use of the term *subsidium* recalls *subsidium charitativum*, Bishop Paul's name for his fund drive. See ZGAE, 1894–1897, *11*:436. The 1622 inventory, fol. 30ʳ/last 2 lines, lists a "Document of Varmia Bishop Paul ordering a fund drive for the recovery of the castles in Rössel, Lidzbark etc."

M 5   During his fund drive, Bishop Paul Legendorf of Varmia wrote to certain priests, demanding that they should contribute the jewels of their churches to his cause.

N 1   On 24 September 1455 Pope Calixtus III ordered sentences of excommunication and interdict to be issued against the Prussian League as rebels against the Teutonic Knights (Theiner, II, 98–101, no. 142). But Pope Pius II, previously designated bishop of Varmia, withdrew that censure on 10 January 1460 (ibid., 134–135, no. 172, from a papal source). Obłąk's reference (p. 48, n. 88) to an anti-Turkish League is now withdrawn. The 1622 inventory, fol. 21ʳ/5–6, makes it clear that Pius II's absolution affected the subjects of the Varmia diocese.

N 4   ODA Q 5, dated 7 and 13 April 1430 (CDW, IV, 350–353, no. 311).

N 5   ODA C 99, dated 3 September 1496, an order by the bishop of Chełmno absolving Balthasar Scaibot from the censure placed on him by the bishop of Varmia. Balthasar Scaibot was a military vassal of the Varmia Chapter, who rebelled against it. He organized all the vassals of the Olsztyn district, and citizens too, against the Chapter, making himself the leader of the rebellion. When the Chapter complained to the pope about George of Schlieben (see M 1, above), Balthasar Scaibot was named as one of his accomplices (SRW, I, 107/3; 108/8 up–4 up; 113, n. 125/6; 160/7 up; 162/8 up–7 up; 167/15; 169/25; 173/11 up). Obłąk's reference (p. 48, n. 90) to ZGAE, 1858–1860, *1*:151, should be disregarded.

N 6   Bishop Francis Kuhschmalz was pressed by the Varmia Chapter to collect the 25,000 marks owed them by the Teutonic Knights (see L 9, above). "Quite frequently requested by his Chapter to recover the 25,000 marks which the Grand Master and his Order were fined [to recompense] the diocese [of Varmia, Bishop] Francis refused to do this, to the harm of his diocese" (SRW, I, 89/6 up–4 up).

N 7   Paul Rusdorf, Grand Master of the Teutonic Knights from 1422 to 1441 (not 1424–1435, as in Obłąk, p. 49, n. 91), replied to Bishop Francis Kuhschmalz's dispatch (see N 6, above).

N 8   After the death of Bishop Paul Legendorf of Varmia on 23 July 1467 (ZGAE, 1858–1860, *1*:148), a three-sided struggle to succeed him ensued. One of the contestants was Andrew Oporowski, who held the diocese for a time and, to gain support, made provision for certain appointments. Various letters written by Oporowski from August to November 1473 (now in ODA L 70; ZGAE, 1858–1860, *1*:163, n. 2) do not refer to Resil.

N 9   N 9's *Pronuntiatio arbitrorum inter Ecclesiam et +* strongly resembles *Quaedam pronuntiatio arbitrorum super concordia facta inter Ecclesiam Varmiensem et Ordinem de a[nno] 1374* in the Chapter's 1622 inventory, Y 4/1, fol. 2ᵛ. The presence of *Quaedam* suggests that this pronouncement of the arbiters is ODA Q 4, dated 28 July 1374 (CDP, III, 156–158, no. CXVIII, and CDW, II, 513–517, no. 496). See B 4, above.

O 1   ODA L 19, dated 30 June 1441, in German, original parchment with seals.

O 2–3   ODA L 2, 8; both dated 5 February 1442.

O 4   For Copernicus' personal involvement in the Greusing case, see Correspondence, Letter 1, n. 5. The 1622 inventory, fol. 8ʳ/7 up–6 up, lists a "Copy of the sale of 8 parcels in Wusen by Philip Greussing for the venerable Chapter of Varmia."

O 5   Reading *Quietantia* with superlinear *-a*, rather than *Quietantie*, as in Obłąk, p. 49/11. ODA K 3 contains a receipt dated 10 May 1465 by Bartholomew Libenwald, whose expenses on behalf of the Chapter were enumerated in his *Memoriale* (SRW, I, 305/313).

O 6   ODA L 6. Peter Polen, a citizen of Olsztyn, became involved in a lawsuit with his fellow-citizen Andrew Melczer. It was referred to a court of arbitration, which settled the dispute on 30 September 1440 (SRW, I, 302, n. 2/1–9). See E 16, above.

O 7   ODA A 20, 1507–1508. Christopher of Delen (or van der Delau) was the burgrave of Olsztyn in 1507–1508 (ZGAE, 1866, *3*:597/7 up–5 up). The description of Copernicus' O 7 was given more fully in the inventory of 1508: "Deposition of the witnesses in the matter of the registration of Christopher of Delau's house" (Obłąk, p. 24/8 up). The 1622 inventory, fol. 22ᵛ/4 up–3 up, lists the "Proceedings of the investigation concerning the clandestine donation of the land in Olsztyn to Christopher of Delen."

O 8   ODA A 2/1–30, ranging in date from 1516 to 1519. See O 4, above.

NOTES

O 9   ODA A 25/1–7, dated 1510–1511. Bogener was known also as Fröhlich.

O 10   ODA J 25, dated 6 July 1517.

O 11   ODA T 3, a collection of 18 documents ranging in date from 1511 to 1517. Tynappel was an agent of the Teutonic Knights who engaged in conflicts with the Varmia Chapter. Obłąk's reference (p. 49, n. 101) to 23 documents, commencing in 1510, is now withdrawn.

P 1   Anselm, the first bishop of Varmia, originally wished to have the Chapter consist of 24 canons, enjoying equal incomes. But he succeeded in establishing only 16 (CDW, II, 474/4–12). A century or so later his successor, Johannes Streifrock, on 24 February 1363 created the eight additional canonries, "which we wish to be called 'minor'" because they were endowed with lower incomes (CDW, II, 349/3; 474/15–18, from Theiner, I, 678, no. 915). The existence of this inequality was acknowledged by Pope Innocent VII on 31 December 1414 (CDW, III, 395/5–6). As soon as the Chapter's finances permitted, equalization would occur (CDW, III, 477/1–6). But this equalization was opposed by the holders of the major canonries, who were therefore sued by their less wealthy colleagues. The reigning pope's decision was rendered on 2 December 1411, and was sent out from Rome a month later on 2 January 1412 (CDW, III, 475–482, no. 470; ODA L 21). It mentions that in the Varmia Chapter "there are major, medium, and minor prebends" (CDW, III, 476/10).

P 2   ODA B 5, the original parchment, dated 27 April 1426 (CDW, IV, 160–165, no. 106). Where *minorum* was required, Copernicus miswrote *minarum*, by attraction to *mediarum*. Before this suppression, the Varmia Chapter consisted of 24 prebends: 16 major, 4 medium, 4 minor (ZGAE, 1858–1860, *1*:127, n. 3). After the suppression, only the 16 major prebends or canonries remained. The 1622 inventory, fol. 28/6–7, listed a "Document carrying out the decision in favor of the canons holding a major prebend, and against the prebendaries holding the medium prebends, 1412."

P 3   Pope Clement VI granted Otto of Russen the next Varmia canonry to become vacant (CDW, III, 620–629, no. 629). This grant, made on 10 November and notarized on 12 December 1344, led to Otto's installation as a canon of Varmia on 15 March 1345 (CDW, III, 629–631, no. 630). On the basis of the income from the property he inherited, Canon Otto created a vicariate, to which his relatives agreed on 10 July 1347 and 1 April 1348 (CDW, II, 110–113, no. 105). Then on 9 November 1349 (not 1347, as in Obłąk, p. 49, n. 104) the bishop of Varmia, on the advice of the Chapter, consented to the establishment of the vicariate (CDW, II, 143–144, no. 142). ODA C 88, cited by Obłąk, is a German translation of the consent; the original Latin parchment was filed with the Wills, under the letter R.

P 4   ODA C 56, a copy of the document, dated 7 October 1366, by which Bishop Johannes Streifrock of Varmia, when he was still the Chapter's custodian, established two vicariates in Frombork Cathedral (CDW, II, 416–417, no. 404). The funds for these vicariates were derived from the rents paid by the peasants living in the villages Dewiten (misread by Copernicus' predecessor as Degeten; Obłąk, p. 24/3) and Vangaiten. In the text of ODA C 56 the latter village is called Hoenberg, but a marginal note states "otherwise known as Wangayten."

P 5   ODA F 10, the original parchment, dated 5 April 1496. This vicariate was established by Elias of Darethen, who became a Varmia canon on 26 May 1486, was the administrator of the Chapter in 1493, and died on 8 November 1498 (SRW, I, 242, n. 133).

P 6   A hospital that had been built near Frombork Cathedral before 1456 was granted to the Order of St. Anthony by Bishop Lucas Watzenrode of Varmia on 7 April 1507 in a ceremony at which Copernicus was present (StC VIII, 49, no. 52). The Order's acceptance on 17 May 1507 (not 1503, as in Obłąk, p. 50, n. 107) incorporated the bishop's grant within itself (ODA L 18; printed in *Jahrbücher des Vereins für mekleburgische Geschichte und Alterthumskunde*, 1868, *33*:27–31). But the Order could not make a success of the hospital, and on 8 August 1519 renounced their rights thereto (ODA C 24; printed, ibid., pp. 33–35). The 1622 inventory lists the Antonites' acceptance of the hospital in 1507, their renunciation of it in 1519 (fol. 13ᵛ/1–3), and the disposal of their property in Varmia (fol. 21ᵛ/5–6).

P 7   ODA L 25, the original parchment in two sheets, dated 11 July 1355 (CDW, II, 222–226, no. 224). The three vicariates were established by Varmia Bishop Johannes Belger of Meissen shortly before his death on 30 July 1355.

[After P 7, Copernicus wrote an entry which he deleted and transferred to Q 6.]

P 8   Martin Achtsnicht was a Varmia canon from 18 February 1479 until 14 August 1502. Soon thereafter he died, on 4 March, the year on his tombstone being unclear, perhaps 1504 (SRW, I, 236, n. 97); 4 March 1504 was not the date of his will (as in Obłąk, p. 50, n. 109). ODA T 23, fol. 1ᵛ–3ᵛ, is a copy of his will, in which he bequeathed to Frombork Cathedral a gilded silver chalice and two silver cups (ZGAE, *8*:518/6 up–4 up).

P 9   Copernicus' *testį* was misread as *testium* (witnesses) by Obłąk (p. 50/8) instead of *testamenti* (will). Henry of Sonnenberg was the first provost of the Varmia Chapter (SRW, I, 213, n. 3). Provost Henry's will, dated 7 May 1314, was approved by Bishop Eberhard of Varmia on 5 January 1320, after Provost Henry's death (between 1 and 3 November, 1317 or 1318; CDW, I, 335, n. 7). His will established a "permanent vicariate ... to be provided for whoever is provost at the time" (CDW, I, 334/5–7). The bishop's approval, printed in CDW, I, 333–336, no. 195, from the parchment original in ODA C 26, had the following memorandum on its verso:

> A.D. 1515, the venerable Chapter of the Frombork Cathedral at a general Chapter meeting as is its custom in compliance with its current statutes [¶ 45; Hipler, p. 260; PII, 515, no. 45/8–10] on the day [7 May] after St. John's-before-the-Latin-gate discussed various matters concerning the Cathedral, but especially the vicariates existing from olden times in the aforesaid Cathedral and to a great extent depleted. Joined together in the income for the vicariates of the sixteen canonical prebends, all and each of them do not yet have twelve marks. The Chapter assigned the vicariate mentioned in this document to be preserved in perpetuity in the name of St. Peter the Apostle in honor of the provostship. The Chapter also resolved that this document should be kept hereafter in the treasury of the castle in Olsztyn, with a copy of the said [document] deposited in the vicarage (CDW, I, 337).

Five years after this decision of the Chapter, Copernicus listed the bishop's approval of Provost Henry's will as P 9 in his own inventory. In revising its statutes on 7 May 1391, the Chapter "diligently attended to holding the Chapter's meetings on the four anniversary days of the late honorable Henry of Sonnenberg, provost of the Varmia Chapter" (CDW, III, 331, no. 21/1–5).

Q 1   ODA J 16, the original parchment, dated 18 December 1422 (CDW, III, 586–590, no. 598), where *ad ecclesiam ... elemosine* (587/15–16) is echoed by Copernicus' Q 1's *elemosine in ecclesia Var[miensi]*.

Q 2   This is "one or more public documents" (CDW, III, 589/9 up) ordered by the Varmia Chapter concerning the establishment mentioned in Copernicus' Q 1.

Q 3   ODA P 50, the original parchment, dated 29 December 1354 (CDW, II, 214, no. 216).

Q 5   Obłąk omitted *areae* (ZGAE, 1974, *37*:196/5), the third word in Copernicus' Q 5, which was registered more fully in the inventory of 1508: "In the case of the land around the house bequeathed by the late Jo[annes] Rex for the vicariate of Saint Wenceslaus" (Obłąk, p. 24/5 up–4 up; for Joannes Rex, see H 6, above). The 1622 inventory lists a "document concerning the purchase of a house to be added to the same vicariate [in the chapel of St. George] by Canon Joannes Rex" (fol. 16/last 2 lines), and also his will (fol. 24ʳ/5 up).

Q 6   In Elditen, on 10 July 1289 the bishop of Varmia granted 110 parcels to a warrior and his heirs. The latter requested a copy when the paper on which the grant was originally written began to disintegrate. That copy (ODA C 1, fol. 26) was made on 26 November 1370 (CDW, I, 136–138, no. 79). Copernicus' Q 6 was recorded more fully in the 1508 inventory: "Document concerning the sale of the rent of 40 marks, light money, in the property of Elditen" (Obłąk, p. 24/11–12). In his haste Copernicus miswrote the numeral: instead of *XL*, he wrote *XI*.

R 1   On 7 May 1375 the Varmia Chapter in Frombork agreed to the establishment of a vicariate in Melsac (CDW, II, 545–546, no. 508). Obłąk's identification (p. 50, n. 114) of Copernicus' R 1 with CDW, III, no. 585, dated 17 April 1422, is now withdrawn.

R 2   ODA F 23, the original parchment, dated 12 May 1375 (CDW, II, 546, no. 509). Five days after the establishment of the vicariate mentioned in R 1, above, in Lidzbark the bishop of Varmia confirmed the arrangement.

R 3   ODA F 18, the original parchment, dated 25 July 1359 (CDW, II, 283–284, no. 286). ODA F 18's *vicariam ... apud sanctum Georgium in melsac ... fundavit* is echoed by Copernicus' R 3's *Fundatio vicariae apud s[anctum] Georgium ibidem* [= Melsac].

R 5   See C 1.

COPERNICUS' BREAD TARIFF

In the year 1531 Copernicus and Giese were designated Guardians of the Chapter's [Counting] Table (*mensae venerabilis Capituli deputati tutores*). While so engaged, Copernicus collected large sums of money owed to the Chapter and deposited them on the Chapter's counting table. At about the same time he devised his *Bread Tariff*. This document lay unnoticed for centuries until it was discovered in the library of the University of Uppsala and published in MCV, 1878, *1*:48–49. Then it was printed for the second time by PI², 213–214; for the third time by Dmochowski, pp. 187–189; and for the fourth time by NM, pp. 52–53, with a photofacsimile in Plate XXIII. Although Copernicus wrote the *Tariff* with his own hand, he himself did not attach his name to it in any way. That was done for him by Felix Reich, who stated tersely at the top of the sheet: "The author [is] Nicholas Copernic, canon of Varmia." The *Tariff* was translated into Polish by Dmochowski, pp. 203–205, and by Wasiutyński, p. 374 (omitting the Table). What follows is the first translation into English.

OLSZTYN BREAD TARIFF, ACCORDING TO THE PRICES OF THE GRAINS, WHEAT AND RYE[1]

From one sack of both grains [wheat and rye], after a careful weighing has been conducted and [the weight of] the bread basket[2] has been subtracted,[3] about 67 pounds of bread are produced. But[4] since the darnel and tares are usually separated from the grain before it is ground in order that the bread may come out cleaner and purer, it was agreed heretofore to subtract one pound for such cleansings, so that at least 66 pounds of bread result from one sack. Furthermore, the ordinary[5] expenditures[6] are 6 shillings 4 pence, namely, the baker's usual 4-shilling wage, 1 shilling for transportation, 1 shilling for salt and yeast, 4 pence for sifting. But the bran and chaff are enough to cover the baking expenses, provided that 8[7] sacks[8] of bran invariably sell[9] for 6 shillings. Hence, the same constant ratio of the price of grain to the production of loaves[10] prevails. Thus, for example, when grain is bought for 33 shillings [a sack], 6 penny[11] loaves will weigh 2 pounds. But when the price is 22 [shillings a sack], 6 loaves ought to weigh 3 pounds, and so on, as in the Table below, beginning with 9 [shillings] and increasing by 3 [shillings].

Price of Grain per sack	Weight of Six penny[12] loaves		Price of Grain per sack	Weight of Six penny[12] loaves	
shillings	pounds	skoters[13]	shillings	pounds	skoters
9	7	16	39	1	$33\frac{3}{13}$
12	5	24	42	1	$27\frac{3}{7}$
15	4	$19\frac{1}{3}$[14]	45	1	$22\frac{2}{5}$
18	3	32	48	1	18
21	3	$6\frac{6}{7}$	51	1	$14\frac{2}{17}$
24	2	36	54	1	$10\frac{2}{3}$
27	2	$21\frac{1}{3}$	57	1	$7\frac{11}{19}$
30	2	$9\frac{3}{5}$	60	1	$4\frac{4}{5}$
33	2	0	63	1	$2\frac{2}{7}$
36	1	40	66	1	0

A document closely related to the *Bread Tariff* was entitled by Felix Reich "Dr. Nicholas Coppernic's Computation for the Baking of Bread" (*Panis coquendi ratio Doctoris Nicolai Coppernic*). This second Copernican bread document was found in the same Uppsala composite volume as the *Tariff*, a few folios behind it (fol. 63^v-64^r, 68^r). Unlike the *Tariff*, which Copernicus wrote out with his own hand in its final form, this second document is a preliminary draft, not in Copernicus' handwriting. It was first published in MCV, I, 49–50; then in PI², 215; and in Dmochowski, pp. 188–189, with a translation into Polish (pp. 204–205). What follows is the first translation into English.

INVESTIGATING THE COMPUTATIONS CONNECTED WITH THE ORDINARY TOP-QUALITY LIGHT BREAD

First, weigh a sack of this year's pure rye and[1] find out how many pounds of rye are contained in one sack.

Then, if the width and depth of any[2] sack in Lidzbark, Olsztyn, and anywhere [else] are determined, a comparison of one with another will make it possible to ascertain accurately enough how much the sacks differ,[3] at least to the extent that is enough for this purpose at the outset.

And because the flour that is made from a sack of rye weighs approximately the same as the grain [from which the flour was made], accordingly take as much of this flour by weight as you wish, and sift it in a suitable way through[4] a flour bolter. Weigh the bran that is left. Its weight will indicate also how much flour remains after being sifted. And if somebody does not mind, both [bran and flour] may be weighed again solely [for the purpose of seeing] whether both restore the weight of the flour before it was sifted. Then in this way we shall come to know how much bran is usually sifted from a sack of rye.

When this is learned, take as much of the sifted flour by weight as you wish, and let loaves of light bread be made from it. It does not matter much whether they are big or small, provided that once more the bread produced from the flour is weighed,[5] and a record is kept of how many pounds of light bread are involved.

Let this be done in Lidzbark, Olsztyn, and elsewhere, if agreeable, and let the findings and weights[6] be assembled and compared. For from these [figures] the true and just price and weight of bread are determined without any uncertainty.

With regard to wheat[7] too, the method described above for rye may be applied. In all these [operations] let the weighing be on the mark, not in excess,[8] as is the practice of merchants, since we are investigating not business[9] but an accurate procedure.

NOTES ON THE BREAD TARIFF

[1] Copernicus' *siligo* (rye) was misunderstood by MCV, I, 50/28-29, and PI², 214/last 2 lines, to be "white wheat" (*weisser Weizen*).

[2] Copernicus' *metreta* was misinterpreted as "grinding" (*zmielenie*) by Dmochowski, p. 203/7, and as "measurement" (*zmierzenie*) by Wasiutyński, p. 374/4.

[3] Copernicus' *deducta* was omitted by Dmochowski and Wasiutyński.

[4] Copernicus' *vero* was misprinted as *fero* by PI², 213/last line.

[5] Copernicus' *communes* was mistranslated as "all" (*wszystkie*) by Dmochowski, p. 203/12.

[6] *Expensi* (MCV, I, 49/4; PI², 214/4), being obviously a misreading, was emended (incorrectly) by Dmochowski, p. 187/11, to *Expensis*, where Copernicus wrote *Expens[a]e*.

[7] MCV, I, 49/7, and PI², 214/7, followed by Dmochowski, p. 187/last line, misread Copernicus' *8* as *s*, and treated it as an abbreviation of *semi* (half). This mistaken conversion of 8 into $1/2$ misled Dmochowski, p. 203/7 up, into "half-bran" (*półotrąb*) and Wasiutyński, p. 374/12, into "half-sack" (*pół korca*).

[8] Copernicus' *mod* (*modiis*) was misread as *modio j* by NM, p. 52/4 up.

[9] Copernicus' *veniant* was mistranslated by Dmochowski, p. 203/7 up (*przychodziło*), and by Wasiutyński, p. 374/12 (*idzie*), as though it were a form of *venio* (come) instead of *veneo* (to be sold).

[10] Copernicus' *panum* was misread as *panem* by MCV, I, 49/9, and PI², 214/9, followed by Dmochowski, p. 188/2.

[11] Copernicus' *obolares* was mislabeled an error by NM, p. 52/last line. Copernicus' *6 panes obolares* (6 loaves, at a penny each) does not mean "sixpenny bread," as in Wasiutyński, p. 374/15 (*chleb 6-fenigowy*). To hold the price of a loaf of bread fixed at one penny was the purpose of Copernicus' *Tariff*. If grain sold for 9 shillings a sack when the harvest was plentiful, six one-penny loaves would weigh $7^1/_3$ lbs, or more than 1 lb each. At the other extreme, with a lean harvest and grain selling for 66 shillings a sack, six one-penny loaves, all together, would weigh only 1 lb. Copernicus evidently regarded a penny as the "just price" for a loaf of bread. The weight of that one-penny loaf, however, would fluctuate with the market price of grain, from more than 1 lb in a good year to $1/_6$ lb in a lean year. Although Copernicus' *Tariff* has nothing in common with a tax imposed by a government to collect money for public purposes, it was mislabeled a "Bread Tax" by PI², 213/3: *Brot-Taxe*, and *taksa chlebowa* by Dmochowski, p. 187, n. 1, p. 203/1, 4; Sikorski, p. 88, no. 330; StC VII, 158, no. 322; and NM, p. 28/11 up, p. 52, no. 20/1. Embracing the widest fluctuations in the price of grain, from 9 shillings to 66 shillings a sack, Copernicus' *Tariff* was intended to last an unlimited number of years. He did not indicate when he wrote it. But in the Uppsala composite volume (H 156) in which it is preserved, it follows immediately after a printed document dated 1531. In that year Copernicus was designated one of the Guardians of the Chapter's Counting Table (PI², 211/4 up-3 up; 257/4 up-3 up; NM, pp. 53-54; p. 55, 11/10-11; p. 61/1-2). The accounts in which he is so designated were overlooked by Schmauch, who conjectured that in 1530 Copernicus was the Chapter's Master of the Bakery (*magister pistoriae*; ZGAE, 1943, 28:69/6 up-3 up). According to the Chapter's statutes, the "Master of the Bakery ... is to have jurisdiction over the peasants belonging to the same [city of Frombork] as far as the collection of rents is concerned, also over the mill located in the said city, and over other places related to the said mill" (Hipler, p. 258, 39/8-11, 14; PII, 513, 39/8-10, 13). How important the products of the mill and bakery were in the daily lives of the canons is made clear by Statute 12: "no canon entering for the first time is to receive any distribution [of the Chapter's annual income] unless he pays the bakery forty marks, in accordance with the previous practice" (Hipler, p. 251, 12; PII, 504, 12). Statute 6 provided that "if anyone dares to contravene [the order of the church procession] ... he is to be punished in the distributions of the loaves" (Hipler, p. 249/3-6; PII, 501, 6/5-7). When a canon died, his heirs could occupy his premises for twenty days, "and could receive the loaves usually distributed to the said deceased canon only for the said twenty days" (Hipler, p. 254, 26/4-6; PII, 508, 26/4-6). "A resident canon who absents himself from the cathedral for more than 30 consecutive days will be deprived of the distribution of bread and beer as well as other things" (Hipler, p. 251, 16/3-5; PII, 505, 16/3-5). Had Copernicus been the Master of the Bakery in 1530, the duties of that office might well have led him to compose his *Bread Tariff*. But no contemporary evidence puts him in that office then. On the other hand, his tasks as a Guardian of the Chapter's Counting Table in 1531 may have induced him to compose the *Tariff* as a set of guidelines for regulating the production of a commodity essential for the daily welfare of those around him. When the Chapter's statutes were revised in 1532, Statute 12 remained in force, while Statute 6 changed the punishment from deprivation of bread to a fine of 2 marks, Statute 16 omitted the reference to beer, and Statute 26 modified the bread provision (ZGAE, 1972, 36:52, 56, 60, 74).

[12] Copernicus' *obolarum* was mistakenly changed to *obolorum* by Dmochowski, p. 188/7. In the last column of the Table, Copernicus omitted *-ol-*, and had to squeeze these two letters in, above the line.

[13] Copernicus' *sc^t* was misread as *Scpl.* by MCV, I, 49/16, and PI², 214/16, followed by Dmochowski, p. 188/9. According to MCV, I, 50/9 up, and PI², 214/28, "Copernicus' pound is divided into 48 scruples." But the traditional pound was divided into 12 ounces, each ounce in turn being subdivided into 24 scruples, so that 1 lb = 288 scruples, not 48 scruples. Copernicus' pound was divided into 48 skoters; see Letter IV, ¶ 2. As the price of grain rises from its minimum of 9 shillings a sack, the weight of six one-penny loaves shrinks more rapidly at first, and more slowly as the grain price approaches its maximum of 66 shillings a sack. Thus, when grain rises from 9 shillings to 12, bread shrinks from $7\frac{1}{3}$ to $5\frac{1}{2}$ lbs, or $\frac{11}{18}$ lb per shilling. It drops only $\frac{1}{18}$ lb per shilling in the intermediate range, when grain rises from 33 to 36 shillings a sack. At the other extreme, as grain rises from 63 to 66 shillings a sack, the shrinkage is only $\frac{1}{63}$ lb per shilling.

[14] Copernicus $19\frac{1}{3}$ was mistakenly changed to $19\frac{1}{5}$ by MCV, I, 49/last line, followed by PI², 214/19, and Dmochowski, p. 188/12.

NOTES ON THE BAKING OF BREAD

[1] The first *et* was omitted by PI², 215/7, and Dmochowski, p. 188/11 up.

[2] Copernicus' *cuiuslibet* does not mean "every," as in Dmochowski, p. 204, 11/3 *(każdego)*.

[3] Misprinted as defferentia in Dmochowski, p. 188/6 up.

[4] Misprinted as *par* in MCV, I, 50/3, and PI², 215/10.

[5] Reading *appendatur* with PI², 215, II/18, and Dmochowski, p. 189/10, not *appendantur*, as in MCV, I, 50/12.

[6] Reading *examinata* rather than *exeuntia*, as in MCV, I, 50/14; PI², 215, 11/5 up; Dmochowski, p. 189/13, p. 205/4: *wyniknie* (results).

[7] MCV, I, 50/3 up–2 up, mistakenly identified Copernicus' *Losebroth* (light bread) as wheat bread. But Copernicus' entire discussion up to this point concerned rye *(siligo)*, which was misunderstood by MCV to be "white wheat" (see Bread Tariff, n. 1, above).

[8] Copernicus uses the German word *ausschlag*.

[9] Copernicus' *mercaturam* does not mean *zysk* (profit), as in Dmochowski, p. 205/11.

COPERNICUS AS A GUARDIAN OF HIS CHAPTER'S COUNTING TABLE

INTRODUCTION

In 1531 Copernicus was chosen by the Varmia Chapter to be a Guardian of its counting table (*mensae venerabilis Capituli deputatus tutor*). Sharing this function with him was his lifelong friend Tiedemann Giese. Together they were responsible for receiving the money owed to the Chapter by purchasers who had bought some of its land on the installment plan. At the same time the Chapter invested in property, such as a mill, from the leasing of which a steady income was anticipated. Such transactions were financed by borrowing from special funds operating within the Chapter's jurisdiction and receiving the income generated by the properties thus acquired. Other separate accounts managed the various bequests received by the Chapter.

All this entailed rather complicated bookkeeping, which was made even more involved by the monetary situation. For Prussian money, the standard currency in Varmia as part of Royal Prussia, had recently undergone a change from its old standard to a new standard, the ratio being 4:3. During Copernicus' guardianship, however, the bookkeeping was not based on the new standard, because the major transactions had been negotiated before the change in the currency. For that reason the accounts were kept on the old standard, light money (*ad levem monetam veteris numeri*). Sums recorded in the new currency were converted into their equivalent on the old standard.

Prussian currency was made of silver alloyed with copper (paper money had not yet been introduced). But for bigger transactions foreign gold coins were more convenient. All these Spanish, English, Hungarian, and German gold pieces had to be evaluated in terms of the light money on the old Prussian standard.

These complexities do not make the Guardians' accounts any easier for us to understand. But they were kept for a small number of intensely interested and thoroughly informed persons, the sixteen canons of the Varmia Chapter. The Guardians' accounts had to be approved by the Chapter's chancellor for submission to the Chapter, the final authority in these matters. Since the canons' financial security was at stake, any cheating or major blunder was not likely to escape their attention.

On 30 April 1532 the surplus of receipts over expenditures was handed to the administrator for safekeeping in the castle at Olsztyn, and the account was subsequently submitted to the Chapter, which approved it and honorably discharged Giese and Copernicus.

This account consists of three interrelated documents, the first and last of which were written by Giese, while the second was written by Copernicus. The opening paragraph of Document I was quoted by PI[2], 211/last 6 lines, but the three documents as a whole were published for the first time in NM, pp. 53–62, with photofacsimiles of Documents II and III in Plate XXIV a–e. What follows is the first translation and systematic analysis of all three documents.

I

Account of the sums collected from the repurchased properties in Baysen, Codien, Reberg, etc. for various offices and deposited on the table of the venerable Chapter by us, Nicholas Coppernic and Tiedemann Giese,[1] canons and designated Guardians[2] of the said table in 1531 A. D. And this whole account is reckoned in light money of the old standard, in accordance with which this repurchase was done for the most part.

RECEIPTS

A. D. 1526 Caspar Damitz of Elbląg, in the name of the mighty lord George of Baisen,[3] governor of Malbork, to amortize the properties in Codin, Reberg, and others located in the Tolkemit district,[4] had to pay 2000 marks. But as offset for certain abandoned [parcels] in Reberg, which, however, were not really abandoned, a hundred marks having been improperly deducted,[5] he paid only 1900 marks.

In the year 1528 the doughty Achatius Czehm,[6] with the approval of the governor George of Baysen, repurchased the properties in Baysen, for which 2000 marks were owed in accordance with the contract when the properties were bought by him previously, and 300 marks for the purchase from the heirs of the late Pregers. In connection with those [properties] this year he paid 1300 marks in accordance with the contract. In the following year 1529, however, and in 1530 he paid 1000 marks on the new[7] standard equal, on the old standard, to 1333 marks 20 shillings.[8]

Up to that time the same [individual] had paid $187\frac{1}{2}$ marks toward the purchase of the [properties] bought from the heirs of Keserynne. But because this money had been spent for the aforementioned[9] repurchases, it is for that reason not reckoned as an entry in the current receipts.

Total of all receipts, in terms of light money of the old standard:
MMMMDXXXIII marks 20 shillings — 4533 marks 20 shillings[10]

EXPENDITURES FROM THE AFORESAID SUMS FOLLOW, AS LOANS

In the year 1526 the venerable Chapter, buying from George of Baysen the Haselau mill, repurchased at that time among other properties, paid him for them 200 marks, which the Chapter decided to charge to those offices to which it meanwhile assigned the entire income to be derived from the leasing of all the Haselau properties until the aforesaid sum of 200 marks is repaid.   200ᵐ

In the year 1527, for the purchase of the Crumse properties, the Chapter contracted a loan of 250 marks. The Chapter decided that to repay those [marks], every year one last of rye should be sold at the castle in Olsztyn, and the money sequestered for this purpose to amortize the principal until [the loan is] fully paid off.

In the same year a loan was contracted for the workshop, from whose products should be repaid   50ᵐ

In the year 1531, for the construction of a tavern or new inn in Frombork, the venerable Chapter took a loan of 461 marks 22 shillings on the new standard, equal to 615 marks 9 shillings 2 pence[11] on the old [standard].

For the salary of the preacher in the Frombork cathedral a loan was arranged for 15 new marks, equal to 20 [old] marks.[12] These should be repaid from the next income of his vicariate.

First total: 1135 marks 9 shillings 2 pence[13]

II

Account[14] of the sums collected from the repurchased properties in Baysen, Codien, Reberg etc. for various offices and deposited on the table of the venerable Chapter by Tiedemann Giese, custodian, and Nicholas Coppernic, canons, and designated Guardians of the said table in 1531 A.D. And this whole account is reckoned in light money of the old standard, in accordance with which this repurchase was done for the greater part.

RECEIPTS

A. D. 1526 Caspar Dambrowitz,[15] a citizen of Elbląg, in the name of the mighty lord George of Baysen, governor of Malbork, to amortize the properties in Codyn, Reberg, and the others in the Tolkemit district belonging to him, had to pay 2000 marks. But on the pretext of offsetting certain abandoned [parcels] in Reberg, which, however, were not really abandoned, 100 marks having been improperly subtracted,[16] he paid 1900 marks.

Total: 1900^m

BAYSEN

In the year 1528 the doughty Achatius Zehme,[17] with the approval of the governor George of Baysen, repurchased the properties in Baysen, for which he owed 2000 marks in accordance with the contract when the properties were bought by him previously, and 300 marks for those purchased from the heirs of the late Pregers. In connection with those [properties] he paid 1300 marks this year [1528]. But in the next two years he paid 1000 marks on the new standard, equal to 1333[18] marks 20 shillings on the old standard.

In addition he paid $187\frac{1}{2}$ marks toward the purchase of the [properties] bought from the heirs of Kaserynne.[19] But because this money had been spent for the aforementioned[20] repurchases, it is not reckoned here again as income.

Total receipts: 4533^m 20^s [21]

Correct A[lexander] Sculteti[22]

EXPENDITURES

And, in the first place, made by the venerable Chapter in the form of a loan

In the year 1526 the venerable Chapter, having bought the Haselau mill from George of Baysen among other properties repurchased at that time, paid him for the said [mill] 200 marks, which the Chapter decided to charge to the offices to which meanwhile it allocated the entire income to be derived[23] from the leasing of all the Haselau properties until the aforesaid sum of 200 marks is paid and recovered. 200ᵐ

In the year 1527 the Chapter contracted a loan of 250 marks for the purchase of the Cruntsche[24] properties. It also decided that every year 1 last of rye should be sold at the castle in Olsztyn, and the money sequestered for this purpose until the loan is completely paid off. 250ᵐ

In the year 1531 the venerable Chapter again took as a loan for the construction of an inn in Frombork 461 marks 22 shillings on the new standard, equal to 615[25] marks 9 shillings 2[26] pence[27] on the old [standard].

First.[28] Of the aforesaid money borrowed by the Chapter, total: 1065ᵐ 9ˢ 2ᵖ [29]

It was decided to charge this loan to the legacies of Nicholas.[30]

EXPENDITURES FOR CERTAIN OFFICES FOR THE REDUCTION OF THEIR INDEBTEDNESS AND ITS PARTIAL ELIMINATION

From the legacies of the late Bishop Nicholas:

In the year 1529, by decision of the venerable Chapter, for the purchase of cannon, handguns, lead, and powder for the defense of the Cathedral, there were given 160ᵐ

There were added 2 marks 53 shillings 5 pence, and for the gunner in charge of the cannon and powder 1ᵐ

In the year 1531 for the defense of the rights of the fishermen against the people of Elbląg, 10 marks, be it noted, equal to 13 old marks 20 shillings,[31] and for a horse in Colmensche 20ᵐ

total: 33ᵐ 20ˢ

Second[32] total: 197ᵐ 13ˢ 5ᵖ [33]

In the year 1527 for the office of the workshop 50ᵐ

For the pulpit or vicariate of the custodian And[rew] Cletz:[34]

In the year 1529, for 4 parcels bought in Wusen from Gregorius Henkel, to the same [individual] there was paid a total of 182ᵐ 26ˢ 4ᵖ

In the year 1530 to pay for $1/4$, a quarter, of the Appelaw properties bought[35] from Alexander Plastevig and his heirs 70ᵐ

Likewise, in payment for the second quarter of the same properties bought[36] from George Sack 116ᵐ

In the year 1531, when the venerable Chapter wanted to repurchase the rent in Vusen for this office from the office of the students, to complete this repurchase there were paid 31ᵐ 6ˢ 4ᵖ

Likewise, for buying 1 mark of the monetary rent for the house of Jo[hannes] Vogt in Frombork from the bequest of the late preacher Caspar Greve, to complete the capital sum there was given 1ᵐ

Likewise, for part of the salary of the preacher in the Frombork Cathedral 20 marks were borrowed, which should be repaid from the income[37] of this office.

Total: 420ᵐ 33ˢ 2ᵖ [38]

For the common [treasury] of the vicars:

In the year 1530, 30 July, to the same community, which wanted to buy certain hereditary funds in Frombork, there were advanced   53^m 20^s

    Third Total: 523^m 53^s 2^p, light [coinage][39]

OTHER COMMON EXPENDITURES

In the year 1529, to Joannes Vogt,[40] who was sent to Toruń[41] in the matter of the Baysen repurchase, for his expenses   4^m

For the expenses of Achatius Zehme[42] incurred in the inn   2^m 37^s

For the exchange of demonetized currency, a loss was incurred in the Koenigsberg mint[43] of   1^m 28^s 8^p

    Fourth[44] Total: 8^m 6^s 2^p [45]

All the totals of expenditures   1794^m 22^s 5^p [46]

                      Correct A Scult

The remaining income   2738^m 57^s 1^p [47]

                      Correct A Sculteti

In the year 1531, 23 October, the money remaining on the table, having been counted again, was found [to be] as follows:

In new[48] shillings: 187 marks 21 shillings, equal to 249 old marks 48 shillings

In unit groats:[49] 721 marks 48 shillings equal, be it noted, to 962 [old] marks 24 shillings[50]

In 3-groat [pieces]: 19$^1/_2$ marks, equal to 26[51] old marks

In ducal groats, at 4 for 16 pence: 7 marks 20 shillings, equal on the old standard to 9 marks 46 shillings, 4 pence[52]

In Polish half[-shillings]: 12 marks 46$^1/_2$ shillings equal, be it noted, to 17 [Prussian] marks 2 shillings on the old standard[53]

In old pennies: 8^m

In demonetized old pennies and groats, mixed: equal to   18^s 4^p

In gold

In Spanish doblons: 20, at 5$^1/_2$^{m} [each], equal to   110^m

In angelots: 68, at 4 marks [each], equal to   272^m [54]

In rose nobles: 2, at 5^m 45^s [each], equal to   11$^1/_2$^{m} [55]

In Hungarian [coins]: 313,[56] [equal to]   939^m

In Rhenish [coins]: 1, [equal to]   2^m

In clemmergulden: 13, at 1$^1/_2$^{m} [each], equal to   19$^1/_2$^{m}

In the [Holy Roman] Empire's [coins]: 34, at 1^m 37$^1/_2$^{s} [57] [each], equal to   46^m 45^s

In horngulden: 43, at 50^s [each], equal to   35^m 50^s [58]

In the lighter horngulden: 3, equal to   2^m 24^s [59]

    Total: 2712^m 18^s 2^p [60]

                      Correct Alex Sculteti

Missing from the above total, as it stands: 26^m 38^s 5^p

                      Alex Scul

After this money was deducted, the grand total of receipts remained: 4506^m 41^s 1^p [61]

                      Correct A. S.

In the first[62] repurchase 1900^m were paid, as above,[63] and in the second, 2300^m,[64] without regard to[65] what had accrued to this total[66] on account of the new standard of coinage. Hence, of the aforementioned 26^m 38^s 5^p, from the first sum 12^m 3^s 1$^{2/3 p}$ are to be subtracted proportionally,[67] but from the second [sum] 14^m 35^s 3$^{1/3 p}$.[68]

Therefore, the first total for distribution to the various offices will be 1887^m 56^s 4$^{2/3 p}$.[69]

The second total, from which the vicars' common expenses, 8^m 6^s 2^p,[70] will also be deducted, will be 2277^m 18^s 0$^{1/3 p}$.[71] This [total], to which 333^m 20^s of the old[72] [standard] are added for 1000^m of the new standard, as above,[73] increases to 2610^m 38^s 0$^{1/3 p}$ [74] for distribution to the offices for the second[75] repurchase, that in Baysen, as follows.

1531

Distribution of the first total of the purchase of the properties in the Tolkemit district, according to the shares contributed to the various offices, 1887^m 56^s 4$^{2/3 p}$, mentioned above, in the following manner:

legacies of the late Bishop N[icholas]	894^m	24^s	5$^{3/8 p}$
workshop	61	21	2$^{1/3}$
mortuary	77	52	5$^{1/8}$
school[76]	198	14	0$^{2/5}$
vicariate of the custodian or pulpit	210	2	0$^{1/4}$
vicariate of Mart[in] Achts[nicht][77]	151	2	1
vicariate of Zacha[rias][78]	61	21	2$^{1/3}$
vicariate of [St.] Barptolomeus	103	50	1$^{1/3}$
vicariate of 15 prebends	129[79]	47	4$^{7/10}$

Second distribution, which is of the second total and is[80] from the Baysen purchase:	2610^m [81]	38^s	0$^{1/3 p}$
legacies of Bishop N[icholas]	1075^m	28^s	1^p
workshop	73	46	4$^{1/2}$
mortuary	93	38	3$^{1/4}$
school	238	21	4$^{1/3}$
pulpit	252	33	0
vicariate of Mart[in] Achts[nicht]	181	36	3$^{1/3}$
vicariate of Zachar[ias]	73	46	4$^{1/2}$
vicariate of Barpto[lomeus]	124	52	4
vicariate of the prebends	212	49	0$^{2/3}$
community of vicars	113	29	3
[participants in canonical] hours[82]	170	15	3$^{1/8}$

The distribution from both[83] of the preceding[84] totals, taken together, amounts to 4498^m 34^s 5^p.[85]

Legacies of Bishop N[icholas]: 1969^m 53^s 2^p.[86] From this [figure] was deducted the amount of the money borrowed by the Chapter, as above:[87] 1065^m 9^s 2^p; also for ordinary expenses: 197^m 13^s 5^p.[88] Balance in this office: 707^m 30^s 1^p.[89]

Workshop: 135ᵐ 8ˢ 1ᵖ,[90] from which 50ᵐ were paid out previously, as above.[91]
Balance: 85ᵐ 8ˢ 1ᵖ

mortuary	171ᵐ	31ˢ	$2 \, 3/8$ ᵖ [92]	
school:	436ᵐ	35ˢ	$4 \, 3/4$ ᵖ [93]	
pulpit:	462ᵐ	35ˢ	$0 \, 1/4$ ᵖ,[94]	minus 420ᵐ 33ˢ 2ᵖ;[95]
balance:	42ᵐ 1ˢ $4 \, 1/4$ ᵖ [96]			
vicariate of Martin Achts[nicht]		332ᵐ	38ˢ	$4 \, 1/3$ ᵖ [97]
vicariate of Zach[arias]		135ᵐ	8ˢ	1ᵖ [98]
vicariate of Barpto[lomeus]		228ᵐ	42ˢ	$5 \, 1/3$ ᵖ [99]
vicariate of the prebends		342ᵐ	37ˢ	1ᵖ [100]
community of vicars		113ᵐ	29ˢ	3ᵖ;
	minus	53ᵐ	20ˢ;	
	balance:	60ᵐ	9ˢ	3ᵖ [101]
[participants in canonical] hours:		170ᵐ	15ˢ	$3 \, 1/8$ ᵖ [102]

Reduction of this foregoing[103] money, which [is] on the table, to the new standard, light money:

legacies [of Bishop Nicholas]: 707ᵐ 30ˢ 1ᵖ,[104] equal, in new marks, to 530ᵐ [105] 37ˢ $3 \, 1/4$ ᵖ [106]

workshop: instead of 85ᵐ 8ˢ 1ᵖ,[107] 63ᵐ	51ˢ [108]	$0 \, 3/4$ ᵖ [109]	
mortuary:	128ᵐ	38ˢ	$3 \, 1/2$ ᵖ [110]
school:	327ᵐ	26ˢ	5ᵖ [111]
pulpit:	31ᵐ	31ˢ	$1 \, 3/4$ ᵖ [112]
vicariate of Achts[nicht]:	249ᵐ	29ˢ	$0 \, 1/4$ ᵖ [113]
vicariate of Zach[arias]:	101ᵐ	21ˢ	$0 \, 3/4$ ᵖ [114]
vicariate of Barpt[olomeus]:	171ᵐ	32ˢ	1ᵖ [115]
vicariate of the prebends:	256ᵐ	57ˢ	$5 \, 1/4$ ᵖ [116]
community of vicars:	45ᵐ	7ˢ	$0 \, 3/4$ ᵖ [117]
[participants in canonical] hours	127ᵐ	41ˢ	4ᵖ [118]

Total: 2034ᵐ 13ˢ $4 \, 1/2$ ᵖ,[119] new money

The venerable Chapter approved this account.

signed, Alexander Sculteti, chancellor

III

In the year 1532, account[120] of the receipts and expenditures[121] of the money from the repurchases of Baisen, Codien, Reberg etc. in relation to the sums which were distributed to the individual offices in the account of the preceding year by Tiedemann Giese, custodian, and Nicholas Copernicus, canon of Varmia, Guardians of the Chapter's [counting] table. And this account proceeds on the basis of the new standard of light money.

Receipts, for the office of the pulpit: the 15 marks, which were lent last year toward the salary of the preacher, were repaid. These 15 marks, when added to the total capital of this office, as reported last year, make the balance
46ᵐ 31ˢ 2ᵖ [122]

Receipts for the legacies of the late Bishop Nicholas:
from the Haselau leases,[123] 5ᵐ
for the horse bought[124] in Colmensze and now resold, 8ᵐ
total receipts: 13ᵐ

EXPENDITURES

On 23 January the venerable Chapter repurchased from George Rautenberg of Elbląg the Claukendorf and Lockerat properties in the Tolkemit district by returning, in accordance with the letter of the agreement, 100 Polish marks of the Polish standard, reckoning 48 groats for each mark. In addition, 106 Hungarian florins[125] and 1 Rhenish[126] [coin] were paid, making 240 marks in Prussian money.[127]

In the matter of the defense of the fishing and netting [rights] against the people of Elbląg, expenses were incurred in Tolkemit, Elbląg, Cracow, Frombork, and for couriers dispatched to various places, as in the special account amounting to $46^m\ 16\frac{1}{2}^s$.

In the matter of the restitution of the Tolkemit district retained by the king, in the year 1525 there had been borrowed from the workshop as a loan for the expenses in Cracow 20 Hungarian florins, which have now been repaid, equal to 45 marks.[128]

When the same district was restored, a silver gilded bowl was given to the commissioners. [Weighing] (uncertain reading) $7\frac{1}{2}$ skoters, it was bought by the Most Reverend bishop of Varmia [Maurice Ferber] at a price of 19 marks for each mark [by weight], equal to $91^m\ 26^s$.[129]

Total expenditures: $422^m\ 42\frac{1}{2}^s$[130]

When the receipts are subtracted from this total of the expenditures, the balance is a total of expenditures of $409^m\ 42\frac{1}{2}^s$.[131] If this total of expenditures is deducted from the total of the capital attributed last year to this office of the legacies, that total being $530^m\ 37^s\ 3\frac{1}{2}^p$,[132] there will remain for the same legacies a total capital of $120^m\ 55^s\ 5^p$.[133]

After these accounts were completed, the venerable Chapter decided that the aforementioned sums collected from the repurchase of Baisen, Codin, Reberg etc. should be allotted and delivered to the same offices to which distributions [had been made], in accordance with the divisions and shares in the accounts of this year and the preceding[134] years, and that thus this account, and the anxiety over its safekeeping and protection, should finally be terminated. And this was done, as follows. The office of the legacies of the late Bishop Nicholas took as its share:

	120^m	54^s	5^p
workshop:	63^m	51^s	1^p
	Received Alexander Sculteti		
mortuary:	128^m	38^s	$3\frac{1}{2}^p$
school:	327^m	26^s	5^p
pulpit:	46^m	31^s	2^p
vicariate of the late Achtesnicht:	249^m	29^s	
vicariate of the late Zacharias:	101^m	21^s	1^p
vicariate of St. Bartolomeus:[135]	171^m	32^s	1^p
united vicariates:	256^m	57^s	$5\frac{1}{2}^p$
[participants in] Our Lady's hours:	127^m	41^s	4^p
community of vicars:	45^m	7^s	$\frac{1}{2}^p$

I, Johannes Faulhaber, received it.[136]

In the aforementioned year, on the last day of April, I, Tiedemann Giese, and Achatius von der Trenck, legates of the venerable Chapter in Olsztyn, dispatched in accordance with a decision of the same Chapter, presented all the aforementioned money, except the sums for the workshop and community of vicars, these sums having been returned as [indicated] above, to the venerable Felix Reich, administrator, to be kept in the castle in the same place. The same temporary administrator will hereafter make an accounting of these funds to the Chapter. And in this way this account is closed.

The venerable Chapter approved this account, gave a receipt for it, and discharged the aforesaid venerable Tiedemann, custodian, and Nicholas Coppernic.

[Signed] with my own hand, Alexander Sculteti, canon and chancellor

NOTES

[1] Document I was written by Giese.

[2] PI², 211/12 up–3 up, cited *tutores* (Guardians) as evidence that in 1531 Copernicus was a *nuntius* (legate), a different official in the service of the Chapter.

[3] Jorg von Basen (as the German chronicler called him) delivered a speech on 25 May 1525 as an emissary of the king of Poland on the occasion of the public announcement in Koenigsberg of the secularization of the duchy of Prussia as a fief of Poland (Acten, V, 771–772).

[4] The town of Tolkemit, with its surrounding villages, was given by King Casimir IV of Poland to a member of the Baisen family in 1457, presumably as a reward for his political and military services in the Thirteen Years' War (1454–1466). In 1476 and 1489 the royal grant was confirmed for the sons of the original grantee. Then on 2 June 1503 the king permitted the bishop of Varmia to buy the Tolkemit district from George of Baisen (1469–1546). In addition, on 10 February 1508 King Sigismund I gave to the Varmia church, bishop, and Chapter three villages, which had formerly been granted by Casimir IV to a monastery in Elbląg but were now deserted. Then on 25 February 1519 Sigismund transferred the Tolkemit district "to the canons and Chapter of Varmia with the obligation of singing in the chapel of St. George one solemn mass with prayers for the peaceful and happy state of the kingdom of Poland, and of performing solemn rites every year for the salvation of the souls of the kings of Poland"; see Summaria, part I, no. 435, 440, 2076; part III, no. 792; part IV, no. 245, 2900. Through Caspar Dam(brow)itz as agent, George of Baisen repurchased from the Chapter certain Tolkemit properties, which he paid off in installments.

[5] Reading *deductis*, not *deductos*, as in NM, p. 54/8.

[6] Zceme, as the German chronicler called him, was also an emissary of the king of Poland in Koenigsberg on 25 May 1525 (see n. 3, above).

[7] Originally mistakenly written as "old," which was deleted, while "new" was inserted above the line. In NM, p. 54, n. 6, *veterum* should be *veterem*.

[8] $1333^m\ 20^s$, old standard $= {}^4/_3\ (1000^m$, new standard).

[9] Presumably in an earlier account.

[10]
Dam(brow)itz	1900^m
Czehm	1300
	1333 20^s
total	$4533^m\ 20^s$

[11] "Should really be $610^m\ 17^s$," says NM, p. 55, n. e, without any explanation. But Giese's computation is correct:

new standard: $461^m = 27{,}660^s;\ 27{,}660^s + 22^s = 27{,}682^s = 166{,}092^p$

old standard: $615^m = 36{,}900^s;\ 36{,}900^s + 9^s = 36{,}909^s = 221{,}454^p;\ 221{,}454^p + 2^p = 221{,}456^p$

$$^4/_3\ (166{,}092^p) = \frac{664{,}368^p}{3} = 221{,}456^p$$

[12] $^4/_3\ (15\text{ new marks}) = \dfrac{60}{3} = 20\text{ old marks}$

¹³ 200ᵐ
   250
    50
   615 9ˢ 2ᵖ
―――――――――――
1135ᵐ 9ˢ 2ᵖ

¹⁴ This account was written by the hand of Copernicus; facsimile in NM, Plate XXIV a–d, and in NCCW IV, Plate XLII.

¹⁵ Miscalled Damitz, in Document I.

¹⁶ Reading *deductis*, not *deductos*, as in NM, p. 55/10 up.

¹⁷ Czehm, in Document I.

¹⁸ Initially Copernicus wrote *1353*, which he deleted.

¹⁹ Keserynne, in Document I.

²⁰ Presumably in an earlier account.

²¹ Dambrowitz 1900ᵐ
   Zahme   1300
                  1333 20ˢ
   total   4533ᵐ 20ˢ

²² Alexander Sculteti reviewed this account in his capacity of Chancellor of the Chapter. For a summary of his later career, see TCT, pp. 382–386.

²³ Reading *obventurum*, not *obventurorum* as in NM, p. 56/11.

²⁴ Crumse, in Document I.

²⁵ Copernicus started to write this number in Roman numerals (*D* ...), which he deleted.

²⁶ Copernicus wrote this number in Roman numerals *(ij)*, which he deleted.

²⁷ NM again (p. 56, n. d) mistakenly insists that this sum "should really be 610ᵐ 17ˢ."

²⁸ Inserted in the left margin by Sculteti.

²⁹ Haselau   200ᵐ
   Cruntsche 250
   Frombork   615   9ˢ 2ᵖ
   total   1065ᵐ 9ˢ 2ᵖ

³⁰ Nicholas of Tungen, bishop of Varmia, left a will, dated 29 January 1489. Its provisions were reported in ZGAE, 1932, *24*:51–52. The decision to charge the loan to Nicholas' legacies was inserted later by Giese.

³¹ 13 old marks 20 shillings = 780ˢ + 20ˢ = 800ˢ = ⁴/₃ (10 new marks) = $\frac{40^m \times 60}{3}$ = 800ˢ. Copernicus' *keutelarum* is a latinization of *Keutel*, related to English "kiddle." In earlier times *keutel (cutel)* denoted nets for catching eels: ... *retibus agwillarum quae* ... *Cutel nominantur* (CDW, I, 268/16 up–15 up; dated 8 July 1310). Later on, the designation shifted to eel boats: ... *navium anguillariorum sive keutelarum* (*Rocznik Olsztyński*, 1972, *10*:301, no. 14/1, dated 5 April 1512). The income derived by the Chapter from the fishermen was used to commemorate the Polish kings (ZGAE, 1972, *36*:66/3 up–68/9).

³² Written in the left margin by Sculteti.

³³ Cathedral   160ᵐ
   armaments   3 53ˢ 5ᵖ
   1531   33 20
   total   197ᵐ 13ˢ 5ᵖ

³⁴ Cletz, a Varmia canon since 4 June 1485, served as Chapter custodian from 1499 until his death in 1515.

³⁵ Reading *emptorum*, not *emptis*, as in NM, p. 57/4; the identification of Appelaw with Haselau in NM, p. 57, n. 7, and p. 96, is dubious.

³⁶ Reading *em[p]torum*, not *emptis*, as in NM, p. 57/6. The "Transaction with George Sack, 1531" was listed in the 1622 inventory, fol. 25/4 up.

³⁷ Copernicus omitted the *t* in *proventibus*.

³⁸ Henkel   182ᵐ 26ˢ 4ᵖ
   Plastewig   70
   Sack   116
   Vusen   31   6   4
   Wogt   1
   preacher   20
   total   420ᵐ 33ˢ 2ᵖ

[39] This third total was written by Sculteti.

Workshop	50^m		
Pulpit	420	33^s	2^p
Vicars	53	20	
Third total	523^m	53^s	2^p

[40] Vogt's house in Frombork was the subject of the second transaction in 1531 for the pulpit account.

[41] Reading *Thorunam*, not *Thoruniam* as in NM, p. 57/21.

[42] See n. 6, above.

[43] At the Koenigsberg mint the old currency of Ducal Prussia had to be exchanged for the new coinage, at a loss to the public; see Copernicus and the Money Question, M ¶ 17.

[44] Written by Sculteti.

[45]
Vogt	4^m		
Zehme	2	37^s	
Koenigsberg	1	28	8^p
total	8^m	6^s	2^p

[46]
1st (sub)total	1065^m	9^s	2^p
2nd	197	13	5
3rd	523	53	2
4th	8	6	2
grand total	1794^m	22^s	5^p

[47]
Total income	4533^m	20^s	
Total expenditure	1794	22	5^p
remaining income	2738^m	57^s	1^p

[48] Copernicus mistakenly wrote *antiquis*, where *novis* is required, as is shown by the following computation:

$187^m 21^s = 11{,}220^s + 21^s = 11{,}241^s$; $249^m 48^s = 14{,}940^s + 48^s = 14{,}988^s$; $^4/_3(11241^s) = \dfrac{44{,}964^s}{3} = 14{,}988^s$

[49] 1 Prussian groat = 3 Prussian shillings

[50] $721^m 48^s = 43{,}260^s + 48^s = 43{,}308^s$; $962^m 24^s = 57{,}720^s + 24^s = 57{,}744^s$; $^4/_3(43{,}308^s) = \dfrac{173{,}232^s}{3} = 57{,}744^s$

[51] Copernicus started to write this number in Roman numerals (*XX*...), which he deleted when he switched to Hindu-Arabic *26*.

26 old marks = $^4/_3 (19^1/_2$ new marks$) = ^4/_3 \times \dfrac{39}{2} = \dfrac{156}{6} = 26$

[52] $7^m 20^s = 420^s + 20^s = 440^s = 2640^p$;

$9^m = 540^s$; $540^s + 46^s = 586^s = 3516^p$; $3516^p + 4^p = 3520^p$;

$^4/_3(2640^p) = \dfrac{10{,}560^p}{3} = 3520^p$

NM, p. 57, n. n, suggests that Copernicus was confused about the value of the ducal groat, and that he should have equated $4 \times 16^p = 64^p$ with $1^m 4^p$! But $1^m 4^p \neq 64^p$; $1^m 4^s = 64^s$. For the decline in the value of the groat, see Copernicus and the Money Question, E ¶ 14–15.

[53] $12^m = 720^s$; $720^s + 46^1/_2{}^s = 766^1/_2{}^s$;

$17^m = 1020^s$; $1020^s + 2^s = 1022^s$;

$^4/_3(766^1/_2{}^s) = \dfrac{3066^s}{3} = 1022^s$

[54] At first Copernicus wrote *262*, which he replaced by *270* in the line; then he deleted both of these incorrect equivalents and wrote the correct equivalent, *272*, above the line.

[55] Initially, Copernicus mistakenly wrote *12¹/₂*, which he deleted.

[56] At 3^m each.

[57] Departing from his usual division of the mark into 60 shillings, here Copernicus equates the mark with 100 shillings, so that the Empire's 34 coins at $1^3/_8{}^m$ each, equal $46^3/_4{}^m = 46^m\ 45^s$. Had Copernicus here as elsewhere assigned 60 shillings to the mark, he would have made an error of $8^m\ 30^s$. Did he commit so gross a blunder, and did it pass unnoticed by his fellow-Guardian Giese and supervisor Sculteti? These two questions are answered affirmatively by NM, p. 58, n. pq!

[58] $43 \times 50^s = 2150^s = 35^m 50^s$

[59] The lighter horngulden was worth 48^s, 2^s less than the horngulden at 50^s; $3 \times 48^s = 144^s = 2^m 24^s$

[60]

	marks	shillings	pence
new shillings	249	48	
1-groat	962	24	
3-groat	26		
ducal	9	46	4
Polish	17	2	
pennies	8		
mixed		18	4
doblons	110		
angelots	272		
nobles	11	30	
Hungarian	939		
Rhenish	2		
clemmergulden	19	30	
Empire	46	45	
horngulden	35	50	
light horngulden	2	24	
total	2712^m	18^s	2^p

[61] $4533^m\ 20^s = 4532^m\ 80^s = 4532^m\ 79^s\ 6^p$

$$\begin{array}{r} - \quad 26 \quad 38 \quad 5 \\ \hline 4506^m\ 41^s\ 1^p \end{array}$$

[62] After *prima*, Copernicus started to write *sol-* (paid), but deleted it when he realized that he needed *redemptione* first.

[63] Copernicus' *utā* (*ut supra*) was misread as utum(?) by NM, p. 58/8 up; 1900^m were paid by Dambrowitz.

[64] Copernicus started to write *233*, which he deleted; Zehme paid $1300^m + 1000^m = 2300^m$.

[65] Initially Copernicus wrote *cum* (together with), which he deleted and replaced by *absque* above the line.

[66] Copernicus inserted *huic summae* in the left margin.

[67] $26^m 38^s 5^p = 1560^s + 38^s 5^p = 1598^s 5^p = 9588^p + 5^p = 9593^p$, to be divided in the proportion $1900:2300$

$1900 + 2300 = 4200$;

$$9593^p \times \frac{19}{42} = \frac{182{,}267^p}{42} = 4339\ ^2/_3{}^p = 723^s\ 1^2/_3{}^p = 12^m\ 3^s\ 1^1/_8{}^p$$

[68] $$9593^p \times \frac{23}{42} = \frac{220{,}639^p}{42} = 5253\ ^1/_3{}^p = 875^s\ 3^1/_3{}^p = 14^m\ 35^s\ 3^1/_3{}^p$$

[69] $1900^m = 1899^m\ 59^s\ 6^p$

$$\begin{array}{r} - \quad 12 \quad 3 \quad 1\ ^2/_3 \\ \hline 1887^m\ 56^s\ 4\ ^1/_3{}^p \end{array}$$ (Copernicus' $^2/_3{}^p$ repeats the subtrahend instead of subtracting it).

[70] This is the fourth (sub)total of the expenditures.

[71] $2300^m = 2299^m\ 59^s\ 6^p$

$$\begin{array}{r} - \quad 8 \quad 6 \quad 2 \\ - \quad 14 \quad 35 \quad 3^1/_3 \\ \hline 2277^m\ 18^s\ \ ^1/_3{}^p \end{array}$$ (according to Copernicus, although this should be $^2/_3{}^p$)

Initially Copernicus wrote $2267^m 18^s\ ^1/_3{}^p$, which he deleted.

296

NOTES

[72] Copernicus inserted *veteris* above the line.

[73] Copernicus *utā* was misread as utum (?) by NM, p. 59/2.

[74] 2277^m 18^s 1/3^p
  +333   20
  ―――――――――――
  2610^m 38^s 1/3^p

[75] Copernicus wrote and deleted *s* after *altera*.

[76] Bishop Nicholas of Tungen's will left 20^m for the cathedral school (ZGAE, 1932, *24*:50/7-8). The office of the mortuary in Frombork Cathedral was concerned, not with the temporary reception of the dead, but with the annual commemoration of donors on the anniversary of their death.

[77] Martin Achtsnicht (Achtesnicht), a Varmia canon for a quarter-century, in his will established a vicariate, to which he left 400 marks. His will also provided that a quarter of the proceeds from the sale of his curia should be used for choral prayer, and a second quarter for the school (ZGAE, 1932, *24*:52/3 up-2 up, 53/15-16). In 1539, on 6 March and 10 November, as an official of the Chapter, Copernicus handled certain sums of money allotted to the vicariate of Achtsnicht (ZGAE, 1927-1929, *23*:798, no. 4; Sikorski, p. 116, no. 452).

[78] In 1508, shortly before his death on 20 January 1509, canon Zacharias Tapiau bequeathed 522^1/$_2$^{m} for the establishment of two vicariates (ZGAE, 1914-1916, *19*:817; *24*:57/6). In January 1511, as a Chapter inspector (*visitator*), Copernicus transferred the "remaining money on deposit in the castle [of Olsztyn] for the vicariates of the venerable Zacharias" to Frombork (Sikorski, p. 28, no. 54; PI^1, 381; PI^2, 256, omitting Zacharias' name).

[79] Initially Copernicus wrote *29*, which he deleted. These 9 items add up to 22^5/$_6$^{p} = 3^s 4^5/$_6$^{p} (rather than Copernicus' ^{2}/$_3$^{p}); 233^s + 3^s = 236^s = 3^m 56^s; 1884^m + 3^m = 1887^m; 1887^m 56^s 4^5/$_6$^{p}.

[80] Copernicus inserted *et est* above the line.

[81] Initially Copernicus wrote *2660*, but replaced the third digit by *1*. The 11 items in the second distribution add up to 31^2/$_3$^{p} = 5^s 1^2/$_3$^{p}; 393^s + 5^s = 398^s = 6^m 38^s; 2604^m + 6^m = 2610^m; 2610^m 38^s 1^1/$_3$^{p} (not 0^1/$_3$^{p}, as in Copernicus).

[82] Copernicus' *horarum* was misread as *horarium* by NM, p. 59/9 up. A canon who participated in certain canonical hours received specified payments in money and in kind (Hipler, pp. 251-252, no. 17; PII, 505, no. 17; ZGAE, 1972, *36*:64, no. 19).

[83] Initially Copernicus wrote *utraque*, which he deleted, while putting *ambobus* above the line.

[84] Copernicus' p̄cedentibus (*praecedentibus*) was misread as *procedentibus* by NM, p. 59/8 up.

[85] 1887^m 56^s 4^2/$_3$^{p}
  +2610   38   0^1/$_3$
  ―――――――――――
  4498^m 34^s 5^p

[86] first distribution   894^m 24^s 5^3/$_8$^{p}
  second distribution 1075   28   1
  ―――――――――――
  1969^m 53^s 0^3/$_8$^{p} (2^p, according to Copernicus)

[87] Copernicus' *utā* (*ut supra*) was misread as utum (?) by NM, p. 59/5 up. The Haselau, Cruntsche, and Frombork loans were charged to Bishop Nicholas' legacies near the first subtotal under Expenditures.

[88] This is the second subtotal under Expenditures.

[89] 1055^m 9^s 2^p 1969^m 53^s 2^p
  + 197   13   5   —1262   23   1
  ―――――――――――   ―――――――――――
  1262^m 23^s 1^p 707^m 30^s 1^p

[90] first distribution   61^m 21^s 2^1/$_3$^{p}
  second distribution 73   46   4^1/$_2$
  ―――――――――――
  135^m 8^s ^{5}/$_6$^{p} (rounded to 1^p by Copernicus)

[91] Copernicus' *utā* was misread as utum (?) by NM, p. 59/2 up. Under Expenditures, immediately after the second subtotal, 50^m were charged to the workshop in 1527.

[92] first distribution   77^m 52^s 5^1/$_8$^{p}
  second distribution 93   38   3^1/$_4$
  ―――――――――――
  171^m 31^s 2^3/$_8$^{p}

[93] first distribution   198^m 14^s 0^2/$_5$^{p}
  second distribution   238   21   4^1/$_3$
  ―――――――――――
  436^m 35^s 4^{11}/$_{15}$^{p} (rounded to ^{3}/$_4$ by Copernicus)

[94] first distribution   210^m 2^s 0$^1/_4^p$
     second distribution   252 33 0
                            ─────────────────
                            462^m 35^s 0$^1/_4^p$

[95] Under Expenditures, second item within the third subtotal, 1529–1531: Wusen, Appelaw, Sack, Vusen, Vogt, preacher.

[96]   462^m 35^s 0$^1/_4^p$
        —420 33 2
        ─────────────────
         42^m 1^s 4$^1/_4^p$

[97] first distribution   151^m 2^s 1^p
     second distribution   181 36 3$^1/_3$
                            ─────────────────
                            332^m 38^s 4$^1/_3^p$

[98] first distribution   61^m 21^s 2$^1/_3^p$
     second distribution   73 46 4$^1/_2$
                            ─────────────────
                            135^m 8^s $^5/_6^p$ (rounded to 1^p by Copernicus)

[99] first distribution   103^m 50^s 1$^1/_3^p$
     second distribution   124 52 4
                            ─────────────────
                            228^m 42^s 5$^1/_3^p$

[100] first distribution   129^m 47^s 4$^7/_{10}^p$
      second distribution   212 49 0$^2/_3$
                            ─────────────────
                            342^m 36^s 5$^{11}/_{30}^p$ (instead of 36^s 5$^1/_3^p$, Copernicus writes 37^s 1^p)

[101] The community of vicars did not share in the first distribution, but in the second distribution received 113^m 29^s 3^p. It invested 53^m 20^s in Frombork on 30 July 1530, according to the last entry just before the third subtotal of expenditures.

        113^m 29^s 3^p
        — 53 20
        ─────────────────
         60^m 9^s 3^p

[102] first distribution   did not participate
      second distribution   170^m 15^s 3$^1/_8^p$

[103] Copernicus' $\overline{pceden}$ (praecedentis) was misread as procedentis by NM, p. 60/12.

[104] The first entry after the addition of the two distributions gave the balance in Bishop Nicholas' legacies as 707^m 30^s 1^p, on the old standard.

[105] Initially Copernicus wrote 503, which he deleted and replaced by 530, above the line.

[106] 707^m 30^s 1^p = 42420^s + 30^s 1^p = 42,450^s 1^p = 254,701^p, old standard; $^3/_4$ old standard = new standard; $\dfrac{254{,}701^p \times 3}{4} = \dfrac{764{,}103^p}{4} = 191{,}025^3/_4^p = 31{,}837^s 3^3/_4^p = 530^m 37^s 3^3/_4^p$ (but 3$^1/_4^p$, according to Copernicus)

[107] The workshop's balance was 85^m 8^s 1^p, old standard.

[108] Initially Copernicus wrote 47, which he deleted and replaced by 51, above the line.

[109] 85^m 8^s 1^p = 5100^s + 8^s 1^p = 5108^s 1^p = 30,648^p + 1^p = 30,649^p, old standard; $^3/_4$ old standard = new standard; $\dfrac{30{,}649^p \times 3}{4} = \dfrac{91{,}947^p}{4} = 22{,}986^3/_4^p = 3831^s 0^3/_4^p = 63^m 51^s 0^3/_4^p$

[110] first distribution   77^m 52^s 5$^1/_8^p$
      second distribution   93 38 3$^1/_4$
                            ─────────────────
                            171^m 31^s 2$^3/_8^p$

171^m 31^s 2$^3/_8^p$ = 10,260^s + 31^s 2$^3/_8^p$ = 10,291^s 2$^3/_8^p$ = 61,746^p + 2$^3/_8^p$ = 61,748$^3/_8^p$, old standard; $^3/_4$ old standard = new standard; $\dfrac{61{,}748^3/_8^p \times 3}{4} = \dfrac{185{,}245^1/_8^p}{4} = 46{,}311^9/_{32}^p = 7718^s 3^9/_{32}^p = 128^m 38^s 3^9/_{32}^p$ (rounded to 3$^1/_2^p$ by Copernicus)

[111] first distribution   198^m 14^s 0$^2/_5^p$
      second distribution   238 21 4$^1/_3$
                            ─────────────────
                            436^m 35^s 4$^{11}/_{15}^p$

$436^m\ 35^s\ 4^{11}/_{15}{}^p = 26160^s + 35^s\ 4^{11}/_{15}{}^p = 26195^s\ 4\ ^{11}/_{15}{}^p = 157170^p + 4\ ^{11}/_{15}{}^p = 157174\ ^{11}/_{15}{}^p$, old standard; $^3/_4$ old standard = new standard; $\dfrac{157{,}174^{11}/_{15}{}^p \times 3}{4} = \dfrac{471{,}524^3/_{15}{}^p}{4} = 117{,}881^1/_{20}{}^p = 19{,}646^s\ 5^1/_{20}{}^p = 327^m\ 26^s\ 5^1/_{20}{}^p$ (rounded to 5^p by Copernicus)

[112] The pulpit's balance was $42^m\ 1^s\ 4^1/_4{}^p = 2520^s + 1^s\ 4^1/_4{}^p = 2521^s\ 4^1/_4{}^p = 15{,}126^p + 4^1/_4{}^p = 15{,}130^1/_4{}^p$, old standard; $^3/_4$ old standard = new standard; $\dfrac{15{,}130^1/_4{}^p \times 3}{4} = \dfrac{45{,}390^3/_4{}^p}{4} = 11{,}347^{11}/_{16}{}^p = 1891^s\ 1^{11}/_{16}{}^p = 31^m\ 31^s\ 1^{11}/_{16}{}^p$ (rounded to $1^3/_4{}^p$ by Copernicus)

[113] The vicariate of Achtsnicht had a balance of $332^m\ 38^s\ 4^1/_3{}^p = 19{,}920^s + 38^s\ 4^1/_3{}^p = 19{,}958^s\ 4^1/_3{}^p = 119{,}748^p + 4^1/_3{}^p = 119{,}752^1/_3{}^p$, old standard; $^3/_4$ old standard = new standard; $\dfrac{119{,}752^1/_3{}^p \times 3}{4} = \dfrac{359{,}257^p}{4} = 89{,}814^1/_4{}^p = 14{,}969^s\ ^1/_4{}^p = 249^m\ 29^s\ ^1/_4{}^p$

[114] The vicariate of Zacharias had a balance of $135^m\ 8^s\ 1^p = 8100^s + 8^s\ 1^p = 8108^s\ 1^p = 48{,}648^p + 1^p = 48{,}649^p$, old standard; $^3/_4$ old standard = new standard; $\dfrac{48{,}649^p \times 3}{4} = \dfrac{145{,}947^p}{4} = 36{,}486^3/_4{}^p = 6081^s\ ^3/_4{}^p = 101^m\ 21^s\ ^3/_4{}^p$

[115] The vicariate of Barptolomeus had a balance of $228^m\ 42^s\ 5^1/_3{}^p = 13{,}680^s + 42^s\ 5^1/_3{}^p = 13{,}722^s\ 5^1/_3{}^p = 82{,}332^p + 5^1/_3{}^p = 82{,}337^1/_3{}^p$, old standard; $^3/_4$ old standard = new standard; $\dfrac{82{,}337^1/_3{}^p \times 3}{4} = \dfrac{247{,}012^p}{4} = 61{,}753^p = 10{,}292^s\ 1^p = 171^m\ 32^s\ 1^p$

[116] According to Copernicus' computation, the vicariate of the prebends had a balance of $342^m\ 37^s\ 1^p = 20{,}520^s + 37^s\ 1^p = 20{,}557^s\ 1^p = 123{,}342^p + 1^p = 123{,}343^p$, old standard; $^3/_4$ old standard = new standard; $\dfrac{123{,}343^p \times 3}{4} = \dfrac{370{,}029^p}{4} = 92{,}507^1/_4{}^p = 15{,}417^s\ 5^1/_4{}^p = 256^m\ 57^s\ 5^1/_4{}^p$

[117] The community of vicars had a balance of $60^m\ 9^s\ 3^p = 3600^s + 9^s\ 3^p = 3609^s\ 3^p = 21{,}654^p + 3^p = 21{,}657^p$, old standard; $^3/_4$ old standard = new standard; $\dfrac{21{,}657^p \times 3}{4} = \dfrac{64{,}971^p}{4} = 16{,}242^3/_4{}^p = 2707^s\ ^3/_4{}^p = 45^m\ 7^s\ ^3/_4{}^p$. This should be 46^m, says NM, p. 60, n. ii, without any explanation. But 46^m new standard = $61^m\ 20^s$, old standard, as against a balance of $60^m\ 9^s\ 3^p$, old standard, in the account of the community of vicars.

[118] The office of canonical hours had a balance of $170^m\ 15^s\ 3^1/_8{}^p = 10{,}200^s + 15^s\ 3^1/_8{}^p = 10{,}215^s\ 3^1/_8{}^p = 61{,}290^p + 3^1/_8{}^p = 61{,}293^1/_8{}^p$, old standard; $^3/_4$ old standard = new standard; $\dfrac{61{,}293^1/_8{}^p \times 3}{4} = \dfrac{183{,}879^3/_8{}^p}{4} = 45{,}969^{27}/_{32}{}^p = 7661^s\ 3^{27}/_{32}{}^p = 127^m\ 41^s\ 3^{27}/_{32}{}^p$ (rounded to 4^p by Copernicus).

[119] Copernicus understated this total by $3^3/_4{}^p$, since it should be $2034^m\ 14^s\ 2^1/_4{}^p$.

[120] This account was written by Giese.

[121] Initially Giese wrote *officiis* here, but deleted it because he needed it two lines below.

[122] The pulpit's previous balance $\quad 31^m\ 31^s\ 1^3/_4{}^p$
$\underline{+15\qquad\qquad}$
$46^m\ 31^s\ 1^3/_4{}^p$ (rounded to 2^p by Giese)

[123] Giese started to write *perag-*, which he deleted.

[124] For 20^m, as the last item in the second subtotal of expenditures in 1531 (Document II). *Colmensze*, as written here, shows that *Colmensche* was written there (as against Colmensehe in NM, p. 56/5 up, p. 93/right/11).

[125] The Hungarian florin was worth $2^1/_4$ Prussian marks (20 Hungarian florins = 45 Prussian marks, according to the third item in this section). Hence, 106 Hungarian florins = $238^1/_2$ Prussian marks. In Copernicus' 1531 inventory of gold coins, the Hungarian piece was equated with 3 Prussian marks (see n. 56, above).

[126] This Rhenish coin is the fifth item in Copernicus' 1531 inventory of gold pieces. There it is equated with 2^m, but here with $1^m\ 30^s\ (= 240^m - 238^1/_2{}^m)$.

[127] This expenditure of 240 Prussian marks (= 106 Hungarian florins + 1 Rhenish coin) does not include the 100 Polish marks.

[128] Hence, the Hungarian florin was worth $2^1/_4$ Prussian marks.

[129] The price $91^m\ 26^s = 5486^s$, at $19^m = 1140^s$ per mark by weight = 96 skoters, or $11^7/_8{}^s$ per skoter,

bought slightly less than 462 skoters = 4 marks by weight, 78 skoters. NM's misreading (mr. IIII j, p. 61/12 up) does not correspond to Giese's Latin text (Plate XXIVd, right column, 5 up–4 up).

[130] $\quad$ 240^m
$\qquad$ 46 $\quad$ 16$^{1/_2 s}$
$\qquad$ 45
$\qquad$ $\underline{91 \quad 26}$
$\qquad$ 422^m 42$^{1/_2 s}$

[131] $\quad$ 422^m 42$^{1/_2 s}$
$\qquad$ $\underline{-\ 13}$
$\qquad$ 409^m 42$^{1/_2 s}$

[132] Initially Giese wrote *L* (fifty), which he deleted.

[133] $\quad$ 530^m 37^s 3$^{1/_2 p}$
$\qquad$ $\underline{-409 \quad 42 \quad 3}$
$\qquad$ 120^m 55^s $^{1/_2 p}$ (this fraction was deleted by Giese, who replaced it by $v = 5$)

According to NM, p. 61, n. e, this result should be 120^m 54$^{1/_2 s}$ 3$^{1/_2 p}$ (which Giese would have written as 120^m 55^s $^{1/_2 p}$).

[134] Giese's $\overline{pcedent\jmath}$ was misread as *procedentis* by NM, p. 61/last line.

[135] Under the heading "For the vicariate of St. Barto[lomeus]," an unidentified individual twice wrote and twice deleted in the Chapter's account books 1508–1547 (Archives of the Varmia Diocese, RF 11, fol. 27^v–28^r): "From the hands of the venerable Tiedemann Giese, Dr. Nicho[las Copernicus] and Felix Reich, on 8 November I received 28^m 21^s" (Sikorski, p. 90, no. 342). As administrator, Reich received 171^m 32^s 1^p for the vicariate of St. Bartholomew on 30 April 1532. This aborted transfer of 28^m 21^s presumably took place (or failed to take place) on 8 November 1532, since the following entry in the account books is dated 19 January 1533.

[136] Written by Faulhaber, a vicar.

COPERNICUS AS A PHYSICIAN

In the summer of 1501 the Varmia Chapter granted Canon Nicholas Copernicus a further leave of absence, "principally because Nicholas promised to study medicine, and as a helpful physician would some day advise our most reverend bishop and also all the members of the Chapter."[1] Copernicus chose to pursue his medical studies at the University of Padua, which he entered in the winter semester of 1501. Remaining in Padua only two years, he did not earn a medical degree, for which three full years were required. The lack of a medical degree, however, did not prevent him from practicing that still unlicensed profession. When his uncle, the bishop of Varmia, fell ill, Copernicus served as his personal physician. On 7 February 1507 the Chapter awarded Copernicus a monetary bonus, over and above his regular income as a canon, for every year he would spend in the bishop's service.[2] This responsibility of looking after the bishop's health required Copernicus to live, not in the Chapter's cathedral town of Frombork, but in the episcopal headquarters in Lidzbark.

The notation "For the episcopal library in the residence in Lidzbark" was written[3] on the title page of a copy of the edition of the *Surgery* by Pietro d'Argelata († c. 1423) which was published in Venice on 12 September 1499. A year and a half earlier, 11 March 1498, an edition of the alphabetically arranged dictionary of medicinal simples by Matteo Silvatico († c. 1342) was also published in Venice, which then controlled Padua and its university. A copy of Silvatico's simples is bound with the Lidzbark copy of Argelata.[4] On fol. 182^v of Silvatico, there is a handwritten recommendation[5] of the imperial pills (mis)attributed to Arnold of Villanova (c. 1240–1311):

These can be taken at any time without previous preparation, diet or precaution, in the morning and evening, before a meal or after it, without syrup, by anybody in good or bad health. They are efficacious in the digestion of every substance and in any[6] illness. They harmlessly eliminate whatever is excessive. They find their way to, and very much strengthen the principal and weak organs. By inducing cheerfulness they postpone gray hair, which comes from corrupt humors. They reinforce whatever has been damaged by sharp, salty humors. They take care of the power of vision above all. They regulate the stomach and keep it in order. They repress catarrh, stop a cough, relieve the congestion and all the [other] malfunctions of the throat and mouth, alleviate gas in the stomach, get rid of opisthotony,[7] sharpen the wits, strengthen and invigorate the nerves, preserve the teeth from decay, guard against the plague, against arthritic itch and the gout, induce sleep, protect weakened bodies against the incursion of illnesses, drive away both forms of colic with phlegm, and purge gently. Finally, anybody who wants to be purged by these pills should take one on the first day, two on the second day, three on the third day, and so on, until the seventh day, or as much as seems to help the patient. The composition of the pills is as follows:

R$_x$ amomum		rhubarb, weighing as much as the aforementioned ingredients
anise		
cardamom		
ginger		
cinnamon		aloe syrup, weighing as much as the aforesaid total
zedoary[8]		
mastic		
nutmeg	1 scruple	
gillyflower		
crocus		Compound everything with the essence of violets or roses, and press it into one mass.
cubeb		
aloe syrup		
good turpeth		When you want to use it, make pills in the shape of a bean or pea.
manna		
agaric		
senna		
five grains of mirabelle		

After the death of his uncle Lucas Watzenrode, Copernicus was consulted from time to time by the bishops of Varmia who succeeded his uncle: Fabian von Lossainen, Maurice Ferber, and Johannes Dantiscus. In addition, on Saturday, 24 February 1532, Copernicus wrote a prescription for a sister of a fellow-canon Achatius Freundt (c. 1480–1533). She was very sick, "throwing up gobs from her stomach, with loss of appetite." Copernicus prescribed that her food should be mixed with

R$_x$ red coral		
cinnamon amomum	1	scruple
Abbot Diarod's preparation	2$\frac{1}{2}$	scruples
white sugar	$\frac{1}{2}$	pound
distilled water	$\frac{1}{2}$	quartern,
the compound to be made in a round form.[9]		

Copernicus' services as a physician were sought also by Duke Albert of Prussia for one of his henchmen, a story that may be followed in the section on Copernicus' Correspondence, Letters 14–15.

Copernicus treated his fellow-canon Felix Reich. "While I was still alive," said Reich in his will, dated 22 November 1538, "I gave him as his personal property" a copy of the *Materia medica* by Dioscorides, the principal ancient authority on pharmacology. Unlike Reich's will, Copernicus' will has not been preserved. Nevertheless, we know that he bequeathed three medical volumes to Fabian Emerich (1477–1559), his successor as physician to the Chapter.[10]

One of them contains Arnold of Villanova's *Breviarium practicae* (Pavia, c. 1485), an encyclopedia of practical medicine.[11] Bound with it is a copy of the first edition of the *Canonica de febribus* (Bologna, 8 March 1487), the systematic treatise on fevers[12] by Michele Savonarola (c. 1384–1468), the grandfather of Friar Girolamo Savonarola, who was burned at the stake on 23 May 1498 in Florence while Copernicus was in Bologna. The marginal notes on fol. 5–13 of Villanova's *Breviarium* have been misattributed to Copernicus.[13] But the facsimile in KMW, 1970, p. 598, clearly shows that they were written by someone else. Since they concern the eye, they may have been written by Emerich, who was especially interested in ophthalmology. The composite Villanova-Savonarola volume was originally owned by a Gdańsk physician, who bequeathed it to a businessman in that city.[14] How Copernicus acquired it has not yet been clarified. It bears Emerich's ownership note that it came to him from Copernicus' estate through the executors of his will in accordance with the testator's command.[15]

The second of the three volumes bequeathed by Copernicus to Emerich contains the commentary on Ibn Sina (Avicenna), Canon I, fen 4 (ed. Venice, 4 February 1485) by Ugo Benzi (1376–1439), sometimes called Hugo of Siena. Benzi's commentary is bound with a copy of the Venice 1486 edition of Michele Savonarola's *Practica medicinae*.[16] The last of the volumes in Copernicus' legacy to Emerich consists of the *Practica*, which was begun on 10 June 1418, "after 36 years of the customary practice," by the Portuguese physician Valescus of Taranta. Copernicus bequeathed to Emerich a copy of the edition which was completed on 19 May 1490.[17] Whether any of the annotations in the three medical volumes left to Emerich were written by the testator Copernicus is uncertain.

By contrast with Copernicus' achievement in astronomy and his analytical mastery of the money question, his activities as a physician did not advance medicine at all, in the judgment of the twentieth century. Quite different was the opinion of those closer to his own time: "In medicine he was honored like a second Aesculapius."[18]

The foregoing survey of Copernicus' medical studies and medical practice justifies an affirmative answer to the question recently asked: "Was the astronomer Nicholas Copernicus a physician?"[19] The propounder of this question took seriously a hoax jocularly pretending that Copernicus introduced the buttering of bread as a remedy against the plague.[20]

BIBLIOGRAPHY

Berg, Alexander, "Der Arzt Nikolaus Kopernikus und die Medizin des ausgehenden Mittelalters," pp. 172–201, in Johannes Papritz and Hans Schmauch, eds., *Kopernikus-Forschungen* (Leipzig, 1943; Deutschland und der Osten, Quellen und Forschungen zur Geschichte ihrer Beziehungen, vol. 22)

Eis, Gerhard, "Zu den medizinischen Aufzeichnungen des Nicolaus Coppernicus," *Lychnos*, 1952, pp. 186–209

Flis, Stanisław, "Kopernikowski inkunabuł medyczny w Olsztynie," KMW, 1970, pp. 589–606

Hand, Samuel B. and Kunin, Arthur S., "Nicholas Copernicus and the Inception of Bread-Buttering," *Journal of the American Medical Association*, 1970, *214*:2312–2315, and cover of Vol. 214, No. 13 [a hoax]

Holz, Max, "Pro cassia fistula doctori Nicolao Koppernic," ZGAE, 1935, 25:233–237

Konopka, Stanisław, "Mikołaj Kopernik wśród lekarzy," pp. 189–209, in Józef Hurwic ed., *Mikołaj Kopernik; Szkice monograficzne* (Warsaw, 1965)

Rytel, Alexander, "Nicolaus Copernicus — Physician and Humanitarian," *Polish Medical History and Science Bulletin*, 1956, *1*:3–11 [unreliable]

Valentin, H., "Herkunft und Echtheit des im Frauenburger Museum vorhandenen Koppernikus-Rezepts," *Süddeutscher Apotheker-Zeitung*, 1931, pp. 152–154

Zimmermann, Walter, "Bemerkungen zu H. Valentin," ibid., pp. 154–155

NOTES

[1] TCT, p. 327; StC VIII, 43, no. 38.

[2] StC VIII, 48, no. 50.

[3] By Copernicus, according to MCV, I, 60/7; PI^1, 336/9 up, and L. A. Birkenmajer, *Stromata Copernicana*, p. 324/6 up.

[4] StC XVI, 370, no. 11.

[5] MCV, I, 61–62; PI^2, 314–315, misattributed this recommendation to Copernicus.

[6] Deleting *a* (PI^2, 314/11 up).

[7] Reading *opisthotoniam*, not stotonomam (as in MCV, I, 61/11–12; PI^2, 314/5 up).

[8] Reading *zeduarii*, as in PI^2, 312/last line, and PII, 245/10 up, not zoduarii (as in MCV, I, 62/12, and PI^2, 315/17 up). In DSB, IV, 121/right/24, zeodary is a typographical error.

[9] MK, p. 577.

[10] Like Copernicus, Emerich practiced medicine without holding a doctoral degree in that subject. The attribution of the doctorate in medicine to Emerich (L. A. Birkenmajer, *Stromata Copernicana*, p. 316/n. 1; repeated in StC XVI, 369/3) lacks documentary support.

[11] Facsimile of the initial page in KMW, 1970, p. 595.

[12] Facsimile of the initial page in KMW, 1970, p. 601, and of the colophon on p. 594.

[13] By Zinner, *Entstehung*, p. 407, no. 24. Other misattributions were corrected by H. Valentin, "Herkunft und Echtheit des im Frauenburger Museum vorhandenen Koppernikus-Rezepts," *Süddeutscher Apotheker-Zeitung*, 1931, pp. 152–154.

[14] Facsimile in KMW, 1970, p. 592.

[15] StC XVI, 368, no. 8 (1).

[16] StC XVI, 369, no. 9.

[17] StC XVI, 369–370, no. 10.

[18] StC XXI, 14/25–26.

[19] Constantin Bart, "L'Astronome Nicolas Copernic a-t-il été médecin?," *Actes du XXXe Congrès international d'histoire de la médecine* (Quebec, 1977), pp. 477–486.

[20] Samuel B. Hand and Arthur S. Kunin, "Nicholas Copernicus and the Inception of Bread-Buttering," *Journal of the American Medical Association*, 1970, *214*: 2312–2315.

LETTERS WRITTEN BY COPERNICUS

INTRODUCTION

The surviving letters written by Copernicus present a startling contrast to the extant correspondence of the three astronomers to whom he is most often compared: Tycho Brahe (1546–1601), Galileo Galilei (1564–1642), and Johannes Kepler (1571–1630). Only seventeen of Copernicus' letters survive. With but two exceptions (Letters III, IV), they are all quite brief and business-like. The light they throw on Copernicus' life and character, and on his intellectual development, flickers by contrast with the steady, strong illumination provided by the far more copious correspondence conducted by Brahe, Galileo, and Kepler. By reason of their paucity, Copernicus' letters are all the more precious, since his other writings are sternly taciturn, unlike the outspoken publications of his three fellow-astronomers.

Copernicus' two longer letters were both written as responses. In Letter III he castigates an astronomical work which was sent to him by a friend eager to know his opinion of it. In Letter IV he replies to an objection directed by a fellow-canon against his treatise on money.

The remaining fifteen briefer letters are all concerned in one way or another with the affairs of the Varmia Chapter, of which Copernicus was a canon. Letter I is addressed to the Chapter itself; twelve others to a Varmia bishop; and the remaining two to the duke of Prussia, to whom Copernicus, whose mother tongue was German, wrote in that language, whereas he composed all his other letters in Latin.

The two letters (XIV, XV) addressed to the duke of Prussia were in the Koenigsberg archives at the time of World War II. After the war (Koenigsberg having been annexed by the Soviet Union and renamed Kaliningrad), they were kept at Göttingen, and now they are in West Berlin. Letter XV was the first of Copernicus' letters to be published (in 1819).

Letter I, addressed to the Varmia Chapter, remained in Frombork until that city was captured on 9 July 1626 (SRW II, 611/21) by Sweden, which confiscated the Chapter's archives. In 1798 Sweden sent Letter I to Prussia, where it was preserved in the Koenigsberg archives until 1930. As part of an exchange it was returned in that year to the Frombork archives, but was lost during World War II. Fortunately, just before that loss, photofacsimiles of it had been published.

The remaining twelve letters, all addressed to a bishop of Varmia, remained in Lidzbark until in 1703 "the young and arrogant king of Sweden, Charles XII, about eighteen years old, with a part of his army entered Varmia and ... moved to Lidzbark. He remained in the bishop's palace ... throughout the entire winter and the spring, until 25 June of the following year 1704. He transferred to Sweden the old manuscripts of the episcopal archives, together with the library" (SRW II, 649). Thus in 1703–1704, in Charles XII's replay of Gustavus Adolphus' scenario, the Lidzbark episcopal archives suffered the same fate as had overtaken the Frombork Chapter's archives three-quarters of a century before.

In Sweden, Copernicus' letters to a Varmia bishop were acquired by the library of the University of Uppsala, which still holds Letters II and IX. In 1810, however, it gave up seven others to the administrator of the Czartoryski Library in Puławy. Six of these (VII, VIII, X, XII, XIII, XVII) are now preserved in the Czartoryski Library in Cracow, which was founded in 1873. But Letter VI, which had been obtained in the same way, was sent from Puławy to Great Britain, where a facsimile of it was published. Although it was brought back to Poland, it has not yet been found. Letter V also is lost; a late copy of it is possessed by the Czartoryski Library.

Letter XVI somehow survived the Swedish seizure of the Lidzbark archives until the three successive partitions of Poland in 1772, 1793, and 1795. The last independent bishop of Varmia then transferred Letter XVI to his private collection. This was dispersed after his death in 1801, and Letter XVI was acquired by the institution which is now called the National Library in West Berlin.

Letter XI has not survived. But in 1618, while it was still in the episcopal archives in Lidzbark, it was copied by Jan Brożek, that very zealous collector of books and documents related to Copernicus. Brożek's copy is preserved in the Jagiellonian Library in Cracow.

Brożek also had "various letters ... written by the hand of Copernicus himself to his uncle Lucas [Watzenrode] and others," including astronomers in Cracow "with whom he conferred about eclipses and observations of eclipses," according to the earliest well-informed biographer of the astronomer; see Erna Hilfstein, *Starowolski's Biographies of Copernicus* (Wrocław, 1980; StC XXI, 15). But these letters, once possessed by Brożek, now seem irretrievably lost.

Also lost is Copernicus' letter of 1 July 1540 to the individual who by a strange twist of fate later became the second and final editor of the *Revolutions*. The loss of this letter is particularly regrettable because it evidently set forth Copernicus' attitude toward the place of hypotheses in science, as can be inferred from the answer it elicited on 20 April 1541, after having been received in March 1541 after a long delay in transit (StC XVI, 453–455).

The surviving correspondence shows that four other letters of Copernicus have been lost. One was an official report to the Varmia Chapter, like Letter I (see n. 9, thereon). The second concerned two vacant Varmia canonries to be filled in accordance with a plan transmitted from one bishop to another by Copernicus (Letter IX). The third informed a relative that Copernicus as proxy had taken possession of the Varmia canonry recently conferred on the relative's son *in absentia* (Letter XIII, n. 3). The fourth and last asked a fellow-physician for help in treating a patient (Letter XIV, n. 1). The nature of these four lost letters suggests that, while any and all additional information about Copernicus will always be gladly received by scholars, they need not anticipate any astonishing discoveries.

I

When Copernicus wrote Letter I, he was serving his Chapter as its administrator, normally stationed in Olsztyn in the southern part of the diocese of Varmia. On account of an emergency, however, he hurried north to Lidzbark, the residence of the bishop of Varmia. After conferring with the bishop, he went on to Melsac in order to investigate certain crimes. While he was in Melsac, he wrote Letter I to his Chapter in Frombork in the northwestern part of the diocese.

Letter I was written at a time of high tension between the diocese of Varmia and its powerful neighbor, the Knights of the Teutonic Order. The pope had recently sent an emissary in an effort to avert the impending war between the Knights and the kingdom of Poland, of which the diocese of Varmia was a part. A military officer in the employ of the Knights was embroiled with a local official under Copernicus' jurisdiction. It was rumored that the Grand Duke of Moscow, on whose support the Knights were counting, had just made peace with Poland.

Venerable and worshipful gentlemen,
honorable masters:

I learned from his Most Reverend Lordship [the bishop of Varmia] yesterday what your Reverences[1] write about preparing the reception.[2] The arrangements are virtually complete for either [contingency], whether it happens to be a fish day or a meat day.[3]

P[hilip][4] Greusing's letter[5] impelled me to leave Olsztyn[6] sooner [than I had intended]. At my invitation the burgrave[7] left with me. In[8] Lidzbark he received more complete information, as a result of which Greusing will be unable to complain that he has been denied justice.[9]

His Most Reverend Lordship also commissioned me to advise[10] your Reverences[11] concerning the reply to be given to the Grand Master [of the Teutonic Knights]. If the letter[12] has not been sent, in the copy transmitted [to you] by his Lordship the following clause is to be added:[13] "that holy justice may not be blocked,"[14] the better to forestall their perverse and quibbling interpretation.

His Lordship has also received the news that [the Grand Duke of] Moscow has signed with the king [of Poland] a permanent peace treaty, the provisions of which his Lordship expects to learn at any moment.[15] Thus the complete[16] confidence of our neighbors has accordingly now collapsed.[17]

I commend myself to your Reverences.[18]

Melsac,[19] 22[20] October 1518

I shall leave from here too as soon as I can.[21]

N. Coppernic

To the Venerable[22] and Worshipful
Officers, Canons, and Chapter
of the Church of Varmia,
most honorable masters

¹ Copernicus' abbreviation *d^es* was improperly expanded as *dignitates* by Baranowski (p. 589/3), who published Letter I for the first time. His reading was repeated by Hipler, p. 165; PII, 143; and Dmochowski, p. 197. But Triller, p. 124, expanded correctly (*dominationes*), and was followed by StC XVIII, 215.

² This reception was to honor the papal emissary, Nicholas Schönberg (1472–1537; for his papal credentials, dated 19 March 1518, see Joachim, II, 171, Document no. 9). Schönberg consulted first with King Sigismund of Poland in his capital city, Cracow, and then proceeded to the headquarters of the Teutonic Knights in Koenigsberg. On 13 July the king informed the bishop of Varmia about Schönberg's movements (MK, p. 536, citing Czartoryski Library MS 1601, p. 135). The bishop promptly dispatched to Koenigsberg an invitation to Schönberg to visit him in his palace in Lidzbark. On 28 July Schönberg replied that he could be expected in Lidzbark "during the following week or the week thereafter" (ZGAE, 1914–1916, *19*:314/10–12). The reception for Schönberg was to take place in Lidzbark, not in Olsztyn (as PI², 98/7–8, mistakenly thought), nor in Melsac (as was thought by MK, p. 537; Wasiutyński, pp. 284–285; and StC XVIII, 45, 88, equally mistakenly).

³ In his letter of 28 July, Schönberg indicated that he might be in Lidzbark before the middle of August. Some two months later, however, on 21 October (Copernicus' "yesterday") the bishop still did not know exactly when to expect Schönberg. The Frombork Chapter having inquired about the reception for him, the bishop wanted to assure them that almost everything was ready, whether he arrived on a Friday or on some other day of the week. But instead of answering the Chapter in a separate communication of his own, the bishop had his reply to the Chapter's query about the reception incorporated in the report which Copernicus was required to make to the Chapter about an administrative matter.

On Christmas Day, 1518, Schönberg was in Frombork, where he wrote two letters (Joachim, II, 197–198, Documents no. 33–34), and he spent the holiday with Bishop Fabian in Lidzbark (ibid., p. 33/last line). At that time Copernicus was not in Frombork in northern Varmia, nor in Lidzbark in central Varmia, but in Olsztyn in southern Varmia, where he was serving as the Chapter's administrator (see above, Administrative Documents, Leases, n. 222). Hence he and Schönberg did not meet in 1518. That is why Schönberg said nothing about 1518 when 18 years later, on 1 November 1536, he wrote from Rome urging Copernicus to release the *Revolutions*. When Copernicus received the letter of Schönberg (who had in the meantime become a cardinal), he filed it for future reference. After finally deciding to publish the *Revolutions*, he inserted Schönberg's letter in the front matter. In his Preface, he called Schönberg the foremost of his friends who exhorted him to publish the *Revolutions*. But, like Schönberg, he too said nothing about their having met in 1518.

⁴ Copernicus' abbreviation *p* was misread as *d* by Triller, p. 124.

⁵ Not "letters" (*Briefe*, PI², 103/11). Although *literae* is plural in form, here it denotes only one letter.

⁶ Copernicus had gone to Olsztyn to serve his Chapter as administrator of its holdings. His three-year term of office had begun on 11 November 1516.

⁷ Copernicus started to write *burgrabius*. But he formed only the first two letters, and then deleted them because he decided to insert *inde* before *burgrabius*. The burgrave of Olsztyn, Christopher Drauschwitz, had witnessed legal documents drafted by Copernicus some five months earlier, on 27 March and 29 May 1518 (see Financial Transactions, II, III, above).

⁸ Not "from" (*von*, as in Schultheiss, p. 43). Copernicus says explicitly "*in heilsberg*," meaning that he and Drauschwitz were in Lidzbark when the burgrave received the additional information.

⁹ On 3 August 1513 Philip Greusing sold to the Varmia Chapter 8 parcels of land as well as his share in both the old and new taverns in a village in the Melsac district (ZGAE, *27*:528/10 up–8 up). Since he transacted this sale through proxies, presumably he was already a military officer of high rank in the employ of the Teutonic Knights. Later on he was involved in litigation with the Varmia Chapter. The trial was scheduled to take place in November 1516 in Koenigsberg. But the Chapter did not want to send its spokesmen there for fear that they would be robbed on the public roads. It therefore asked the Grand Master for a postponement until January 1517 (StC VIII, 69, Document no. 109). On 10 November 1517 Greusing ordered the officials of certain villages to stop making payments to the Varmia Chapter and instead to start giving those payments to himself (ibid., 81, Document no. 147). In this connection on 25 November 1517 the bishop of Varmia sent an instruction to the Chapter's chancellor regarding Greusing (ZGAE, 1905, *15*: p. 226, n. 3). Greusing's conflict with Drauschwitz, the burgrave of Olsztyn, was reported by Copernicus to the Chapter. His report was forwarded by the Chapter to the bishop of Varmia on 6 October 1518, together with an accompanying letter that said in part:

> Your Most Reverend Lordship will learn from our administrator's letter, which we are enclosing herewith, how well founded is Greusing's complaint against our burgrave in Olsztyn. Nevertheless, we are writing to the administrator that, if any fault or delay is due to our burgrave, it is to be corrected as soon as possible by the administrator's meting out justice to him without delay (ZGAE, 1943, *28*:78, no. 7/1–6).

Greusing's complaint about being denied justice was contained in the letter he sent to Copernicus in Olsztyn. It was the receipt of Greusing's letter that impelled Copernicus to leave Olsztyn sooner than he had intended. During the war of 1520–1521, Greusing was captured. On 24 July 1520 he was released on his word of honor to celebrate a religious holiday on 15 August, and then return by 1 September. He did not return (AcTom V, 285/8 up, 320/20). When he was recaptured, he was tortured to death in Malbork (Hubatsch, p. 82/15 up–12 up).

[10] Copernicus' *commonere* has previously been misread as *commovere*.

[11] Again Copernicus' abbreviation *d* should be expanded as *dominationes* not *dignitates* (see n. 1, above).

[12] Not "letters" (*Briefe*, as in PI², 103/18; cf. n. 5, above). The letter in question was to be sent to the Teutonic Knights. A draft had been submitted to the Chapter by the bishop. He now instructed Copernicus, in writing to the Chapter, to advise it to insert an added clause in the reply to the Knights, if there was still time to do so.

[13] Copernicus started to write *addatur*. But he formed only the first five letters, and then deleted them, because he realized that he had omitted *ut*, which he needed before *addatur*.

[14] This legal formula may imply the death sentence, in particular, by hanging (Triller, pp. 124–133). The culprit (Caspar Paipo), whose case is traced by Triller, may be connected in some way with the Greusing-Drauschwitz dispute.

[15] That moment never came. Basil III, Grand Duke of Moscow, did not sign a permanent peace treaty with King Sigismund I of Poland. The alleged news (*novitates*) about the treaty turned out to be an inaccurate rumor about a temporary truce between the two rulers.

[16] Between *tota* and *confidentia* Copernicus started to write a word (perhaps *iam*), which he abandoned after forming the first letter (*j*).

[17] On 10 March 1517 Albert, Grand Master of the Teutonic Knights ("our neighbors," as Copernicus calls them), had concluded a military alliance with Basil III against Poland (Joachim, I, 299–302, Documents no. 130, 131). But that alliance fell apart when Moscow and Poland agreed to a truce.

[18] Copernicus' abbreviation *d.* should be expanded to *dominationibus*, not *dignitatibus* (see nn. 1, 11, above).

[19] After receiving Greusing's letter in Olsztyn, Copernicus left with the local burgrave. He was in Lidzbark on 21 October, and on 22 October he wrote Letter I to the Chapter from Melsac.

[20] Not "2" (as in PII, 141/14), nor "21" (as in PI², 103/2, and MK, p. 536/3 up). The presence of this error in PI², 103, was mistakenly denied by Triller, p. 125, n. 3.

[21] Copernicus had left Olsztyn in a hurry, tarried briefly in Lidzbark, and gone to Melsac where he wanted to spend as little time as possible. It was in the vicinity of Melsac that Paipo's crimes were perpetrated. In a letter Paipo had asked the Varmia Chapter to try him in connection with a murder. The Chapter set a date for the trial in replying to him from Frombork on 16 April 1517. The Chapter's reply is said to be in Copernicus' handwriting, with the characteristic spellings "Frawemburg/Frawemborg" (NM, pp. 20/4–6, 5 up; 36/2; 65, no. 4; 74, no. 4; "signature" in StC VIII, 76, no. 132, mistranslates *autograf* in StC VII, 92, no. 132). This document is described, however, not as the original sent by the Chapter to Paipo, but as a copy made by Copernicus "for his own knowledge and use" (NM, p. 70/7 up–5 up). In that case, the spellings would be characteristic for whoever wrote the original, not for Copernicus, who allegedly copied the original. But the copy (which is photoduplicated in NM's Plate X) is clearly not in Copernicus' handwriting, of which there is an ample sample in NCCW, I. Moreover, Copernicus was not in Frombork on 16 April 1517, when he was in Olsztyn (see Leases, 1517). Two other documents related to Paipo's case are likewise described by NM as being in Copernicus' handwriting. But the photocopies of these two documents (NM's Plates XI, XII) show that their handwriting is different not only from Copernicus' but also from that of the scribe who wrote the Chapter's reply to Paipo. Despite all these discrepancies, the three documents have been uncritically accepted as products of Copernicus' hand and brain (Triller, pp. 181–182). For a critical rejection, see *Isis*, 1973, *64*:551.

[22] PII, 143/3 up's Veneralibus is a misprint for *Venerabilibus*.

Letter I, addressed to the Varmia Chapter, was removed from its archives in Frombork in 1626 to Sweden, which sent it in 1798 to Prussia, where it was preserved in the Koenigsberg archives, Repository 128. In 1930 it passed into the Frombork diocesan archives, as part of an exchange between these two collections (Forstreuter, p. 268). During World War II Letter I was lost, but photofacsimiles of it had been published by Wasiutyński (facing p. 280) and by Schmauch (*Kopernikus-Forschungen*, facing p. 216; ZGAE, 1943, *28*:facing p. 1); see also NCCW IV, Plate V. The text was first printed by Baranowski, p. 589 (to whom the autographs of twelve Copernican letters were lent in 1853; see *Kurier Warszawski*, 1853, p. 863); then by Hipler, pp. 165–166; PII, 143; Dmochowski, p. 197; ZGAE, 1972, *36*:124; and StC XVIII, 215.

Previous translations include Baranowski, p. 589 (reprinted by Polkowski, I, 74–75, and with revisions by Wasiutyński, p. 285); Dmochowski, p. 212; Sikorski, 1973, p. 129 (partial); StC XVIII, 215–216 (these five into Polish) as well as three into German: PI², 103–104; Schultheiss, pp. 43–44, no.2; and Triller, pp. 124–125.

II

Reinhold Feldstedt, who had married a first cousin of Copernicus, and owed him some money, wanted to make a repayment. But he could not reestablish direct contact with Copernicus because war between the Teutonic Knights and the diocese of Varmia had broken out on 1 January 1520. As his Chapter's administrator in Olsztyn, Copernicus was in command of the southern front, while Feldstedt was in his birthplace Gdańsk to the north of the combat area. Among the Varmia canons who took refuge there to escape the fighting was Heinrich Snellenberg. Feldstedt gave him 100 marks for Copernicus. But Snellenberg turned over only 90 marks to Copernicus, who repeatedly asked for the remaining 10, only to be put off by one pretext after another. Finally, all other remedies having been exhausted, Copernicus wrote Letter II to the bishop, asking him to block Snellenberg's income until he paid Copernicus the 10 marks he had been holding back.

My lord, Most Reverend Father
in Christ, my gracious lord:

Some time ago, during the war,[1] the venerable Heinrich Snellenberg[2] received from Reinhold Feldstedt[3] 100 marks of the money Feldstedt owed me.[4] Not long afterward Snellenberg paid[5] 90 of those marks. He remained obligated to me for 10 marks. I often asked him for them. Up to the present time I have not been able to recover them. But, putting me off, he always promised to pay up at the next distribution of the proceeds. Several months having passed, then, it happened that in my presence the venerable administrator[6] counted out a certain share of the money to him. I asked him to pay me then out of that money in accordance with his promises, while I proposed to give him a receipt[7] in full in my own handwriting. Then he again imposed on me with a new objection, and he forced me first to obtain his receipt from Reinhold Feldstedt.[8]

Now the venerable administrator arrived yesterday[9] and distributed the bulk of the proceeds. Holding Snellenberg's receipt, I sought him out, and even so I did not succeed. He said that he wanted to keep all the money [coming to him] from[10] the administrator. If he owed me anything, I should claim it in a legal action in the court of a judge.

I therefore see that I cannot act otherwise, and that my reward for affection[11] is to be hated, and to be mocked for my complacency. I am forced to follow his advice, the advice by which he plans to frustrate me or cheat[12] me if he can.

I have recourse to your Most Reverend Lordship, whom I ask and beseech to deign to order on my behalf the withholding of the income from his benefice[13] until he satisfies me, or a kind provision in some other way for me to be able to obtain what is mine.

I pledge my services with the utmost promptness to your Most Reverend Lordship. May divine goodness preserve you in a completely prosperous long life[14] and happy rule.

Frombork, 29 February 1524

Your Most Reverend Lordship's

Nic. Coppernic[15]

To my lord, Most Reverend Father in Christ,
Maurice [Ferber],[16] by the grace of God bishop
of Varmia, my most honorable and beloved superior

---

[1] War between the Teutonic Knights, on one side, and the diocese of Varmia, supported by Poland, on the other, began on 1 January 1520 and ended with a truce on 15 February 1521.

[2] On 26 September 1501 Snellenberg took possession in person of a Varmia canonry that had become vacant by reason of its occupant's death on 7 May 1499. Snellenberg held the canonry until his death on 24 February 1539 (Hipler, p. 171, n. 1). His name was not Snellenburg (as in Koestler, pp. 183, 622).

[3] Reinhold Feldstedt was a wealthy citizen of Gdańsk (Danzig), from whom the king of Poland on 17 August 1502 borrowed thousands of Hungarian florins, the security being the town of Dirschau and its surrounding villages (Summaria, III, no. 618). A family connection between Copernicus and Feldstedt was established on 21 January 1504, when Feldstedt married Cordula Allen, a first cousin of Copernicus (*Kopernikus-Forschungen*, Table, pp. 136–137:III, 12). On 8 May 1504 Bishop Lucas Watzenrode of Varmia, from whom the king had also borrowed heavily with Dirschau as security, was permitted to cede that town and its villages to Feldstedt (Summaria, III, no. 615, 1501). A loan of 40 Hungarian florins by Feldstedt to the royal notary was guaranteed by the king on 12 March 1505 (ibid., no. 1999). The foregoing are only three of the many Feldstedt financial transactions recorded in Summaria. When Feldstedt died on 24 November 1529, Copernicus was appointed one of the three legal guardians of his orphaned children. The guardianship was terminated on 10 March 1536 (Hipler, pp. 284–285; PI², 264–265).

[4] Feldstedt resorted to an intermediary because he could not repay Copernicus directly. For during the war Copernicus was confined to Olsztyn, where he was in command of the southern front. If it were true that Snellenberg in "the years 1520–1523 stayed with Copernicus in Olsztyn on account of the war" (Hipler, p. 171, n. 1; PI², 216/10–11), then Snellenberg too would not have been able to receive Feldstedt's repayment. But in fact Snellenberg was not in Olsztyn throughout the war. For instance, he was in Tolkemit collecting "more than was due and in excess of the share belonging to him" (PII, 415/23–25). He was apparently also in Elbląg discussing the disposition of the Chapter's light artillery (PII, 411/16). Both Elbląg and Tolkemit are in the north, not far from Gdańsk, where some of the canons were gathered (PII, 415/29). It may have been there that Snellenberg met Feldstedt and received the money owed to Copernicus. Why it was owed, has not yet been clarified.

[5] Copernicus started to write *solvisset* with a second *l*, noticed his error, struck it out, and then wrote the word correctly.

[6] When Giese became Custodian of the Chapter on 12 November 1523, he was at the same time named its general administrator (PI², 26: *General-Official*). The administrator's duties were specified in Sections 33–36, 38, 40, 63 of the Chapter's statutes (PII, 511–514, 521). There is no documentary basis for the statements that "Copernicus agreed with ... Giese that he [Giese] could transmit to him [Copernicus] a certain part of the money" and that "Now for a second time he [Snellenberg] approached Copernicus" (as in StC VIII, 122, no. 256).

[7] The final syllable of *quitaturum*, being slightly detached from the word's first eight letters, was misread as a separate word *nunc* by Prowe, when he published Letter II for the first time in his *Mittheilungen aus schwedischen Archiven und Bibliotheken* (Berlin, 1853), p. 9, no. 2/16. The misreading was repeated by Baranowski, p. 633/13; Hipler, p. 171/10; PII, 144/12, and Dmochowski, p. 196/9; but was eliminated by StC XVIII, 216, no. 2/10.

⁸ When Feldstedt gave Snellenberg 100 marks for Copernicus, Feldstedt prudently obtained from Snellenberg a signed receipt for the full amount. After turning over only 90 marks to Copernicus, Snellenberg, in order to delay (or avoid) paying the remaining 10 marks, insisted that Copernicus in Frombork must obtain from Feldstedt in Gdańsk the receipt which Feldstedt had obtained from Snellenberg during the war. Snellenberg's greedy eyes saw the possibility that Feldstedt might have lost or mislaid his receipt.

⁹ 28 February 1524.

¹⁰ Instead of *a*, Copernicus wrote *ad* because *d* is the first letter of the next word. Noticing his error, he deleted the erroneous *d*.

¹¹ Prowe's misreading *dilatione* (*Mittheilungen*, p. 9/11 up), followed by Baranowski, p. 633/17 up; Hipler, p. 171/19; PII, 144/7 up; and Dmochowski, p. 196/18, was corrected to *dilectione* by StC XVIII, 216, no. 2/16.

¹² StC XVIII, 216, no. 2/18's *defraudere* is a misprint for *defraudare*.

¹³ Copernicus' *mihi decernere arrestum fructuum sui beneficii* does not mean "award me the use of the proceeds of his benefice" (Schultheiss, p. 50: *mir den Genuss der Erträgnisse aus seinem Benefizium zusprechen*). In requesting the withholding of Snellenberg's income, Copernicus may have relied on an earlier episode in his fellow-canon's career. On 6 April 1506 the Varmia Chapter resolved:

> Canon Heinrich Snellenberg, having spent his limit of three years in an educational center, arrived for the purpose of residing [here]. He brought back no documentary proof of his studies, as required by [our] statutes. He was therefore debarred for several weeks from sharing in the distribution [of the canons' income]. Finally, on the initiative of his Reverend Lordship [the bishop], the venerable Chapter generously decided to let him enjoy the distributions one by one, provided that in other respects he conforms to the statutes while he is in residence (PI¹, 209).

Hipler (p. 171, n. 1/5-6) misdated Snellenberg's return to Frombork "6 April 1507" (a year later than P's "1506"); besides, "6 April" was the day on which the Chapter acted, not the day on which Snellenberg returned to Frombork. Some three decades later, Snellenberg, along with three other canons, was put on the royal chancellery's list of candidates for the vacant bishopric of Varmia. When these nominations became known in Frombork, on 21 August 1537 Copernicus was proposed in place of Snellenberg, whose designation would be regarded by everybody as ridiculous (Hipler, p. 171/n. 1; PI², 216).

¹⁴ StC XVIII, 216, no. 2/22-23's *longaeuum* is a misprint for *longaevam*.

¹⁵ A facsimile of this signature is provided by PII, Table Vb.

¹⁶ Maurice Ferber had been elected bishop of Varmia in 1523.

Letter II, addressed to the bishop of Varmia, was kept in the diocesan archives in Lidzbark until 1703-1704, when it was taken to Sweden. There in 1715 it was acquired by the library of the University of Uppsala, where it is still preserved. Facsimiles: Wasiutyński, facing pp. 328, 329; Plate VI. The text was first published by Prowe, *Mittheilungen*, p. 9; then by Baranowski, p. 633; Hipler, pp. 170-172; PII, 144-145; Dmochowski, pp. 195-196; and StC XVIII, 216. Previous translations include Baranowski, p. 633 (reprinted by Polkowski, I, 75-76); Dmochowski, pp. 211-212; StC XVIII, 216-217 (these three into Polish); and Schultheiss, pp. 49-50, no. 6 (into German).

III

In the chronological order of Copernicus' correspondence, his *Letter against Werner* belongs here. On account of its contents, however, it was discussed above, in the section comprising his Minor Astronomical Writings.

IV

The final draft ("E") of Copernicus' essay on the money question was written shortly before 17 July 1526. For on that day a regulation was established to which E refers as the "regulation to be established now". As for Felix Reich, he showed no demonstrable interest in the currency problem before he became a Varmia canon on 29 October 1526. But the urgency of that issue, and the paucity of canons qualified to speak for the Chapter on it, induced Reich to try to master all aspects of the monetary puzzle. With that end in view, he made a collection of the relevant currency writings, including Copernicus' E.

While studying E, Reich imagined that he detected an error in it. As a beginning student of monetary matters, he was familiar with the traditional pound divided into 12 ounces. Hence for him $^2/_3$ lb = 8 oz. But in E, $^2/_3$ lb ≠ 8 oz. Reich was unfamiliar with the local mintner's pound divided into 48 skoters (= 16 oz). Without indicating which pound it was using, Copernicus' E equated $^2/_3$ lb with 32 skoters. Reich, knowing that there were 3 skoters to an ounce, equated $^2/_3$ lb with 8 oz (= 24 skoters). Hence (Reich concluded) E's equation, $^2/_3$ lb = 32 skoters, was wrong. In a letter to Copernicus, which has not been preserved, Reich contended that $^2/_3$ lb = 8 oz = 24 skoters ≠ 32 skoters. In Letter IV Copernicus replied that the traditional 12-oz pound was $^1/_4$ lighter than the mintner's 16-oz pound: $^2/_3$ of a 12-oz lb = 8 oz, but $^2/_3$ of a 16-oz lb = 32 skoters = 10 $^2/_3$ oz ≠ 8 oz.

In his hurry to answer Reich, Copernicus omitted the year from the date of Letter IV, the only one of his surviving letters to suffer from such an omission. The gap may be filled, however, by using another topic discussed in Letter IV. On 25 July 1526 the king announced a tax called *contributio*. This was approved by the Varmia Estates on 1 September 1526, with Copernicus present (SRW, II, 495). When the royal tax was discussed half a year later by the West Prussian Estates, however, Copernicus was not present. But a fellow-canon, who did attend that meeting, reported to Copernicus about it after returning to Frombork. In conveying to Reich his reaction to that report, Copernicus implies that the meeting of the West Prussian Estates had taken place quite recently (March, 1527). Hence, his incomplete date for his letter to Reich (*octava Paschae*; eighth day after Easter Sunday) means 28 April [1527]. For in that year Easter Sunday fell on 21 April, and the eighth day thereafter (by the Roman method of counting intervals which was still in vogue then) was the following Sunday, 28 April 1527.

Reverend Sir, dearest friend:

The ability to shed light on subjects which by their very nature are enveloped in a thick fog is not unimportant. Yet it may also happen that somebody who has a correct understanding may not be able to explain what he knows. Something of this sort, I am afraid, sometimes happens to me too. The analysis of Prussian money, moreover, is of this [foggy] nature on account of the variety of the mixture, not to say confusion, of that money. Hence I am not at all surprised if what I wrote is not comprehended instantly by everybody. I shall therefore try to clarify what your Reverence complains was not understood.

We find, I say [E¶9], that $1/2$ pound of silver cost 2 marks, 8 skoters, when 3 parts of pure silver were mixed with a 4th part of copper, and 112 shillings were made from $1/2$ pound of that alloy. Such coinage therefore possessed the qualities required of sound money in intrinsic value and face value, as is certified by a scrutiny of what follows.

For, as I say, in the 112 shillings weighing $1/2$ pound, according to the prescribed proportion of the alloy, pure silver constituted $3/4$.[1] It follows that of this total [of 112 shillings] $1/3$ (amounting to 37 shillings plus $1/3$ shilling or 2 pence)[2] will contain the pure silver constituting 1 of the aforementioned quarters, or $1/4$ of $1/2$ pound.[3] Therefore, if you add $37 1/3$ shillings to 112 shillings, the total will be $149 1/3$ shillings, weighing $2/3$ pound (for *bes* means $2/3$ of any total, just as *dodrans* means $3/4$), or the total weighs $1/2$ pound plus $1/6$ pound,[4] which is equal to $2/3$.[5]

Here, however, I matched $2/3$ with 32 skoters,[6a] as our whole pound contains 48 skoters.[6b] I should not have said "8 ounces." For, the pound, used especially by pharmacists, which is divided into ounces, is different, being lighter by $1/4$. Therefore, the aforesaid total of $149 1/3$ shillings fills out the $1/2$ pound of pure silver. For since the total weighs $2/3$ pound, if you subtract $1/4$ of it, as is required[7] by the proportion of the alloyed copper, amounting to $1/6$ pound, the remainder is $1/2$ pound.[8] We therefore have as the intrinsic value of this currency $1/2$ pound of pure silver distributed over 149 shillings. But the price is 140 shillings,[9] namely, 2 marks, 8 skoters, as was said.[10] Hence approximately 9 shillings are due to the market value or face value, and in general about $1/15$ of the intrinsic value.[11] In this way, I believe, the matter is cleared up.

If any other difficulty emerges, I offer my services to the best of my ability, provided that something beneficial can be accomplished. I am afraid, however, that unless something different from the previous provisions is adopted, matters will go [from bad] to worse. For they will not stop minting money in this way. Why should those men stop who anticipate[12] profit, but no loss, from whatever occurs?

From Canon Achatius' report,[13] I have learned that the tax (*contributio*) is being discussed.[14] I therefore realize that nothing will be done at this time about the currency. For it is wrong to burden the subjects with a double loss. We will accordingly pay the tax. The money, on the other hand, will remain untouched. Rather, it will not remain untouched, but we will make it even worse and give the king, our master, a lot of money, that is, chaff. But where will the grain be?

I do not know whether it would not have been more seemly, more magnificent, and more regal, I will even say more useful, to drop the tax[15] and improve the coinage now, and if that did not provide enough, to proceed to the tax afterwards. For if I am not mistaken,[16] this procedure would have brought greater benefit and profit by increasing the public income. In other words, it would have yielded a permanent advantage, whereas the other is only yearly.[17] But whatever the situation may be, I admit that I can be mistaken, being only one man with one mind, unaware of or uninformed about what others regard as more useful.[18]

I wish your Reverence the best of health and happiness. Convey to his Reverend Lordship, our[19] superior, my respect and readiness to serve.[20] Frombork, 8th day after Easter Sunday[21] [28 April 1527]

N. Coppnic

To Felix Reich[22]
[On the verso, below the fold] About money

---

[1] Hence the silver contained in 112 shillings of the alloy weighed $^3/_4 \times ^1/_2$ lb = $^3/_8$ lb.

[2] Reading *denarii 2* (with Hipler, p. 195/last line; Dmochowski, p. 48/4; KMW, 1975; and StC XVIII, 226/13 up). The numeral 2 was omitted by PII, 155/8. There were 6 pence in the Prussian shilling, so that 2 pence = $^1/_3$ shilling.

[3] Here Copernicus originally added: "or $^1/_8$ lb," which he later deleted, because to say "$^1/_4 \times ^1/_2 = ^1/_8$" would have been insultingly obvious. Copernicus performed this computation because he wanted to know in how many shillings $^1/_8$ lb silver was contained. Having obtained that number of shillings ($37^1/_3$), he could add it to 112, in order to find out in how many shillings of the alloy $^1/_8 + ^3/_8 = ^1/_2$ lb silver was contained. Having obtained this total ($37^1/_3 + 112 = 149\ ^1/_3$), he could then compare the number of shillings containing $^1/_2$ lb silver with $^1/_2$ lb silver's market price, 2 marks, 8 skoters = 140 shillings (1 mark = 60 shillings = 24 skoters; 2 marks, 8 skoters = 120 + 20 = 140 shillings).

[4] 112 shillings weigh $^1/_2$ lb; $^1/_3 \times 112$ shillings (= $37^1/_3$ shillings) weigh $^1/_3 \times ^1/_2 = ^1/_6$ lb.

[5] $^1/_2 (= ^3/_6) + ^1/_6 = ^4/_6 = ^2/_3$.

[6a-b] Reading *sco* (KMW, 1975), not *sct* (Hipler, p. 196/7, 8), nor *sol* (PII, 155/15, 16).

[7] Reading *deposcit* (KMW, 1975; Hipler, p. 196/14; Dmochowski, p. 48/20; StC XVIII, 226/3 up), not *deposcis* (PII, 155/21).

[8] $^1/_4 \times ^2/_3 = ^2/_{12} = ^1/_6$; $^2/_3 = ^4/_6$; $^4/_6 - ^1/_6 = ^3/_6 = ^1/_2$.

[9] Reading *solidi* (KMW, 1975; Hipler, p. 196/16; Dmochowski, p. 48/23; StC XVIII, 226/last line), not *solidis* (PII, 155/24).

[10] In ¶2, 1st sentence, where *mr. ij et sco viii* shows that what Copernicus wrote here (*mr i sco viii*; KMW, 1975; Hipler, p. 196/17; PII, 155/24) is wrong.

[11] $^1/_{15} \times 140 = 9^1/_3$; here Copernicus puts a numerical value ($\simeq ^1/_{15}$) on what he had previously expressed in words: *paululo* (M ¶ 3); *gaer wenigk mynder* (D ¶ 3; modern spelling: *gar wenig minder*); *paulo* (E ¶ 5; see n. vii, above, in the section on the Money Question). In Copernicus' opinion, a coin's face value should equal $^{16}/_{15}$ of its intrinsic value.

[12] StC XVIII, 227/7's *expecant* is a misprint for *expectant*.

[13] Achatius von der Trenck and Copernicus were the two Varmia canons who witnessed an undated document drafted in October 1523 (ZGAE, 1972, 36:184/3 up). There is no documentary basis for dating Trenck's canonry from 1517 (as Sikorski does, p. 144/8/3, p. 147). Trenck held Varmia canonry no. 8. The dates of his predecessor in that canonry are not yet known. Trenck occupied his canonry until his death on 13 March 1551 (ZGAE, 1866, 3:360/last line). His name was not "Trank" (as in Koestler, p. 183).

[14] On 25 July 1526 King Sigismund I issued a *General Letter concerning the Public Taxes to be Collected in the Lands of Prussia* (*Corpus iuris polonici*, IV, 242; AcTom VIII, 127–128). This tax (*contributio*, the term used in the *General Letter*, is repeated here by Copernicus) was approved by the Varmia Estates on 1 September 1526, with Copernicus present (SRW, II, 495–496). But when the tax was discussed by the West Prussian Estates, Copernicus was not present, while his fellow-canon Achatius von der Trenck was present and told Copernicus about the discussion after returning to Frombork from the meeting.

[15] Copernicus is entirely outspoken in his opposition to the tax. His negative attitude was misrepresented in a ceremonial oration by Jacob Lilienthal (1802–1875), *Braunsberg in den ersten Decennien des siebzehnten Jahrhunderts* (Braunsberg, 1837):

> In the year 1526 begin the subsidies paid to the king of Poland under the designation land taxes and excise taxes, proposed by Bishop Maurice [Ferber] and the two representatives of the Chapter, Tiedemann Giese and Nicholas Copernicus (pp. 40–41).

Lilienthal's misstatement that Copernicus proposed the tax was repeated by Hipler (p. 279, no. 70); PI², 211;

and StC VIII, 130, no. 276, n. 1. But the episcopal archive's official minutes of the meeting record Copernicus as present, and say nothing about his having proposed the tax (SRW, II, 495).

[16] StC XVIII, 227/15's *fallar* is a misprint for *fallor*.

[17] The first 11 sections of the *General Letter* concern payments for a single year, but the sections from 12 on envisage taxes extending over a three-year period (*Corpus iuris polonici*, IV, 244).

[18] Copernicus had not attended a meeting of the West Prussian Estates since 21 March 1522 (see above, Money Question, after n. 4).

[19] StC XVIII, 227/20's *nostra* is a misprint for *nostro*. Copernicus' request that Reich convey his respects to the bishop indicates that when Reich sent his criticism of E to Copernicus, he was studying E in the bishop's palace in Lidzbark.

[20] PII, 156/14's *servilia* is a misprint for *servitia*.

[21] In his eagerness to answer Reich as quickly as possible, Copernicus forgot to include the year when he dated Letter IV the "8th day after Easter Sunday" (*octava pasce*). Because of this gap, when Letter IV was printed for the first time (Baranowski, pp. 590–591), it bore no date. When Hipler (pp. 195–197) printed it for the second time, he pointed out that it dealt with E and therefore was written after E. Emphasizing that Reich's activity as a canon representing his Chapter at meetings of the West Prussian Estates began in 1528, Hipler dated Copernicus' letter to Reich "8 April 1528 (?)." This date was repeated without a question mark by PI², 203/8; but when PII, 154–156, printed Letter IV for the third time, it was dated "between 1526 and 1528" (PII, 157/1). When Dmochowski (pp. 47–49) printed Letter IV for the fourth time, he refrained from specifying any particular year (although he implicitly accepted 1528 on pp. cxli, cxlvi). StC VII, VIII (Document no. 287) opted for "19 April 1528." In 1528, Easter Sunday fell on 12 April. Hence, the "8th day after Easter Sunday" in that year was 19 April, according to the Roman method of counting, which was still used at that time. According to this method, the first and last days of the interval were both included, so that the 19th was reckoned as the 8th day after the 12th. StC XVIII, 103, 132, 182, 204 (2b), 226, accepted StC VII, VIII's "19 April 1528" in printing Letter IV for the fifth time.

It will be recalled that (unlike PI², 203/8) PII, 157/1, assigned a date "between 1526 and 1528" to Letter IV. Since it discusses the tax imposed by King Sigismund's *General Letter*, it was written after the *General Letter* was issued on 25 July 1526. In that year Easter Sunday fell on 1 April, and the "8th day after Easter Sunday" was 8 April. On that day Copernicus had not yet heard about the tax (which was announced $3^1/_2$ months later), and therefore he did not write to Reich about it on 8 April 1526, the date proposed by Sommerfeld.

On 1 September 1526 Copernicus was present at the meeting of the Varmia Estates which approved the tax. But he did not attend the later meeting of the West Prussian Estates which discussed the tax. That meeting was attended by Canon Achatius von der Trenck, who told Copernicus about it, as Copernicus mentioned in his letter to Reich. After becoming a canon of the Frombork Chapter on 29 October 1526, Reich began an intensive study of the coinage question. At that early stage of preparing himself for his chosen role as spokesman of his Chapter on currency questions, he still did not know about the 16-ounce pound, which provoked his (lost) letter to Copernicus in the spring of 1527. Copernicus replied on 28 April 1527, for in that year Easter Sunday fell on 21 April, and *octava pasce* on 28 April.

Scholars who dated Copernicus' letter to Reich in 1528 did so because the letter was written after E, which they misdated in 1528. This mistake ignores E's reference to the "regulation to be established now," the regulation which was established on 17 July 1526. With the recognition that E was written before 17 July 1526 (and not in 1528, before 29 March, as in StC VIII, 133, no. 284), the only reason for dating Copernicus' letter to Reich on 19 April 1528 disappears. On the other hand, 28 April 1527 fits into the known sequence of events:

1526

composition of E	before 17 July
issuance of the *General Letter*	25 July
meeting of the Varmia Estates	1 September
beginning of Reich's canonry	29 October

1527

meeting of the West Prussian Estates	March
Reich's (lost) letter to Copernicus	mid-April (?)
Copernicus' letter to Reich	28 April

Wasiutyński (p. 588, n. 165) saw that Letter IV should be dated in 1527. Polkowski, I, 76–78, did not assign the date 12 April 1526 to Letter IV (despite StC VIII, 134, no. 287).

[22] According to KMW, 1975, p. 260, note m, "To Felix Reich" was written in by Reich himself. Here, however, the initial of the surname is lower-case. On the other hand, in his notarial instrument (NM, Plate VII/10 up) Reich began his surname with an upper-case R.

Copernicus wrote Letter IV in such a hurry that he signed it only with his initials "N. C.," and omitted the name of the addressee, Felix Reich. After Reich's death on 1 March 1539, his collection of monetary writings was bequeathed to Copernicus. It was then that he completed his signature by writing the rest of his surname after the initial "C" in Letter IV. He also added Reich's name as addressee. On the verso, in the upper right-hand corner after Letter IV was folded, he noted that the subject treated in it was money.

In this form, written by Copernicus and slightly amended by him after Reich's death, Letter IV remained in the local archives until they were removed to Sweden. In compliance with the claim advanced in 1801 by the kingdom of Prussia, together with other Varmia documents, Letter IV was sent to Koenigsberg. Karl Faber, superintendent of the secret archives there, first called attention to Letter IV in an article published in 1819, wherein he mentioned the signature as originally curtailed: "N. C." This is how Letter IV was printed for the first time, in 1854 by Baranowski (pp. 590–591), who never disclosed the location of his sources.

Hipler, however, guided by Faber's article, found Letter IV in the Koenigsberg archives. Noting the amplification of Copernicus' signature and the insertion of the addressee's name, Hipler mistakenly concluded that the Koenigsberg document was not the original letter sent by Copernicus to Reich, but a copy dating from the mid-sixteenth century (Hipler, p. 195). In connection with Hipler's printing of Letter IV for the second time, his erroneous contention that the Koenigsberg document was a copy infected subsequent discussions of Copernicus' letter to Reich. Thus, when Prowe printed Letter IV for the third time (PII, 154–156), he called the Koenigsberg document a copy, which he characterized as "contemporary," by contrast with Hipler's mid-sixteenth century dating. Contributing to Prowe's belief that the Koenigsberg document was a copy was his observation that Copernicus' surname was later retraced in a different ink on the verso. When Dmochowski (pp. 47–49) printed Letter IV for the fourth time, he took for granted that it was a copy, and that Reich was the copyist (p. 46).

While Letter IV reposed undisturbed in Koenigsberg, with shrewd foresight it was photographed by Hans Schmauch. For in the turmoil attending World War II, Letter IV disappeared and has not been recovered since. On the other hand, Schmauch's photograph survives, in the Ermland House, Münster, Westphalia.

StC VII, 145, no. 287, correctly identified the Koenigsberg document as Copernicus' autograph letter to Reich. But, in a curious divergence StC VIII, 134, no. 287, reverted to the Hipler-Prowe-Dmochowski blunder of calling the Koenigsberg document a sixteenth-century copy. This blunder was soon rectified in 1975, when Professor Marian Biskup reproduced Schmauch's photograph in KMW, showing the Koenigsberg seal on the verso of Letter IV. Thus, throughout its long history, Letter IV was never copied by hand, but only by a camera. It was printed for the fifth time by Biskup (pp. 259–260), and for the sixth time in StC XVIII, 226–227, where it was recognized as an autograph (p. 204); see also NCCW IV, Plate XLV.

Previous translations include those into Polish by Baranowski, pp. 590–591 (reprinted by Polkowski, I, 76–78); by Dmochowski (pp. 89–91); by StC XVIII, 227–228, and a partial translation by Wasiutyński (pp. 357–358); complete translations into French by Le Branchu (I, 25–27) and into English, by Moore; and a partial translation into German by Schultheiss, pp. 44–45, no. 3.

V

As an unmarried canon, whose time was largely taken up by the affairs of his Chapter and by his absorbing studies, Copernicus needed a housekeeper. Naturally, tongues wagged about what might be going on at night in a house occupied by a bachelor and his female housekeeper. Letter V concerns Copernicus' former housekeeper, who had left his service of her own free will to be married, "through no plan or action" of Copernicus. When the marriage turned out to be a failure because of the husband's admitted impotence, Copernicus tried to stop the couple from separating. They did so nevertheless, and argued their case twice before officials of the Chapter. On a later occasion, the woman was returning from the Koenigsberg fair with her new employer, a matron from Elbląg. In Frombork, although construction of a tavern or new hotel was being financed in 1531, the two females spent the night in Copernicus' house. He regarded this act of hospitality as a small return for his former housekeeper's faithful services. But the bishop took quite a different view of the matter, admonishing Copernicus in his own handwriting (in order to conceal the message from his secretary). Replying in Letter V, Copernicus promises the bishop that in the future nobody will have any proper reason to suspect him of misconduct.

My lord, Most Reverend Father
in Christ, my noble[1] lord:

With due expression of respect and deference, I have received your Most Reverend Lordship's letter. Again you have deigned to write to me with your own hand,[2] conveying an admonition at the outset. In this regard I most humbly ask your Most Reverend Lordship not to overlook the fact that the woman[3] about whom your Most Reverend Lordship writes to me was given in marriage through no plan or action of mine.[4] But this is what happened. Considering that she had once been my faithful servant,[5] with all my energy and zeal I endeavored to persuade them to remain with each other as respectable spouses. I would venture to call on God as my witness in this matter, and they would both admit[6] it if they were interrogated.[7] But she complained that her husband was impotent,[8] a condition which he acknowledged in court as well as outside. Hence my efforts were in vain. For they argued the case before his Lordship the Dean [of the Chapter], your Very Reverend Lordship's nephew, of blessed memory,[9] and then before the Venerable Lord Custodian[10] [of the Chapter]. Hence I cannot say whether their separation came about[11] through him or her or both by mutual consent.

However, with reference to the [present] matter, I will admit to your Lordship that when she was recently passing through here from the Koenigsberg fair with the woman from Elbląg who employs her,[12] she remained in my house until the next day. But since I realize the bad opinion of me arising therefrom, I shall so order my affairs that nobody will have any proper pretext to suspect[13] evil of me hereafter, especially on account of your Most Reverend Lordship's admonition and exhortation. I want to obey you gladly in all matters, and I should obey you, out of a desire that my services may always be acceptable.
Frombork, 27 July 1531

Your Most Reverend Lordship's

most devoted
Nicholas Copernicus

To his lordship, Most Reverend[14] Father
in Christ, Maurice [Ferber],[15] by the
grace of God bishop of Varmia,
my gracious and[16] most honorable lord

[1] The manuscript reads *Generose* (as in AcTom XIII, 248, no. 265/1), not *Gratiose* (as in StC XVIII, 228, no. 3/1).

[2] Neither of Bishop Ferber's autograph letters to Copernicus has been preserved.

[3] This woman has not yet been identified.

[4] Copernicus might have planned or acted to have the woman married off, had he been responsible for making her pregnant. But of her own free will she left to marry a husband (who turned out to be impotent). By contrast with two of his bishops (see Letter XI, n. 9), Copernicus fathered no illegitimate children, and no legitimate children, by contrast with one of his fellow-canons.

[5] Hence the statement (in StC VIII, 145, no. 315) that "she only served in the house of her husband" is wrong.

[6] The manuscript's defective reading faterent was corrected to *faterentur* in AcTom XIII, 248, no. 265/9–10, and StC XVIII, 228, no. 3/10.

[7] StC XVIII, 228/10's *inquirentur* is a misprint for the manuscript's *inquirerentur* (as in AcTom XIII, 248, no. 265/8).

[8] StC VIII (145, no. 315) translates *illa maritum impotentem lamentaretur* as "she complained of the violence of her husband." Would he have "acknowledged in court as well as outside" that he was violent, in connection with a lawsuit seeking to terminate his marriage?

[9] The first arguments were heard before 17 May 1530. For on that day Johannes Ferber (born 1496) died. He had served as the Chapter's Dean since 19 February 1522.

[10] Tiedemann Giese (1480–1550) served as Chapter Custodian from 1523 to 1538. The second hearing took place at some time between 17 May 1530 and 27 July 1531, the date of Letter V.

[11] The manuscript reading *recesserent* was emended to *recesserint* by AcTom XIII, 248, no. 265/12, and StC XVIII, 228/5 up.

[12] The Elbląg employer has not yet been identified. StC VIII (145, no. 315) refers to "her housekeeper from Elbląg," implying that the woman from Elbląg was the employee rather than the employer.

[13] The manuscript reading *suspiceretur* was emended to *suspicaretur* by AcTom XIII, 248, no. 265/6 up and StC XVIII, 229/1.

[14] The manuscript reads *Reverendissimo* (as in StC XVIII, 229/6), not *Reverendo* (as in AcTom XIII, 248, no. 265/2 up).

[15] Maurice Ferber was elected bishop of Varmia on 16 April 1523, the notary being Felix Reich (ZGAE, 3:539/5), and served until his death on 1 July 1537.

[16] The manuscript reads *et* (as in AcTom XIII, 248, no. 265/last line) rather than *atque* (as in StC XVIII, 229/7).

Letter V has not been found, but a late copy is preserved in the Czartoryski Library in Cracow, manuscript 284, p. 169. Baranowski made a transcription of the text, which he refrained from publishing because it reveals Copernicus' recognition of the "bad opinion" arising from his conduct. The text was printed for the first time in AcTom XIII, 248, no. 265, and then in StC XVIII, 228–229. The only previous translation is in StC XVIII, 229 (into Polish).

VI

Johannes Dantiscus (1485–1548), a highly talented but thoroughly hypocritical writer, served with distinction on the secretarial staff and in the diplomatic corps of the king of Poland. Despite Dantiscus' entirely secular outlook and vigorous sexual promiscuity, he schemed to become bishop of Varmia. In order to make himself eligible for this very remunerative post, he had first to acquire a Frombork canonry.

His earliest opportunity was linked with the grave illness of Canon Andrew Copernicus, the astronomer's older brother. Afflicted with an advanced stage of leprosy, Canon Andrew needed a coadjutor. Acceding to the king's wishes, the bishop nominated Dantiscus, who was then the king's secretary. On 1 July 1514 the king instructed his ambassador at the papal court to seek the pope's approval of the appointment of Dantiscus as Canon Andrew Copernicus' coadjutor (AcTom III, 124/10–15; PI², 32). But an even more influential combination pulled the strings behind the scene in favor of a rival candidate, who became Canon Andrew Copernicus' coadjutor on 15 June 1516. Despite the king's strenuous protests to the pope and cardinals in 1517 (AcTom IV, 169–170), Dantiscus' rival acquired the vacant canonry about November 1518, after Canon Andrew's death (Sikorski, p. 146).

While the tug of war over Canon Andrew Copernicus' coadjutor and successor was still being waged, another vacancy occurred with the death of a Varmia canon about September 1515 (Sikorski, p. 145). Dantiscus was named by the king to fill the vacancy. But the crown's right to do so was contested by the Chapter, which in the end was supported by the papal curia (PI², 251). Thus, Dantiscus' second attempt to become a Varmia canon came to naught, like his first effort.

Dantiscus tried again when a Varmia canon died in Rome in the early months of 1528. On 14 June of that year the bishop wrote to the Chapter, in part, as follows:

> I recently had a conversation in Malbork with our venerable brother, Dr. Nicholas Copernicus. ... [see above, Money, at. n. 108] A Varmia canonry is vacant on account of the death of our late brother Eberhard Ferber, who passed away in Rome. I have heard that in the present disturbed state of affairs throughout Germany the benefices which fall vacant in the Roman Curia are filled in accordance with the German bishops' judgment and wishes. Hence I too was ready to confer this same canonry on Johannes Flaxbinder [Dantiscus], provided that His Royal Majesty approved and could obtain from the papacy permission for us to do so (PI², 252).

The king undoubtedly approved, but again a rival candidate had stronger backing in Rome, and Dantiscus was foiled for the third time.

With unwavering support from the crown, however, Dantiscus finally obtained a Varmia canonry in 1529, after 20 February (Sikorski, p. 145). The tide continued to run in his favor while he was still abroad, for on 4 May 1530 he was named bishop of Chełmno. This diocese in Royal Prussia had no Cathedral Chapter capable of opposing the king's will, by contrast with nearby Varmia. Although Chełmno in itself was relatively poor, it was useful as a stepping-stone to more prosperous Varmia.

The health of the Varmia bishop being precarious, in 1532 he decided that he needed a coadjutor. Since the coadjutor usually managed to arrange things in such a way that he was sure to fill the vacancy when it occurred, the choice of a coadjutor was practically tantamount to naming the next bishop. In 1532 the Varmia bishop and Chapter were agreed that Copernicus' closest friend Giese should be the coadjutor of the reigning bishop and therefore presumably the next bishop of Varmia. For, Giese could be counted on to defend staunchly the special privileges of the Varmia diocese against encroachments on the part of the king. The monarch therefore relied on his faithful follower Dantiscus to oppose Giese.

Dantiscus entered his Chełmno bishopric in September 1532. At once he launched a strenuous campaign to break down the stubborn resistance to himself among the Varmia canons. For if he was to become the coadjutor of their bishop, and later their bishop himself, he would need their support. With this end in view, early in April 1533 he invited Copernicus to attend his formal installation as bishop of Chełmno at the episcopal seat Lubawa on Sunday, 20 April 1533.

Dantiscus' invitation has not been preserved, but Copernicus' reply on 11 April 1533 has survived. It is a polite but firm refusal to enhance Dantiscus' prestige by appearing in person as a Varmia canon at Dantiscus' installation as bishop of Chełmno. This refusal required courage on the part of Copernicus. For if Giese lost, and Dantiscus won, in their current competition for the coadjutorship, Dantiscus would soon be bishop of Varmia, and therefore Copernicus' immediate superior. Dantiscus was not especially noted for his magnanimity. On the contrary, he could reasonably be expected, after his elevation, to punish those who had opposed him when he was climbing the ladder of ecclesiastical preferment rung by rung. Such was indeed the subsequent course of events. Yet on 11 April 1533 Copernicus boldly defied the future bishop of Varmia by refusing to attend his formal consecration as bishop of Chełmno.

This ceremony was scheduled to take place in Lubawa on Sunday, 20 April 1533. Unaware of the nature of this occasion (since he mistakenly believes that Dantiscus had already been installed), Koestler erroneously supposes that what Copernicus received was "Dantiscus' invitation to visit him at Loebau Castle" (p. 181). This was an invitation, not to a social visit, however, but to an official consecration. Dantiscus' overtures to Copernicus "for some unfathomable reason were primly rejected," says Koestler. The reason is readily fathomable by anyone who understands the historical context. That description obviously does not fit Koestler, who says:

> One would have thought that the arrival of such an illustrious humanist [as Dantiscus], in the provincial backwoods hidden by the "vapors of the Vistula," would become a joyous event in the lonely Copernicus' life.

Far from being "a joyous event," Dantiscus' arrival cast a deep shadow on the life of Copernicus. If he was lonely (a condition he prized, since it gave him the privacy he needed to think and write), he certainly would not enjoy the company of an unprincipled schemer like Dantiscus. Despite Copernicus' defiance of Dantiscus on 11 April 1533, Koestler (p. 175) characterizes Copernicus as "submissive" to Dantiscus:

> This submission to authority — to ... Dantiscus on the one hand, to Ptolemy and Aristotle on the other — is perhaps the main clue to Copernicus' personality.

The main clue to Koestler's inadequacy as a writer about Copernicus is his weird notion that Copernicus submitted to the authority of Ptolemy, whose astronomy he overthrew after it had ruled supreme for 1400 years.

Your Lordship,[1] Most Reverend Father in Christ:

I have received your Most Reverend Lordship's letter,[2] from which I quite understand your Most Reverend Lordship's kindliness, graciousness, and good will toward me. You do not disdain to communicate to me[3] the [esteem] I have gained not only with you but also with other good men[4] of every sort. This is surely attributable, not to my services but rather, in my opinion, to your Most Reverend Lordship's well-known generosity. Would that I might some day chance to be able[5] to be so deserving! I am of course more delighted than it is in my power to say that I have found such a patron and protector.

Your Most Reverend Lordship, however, requests that I join you on the 20th of this month. Although[6] I would do so with the greatest pleasure, since I have no insignificant reason[7] for attending so eminent a friend and patron, yet the misfortune has befallen me that at that very time Canon Felix[8] [Reich] and I are required by certain business and by compelling reasons[9] to remain at our stations. I therefore ask your Most Reverend Lordship to excuse my absence at that time. At any other time I am unreservedly ready, as I should be, to call on your Most Reverend Lordship and do whatever pleases you, to whom I am indebted in very many other ways, provided that your Most Reverend Lordship so indicates to me at some other time. I acknowledge that hereafter I should not so much satisfy your requests as execute your commands.

Frombork, Good Friday [11 April] 1533

Your Most Reverend Lordship's

most faithful
Nicholas Copernicus

To the Most Reverend Father[10] in Christ,
Johannes [Dantiscus], bishop designate[11]
of Chełmno, my most honorable
patron and protector

---

[1] PII, 157/1's *Domino* is a misprint for *Domine*.

[2] Dantiscus' letter to Copernicus has not survived. The same may be said about Dantiscus' letter to Copernicus' fellow-canon Felix Reich, who was also invited at the same time to attend the installation of Dantiscus as bishop of Chełmno.

[3] PII, 157/5 omits *mihi* after *eadem*.

[4] Koestler (p. 181) mistranslates:

... good will towards me; which he [Dantiscus] has condescended to extend not only to me, but to other men of great excellence.

Koestler's version makes Copernicus align himself, immodestly, with "other men of great excellence." Copernicus, however, diplomatically aligns Dantiscus with "other men of great excellence." One such man was Rainer Gemma of Friesland (1508–1555), a professor at the University of Louvain. He "discussed the motion of the earth and the heavens" with Dantiscus at some time before 11 March 1532, while Dantiscus was traveling to the imperial court at Brussels on a diplomatic mission for the king of Poland (AcTom XIV, 204/4). On 20 July 1541 in a letter to Dantiscus, without mentioning Copernicus' name, Gemma urged the bishop to speed up the printing of the astronomer's masterpiece: "I believe you know to whom I refer, for you once mentioned this famous writer to me" (Fernand van Ortroy, "Bio-bibliographie de Gemma Frisius," *Académie royale de Belgique, Classe des lettres et des sciences morales et politiques, Mémoires, Collection in 8°*, XI, fascicle 2, 1920, p. 410). In his (lost) letter to Copernicus in 1533, Dantiscus evidently recalled Gemma's praise of Copernicus during his recent conversation with Dantiscus. In other words, the bishop relayed to Copernicus the high regard felt for him by Gemma and others. They did not send such compliments directly to Copernicus. His

quae cum apud ipsam obtinui, etiam apud alios quoscumque bonos viros eadem mihi propagare non dedignatur

does not mean

these favourable sentiments I have learned both from yourself and from some other worthy men to whom you have deigned to communicate them (*Edinburgh Philosophical Journal*, 1821, 5:64, anonymous translation).

⁵ The present whereabouts of Letter VI are not known. In 1820 it was sent from the Czartoryski Library in Puławy to Great Britain, where a facsimile of it was published in the *Edinburgh Philosophical Journal*, 1821, 5:opposite p. 63. In the facsimile's line 6 *possem* was misread as *possent* in the accompanying transcription (p. 63/12 up–11 up). When the autograph of Letter VI was sent to Great Britain, a handwritten copy was made and retained by the Czartoryski Library. On the basis of this copy Baranowski published the text of Letter VI for the second time. Although the copy reads *possem* even more clearly than the autograph does, Baranowski (p. 583/9) printed *possum*, which was repeated by Hipler (p. 197, no. 9/8) and PII, 158/2. The correct reading finally appeared in AcTom XV, 273, no. 193/6, and StC XVIII, 229/2 up.

⁶ The Edinburgh transcription correctly read *etsi* (p. 63/9 up). Baranowski, however, who did not know about the Edinburgh transcription, omitted *si* (written separately from *et* in the Czartoryski copy, as it was also in Copernicus' autograph). Baranowski's amputated *et* (p. 583/13) was repeated by Hipler (p. 197, no. 9/12) and PII, 158/5. The correct reading appears in AcTom XV, 273, no. 193/8, and StC XVIII, 230/2.

⁷ Copernicus' "no insignificant reason" refers to the official ceremony when Dantiscus would be formally installed as bishop of Chełmno.

⁸ Copernicus' *d[ominum] felicem* was misinterpreted as "my excellent protector" (*Edinburgh Philosophical Journal*, 1821, 5:64/14 up–13 up).

⁹ Copernicus' "compelling reasons" (*causae necessariae*) are echoed by Reich's *magna causa* in the letter in which he, like Copernicus, declined to attend Dantiscus' consecration. Reich's letter, like Copernicus', was dated 11 April 1533. He explained to Dantiscus that

> the day designated [for the consecration] comes quite inopportunely for us. Together with the others, we have been required for an important reason (*magna causa*) to be in our Frombork Cathedral on St. George's day [23 April]. ... I have agreed with the venerable man Nicholas Copernicus that we shall hereafter of our own accord or without any invitation call on your Most Reverend Lordship, and indeed quite soon, especially if it can be known when we can find you at home or arrive at a time that is not inconvenient, so that your Most Reverend Lordship may understand that what we now lack is not the will to obey, but the possibility [of doing so; MK, p. 391, no. 4; AcTom XV, 272, no. 192].

¹⁰ The Puławy copy's *et* after *Patri* was omitted by Baranowski (p. 583/4 up), Hipler (p. 198/7), and PII, 158, no. 5/2 up, but was reinstated by AcTom XV, 273, no. 193/2 up, and StC XVIII, 230/13.

¹¹ Letter VI "is dated 11 April 1533, that is, a few months after Dantiscus had been installed in his bishopric," says Koestler (p. 181). Copernicus' use of *Electo Culmensi*, however, shows that Dantiscus was still only bishop "designate of Chełmno" on 11 April 1533. Overlooking Copernicus' reference to Dantiscus' official status on 11 April 1533, Koestler misunderstands the nature of Dantiscus' invitation, which asked Copernicus to assist in celebrating on 20 April a significant advance in Dantiscus' campaign against Giese, Copernicus' closest friend. Dantiscus did not invite Copernicus to make a merely social call on 20 April 1533, "to visit him at Loebau Castle," as Koestler erroneously supposes. Hence, Copernicus' refusal to go to Loebau on 20 April 1533 does not show his "inability to loosen up and enter into a human relationship," as Koestler mistakenly concludes (p. 182). On the contrary, Copernicus' refusal shows his unswerving loyalty to his best friend and to the best interests of his Chapter. That is why in his refusal he speaks not only for himself but also for his fellow-canon Felix Reich. Was Reich also unable "to loosen up and enter into a human relationship"?

Some five years later the struggle between Dantiscus and Giese ended in a compromise: Dantiscus was named bishop of Varmia, and Giese bishop of Chełmno. On or shortly before 4 August 1538 Copernicus together with Reich went to Lidzbark, Dantiscus' new episcopal seat (Schmauch, pp. 67, 97, no. 27). They advised against holding Giese's installation ceremony in the Frombork Cathedral and, as representatives of the Varmia Chapter, accompanied Dantiscus on his extensive tour of the diocese to receive the oath of allegiance (PI², 325). The situation on 4 August 1538, when Copernicus and Reich went to Lidzbark together, was quite different from the situation on 11 April 1533, when they both refused to go to Chełmno. By 1538 Dantiscus had triumphed; opposition to him was now out of the question. On the other hand, there still was hope of defeating him in 1533, as he had been defeated three times before.

The *Edinburgh Philosophical Journal* (mistakenly) referred to Dantiscus as bishop of Chełmno on 11 April 1533, perhaps because its facsimile of Copernicus' letter (Plate III, opposite p. 63) does not reproduce the address. This was not written on a separate envelope in which the letter was inserted, as we ordinarily do nowadays. Instead, Copernicus wrote the address in a special place on the verso of the letter. This was then folded so as to show only the address, and it was sealed. It is in the address that Copernicus refers to Dantiscus as *Electo Culmensi*. Through a misreading (as *t*) of the Polish letter *ł* (pronounced like our *w*), the place where the copy of Letter VI was then kept was given as Putawy (p. 63/12) instead of Puławy.

Letter VI was removed from Lidzbark to Sweden in 1703–1704. Then in 1810 it was acquired for the Czartoryski library in Puławy. Only a decade later, in 1820 it was taken to Great Britain by Karol Sienkiewicz (1793–1860), who was sent abroad as a purchasing agent for the Puławy Library; see his *Dziennik podróży po Anglii 1820–1821* (Wrocław, 1953), p. 41. Although intended for the collection of Lord George Spencer (1758–1834), Letter VI was taken instead to Edinburgh by Sienkiewicz (*Dziennik*, pp. 46, 90, 243). His associate William Day is evidently the unidentified "friend" by whom Letter VI was shown to the editors of the *Edinburgh Philosophical Journal* (1821, 5:63/9). Day may also be the anonymous translator of Letter VI (*Journal*, p. 64). Sienkiewicz brought Letter VI back with him to Poland. There it formed part of the library of Władysław Czartoryski (1828–1894), whose father had thought of giving it to Lord Spencer. A copy made before 1820 in Puławy is now preserved in the Czartoryski Library in Cracow (MS 1596, pp. 357–358; Catalogus Czartoryski, II, 255), which also has an earlier copy. There is another eighteenth- or early nineteenth-century copy in the Library of the Ukrainian Academy of Science in Lvov (Ossolineum Collection, MS III, 546, p. 234); see Wojciech Kętrzyński, *Katalog rękopisów biblioteki Zakładu Narodowego Imienia Ossolińskich* (Lvov, 1886), II, 618, no. 42. A facsimile of Letter VI was published in the *Edinburgh Philosophical Journal*, 1821, 5:63–64, accompanied by a transcription (the first publication of the Latin text of Letter VI) and a translation into English (perhaps by William Day). The text was then reprinted by Baranowski, p. 583, from the autograph in the library of Władysław Czartoryski (Polkowski, I, 78, n. *); by Hipler, pp. 197–198; PII, 157–158; AcTom XV, 272–273, no. 193; and StC XVIII, 229–230. Letter VI was translated into Polish by Hoffmanowa, pp. 81–82; by Baranowski, p. 583 (reprinted by Polkowski, I, 78–79, and with revisions by Wasiutyński, pp. 385–386); and by StC XVIII, 230.

VII

Your lordship, Most Reverend Father
in Christ, your most gracious lordship:

I have received your Most Reverend Lordship's letter, which is full of kindness and love. In it you remind me of the friendship and pleasant relationship which I formed with your Most Reverend Lordship when I was still young,[1] and which I perceive still continues in a flourishing state on your part. And so, numbering[2] me among your friends, you have seen fit[3] to invite me to the wedding of your kinswoman.[4]

Surely, Most Reverend Lordship, I should have heeded your Most Reverend Lordship and presented myself at some time to so eminent a superior and benefactor of myself. At present, however, being occupied in a task imposed[5] on me by his Most Reverend Lordship, [the bishop] of Varmia, I cannot leave.[6] Therefore deign to excuse my personal[7] absence and preserve your old attitude toward me, even though I shall not be present.[8] For, a meeting of the minds usually counts for even more than a meeting of the bodies.

For your Most Reverend Lordship, to whom I pledge my services, I wish constant good health[9] amid every happiness.
Frombork, 8 June 1536

Nicholas Copernicus

To his lordship, Most Reverend Father
in Christ, Johannes [Dantiscus],
bishop of Chełmno, his most gracious lordship

[1] "There is no evidence that they [Dantiscus and Copernicus] had ever been friends," says Koestler (p. 184), after quoting (p. 182) this very letter in which Copernicus tells Dantiscus that

> he reminds me of that familiarity and favour with Your Rev. Lordship which I contracted in my youth; which I know to have remained just as vigorous up to now. And since I am thus to be numbered among his intimates ... [In Koestler's "translation" Dantiscus is both "he" and "Your Rev. Lordship"].

Copernicus and Dantiscus may have met in 1512. On 19 January of that year Copernicus was in Sztum in the service of his uncle Lucas Watzenrode, bishop of Varmia (ASPK, V, part 3, page 145). A week later, on 26 January, Dantiscus was at Malbork, a few miles due north of Sztum, as spokesman for the king of Poland at a meeting of the West Prussian Estates (ibid., p. 156).

[2] When Baranowski published Letter VII for the first time, at p. 584/8 he misread *commemorando*, which was repeated by Hipler, p. 198, no. 10/8, and PII, 159/4–5. The correct reading *connumerando* was introduced by StC XVIII, 230–231.

[3] Baranowski's misprint dignita (p. 584/9) was repeated by Hipler, p. 198, no. 10/8, with a *sic*!, and by PII, 159/5, without a *sic*! StC XVIII, 231/1, restored Letter VII's reading *dignata*.

[4] This young lady has not yet been identified.

[5] StC XVIII, 231/4's iniuxit is a misprint for *iniunxit*.

[6] Copernicus perseveres in the defiant attitude expressed in Letter VI. His defiance of Dantiscus over those three years refutes Koestler's uninformed remark about Copernicus' "submission to authority — to ... Dantiscus" (p. 175).

[7] Baranowski (p. 584/15) misread Letter VII's *ipam* as *istam*, which was repeated by Hipler, p. 198/7 up, and PII, 159/10. The correct reading *ipsam* was introduced by StC XVIII, 231/5.

[8] Baranowski's misprint absentt (p. 584/16) was repeated by Hipler (p. 198/6 up), but was emended to *absente* by PII, 159/11. Letter VII's *absentis*, an inappropriate genitive form affected by the two approaching genitives *animorum*, *corporum*, was printed by StC XVIII, 231/6.

[9] StC XVIII, 231/9's valero is a misprint for *valere*.

Letter VII was removed from Lidzbark to Sweden in 1703–1704. In 1810 it was acquired by the Czartoryski Library in Puławy, and is now in the Czartoryski Library in Cracow (MS 2713, pp. 1–2), which also has two copies: MS 54, p. 311, and MS 284 (the reference to MS 240, p. 170, in StC VIII, 159, no. 355, is wrong). While Letter VII was in Puławy, it was translated into Polish by Niemcewicz, IV, 64–65; 2nd ed., IV, 54 (not IV, 24, as in Hipler, p. 163/11, followed by PII, 142/5). Later translations into Polish were made by Baranowski, p. 584 (reprinted by Polkowski, I, 79, and with revisions by Wasiutyński, pp. 396–397), and by StC XVIII, 231. The Latin text of Letter VII was first published by Baranowski, p. 584; then by Hipler, pp. 198–199; PII, 158–159; and StC XVIII, 230–231. Facsimiles of Letter VII: Wasiutyński, facing p. 409; NCCW IV, Plate VII.

VIII

A member of the court of King Ferdinand of Hungary reported on recent and impending national and international political developments to Dr. Johannes Tressler of Wrocław, canon of the local Cathedral Chapter and also its Custodian. On 27 June 1537 Tressler relayed this report to Copernicus in Frombork. Copernicus in turn promptly passed the information on to his bishop. This was how news was communicated from person to person before the advent of newspapers.

Your lordship, Most Reverend Father
in Christ, your most gracious lordship:

A delivery to me by your Most Reverend Lordship's courier[1] was the favorable moment that reminded me to give him something by way of a letter from me too for your Most Reverend Lordship.

These [past few] days[2] I received news from Wrocław,[3] which I send on to your Most Reverend Lordship, although I am afraid that it is already stale as far as your Lordship is concerned because the letter[4] was dated 27 June.[5] Nevertheless, the private written message to me is that a communication arrived from the court of His Royal Majesty Ferdinand,[6] containing the following[7] information:

The Persian shah,[8] prompted by the emperor,[9] pope,[10] and king of Portugal,[11] is attacking[12] Turkey with great strength in order to compel it to leave Italy and withdraw its expeditionary force.[13] A truce between[14] the king of France and the emperor is said to have been signed,[15] with the widow of the duke of Milan being given, together with the duchy, to a son of the king of France.[16] Our [troops], that is, King Ferdinand's, are doing well at Košice. The man who had taken Košice by treachery has been captured, many of the enemy have been slain, and the fortress where the whole Košice calamity began has also been taken.[17] The Bohemians and Moravians are already on the march. In like manner the Silesians are crossing into Hungary on all sides. Perhaps, by the grace of God, they will recover Košice and other [places]. It is also said that in our affairs the Weyda[18] is suing for peace by proposing certain conditions. Whether they will be accepted, we still do not know.

This information in the letter, I pass on, just as I received it, to your Most Reverend Lordship, to whom I pledge my services and myself.
Frombork, 9 August 1537

Your Most Reverend Lordship's

most devoted
Nicholas Copernicus

To his lordship, Most Reverend Father in Christ,
Johannes [Dantiscus], by the grace of God
bishop of Chełmno, and administrator of the
Church[19] in Pomesania,[20] my most gracious lord

---

[1] What Dantiscus' courier delivered to Copernicus has not yet been determined.

[2] Copernicus' *his diebus* means that he received this news shortly before 9 August 1537, the day on which he wrote Letter VIII. Koestler's notion (p. 183) that the news "had reached Copernicus ... two full months earlier" is a gross error. Koestler's "two full months earlier" is apparently an effort to define the interval between 27 June (the day on which the news was dispatched to Copernicus) and 9 August (the day on which he transmitted it). This interval is 43 days, more than two full weeks less than "two full months." Of these 43 days, perhaps

40 were consumed in transporting the news from Wrocław to Frombork. Then, shortly after Copernicus received the news, he passed it on to Dantiscus. To someone accustomed to instantaneous electronic transmission of world news, like Koestler, and unaware that such communication facilities did not yet exist in the sixteenth century, Copernicus' news was "outdated" and also "indifferent political gossip." Actually, it concerned quite recent (and impending) international developments of the highest importance to people living in Silesia and Varmia. In relaying this recently received information to the bishop of Chełmno, Copernicus acted as he had in relaying to his Chapter the (alleged) information he received from the bishop of Varmia (see Letter I, above).

[3] For Copernicus' official Wrocław position, see above, Administrative Documents, Copernicus' Scholastry. He enjoyed the income from this sinecure for thirty-five years, during which he never set foot in Wrocław, his duties there being performed for him by a vicar. Early in 1538 he gave up the scholastry, his successor being nominated by King Ferdinand on 4 February 1538 "after the resignation of Nicholas Copernicus" (PI², 263). From the bishop of Wrocław, Copernicus had received the following notification:

> Inasmuch as I hold the right of advowson in the collegiate Church of the Holy Cross in Wrocław, I agree that Dr. Nicholas Copernicus, possessor of the scholastry in the said church, may resign it in the hands of His Holiness our Pope or the bishop of Wrocław, in favor of Dr. Johannes Rupoldus, canon of Wrocław (PI¹, 314–315).

This notification was shown by Copernicus to Dr. Tressler when they were conferring about Dantiscus' illness in 1538, in late April or early May. Shortly thereafter, in writing to Dantiscus on 16 May, Tressler proposed the notification as a model to be followed by Dantiscus. For, after being elected bishop of Varmia on 20 September 1537, Dantiscus was prepared to relinquish his canonry in the Varmia Chapter. However, he did not look with favor on Tressler's candidacy. On the contrary, his choice was the future spearhead of the Polish Counter-Reformation, Stanislaus Hosius (Stanisław Hozjusz, 1504–1579). In like manner Rupoldus, Copernicus' candidate for the Wrocław scholastry, was passed over in favor of Dr. Johannes Benedict Solfa, the personal physician of the king of Poland.

On 1 June 1542 Tressler and Giese were two of the three persons charged by the pope with helping Copernicus' coadjutor obtain the benefits accruing from that appointment, an arrangement accepted by Copernicus on 28 June 1542. All these indications point to Tressler as Copernicus' informant in Wrocław, and author of the letter of 27 June 1537 to Copernicus. Tressler (or Dressler, as his name also appears in the archives) became a canon of the Wrocław Cathedral Chapter on 26 August 1519, and some four years later its Custodian. As for Johannes Rupoldus (?–1544), he became a Wrocław Cathedral Canon by 1524, and on 15 June 1538 was nominated by King Ferdinand to a vacant canonry in Wrocław's Church of the Holy Cross. In addition, he was a Varmia canon (Zimmermann, pp. 192, 233, 462).

[4] Copernicus' *litterae*, although plural in form, refers only to a single letter, not "letters" (as in Koestler, p. 183).

[5] The news dispatched from Wrocław by Tressler on 27 June reached Copernicus in Frombork early in August, and on 9 August he promptly passed it on to Dantiscus.

[6] Ferdinand I (1503–1564) was elected king of Bohemia in 1526, later in the same year king of Hungary, and in 1531 king of the Romans. Ferdinand's was a traveling court, not permanently established in any one place, but moving back and forth between Vienna, Prague, and other urban centers.

[7] Copernicus inserted *haec* above the line.

[8] Persia was urged by the European powers to join in a two-front war against their common enemy, the Turks. But Shah Tahmasp I (1514–1576) refused on the ground that the Sultan

> has set out in a religious war against the Europeans. If we attack his country under these circumstances, we shall not succeed. And if he had killed my brother or son, I still would not invade his land now that he has marched against the infidels. We do not want to barter religion for worldly wealth (*Die Denkwürdigkeiten Schah Tahmasp's des Ersten von Persien*, ed. Paul Horn [Strassburg, 1891], p. 36).

[9] Charles V (1500–1558), older brother of Ferdinand I, was elected Holy Roman Emperor in 1519.

[10] Paul III (1468–1549) was elected pope in 1534.

[11] John III (1502–1557) ascended the throne of Portugal in 1521; Antonio H. R. de Oliveira Marques, *History of Portugal*, 2nd ed. (New York, 1976), pp. 215–216.

[12] Copernicus' *investat* was emended to *infestat* by Baranowski (p. 585/13), followed by StC XVIII, 231/5 up. But Hipler (p. 199, no. 11/11), using Baranowski's facsimile of Letter VIII, printed *investat*, and so did PII, 160/6. Although there is no such Latin form as *investat*, that is what Copernicus wrote: the letter *v* for the sound of *f*. By the same token, in Letters XIV and XV, which Copernicus composed in German, his native language,

he spelled vlessige, vleissig (PII, 166/2, 6 up; 167/4 up; NCCW IV, Plates XIII, XIV), where modern German prescribes *fleissige, fleissig*.

[13] Suleiman (1494-1566), who became the Ottoman sultan in 1520, ordered an invasion of Italy in July 1537 and withdrawal in the following month; see Roger Bigelow Merriman, *Suleiman the Magnificent* (Cambridge, Mass., 1944), p. 220.

[14] Instead of copying *Inter* from Tressler's letter, Copernicus wrote *Iter* at first, then struck it out and wrote the correct word.

[15] "A truce ... is said to have been signed" was the message passed from the informant in King Ferdinand's court to Tressler, and by Tressler to Copernicus on 27 June 1537. This message is characterized by Koestler (p. 183) as "a rumour about an armistice between the Emperor and Francis I, which happened to be unfounded." Emperor Charles V and King Francis I of France signed a truce on 30 July 1537; see Gabriel-Henri Gaillard, *Histoire de François Premier*, 2nd ed. (Paris, 1769), V, 44.

[16] Francesco II Sforza (1495-1535), duke of Milan, was survived by his widow Christina, who bore him no children, so that the Sforza line died out. The vacant duchy of Milan was claimed by Francis I (1494-1547) for his second son Henry, while the emperor pretended to oscillate between him and his younger brother Charles, Francis' third son (Gaillard, IV, 275).

[17] In the struggle between King Ferdinand and his rival, John Zápolya, king of Hungary, for control of that country, Košice was of crucial importance. On 3 December 1536 it was handed over to Zápolya, as Ferdinand was informed. On 5 January 1537 a local commander wrote to Ferdinand, urgently asking for help in dealing with the situation in Košice; see Eudoxio de Hurmuzaki, *Documente privitóre la istoria Românilor*, II, 1 (Bucharest, 1891), pp. 118-119. On 27 January word was sent to Ferdinand that Košice had risen against Zápolya, a rumor which turned out to be false.

[18] Copernicus wrote the initial *W*, struck it out, and then copied the full word *Weyda*, which was used in King Ferdinand's court to designate Zápolya. His truce with Ferdinand was extended to 1 April 1537, but the surrender of Košice, in violation of the truce, gave Zápolya a great strategic advantage. In the peace negotiations early in June 1537, Ferdinand's spokesman insisted on the restoration of Košice, but the other side had to await instructions from Zápolya. This information was sent to Ferdinand on 10 June, and formed the background for the letter to Dr. Tressler. Ferdinand later made peace with Zápolya on 10 June 1538, Košice being restored to him in a secret agreement. The foregoing analysis was extracted from the classical article by Arpád Károlyi in *Századok*, 1878, *12*:613-617, 687-724, 790-840, by Professor Janos M. Bak, University of British Columbia, whose invaluable help is gratefully acknowledged.

[19] PII, 160/2 up's *ecclesia* is a misprint for *ecclesiae*.

[20] The last Catholic bishop of Pomesania defected to the Lutherans in 1524. Although Pomesania as a whole ceased to be a Catholic diocese, the districts adhering to Rome were administered by the neighboring bishops of Chełmno, including Dantiscus; see Hans Schmauch, "Die Verwaltung des katholischen Anteils der Diözese Pomesanien durch den Culmer Bischof," *Mitteilungen des westpreussischen Geschichtsvereins*, 1936, *35*:113-115. StC XVIII, 232's *Pomorskiego* repeats Baranowski's confusion of Pomerania with Pomesania (p. 585/3 up).

Letter VIII was taken from Lidzbark to Sweden in 1703-1704. Then in 1810 it was acquired by the Czartoryski Library in Puławy. It is now in the Czartoryski Library in Cracow (MS 2713, pp. 7-8), which also has a copy (Catalogus Czartoryski, I, 9: MS 55, pp. 197-198). While Letter VIII was in Puławy, it was translated into Polish by Niemcewicz, IV, 65-66; 2nd ed., IV, 55 (not IV, 24, as in Hipler, p. 163/11, followed by PII, 142/5). Later translations into Polish were made by Baranowski, p. 585 (reprinted by Polkowski, I, 80, and Wasiutyński, pp. 402-403) and by StC XVIII, 232. The Latin text of Letter VIII was first printed by Baranowski, p. 585; then by Hipler, pp. 199-200; PII, 159-160; and StC XVIII, 231-232. Facsimiles of Letter VIII: Baranowski (last page); Wołyński, Plate IV; *Filomata*, 1973, p. 255; NCCW, I, Plate IV, and NCCW IV, Plate VIII.

IX

Maurice Ferber, bishop of Varmia, died on 1 July 1537. The rivalry between Dantiscus and Giese to succeed him was resolved by a deal. On 20 September 1537 the Varmia Chapter elected Dantiscus by unanimous vote, including Copernicus'. In becoming the new bishop of Varmia, Dantiscus left his former bishopric of Chełmno vacant. That vacancy was filled by the king's appointment of Giese on 22 September 1537. A few months later, on the same day (11 January 1538) the pope proclaimed Giese bishop of Chełmno and Dantiscus bishop of Varmia. In moving up to the bishopric of Varmia, Dantiscus gave up his Varmia canonry. He wanted Giese to do the same, and charged Copernicus with the task of accomplishing this result. But Giese had no intention of surrendering his Varmia canonry. For he would need it to be eligible for the Varmia bishopric, to which he was elected on 25 January 1549 (ZGAE, 1894–1897, *11*:67), with papal confirmation on 20 May 1549 (ZGAE, 1858–1860, *1*:346), after Dantiscus' death on 27 October 1548. A decade earlier, therefore, the Giese–Copernicus strategy sought some plausible pretext for postponing the presentation of Dantiscus' proposal to the Chapter. Giese, elected bishop of Varmia in 1549, died on 23 October 1550 (not 1548, 1549, as in PI², 26).

My lord, Most Reverend Father
in Christ, most gracious lord:

With regard to that matter of the canonries, which your Most Reverend Lordship entrusted to me, I have received the plan[1] and shared[2] it with the Most Reverend [bishop] of Chełmno.[3] The matter did not seem to be advanced enough to be referred to the Chapter unless the case of the Cantor,[4] which has intervened, is decided first. When that is done, there will be a better opportunity to bring the proposal up, unless another plan is conceived in the meantime by your Most Reverend Lordship, to whom I wish my services to be acceptable.
Fromork, the fifth day after
Easter Sunday [25 April][5] 1538

Your Most Reverend Lordship's

most devoted
Nicholas Copernicus

To his lordship, Most Reverend Father
in Christ, Johannes [Dantiscus]
by the grace of God bishop of Varmia,
my most kindly and most honorable lord

---

[1] Dantiscus' plan was proposed in a letter to Copernicus that has not been preserved.

[2] StC XVIII, 232/3 up's *communicatio* is a misprint for *communicato*.

[3] Copernicus' letter transmitting Dantiscus' plan to Giese has not been preserved, nor has Giese's response. Before *accepto*, Copernicus opened a parenthesis which he forgot to close.

[4] When Copernicus wrote Letter IX, the Cantor of the Varmia Chapter was Canon Johannes Timmermann or Zimmermann (1493–1564). The nature of the case involving him has not yet been clarified. He remained Cantor

until 11 March 1539, when he resigned as he was becoming the Chapter's Custodian, a more lucrative post made vacant by the death of Felix Reich on 1 March 1539.

[5] Not "17 March," as in PII, 141/11 up. But PII, 161, has the correct date (25 April), not "the erroneous date 15 April" (as is erroneously charged by StC VIII, 170, no. 393, and StC XVIII, 205, no. 7). That erroneous date (15 April) was in Polkowski, I, 82. Letter IX was received in the episcopal palace in Lidzbark on 26 April, as is indicated by a notation below the address. That notation was "appended by the hand of Dantiscus," according to Prowe, *Mittheilungen*, p. 10, followed by MK (p. 400, no. 4), but more likely by the hand of a secretary, when the mail arrived and before the letter's seal was broken.

Letter IX remained in the episcopal archives in Lidzbark until 1703–1704, when it was taken to Sweden. In 1715 it was acquired by the library of the University of Uppsala, where it is still preserved. The Latin text was first published by Leopold Prowe, *Mittheilungen aus schwedischen Archiven und Bibliotheken* (Berlin, 1853), p. 10; then by Baranowski, p. 634; Hipler, pp. 200–201; PII, 161; and StC XVIII, 232–233. Facsimiles: Prowe, *Mittheilungen*, Plate II; Wolyński, Plate V; StC XVI, 386; and NCCW IV, Plate IX. Letter IX was translated into Polish by Baranowski, p. 634 (reprinted by Polkowski, I, 81–82; not 82–83, as in StC VII, 179, no. 393; VIII, 170, no. 393; XVIII, 205, no. 7), and by StC XVIII, 233.

X

During the Protestant Reformation the private lives of the Catholic clergy came under severe criticism. As part of a campaign to eliminate lax practices, previously tolerated, Dantiscus warned the Varmia canons regarding their housekeepers: only a female relative would be permitted to work in a canon's house. Copernicus did not respond at once to the bishop's warning, since "it was not easy to find a proper female relative forthwith." He intended to make the change by Easter 1539. But long before then Dantiscus wrote him a peremptory individual note, demanding earlier compliance. On 2 December 1538 Copernicus sent Dantiscus Letter X, promising to conform within a month.

My lord, Most Reverend Father in Christ,
most gracious lord, to be heeded by me
in[1] everything:

I acknowledge your Most Reverend Lordship's quite fatherly, and more than fatherly admonition,[2] which I have felt even in my innermost being.[3] I have not in the least forgotten the earlier one, which your Most Reverend Lordship delivered in person and[4] in general.[5] Although I wanted to do what you advised, nevertheless it was not easy to find a proper female relative forthwith,[6] and therefore I intended to terminate this matter by the time[7] of the Easter holidays. Now, however, lest your Most Reverend Lordship suppose that I am looking for an excuse to procrastinate, I have shortened the period to a month, that is, to the Christmas holidays, since it could not be shorter, as your Most Reverend Lordship may realize. For as far as I can, I want to avoid offending all[8] good peo-

ple, and still less your Most Reverend Lordship. To you, who have deserved my reverence,[9] respect, and affection in the highest degree, I devote myself with all my faculties.
Gynopolis[10] [Frombork],
2 December 1538

Your Most Reverend Lordship's

most obedient
Nicholas Copernicus

To his lordship, Most Reverend Father
in Christ, Johannes [Dantiscus],
by the grace of God bishop of Varmia,
his most gracious lordship

---

[1] Baranowski, who printed Letter X for the first time (p. 586), misinterpreted Copernicus' *i* as *et*, which was repeated by Hipler (p. 201, no. 14/2) and by PII, 161, no. 9/2. The correct reading *in* was introduced by StC XVIII, 233, no. 8/2.

[2] Dantiscus' admonition has not been preserved. He did not act alone, but in concert with Felix Reich. For according to a secret agreement between these two wily schemers, when Dantiscus sent his admonition to Copernicus, at the same time he dispatched a copy of it to Reich. This was accompanied by a letter from Dantiscus to Reich, the relevant part of which was to be read aloud to Copernicus by Reich in order to intensify the pressure on Copernicus. But in examining Dantiscus' letter, Reich found "certain little words" objectionable. As a notary by profession and a stickler for stylistic precision in legal and quasi-legal documents, Reich concluded that Dantiscus' letter had gone too far. He therefore decided not to read any part of it to Copernicus.

While Reich was more than willing to cooperate with Dantiscus in the bishop's move against Copernicus, he was motivated by considerations of internal Chapter politics rather than by feelings of antagonism toward the astronomer. His basic friendliness toward Copernicus was expressed in the will which he drew up about this very time, on 22 November 1538. As a man of considerable wealth, he could have provided for all fifteen of his fellow-canons. In fact, however, he chose only three, the second being Copernicus, to whom he bequeathed four gold coins (of the type known as David's Harp) and two books. One of these was a copy of Dioscorides' *Materia medica* ("which I gave him as his personal property while I was still alive," says Reich, who was one of Copernicus' patients; there were some eight editions of Dioscorides before the date of Reich's will). The other book consisted of "various works of Chrysostom and Athanasius, translated by Erasmus and collected in a single volume." Erasmus contributed to a five-volume edition of Chrysostom in folio (Basel, 1530), while his translation of Athanasius in folio was printed in Paris 1519 and Strasbourg 1522 (Reich's will was published in KMW, 1972, pp. 656–666, with the passage concerning Copernicus at p. 657/3–6). The copies of Dioscorides, Chrysostom, and Athanasius given or bequeathed by Reich to Copernicus have not been located, as may be inferred from their absence from Paweł Czartoryski's "The Library of Copernicus," StC XVI, 355–396.

In response to Dantiscus' admonition, Copernicus sent him Letter X on 2 December 1538. On that same day Reich replied to Dantiscus, in part as follows:

> As far as the venerable Nicholas Copernicus is concerned, I approve of your Most Reverend Lordship's pious initiative and fatherly admonition. I hope that he will take it to heart, so as not to need my admonition. He will be overcome with shame, I am afraid, if he learns that I am privy to this matter. Had I not been prevented by the insertion of certain little words, I would perhaps have read to him your Most Reverend Lordship's letter insofar as it touches on that business, for this was agreed between us (MK, p. 393).

[3] Copernicus' *intimo corde* does not mean "with thankful heart" (*dankbaren Herzens*, as in PI², 364/11).

[4] Baranowski (p. 586/5) misinterpreted Copernicus' $\overline{pns}$ as *prius*, and also omitted the following *et*. In this double misreading he was followed by Hipler (p. 201, no. 14/5) and PII, 162/2. StC XVIII, 233, no. 8/5, introduced the correct reading *praesens et*.

333

⁵ For the purpose of receiving the oath of allegiance from his newly acquired diocese of Varmia, Dantiscus made a tour of the bishopric. He was accompanied by two representatives of the Chapter, Copernicus and Reich. They joined him in his episcopal residence not later than 4 August 1538. It was during the course of this tour that Dantiscus delivered his earlier admonition "in person and in general." Koestler imagines that Copernicus made a promise to Dantiscus prior to November 1538, and that "in November, Dantiscus reminded Copernicus of his promise" (p. 184). But nothing is known about any such promise before November, or any such reminder in November. Copernicus made a promise in Letter X, which is dated 2 December 1538. He kept his promise. There is no evidence that Copernicus ever broke a promise.

⁶ Copernicus' *protinus* is an indispensable part of his statement: since it was hard to find a suitable housekeeper at once, he needed more time. Yet *protinus* is omitted by PI², 364/16, and by Koestler (p. 185).

⁷ Baranowski (p. 586/8–9) read *intra*, which was repeated by Hipler (p. 201, no. 141/8) and by PII, 162/5, but was corrected by StC XVIII, 233, no. 8/7, to *citra*. Originally Copernicus wrote *ultra*, with the intention of terminating the matter "after" the Easter holidays. On second thought, however, fearing that he would be suspected of "looking for an excuse to procrastinate," he deleted *ultra* and wrote *citra* above it. The *l* rising above the line was misinterpreted as *i* by Baranowski and combined with Copernicus' *ci*, understandably confused with *n*, to produce *intra*. This preposition was not used here by Copernicus, who first wrote *ultra* and then shifted to *citra*.

⁸ Copernicus' *oib̄ᵌ* was misinterpreted by Baranowski (p. 586/14) as *actis*, which was repeated by Hipler (p. 201, no. 14/13) and PII, 162/10. The correct reading *omnibus* was introduced by StC XVIII, 233, no. 8/12.

⁹ In writing *reveretur* (under the influence of the following two subjunctive forms *honoretur, ametur*), Copernicus forgot that the required subjunctive form was *revereatur*. This was introduced without comment by Baranowski, p. 586/15, followed by Hipler (p. 201, no. 14/14) and PII, 162/11. StC XVIII, 233, no. 8/13, reverted to Copernicus' faulty *reveretur* without comment.

¹⁰ The German name of Frombork is Frauenburg (Citadel of Our Lady), which Copernicus matched in Greek as "Gynopolis" in Letter X and also in *Revolutions*, III, 13, and IV, 7, 16 (NCCW, I, fol. 126ᵛ/3 up; II, 145/13–14, 191/29–30). In writing to Dantiscus, a well-known neo-Latin poet and classical scholar, Copernicus felt no hesitation in identifying his place of residence as Gynopolis. "This might perhaps provide a clue to [an] apparently senseless mystification," exclaims Koestler (p. 123), who labels Copernicus "The Mystifier" (p. 121). The alleged mystification is discussed in TCT, p. 291.

Letter X was removed from Lidzbark to Sweden in 1703–1704. Then in 1810 it was acquired by the Czartoryski family, and is now preserved in the Czartoryski Library in Cracow (Catalogus Czartoryski, II, 225: MS 1596, pp. 519–520). Facsimiles: Wołyński (Plate VI); Wasiutyński (facing p. 404); NCCW IV, Plate X. Letter X was translated into Polish by Hoffmanowa, pp. 82–83; Baranowski, p. 586 (reprinted by Polkowski, I, 82, with the year misprinted as 1528, and with revisions by Wasiutyński, p. 410); and StC XVIII, 233–234.

XI

My lord, Most Reverend Father
in Christ, most gracious lord:

I have now done what I should not or could not in any way¹ have failed² to do.³ I hope that what I have done in this matter quite accords with your Most Reverend Lordship's warnings.

But you wish to find out from me how long my uncle lived⁴ — my uncle, the late Lucas Watzenrode of blessed memory, a predecessor of your Most Reverend Lordship. He lived 64 years, 5 months;⁵ he was bishop for 23 years;⁶ he died on 30⁷ March 1512⁸ A. D. With him came to an end the family whose coat of arms is conspicuous at Toruń in their many works and in the ancient monuments.[9] I now pledge my obedience to your Most Reverend Lordship.

Frombork, 11 January 1539

Your Most Reverend Lordship's

most obedient
Nicholas Copernicus

To his lordship, Most Reverend Father
in Christ, Johannes [Dantiscus],
by the grace of God bishop
of Varmia, my most gracious lord

---

[1] Letter XI has not survived. But while it was still in the Lidzbark episcopal archives, it was copied in 1618 by Brożek, while he was searching for documents pertaining to Copernicus. On his return to Cracow, he stitched his copy of Letter XI into his copy of the third edition of *Revolutions*, which is preserved in the Jagiellonian Library in Cracow (shelf mark 311204). Brożek's copy was used as the basis of the first printing of Letter XI by Baranowski. At the end of the first line of Brożek's copy, only the letter *m* survived the stitching process. This letter's first stroke is separated by a slight space from the rest of the letter. Interpreting these strokes as *iu-*, Baranowski (p. 592/3) read *iure*, which was repeated by Hipler (p. 202, no. 15/3) and PII, 163/3. However, StC XVIII, 234, no. 9/2's *m[odo]* is undoubtedly correct.

[2] Brożek wrote obmittere, a non-existent form which was corrected to *omittere* by Baranowski, p. 592/3, followed by Hipler (p. 202, no. 15/3) and PII, 163/3. StC XVIII, 234, no. 9/2, reverted to Brożek's non-existent form.

[3] In Letter X, on 2 December 1538, Copernicus promised to change his housekeeper by the time of the coming Christmas holidays. Now in Letter XI, on 11 January 1539, he informs Dantiscus that he has kept his promise.

On that very day Felix Reich also wrote to Dantiscus. Some ten weeks earlier, on 1 November 1538, Reich had informed Dantiscus:

> Through God's mercy and Dr. Nicholas [Copernicus'] efforts, my loss of blood was stopped in time (MK, p. 392, no. 7).

But no trace of gratitude to his healing physician, no feeling of loyalty to his fellow-canon, is perceptible in the communication which Reich sent to Dantiscus on 11 January 1539:

> Most Reverend Father in Christ,
> my lord and most gracious superior:
>
> Your Most Reverend Lordship behaves so generously toward me — [an attitude] for which I am intensely grateful — as to procure everything that can conduce to restoring my health. I have therefore indicated to [Achatius von der Trenck], the venerable Administrator [of the Chapter] in Olsztyn, what I want him [reading *eum* instead of *cum*] to request from your Most Reverend Lordship in my name. I hope that he will obtain it without inconvenience, provided it is in the power of your Most Reverend Lordship, whom I also thank for the light beer which was sent recently.
>
> With regard to another matter, I observe that your Most Reverend Lordship is anxious to remove the serious scandal in the church. Hence I have no doubt that a prosecution too will be forthcoming, and that other upright canons are waiting for this [move] with keen desire. I therefore now, without being asked and of my own volition, step forward to impart gladly whatever I can muster [reading *adhibere* instead of *adhiberi*], as long as I can, by way of advice or service for so pious a proceeding.
>
> It therefore seems to me that your Most Reverend Lordship should send [reading *mittat* rather than *mittatur*] in a sealed letter at the earliest possible opportunity individually to each of these three of our brothers [one of whom was Copernicus] an order in accordance with the writ dictated by me and written down — since I myself can barely write — by the faithful hand of Fabian

[Emerich (1477–1559), the Chapter's notary], acting as secretary, who may be safely entrusted with everything. And in addition the females should likewise be warned by the local priest under your authority in accordance with the writ which I sent previously. Care should also be taken to omit from the letters to the other two, who do not have legal husbands, what is in that earlier letter concerning Nicholas [Copernicus'] cook, who does have a legal husband. The impending commencement of the proceeding against the women too will strike terror to no small degree. Yet an appropriate limit must be granted to them so that they may in all likelihood be able to secure other homes for themselves within the boundaries of the warning. Have the venerable Achatius, the Administrator in Olsztyn, inform your Most Reverend Lordship about what is generally said in this regard. Yet he is privy to none of the matters about which I am writing.

Whatever the situation may be, may Your Most Reverend Lordship act firmly. God Almighty will strengthen your arm so that you may conduct to a happy ending what you initiated out of zeal. As much as we can, all of us will help make a success of this affair. However, your Most Reverend Lordship must take care nevertheless in commencing the proceeding with the force of law not to introduce in your future letters anything contrary to formal and customary legal style, as it is called. For it often happens that even the tiniest clause may spoil an entire case, so that it is declared null and void if it comes before a higher judge.

I commend myself to the grace and favor of your Most Reverend Lordship. May you be safe and sound for a long time!
Frombork, 11 January 1539

Your Most Reverend Lordship's

F[elix] R[eich]

To my lord, Most Reverend Father in Christ,
Johannes [Dantiscus], by the grace of God
bishop of Varmia, my most worshipful
and gracious lord

The foregoing letter of Reich to Dantiscus, dated 11 January 1539, is preserved in the Czartoryski Library in Cracow (Catalogus Czartoryski, II, 255; MS 1597, pp. 377–378), and was first printed in KMW, 1972, pp. 375–376. It was received by Dantiscus on 15 January 1539, as is indicated by a handwritten notation on it. The legal documents discussed in it were drafted by Dantiscus and his staff, evidently in a great hurry, and sent to Reich for his scrutiny. After receiving them, on 23 January 1539 he replied to Dantiscus as follows (KMW, 1972, pp. 378–379):

Most Reverend Father in Christ,
most gracious lord:

I am sending back all the letters because in one a serious scribal error must be corrected, and that cannot be done here. For the scribe wrote "Henry" instead of "Alexander" [Scultetus, one of the two other Varmia canons under attack, in addition to Copernicus]. Moreover, in my previous letter I warned about the banishment of ten miles and outside the diocese, since your Most Reverend Lordship does not have the power to banish anyone beyond your own diocese, which in certain places (as here [Frombork]) does not extend farther than one mile. Consequently it would have been necessary to delete this reference to ten miles as the distance to which the women are relegated. It is my advice that this too should be done now. Finally, "innocent" is written elsewhere instead of "behaving innocently." And in case there are any other [defects], I am for that reason sending all the documents back as a group, so that also as a group they may hereafter be put in final form one by one.

It is also necessary that the letters to the canons should be tied up separately, and that in like manner the letters to the cooks should be tied up separately, sealed in one envelope, and addressed to the priest. For if this were not done, a great mishap could occur. For if open letters to the cooks fell into the hands of those three canons, I have no doubt that, being intercepted, the letters could not accomplish their purpose. Your Most Reverend Lordship will therefore instruct your courier on his return to deliver to the priest the documents pertaining to the women first and ahead of everything [else], and afterwards the documents concerning the canons to any canon. The latter will undoubtedly give each one his own and, if need be, accost each one [uncertain reading] most circumspectly. Otherwise he will heap no small suspicion on me.

I suppose your Most Reverend Lordship has a sound reason for writing very briefly to

Nicholas [Copernicus], and it does not matter a great deal. Undoubtedly they all [Copernicus and the two other canons under attack] coordinate everything with one another.

As regards the beverage I shall request from the generosity of your Most Reverend Lordship, I have previously designated as my agent [*sollicitatorem*] the venerable Administrator in Olsztyn. In nearly their final decision my physicians now very discreetly allow wine in unlimited quantities to strengthen my heart, a pure, clarified, and gentle wine, however, not harsh or too sweet or too strong [illegible reading], but a mild Hungarian wine, muscatel, malmsey, or suchlike. One flagon of a wine of this category will be enough for me, I believe. The Piotrków beer which your Most Reverend Lordship gave me, I now for the first time feel is very wholesome. At present, therefore, I drink it every day. No small quantity is still left, however, so that there is absolutely no need for concern on the part of your Most Reverend Lordship, to whom I wish to be of the greatest service. Better organized and clearer and more comprehensive, I cannot be. May Christ keep your Most Reverend Lordship safe.

Frombork, 23 January 1539 at night

Your Most Reverend Lordship's

                                                                              F[elix] R[eich]

To my lord, Most Reverend Father in Christ,
Johannes [Dantiscus], by the grace of God
bishop of Chełmno [!], my lord and
most gracious superior

Reich had addressed Dantiscus correctly as bishop of Varmia in his previous letter of 11 January 1539. But now, twelve days later, he calls Dantiscus bishop of Chełmno, the office vacated by Dantiscus a year before, when he moved up to the bishopric of Varmia (PI², 323/6 up–3 up). On 23 January 1539 Reich writes at night for fear that he will be unable to function on the following day. These two manifest signs of moribundity conform to Reich's self-description: on 11 January he can barely write; by 23 January his physicians have made their nearly final decision; he is down to his last flagon of wine; he can't be better organized, clearer, and more comprehensive.

As is indicated by a handwritten notation on Reich's letter of 23 January, it was received by Dantiscus on 27 January. On that same day a shipment that had been ordered by Dantiscus was delivered to Reich, who promptly sent off the following letter:

My lord, Most Reverend Father
in Christ, most gracious lord:

Today I received the wine and the Masovian beer, which however smells strongly of lavender, and that is incompatible with my illness. Nevertheless, I am immensely grateful for everything. But I beg your Most Reverend Lordship not to be concerned about additional beer, since the previous [supply] has not yet been exhausted. As soon as it gives out, I shall not refrain from calling once more on your Most Reverend Lordship's generosity.

Together with the wine from Olsztyn, through the effort of the venerable Administrator in Olsztyn your letter to the Chapter was also delivered to me. I am afraid, however, that it may contain something about the proceedings against the canons' cooks and against the canons themselves. I do not dare deliver the letter to the Chapter lest its members cause a disturbance in this affair. It would be expedient to have the matter settled as soon as possible, and have the letter sent back. Whatever your Most Reverend Lordship may then order [reading *jusserit*, not lusserit] regarding the letter to the Chapter, I shall see to it that the order is carried out. Meanwhile I beg you not to be angry because without any instructions I withheld the letter in accordance with my own judgment. The loss of time will be slight. For, momentous matters cannot be handled now by only three members of the Chapter [the others being temporarily absent].

I commend myself to the customary grace and favor of your Most Reverend Lordship forever.

Frombork, 27 January 1539

Your Most Reverend Lordship's

                                                        Felix Reich
                                      with sick and trembling hand

To my lord, Most Reverend Father in Christ,
Johannes [Dantiscus], bishop of Varmia,
my lord and most gracious superior

The foregoing letter is preserved in the Czartoryski Library in Cracow (Catalogus Czartoryski, II, 255; MS 1597, pp. 493–494), and was printed for the first time in KHNT, 1978, 23:184; facsimile: StC XVIII, Plate 8 (opposite p. 267). A handwritten notation indicates that Reich's letter of 27 January was received on 30 January. A month later, on 1 March 1539, this ungrateful, devious, and disloyal "friend" of Copernicus died.

Tiedemann Giese, the bishop of Chełmno, a true friend of Copernicus, reacted in an entirely different way. On 16 March 1539 he wrote to Dantiscus in an effort to dissuade him, for the good of the church, from proceeding against Copernicus and the other two canons, who should not be excommunicated. This new source (Czartoryski Library, Cracow, Ms 1597, p. 619) was cited in KHNT, 1978, 23:181, n. 5.

[4] Dantiscus' letter asking Copernicus for information about Watzenrode has not been preserved. The reason why Dantiscus wanted this information is made clear by Letter XVII.

[5] Usually such a statement includes the number of days also. By omitting the number of days, and stating that Watzenrode died on the 30th day of the month, Copernicus implies that his uncle was born on the 29th day of the month. The month in question was October, five months earlier than March, the month in which Watzenrode died. Since he lived 64 years, 5 months, 0 days, he was born on 29 October 1447 (not "29 November 1447," as in PI[1], 73).

[6] Watzenrode was elected bishop of Varmia on 19 February 1489 (PI[1], 82), and completed 23 years as bishop in February 1512, before his death in March of that year.

[7] Copernicus dates his uncle's death one day later than does his uncle's chancellor, Paul Deusterwald (c. 1470–c. 1520). Not long after becoming bishop of Varmia, Watzenrode instituted in September 1489 an official chronicle to record his actions. Under his personal supervision, this chronicle (*Memoriale actorum Curiae Warmiensis*, as it was called) was written by several men close to him. From 1496 on, the chronicler was Deusterwald (SRW, II, 2, 45, 52, 55). In 1512 Deusterwald accompanied Watzenrode to the wedding of King Sigismund I in Cracow on Sunday, 8 February. On the return trip north, Watzenrode died. The bishop's final agony was witnessed by Deusterwald, who wrote in the *Memoriale:* "I recorded what I saw with my own eyes was done in connection with his death" (ibid., 171/13–14). Deusterwald (p. 170/4) dates Watzenrode's death on 29 March, as against Copernicus' 30 March. But Copernicus was not in Toruń when his uncle died there. Hence his date (30 March) is second-hand at best. It should not be accepted (as it was in ZGAE, 6:311), but rejected in favor of 29 March, as given by Deusterwald, who was an eyewitness and the bishop's chronicler at the time of Watzenrode's death. The *Memoriale* section dealing with the death of Watzenrode is reprinted in PII, 477–480 (with "29 March" at p. 438/12–17, and Deusterwald's presence at p. 479/8 up–7 up).

After Watzenrode's death, Deusterwald moved into his previously acquired priesthood of the Church of St. Nicholas in Elbląg (SRW, I, 408, n. 106/14–15). He stayed there for the rest of his life. The year in which he died still remains to be determined, but his death is known to have occurred on 27 August (SRW, I, 281/4 up). That was also the day on which a Varmia canon, Balthasar Stockfisch, died. Now, Stockfisch's successor as Canon no. 10 took office in 1521 (Sikorski, p. 144). On 15 February 1521 Stockfisch is referred to as the "late Balthasar" (*defuncti Balthasaris*; Hipler, p. 340/9; PII, 406/13–14). Hence he died no later than 27 August 1520. But his name follows Deusterwald's in the list of those who died on 27 August. Therefore Deusterwald died on that day in the year 1520 at the latest. He did not die in 1522 (as in *Zeitschrift des westpreussischen Geschichtsvereins*, 1902, 44:72, no. 50), although it was not until 19 February 1522 that the vacant priesthood in Elbląg was filled (SRW, II, 248, n. 147). When King Sigismund wrote to the people of Elbląg in 1526 (AcTom VIII, 112, no. 87/7 up–6 up, no month, no day), he referred to "Paul, your former priest, who died three years ago." If the king meant Deusterwald, he may have been somewhat uninformed.

Long after Deusterwald died, a request for a copy of the *Memoriale*, or information about it, was addressed to the City Council of Elbląg, which replied that it had received none of Deusterwald's books after his death (SRW, II, 2, n. 1). The date of the Council's reply (19 January 1538) was misequated with the date of Deusterwald's death by Max Perlbach, *Prussia scholastica* (Leipzig, 1895 = MHW, VI), p. 184, no. 14/11, who was uncritically followed by Oko, p. 794, and by KMW, 1972, p. 643, n. 68. Perlbach also had Deusterwald enter his Elbląg priesthood in 1522, the year in which his successor took office after an interval following Deusterwald's death.

The date of Deusterwald's birth is also uncertain. After entering the University of Leipzig in the summer semester of 1486, he received the bachelor's degree on 23 February 1493 and the master's degree on 28 December 1494 (*Die Matrikel der Universität Leipzig*, ed. Georg Erler, II, 333, 346; III, 123; *Codex diplomaticus Saxoniae regiae*, II[1], 16, 17, 18, Leipzig, 1895–1902). Since Deusterwald took nearly seven years to earn the bachelor's degree, and less than two more years for the master's degree, he may have matriculated at an early age, say, 16. Hence he was born about 1470, and died when he was about 50 years old.

[8] Brożek (and presumably Copernicus also) wrote the year of Watzenrode's death in Roman numerals as MDXII, which was misprinted by Baranowski (p. 592/9) in Arabic numerals as 1522. This error was repeated

in PII, 163, although PII, 477, reads MCCCCCXII. The proper year (1512) appeared in Hipler (p. 202, no. 15/8; p. 316/12), often cited by P, whose mistake was corrected by MK, p. 400, no. 5.

[9] Copernicus' statement that the Watzenrode family came to an end with Lucas' death has been questioned (StC XVIII, 118) on the ground that Lucas had an illegitimate child, "Philip Teschner, the bastard son of Bishop Lucas and a certain girl" (PI[1], 111). She belonged to a highly respectable Toruń merchant family, whose surname was bestowed on the illegitimate child. Hence he was not legally a continuator of the Watzenrode family, and therefore Copernicus' statement that the Watzenrode family came to an end with Lucas is unexceptionable.

Letter XI has not survived. Brożek's copy was the basis of the four printed texts discussed in n. 1, above. Letter XI was translated into Polish by Baranowski, p. 592 (reprinted by Polkowski, I, 83, and in part by Wasiutyński, p. 413), and by StC XVIII, 234.

XII

Canon Felix Reich's health began to fail toward the end of 1538. In the hope of having the impending vacancy filled by his own favorite candidate, Canon Stanislaus Hosius (Stanisław Hozjusz, 1504–1579) wrote to Rome. But Queen Bona Sforza of Poland intervened on behalf of Raphael Konopacki (c. 1510–c. 1570), son of George Konopacki (c. 1480–1543), governor of Pomerania since 1518.

This was not the first time Queen Bona acted to have Raphael Konopacki named a Varmia canon. For as soon as she learned that Maurice Ferber, bishop of Varmia, had died on 1 July 1537, she wrote a peremptory letter to his presumptive successor, Johannes Dantiscus, demanding that he vacate his Varmia canonry in favor of Raphael Konopacki. The text of the queen's letter is published here for the first time, from a photocopy kindly supplied by the Czartoryski Library in Cracow (Catalogus Czartoryski, II, 255:MS 1596, p. 447).

Bona dei gratia Regina Poloniae, magna dux Lithuaniae, Russiae, Prussiae, Masoviae etc. domina

R[everendissime] in Christo pater, sincere nobis dilecte: Intelleximus dominum episcopum varmiensem proximis transactis diebus vita functum esse. Quo fit ut P[aternitas] tua ad sedem episcopatus illius ascensura est, quod felix faustumque sit. Favemus illi ex animo hanc provectionem, quam administraturam credimus in laudem dei optimi maximi, principibus vero suis et reipublicae illi honorem dignitatem et pacem atque commodum.

Et quia habet P[aternitas] tua canonicatum varmiensem, et indecorum est episcopum esse et canonicum in una atque eadem ecclesia, cupimus a P[aternitate] tua plurimum ut de ipso canonicatu Raphaeli Konopacki filio Magnifici Palatini Pomeraniae provideat eumque illi conferat. Qua re nobis multum gratificabitur et ipsum Palatinum eiusque totam familiam reddet sibi [*de-* (deleted)] ecclesiae suae devinctam. Quod non parum profuturum existimamus P[aternitati] tuae atque eius ecclesiae. Scribat autem nobis per hunc nuncium et voto nostro respondeat.

Quia vero P[aternitas] tua successit in episcopatum, succedat etiam in tributum.

Dominus episcopus varmiensis Mauricius, qui nuper mortuus est, dare solitus erat Serenissimo Regi, filio nostro carissimo, equos bonos Germanicos [marginal insertion: et de equirea sua.] Volumus ut P[aternitas] tua idem faciat. Gratiam principis sui cumulatiorem habebit.

Datum Cracoviae die xi mensis Julii
Anno domini MDXXXVII

                                        Ad mandatum suae Majestatis
                                                        Reginalis propriae

In Christo patri domino Joanni [Dantisco]
Episcopo varmiensi, sincere nobis dilecto

                                                        [Received] 19 July

Bona, by the grace of God, Queen of Poland, Grand Duchess of Lithuania, sovereign of Russia, Prussia, Masovia etc.

Most Reverend Father in Christ,
my dearly beloved:

I have learned that the bishop of Varmia died a few days ago. As a result, your Paternity will ascend to the seat of that bishopric. Good luck and happiness! I wholeheartedly approve of your promotion, which I believe will redound to the glory of God on high, but to the honor, dignity, peace, and advantage of your rulers and your commonwealth.

Your Paternity possesses a Varmia canonry, and it is unbecoming to be the bishop and a canon in one and the same church. I therefore very much want your Paternity to dispose of that canonry for Raphael Konopacki, son of [George Konopacki] the illustrious governor of Pomerania, and bestow it on him. By so doing, you will please me very much, and bind the governor and his entire family to your church. I believe this will benefit your Paternity and your church not a little. However, write to me by this courier and comply with my wishes.

Because your Paternity has succeeded to the bishopric, succeed also to the assessment. Bishop Maurice [Ferber], who died recently, used to give His Most Serene Majesty, my very dear son [who had been designated co-regent on 20 February 1530], good German horses, even from his own stable. I want your Paternity to do the same. You will have the deepest thanks of your ruler.

Done at Cracow, 11 July A.D. 1537

                                                   By order of Her Royal Majesty
                                                                  herself

[A photofacsimile of the recto of this letter was published in StC XVIII, which misdated the letter 11 June 1537 (Plate 6, pp. 153, 271, 293). The correct date (11 July 1537) appears on the letter's verso, which was not reproduced by StC XVIII.]

Despite Queen Bona's intervention on behalf of Raphael Konopacki, he did not obtain Canonry no. 15 when it was vacated by Dantiscus. Instead, on 27 July 1538, it fell into the hands of Dantiscus' staunch ally, Hosius.

When a vacancy in Canonry no. 2 loomed on account of Reich's failing health, the queen interceded again for Raphael Konopacki, and this time she was successful. Inside the Chapter, in this affair her agent was Copernicus, a relative of the Konopacki family through George Konopacki's marriage to a granddaughter of Copernicus' grandmother.

Canon Felix Reich died on 1 March 1539 and was buried on the following day. On 3 March Canon Paul (not "Stanisl.," as in PI², 249/10 up) Płotowski, Provost of the Varmia Chapter, sent the following message to Bishop Dantiscus:

> I know that a long time ago Hosius wrote to the papal curia, and I believe that Her Sacred Majesty confirmed the position for someone else. Raphael Konopacki has Her Sacred Majesty's later nomination to your church, Reverend Father. I knew nothing about it, except that today the venerable Dr. Copernicus brought it to me after the burial of Felix (may he rest in peace) and told me that he is one of the governor's proxies (MK, p. 393).

On the same day (3 March 1539) Copernicus sent Letter XII to Dantiscus, who in 1530 had been pitted against George Konopacki in a bitter struggle. This time Copernicus was on the winning side, since the combined pressure of the queen and governor prevailed over the Dantiscus–Hosius–Płotowski faction. For, after receiving Dantiscus' approval, on 11 March 1539 Copernicus took possession of the canonry for Raphael Konopacki, who was still absent.

My lord, Most Reverend Father
in Christ, most gracious lord:

Canon Felix [Reich] of blessed memory, the [Chapter] Custodian,[1] having passed away, was buried yesterday [2 March 1539]. By right of proxy I have claimed the vacant canonry and prebend for Raphael Konopacki. I have acted to have the case advanced,[2] and the office granted to him according to the terms of the papal letter, on the strength of his nomination by the Most Serene Queen of Poland etc.[3] It remains[4] for your Most Reverend Lordship to deign to express your approval of the foregoing, and to indicate that[5] approval to your Chapter, so that possession of the vacant prebend may be given to him[6] or to me as proxy. In this matter your Most Reverend Lordship on the basis of the papal letter will render to the governor of Pomerania, his son, and me an extraordinary service which we shall strive to deserve at the hands of your Most Reverend Lordship.

Frombork, 3 March 1539

Your Most Reverend Lordship's[7]

servant
Nicholas Copernicus

To my lord, Most Reverend Father in Christ,
Johannes [Dantiscus], by the grace of God
bishop of Varmia, my most gracious lord

---

[1] By a slip of the pen, Copernicus wrote Custus instead of *Custos*.

[2] Letter XII was found in the Polish Library in Paris in 1856, and the Latin text was printed for the first time by Dominik Szulc, in *Biblioteka Warszawska*, 1857, part IV, page 782, where *rea provideri* was misread as *ea promoveri*. These two misreadings were repeated by Hipler, p. 202, no. 16/last two lines, and PII, 164/2. StC XVIII, 235/6, read *provideri* correctly, while retaining the erroneous *ea*.

³ After *poloniae* Copernicus wrote an abbreviation which baffled Szulc so much that he omitted it, and in this respect he was followed by Hipler and P. StC XVIII, 235, no. 10/6, however, printed *etc.* correctly.

⁴ Copernicus' *sup est* was misinterpreted by Szulc (p. 782/15) as *supra est*, which was repeated by Hipler, p. 203/2, and PII, 164/4. StC XVIII, 235/6, introduced the correct reading *superest*.

⁵ Szulc (p. 782/18) read correctly Copernicus' *eumque*. But it was mistakenly changed to *cumque* by Hipler, p. 203/3, and PII, 164/5.

⁶ Copernicus' *illi* refers to Raphael Konopacki, not to the Chapter (as in StC VIII, 176/last line). After *illi* Copernicus wrote *pr-* as the beginning of *procuratori*, but deleted these two letters when he realized that he needed *vel mihi* before *procuratori*.

⁷ Szulc (p. 782) omitted Copernicus' $\overline{E\ R\ d\ vre}$, which was omitted also by Hipler, p. 203, and PII, 164.

Letter XII was removed from Lidzbark to Sweden in 1703-1704. Then in 1810 it was acquired by the Czartoryski family, and after a temporary sojourn in Paris it is now preserved in the Czartoryski Library in Cracow (Catalogus Czartoryski, I, 59:MS 307, pp. 123-124); facsimiles: MK, facing p. 388; NCCW I, Plate VI; and NCCW IV, Plate XI. Four printings of the Latin text are mentioned in n. 2, above. Translations into Polish were made by Hoffmanowa, pp. 83-84; by Szulc, p. 782 (reprinted by Polkowski, I, 83-84, and also in Polkowski, *Żywot Mikołaja Kopernika* [Gniezno, 1873], p. 217, and by Wasiutyński, p. 413); and by StC XVIII, 235.

XIII

Upon receipt of Letter XII, Dantiscus responded promptly in a letter of approval (which has not been preserved). Letter XIII acknowledges receipt of that (lost) letter.

My lord, Most Reverend Father
in Christ, most gracious lord:

With the approval and consent of your Most Reverend Lordship, today on behalf of Raphael Konopacki I obtained from the Chapter possession of the canonry and prebend made vacant by the death of Felix [Reich] of blessed memory. I thank your Most Reverend Lordship therefor, and I have no doubt that Raphael himself, together with his father,¹ the governor of Pomerania, will acknowledge this² kindness of your Most Reverend Lordship as they should.³ I⁴ also want to devote myself to⁵ your Most Reverend Lordship, to whom I pledge my services.
Frombork, 11 March 1539

Your Most Reverend Lordship's

                most obedient
                Nicholas Copernicus

To my lord, Most Reverend Father
in Christ, Johannes [Dantiscus],
by the grace of God bishop of Varmia,
my most gracious lord   [received 13 March]

¹ Letter XIII was printed for the first time by Baranowski, who misinterpreted (p. 588/9) Copernicus' pre as *patrono*, which was repeated by Hipler, p. 203, no. 17/8, and PII, 164/last line. StC XVIII, 236/1, introduced the correct reading *patre*.

² After *hanc* Copernicus started to write *benef(actionem)*, but deleted those five letters and switched to *benevolentiam*.

³ In a (lost) letter Copernicus informed George Konopacki that he had taken possession of the canonry for Raphael. Then on 20 April 1539 George Konopacki wrote to Dantiscus as follows (ZGAE, 1972, *36*:194):

> Most Reverend and Gracious Lord:
>
> You should not be unaware that the distinguished and venerable Nicholas Copernicus, doctor of laws and canon of Varmia, wrote me a letter which he dispatched and I received. From it I learned in detail that with God as his guide he has taken possession for my son Raphael of the canonry and prebend of the Church of Varmia (which were made vacant by the death of Felix Reich of blessed memory), and that it only remains for Raphael to take personal possession. If he encounters any obstacle in this matter (may he not!), I most humbly beseech your Most Reverend Paternity (for I have understood that you always support my affairs and those of my son with special kindness, and we look upon you as the anchor of our hope) to deign to assist him with the same utmost kindness and good-will and help. For my part I shall act with all untiringly preserved promptness to satisfy and comply courteously with any wishes whatsoever of your Most Reverend Paternity.
> Świecie, 20 April 1539
>
> George Konopacki, governor of Pomerania and
> prefect of Świecie

The corresponding letter which Raphael Konopacki presumably wrote to Dantiscus has not been preserved.

⁴ Instead of *ego*, Baranowski, p. 588/9, misprinted *ago*, which was repeated by Hipler, p. 203, no. 17/9, with a *sic*!, and by PII, 164/last line, without a *sic*! StC XVIII, 236/1, introduced the correct reading *ego*.

⁵ Baranowski, p. 588/10, misinterpreted Copernicus' *eidē* as *eximie*, which was repeated by Hipler, p. 203, no. 17/9, and PII, 165/1. StC XVIII, 236/2, introduced the correct reading *eidem*.

Letter XIII was taken from Lidzbark to Sweden in 1703–1704. Then in 1810 it was acquired by the Czartoryski family, and is now preserved in the Czartoryski Library in Cracow (Catalogus Czartoryski, II, 255 :MS 1596, pp. 557–558). Facsimiles: Wołyński, Plate VII; NCCW IV, Plate XII. The Latin text was first printed by Baranowski, p. 588 (not 558, as in StC XVIII, 206, no. 11); then by Hipler, p. 203; PII, 164–165; and StC XVIII, 235–236. Translations into Polish were published by Baranowski, p. 588 (reprinted by Polkowski, I, 84, and also in his *Żywot*, p. 217) and by StC XVIII, 236.

XIV

In 1541 Duke Albert of Prussia wrote

To Nicholas Copernicus,
Canon of Frombork
6 April

Through the puissant and honorable lord Johann of Werden, who is dearly beloved by me, you have made me a most courteous offer, should I want to use you personally in my own illnesses or those of others, to show your cooperativeness in such cases and your willingness to come here to me to serve at my pleasure. Accordingly I do not want to conceal from your graciously disposed self that at the present time Almighty Sempiternal God is inflicting on one of my counselors and subordinates an affliction and severe illness which does not get better, but gets worse the longer [it lasts], although I would gladly have every human and possible remedy employed to the extent permitted by dear God's mercy. I sincerely ask you, still not burdened by your offer, to come to me here with the bearer of the present [letter], and with kindness matching my heartfelt trust in you, to impart to the aforementioned good man your faithful advice and opinion whether he may ever be relieved of his painful malady through the bestowal of divine mercy and your cooperation. With all good will I am eager to obtain this from you.

Koenigsberg, 6 April [PI², 469; StC XVIII, 264]

On the same day, by the same courier, the duke wrote

To the Varmia Chapter,
6 April [1541, not 1539 as in PI², 470/7]

I am acting to inform you, kind friends, that through the puissant and honorable burgrave Johann of Werden, mayor of Gdańsk, commander in Steinburg, whom I especially love, I recently had many negotiations conducted with your worthy, honorable, and learned colleague, Nicholas Copernicus, canon of Varmia. Should I ever need him personally, he would freely come to me, and in the case for which I would summon him, he would make use of his best knowledge, for which I would then thank him kindly for his help. Now my honorable commander in Tapiau and dearly beloved counselor George of Kunheim has been smitten by a grave illness which gets worse daily, and in which I would like to see him advised and assisted as soon as possible with God's help. Accordingly I sincerely and courteously ask you to please me by arranging with the aforementioned Copernicus, to whom I am also writing by this same courier, that in view of this letter he will not decline to come here to me and, together with my other physicians, to the extent permitted by God's mercy and his own knowledge, promptly give his best help to my aforementioned counselor [George] of Kunheim so that he may be restored to health. I wish [to convey] my sincere indebtedness to you, for whom I feel all goodwill. I want to express all the worthy honor I feel also for Nicholas Copernicus.

Koenigsberg, 6 April [PI², 470]

Without delay, on 8 April 1541, the Officers, Canons, and Chapter of the Varmia diocese replied to the duke as follows:

In accordance with your Princely Grace's thinking and sincere communication about the oppressive and gravely weakened physical condition which by God's will has befallen our especially and dearly beloved friend, the honorable and reputable lord George of Kunheim, your Princely Grace's counselor and dear and loyal commander in Tapiau, we have conferred and agreed with our colleague and dearly beloved older brother, the honorable and worthy Nicholas Copernicus, that to the extent required to furnish suitable and satisfactory services to your Princely Grace, in accordance with your Princely Grace's sincere desire, without any troublesome excuse, at his advanced old age, he will gladly comply and, as indicated in your Princely Grace's letter, accompany your Princely Grace's courier to your Princely Grace etc. For this purpose we have also exempted his honorable self from several obligations in our church etc. [PI², 470-471].

Together with the courier and his fellow-canons' consent, Copernicus promptly set out from Frombork for Koenigsberg, where the duke soon wrote to the Varmia Chapter as follows:

Albert, by the grace of God
margrave of Brandenburg,
duke of Prussia, etc.

My greetings and good will, first of all, to you, especially beloved, worthy, honorable, estimable, and learned gentlemen. I have your friendly statement that at my sincere request you had a discussion and came to an agreement with my dearly beloved honorable and learned Nicholas Copernicus, doctor of medicine etc., your colleague and dear, friendly older brother. You also conducted matters so far that he proceeded to come to me, for the purpose of serving in a satisfactory and suitable manner without any troublesome excuse in his advanced old age in accordance with my sincere desire and also in view of my letter, in the company of my servitor, whom I dispatched for this purpose. Wishing to be of service to me and my counselors in considering how you might save them from harm and not let it issue from you, you accordingly let him come here, together with a further indication of your offers. To you, who are animated by such feelings as you have for my associates and myself, I now express my deepest thanks. I thoughtfully bear in mind that you do not gladly spare the aforementioned Nicholas Copernicus at the present time, yet you sent him to please me and alleviate the infirmity of George of Kunheim, my dear, loyal, and honorable commander in Tapiau and counselor.

In this connection I must conclude that in citing the reasons you advanced, you did not anticipate that it is a question of a further demonstration of your aforementioned readiness to be of service, which also extends so much further in gratifyingly pleasing me. The illness of my aforementioned counselor George of Kunheim, however, has reached the stage where the doctor feels the need to remain here a while in order to deploy his God-given skill for the patient's sake. My absolutely sincere plan and request therefore go forth to you as friends to please me, with regard to my aforementioned faithful commander, by not penalizing your Copernicus for his absence at the present time, and by allowing him to remain here in his quarters somewhat longer (recalling that it is quite Christian and praiseworthy in such a case to act as a fellow-sufferer, one with another), and to regard him not otherwise than as if he were personally present among you.

To all of you I am in all sincerity in a state of indebtedness. My counselor, frequently mentioned above, who enjoys serving a worthy Chapter without any applause, will likewise gladly incur a debt, provided that the Almighty helps him to recover his health. Koenigsberg, 13 April 1541 [PI^2, 471–472]

Two days later, "the Officers, Canons, and Chapter of the diocese of Varmia, always ready to oblige," replied to the duke as follows:

We have with great appreciation received the letter which your Princely Grace kindly sent us regarding the agreeable departure of our colleague and much beloved older brother, the honorable and worthy Nicholas Copernicus etc., and we have read your additional comments with considerable sympathy. We would gladly see our aforementioned dear, affectionate colleague still among us in accordance with our church's custom and rule on the occasion of this joyous holiday of the splendid invincible resurrection of Christ

from death to life. However, since the honorable lord George [here the Chapter repeats the duke's description of the patient's condition], we cannot and would not obstruct your Princely Grace's sincere plan and wish. On the contrary, in a spirit of service we want to grant whatever your Princely Grace desires in this case. For to this and other possible requests of your Princely Grace for service, we are always favorably disposed.
Frombork, Good Friday [15 April] 1541 [PI², 472]

At the duke's insistence and with the Chapter's consent, Copernicus remained in attendance on his patient in Koenigsberg more than three weeks. When he left for Frombork, he carried with him the following letter from the duke

To the Chapter in Frombork, 3 May

In accordance with my sincere plan and desire, you kindly put at my disposal for a time a fellow-member of your Chapter, the worthy, honorable, and learned doctor Nicholas Copernicus, whom I especially love. I thank all of you most sincerely therefor. Even against his will I kept him from you for some time. I sincerely ask you not to let this harm him, and not to penalize him for it, but much rather to be willing kindly to excuse him for being absent so long at my earnest insistence. Just as I would expect all this from you, so in a similar situation and in a much more important situation I am at all times ready and willing to be indebted to all of you in all sincerity.
Koenigsberg [PI², 473]

On 5 May the Chapter replied to the duke in a document which has become somewhat illegible. Nevertheless its contents are indicated by its heading: The Officers etc. write with reference to Nicholas Copernicus' absence [PI², 473]. Presumably the Chapter assured the duke that it did not plan to penalize Copernicus in any way for absenting himself so long from Frombork. Before leaving Koenigsberg on 3 May, Copernicus had communicated with the Polish royal physician about Kunheim. Six weeks later, having received no information about this correspondence, the duke sent the following letter

To Doctor Nicholas Copernicus, 14 June

You wrote to the physician of His Majesty the King of Poland, my gracious lord and dearly beloved uncle, with regard to the illness of my dear, loyal counselor George of Kunheim, my honorable commander in Tapiau, in order to learn his opinion about this case. Since, for the sake of the aforementioned George of Kunheim, I should like to know that doctor's advice and opinion about George's illness, I want to ascertain whether he has by now revealed his judgment to you. My entirely sincere thought is that, if this has happened, you may wish to inform me about it conveniently by this courier of mine. With all respect I wish to obtain this from you and familiarize myself with it, no passages being omitted.
Koenigsberg [PI², 473–474; StC XVIII, 264]

By the same courier Copernicus replied to the duke on the following day:

XIV, JUNE 15, 1541

Serene and honorable Prince, gracious lord:

My assiduous and wholehearted services are at all times in readiness for your Princely Grace. I refer to and acknowledge your Princely Grace's letter and communication [of 14 June 1541]. I wrote to Jan Benedict [Solfa], physician to His Majesty, king of Poland, to ask him to the best of my ability how help may be provided to the honorable and puissant lord George of Kunheim, your Princely Grace's officer, in his illness.[1] I hoped that the same courier would bring a reply.[2] Up to the present time I have not received any letter from the aforementioned doctor. That surprises me.[3] I was unable to write to your Princely Grace anything essential with regard to the case. I am therefore[4] inclined to communicate by a non-scheduled courier with the same physician again in reference to the same matter. What I learn from him, I will without delay present to your Princely Grace, to whom I humbly pledge my assiduous and indefatigable services.
Frombork, 15 June 1541

Your Princely Grace's

constant servant
Nicholas Copernicus

To the serene and honorable Albert,
by the grace of God margrave of Brandenburg,
duke of Prussia and Wendland,
burgrave of Neuenburg, and prince of Rügen,
my gracious lord

---

[1] Copernicus' letter to Solfa has not been preserved. If it was sent from Koenigsberg on or shortly before 3 May, it would have reached Cracow about the middle of that month. Jan Benedict Solfa was admitted to the University of Cracow in the summer semester of 1505 (*Album studiosorum*, II, 94, right hand column, 6 up: Johannes Benedicti). At Cracow he received the bachelor's degree in 1507, and the master's degree in 1512 (*Statuta necnon liber promotionum ... in universitate ... Jagellonica*, ed. J. Muczkowski [Cracow, 1849], pp. 146, 153). Unlike Copernicus, who went to the University of Padua for his medical training, Solfa studied medicine at the University of Bologna. Again unlike Copernicus, who did not finish the medical curriculum, on 11 January 1516 Solfa was awarded the M. D. degree by the University of Bologna, where he called himself simply Johannes Polonus (Giovanni Bronzino, *Notitia Doctorum sive catalogus doctorum qui in collegiis philosophiae et medicinae Bononiae laureati fuerunt ab anno 1480 usque ad annum 1800* [Milan, 1962], p. 14, citing the *Liber secretus 1504–1575*, fol. 26^r). On 8 May 1523 Solfa was named royal physician at a salary of 100 marks a year (Summaria, IV, 2, p. 278, no. 13546).

[2] When a regular courier delivered a letter, he would usually wait long enough to find out whether the recipient's reply would be ready in time for the return trip. Had Solfa received Copernicus' letter in mid-May, and replied at once, his answer would have reached Frombork by early June.

[3] When the duke's courier arrived in Cracow, Solfa may have been out of town, or he may have been unable to reply at once, or he may have decided he needed time to think about Kunheim's case. Whatever the reason for his delay, his response reached Copernicus on 20 June 1541, five days after Copernicus informed the duke that he had not yet heard from Solfa.

[4] PII, 166/12's *noch* is a misprint for *nach*.

XV

The day after Copernicus received Solfa's letter, he wrote to the duke as follows:

Serene, honorable Prince, gracious lord:

Just yesterday I received from Jan Benedict [Solfa], the physician of His Majesty the king of Poland, a letter and an answer to my message about the honorable George of Kunheim, commander in Tapiau etc.[1] But since no mention[2] is made therein of any other special[3] or extraneous matters, I have forwarded the original letter to your Princely Grace. From it your Princely Grace will learn this doctor's[4] opinion and advice. If I knew anything better to contribute thereto that would be helpful in restoring that good man, your Princely Grace's officer, to health, no labor, exertion, and trouble would be vexatious[5] to me that would be beneficial to your Princely Grace, to whose service I am devoted.
Frombork, 21 June 1541

Your Princely Grace's

obedient servant
Nicholas Copernicus

To the serene and honorable prince,
Albert, by the grace of God margrave of Brandenburg,
duke of Prussia and Wendland, burgrave of Neuenburg,
and prince of Rügen, my gracious lord

---

[1] Solfa's reply to Copernicus has not been preserved.
[2] Reading *berurth* (with PII, 167/6), not *beruth* (as in StC XVIII, 237, no. 13/5).
[3] Reading *besonderen* (with StC XVIII, 237, no. 13/5) rather than *besonderlichen* (as in PII, 167/5).
[4] Reading *doctoren* (with StC XVIII, 237, no. 13/7) rather than *doctoris* (as in PII, 167/7).
[5] Reading *verdrislich* (with StC XVIII, 237, no. 13/11), not *vordrislich* (as in PII, 167/12).

On the following day the duke wrote

To Nicholas Copernicus, 22 June
I have received your letter, together with which the royal physician's letter was forwarded. I thank you most sincerely for showing such zeal and forwarding the letter. Since I understand that something is said in the letter about my commander's illness, I have kept the letter in the expectation that you will be satisfied by my doing so. In case you cannot do without the letter or cannot let me have it, and wish to so inform me, I want to send it back again to you with thanks, for I am disposed to hold you in my especially good graces.
Koenigsberg [PI², 475; StC XVIII, 265]

Since this letter of 22 June 1541 ends the surviving correspondence between the duke and Copernicus, presumably the latter did not ask to have Solfa's letter returned. The royal physician's advice, like Copernicus' treatment of the patient, did not succeed in prolonging his life to any considerable extent, for he died on 29 September 1543, some four months after Copernicus. A letter written by the duke on the occasion of Kunheim's death has been preserved (*Altpreussische Monatschrift*, 1902, 39:143–145). Kunheim had served Albert by being one of his emissaries to arrange the truce of 1521 (19 March, 9 April), and the peace of 1525 (9 April), approving Poland's sovereignty over Prussia (Summaria, IV, 1, p. 207, no. 3648; p. 209, no. 3684; p. 278, no. 4723; IV, 2, p. 313, no. 14211), and the infeudation of Ducal Prussia (AcTom VII, 237).

Letters XIV and XV were among the documents sent by Sweden in 1798 to Koenigsberg. They remained there until World War II, when they were removed to Göttingen. Now they are in West Berlin, Geheimes Staatsarchiv, Stift für Preussischer Kulturbesitz, HBA C1a, Kasten 497. The German text of Letter XIV was first printed by Baranowski, p. 636; then by Prowe, *Nicolaus Coppernicus in seinen Beziehungen zu dem Herzoge Albrecht von Preussen* (Thorn, 1855), pp. 30–31; Hipler, pp. 204–205; L. Prowe, *Copernicus als Arzt* (Halle, 1881), p. 15; PI², 474, and PII, 166; and StC XVIII, 236. Facsimiles of Letter XIV: PII, Plate 3; *Kopernikus-Forschungen*, p. 24; *Die Mittelstelle*, 1943, 2:53; NCCW IV, Plate XIII. Translations into Polish were published by Baranowski, p. 636 (reprinted by Polkowski, I, 85), and by StC XVIII, 237.

The German text of Letter XV was first printed in *Beiträge zur Kunde Preussens*, 1819, 2:266, marking the first time that any letter written by Copernicus was published; thereafter the text of Letter XV was published by Baranowski, p. 637; Prowe, *Nicholaus Coppernicus in seinen Beziehungen zu dem Herzoge Albrecht von Preussen* (Thorn, 1855), p. 31; Hipler, pp. 205–206; Prowe, *Copernicus als Arzt* (Halle, 1881), p. 15; PI², 474–475; PII, 167; Wasiutyński, p. 463 (partial); and StC XVIII, 237. Facsimiles of Letter XV: *Beiträge zur Kunde Preussens*, 1819, 2; *Denkschrift zur Enthüllungsfeier des Copernicus Denkmals zu Thorn* (Thorn, 1853), 189; Wołyński, Plate IX; Wasiutyński, facing p. 464; *Kopernikus-Forschungen*, pp. 31–32; Sikorski, 1973, opposite p. 161; and NCCW IV, Plate XIV. Translations into Polish were published by Baranowski, p. 637 (reprinted by Polkowski, I, 86), L. A. Birkenmajer, *Wybór pism*, 1920, p. 117; 1926, p. 87; and by StC XVIII, 237–238.

XVI

On 8 June 1541 Bishop Dantiscus sent the following letter to the Varmia Chapter:

Venerable lords, sincerely
beloved brethren:

What the most illustrious lord, the duke [of Prussia], our neighbor, replied to my most recent letter, and what this day I [proposed to] answer, you will learn from the accompanying [documents]. The draft of my response to him will accordingly have to be changed in certain places. How this should be done, we must decide by mutual agreement. Therefore think it worthwhile that my venerable brothers, the provost [of the Varmia Chapter, Paul Płotowski], together with Dr. Nicholas [Copernicus], should meet me here tomorrow before lunch time, so that we may consult with one another about the matters which affect my affairs as well as those of the venerable brethren. I faithfully wish them good health.
Braniewo, 8 June 1541

[ZGAE, 1930–1932, 24:260, overlooking the previous publication of this letter in an incomplete form by Prowe, *Nicholaus Coppernicus in seinen Beziehungen zu dem Herzoge Albrecht von Preussen* (Thorn, 1855), p. 12, and by Hipler, pp. 288–289, with the wrong month, July instead of June. StC VIII, 195, no. 460, translates *prandii* as "breakfast." Dantiscus would hardly have asked Copernicus and his provost to ride five miles from Frombork to Braniewo and then have a discussion with him about important matters on an empty stomach before breakfast.]

Copernicus was invited presumably because, having just spent more than three weeks (from about 10 April to 3 May) in Koenigsberg, rendering medical services to a favorite henchman of the duke, he could be expected to provide useful current information about the situation in the duchy. During his meeting with Dantiscus on 9 June 1541, Copernicus informed his host that he had finally made up his mind to let his *Revolutions* be printed. Dantiscus, a prolific poet, promptly composed an epigram for the forthcoming volume. Laudatory poetry of this kind was a familiar feature of books printed during this period, and was often called *titulus* in Latin. Dantiscus sent his *titulus* to Copernicus not long after their meeting on 9 June 1541. Copernicus acknowledged its receipt in the following letter, dated 27 June 1541.

My lord, Most Reverend Father
in Christ, most gracious lord:

I have received your Most Reverend Lordship's very gracious and quite friendly letter.[1] Together with it you did not disdain to transmit also a truly elegant and relevant epigram[2] for the reader of my [*Six*] *Books* [*on the Revolutions of the Heavenly Spheres*],[3] not because I deserve it, but because your Most Reverend Lordship is accustomed to honor scholars with your extraordinary friendliness.[4] I shall therefore place your Most Reverend Lordship's epigram[5] in the forefront of my work, provided that the work is worthy to deserve being so highly embellished by your Most Reverend Lordship. Yet people who know more[6] than I do, and to whom I should listen, say over and over again that my work is not negligible.[7] I really wish to merit, as far as I can, the extraordinary kindness and fatherly affection with which your Most Reverend Lordship does not cease to favor me, and to serve and obey you, as I should, in all matters within my power.

Frombork, 27 June 1541

Your Most Reverend Lordship's

most obedient
Nicholas Copernicus

To my lord, Most Reverend Father
in Christ, Johannes [Dantiscus],
by the grace of God bishop of Varmia,
my most gracious lord

---

[1] Dantiscus' letter to Copernicus has not been preserved.

[2] Although the accompanying letter has not been preserved, Dantiscus' epigram has survived because it was printed by Rheticus in 1542; by Hipler (pp. 105-106) in 1873; and by Stanislaus Skimina, *Ioannis Dantisci poetae laureati carmina* (Cracow, 1950; Corpus antiquissimorum poetarum poloniae latinorum, VII), pp. 209-210; in addition, Skimina published half of it in *Twórczość poetycka Jana Dantyszka* (Cracow, 1948), p. 76. Unaware that Dantiscus' epigram had been printed three and a half times, Koestler asserts that it

> is lost. After thanking Dantiscus for his "extraordinary benevolence," Copernicus dropped his epigram into the waste bin, as he had done with Dantiscus' earlier invitations. He really was an old sourpuss (p. 188).

This characterization of Copernicus is based by Koestler on his own incompetence.

[3] Although composed for Copernicus' *Revolutions*, Dantiscus' epigram was printed at the end of the Dedication of Copernicus' *Sides and Angles of Triangles*. This little work, essentially an extract from the *Revolu-*

tions, was published under the editorship of Rheticus in Wittenberg, the intellectual center of anti-Catholic Lutheranism. Hence, Rheticus presented Dantiscus' epigram anonymously, giving no indication that its author was a Roman Catholic bishop. Rheticus, a fervent astrologer, liked Dantiscus' epigram because at lines 11–14 it advised the reader that

> You must first master the doctrine which these principles
> Set briefly before you, if you wish to know
> What fates govern future events, what disasters
> Are brought to people by hostile stars.

This professed ability of astrology to penetrate the dark veil covering human destiny was dear to the heart of Rheticus but utterly foreign to the thinking of Copernicus. Yet he gave the epigram to Rheticus in order to fulfill his promise to Dantiscus. Knowing Copernicus' aversion to astrology, however, Rheticus kept the epigram out of the *Revolutions* by publishing it a year earlier in *Sides and Angles of Triangles*. At the same time Rheticus suppressed Dantiscus' name in order to avoid offending the Lutheran leaders, many of whom, like himself, were addicted to astrology and were therefore pleased by the anonymous epigram.

The poem printed by Rheticus is not the epigram supplied by Dantiscus to Copernicus, according to StC XVIII, 198, no. 38. This denial contends (p. 156) that what Rheticus printed is unsuitable for an astronomical work such as the *Revolutions*. But in the 18-line poem printed by Rheticus, lines 4–10 declare:

> These writings show you the way to the heavens,
> If you want to grasp with your mind the boundaries
> Where the very beautiful universe expands its immense spaces,
> Or the region of the heavens where the planets wander,
> And the changes which their perpetual courses undergo
> [And the reasons]
> Why the moon enshrouds its brother [the sun] with blinding darkness, and
> Why that [brother] refuses to let the moon borrow its light.

As these seven lines show, the poem printed by Rheticus was composed for an astronomical work, and Copernicus himself judged Dantiscus' epigram "relevant for the reader of my Books" (*ad lectorem librorum meorum ... ad rem*).

The denial, however, asserts that Dantiscus' epigram was intended for Copernicus' *Sides and Angles of Triangles*. This purely trigonometrical work, which says nothing about eclipses and planetary orbits, had not even been conceived in anyone's mind in June 1541, when Dantiscus wrote his epigram for the *Revolutions*.

The impression acquired by Dantiscus from Copernicus at their meeting on 9 June 1541 about the forthcoming *Revolutions* was faulty in two respects. The bishop received the misapprehension that the *Revolutions* was brief, and that it was intended for young students. Hence Dantiscus addressed his epigram to studious youth (*studiosa juventus*, in line 1 here, and likewise in poem VII, p. 36, Skimina's *Carmina*), and he also referred to the book's brief principles (*breviter tradunt haec elementa*, in line 14). Actually, *Revolutions* was written for mature scholars, not young students, and is far from brief, the autograph manuscript in Copernicus' own handwriting being 212 folios in length (NCCW, I).

The denial faces two undeniable facts: (1) Dantiscus wrote an epigram; (2) Rheticus published a poem. In denying the identity of these two pieces, the denial contends that (1) Dantiscus' epigram is lost; (2) although Dantiscus composed the epigram for *Sides and Angles*, the poem printed in *Sides and Angles* by Rheticus was probably composed by himself. "The Twelve Zodiacal Signs and the Beer of Wrocław, a Joke" is ascribed to Rheticus. But even this jocular comparison of the succession of the celestial signs with making the rounds of the Wrocław beerhalls was versified by one of his friends (Burmeister, I, 88). "Of Rheticus' own poetry, nothing has come down to us" (Burmeister, I, 34).

We may therefore safely conclude that (1) the poem in *Sides and Angles* was not written by Rheticus; (2) it was written by Dantiscus (3) for the *Revolutions*, not for *Sides and Angles*.

[4] Copernicus' *benevolentia singulari* is an echo of *benevolentiam singularem* in Suetonius' *Caligula*, ch. 3.

[5] Here *titulum* refers to Dantiscus' epigram, not to his title, as in NCCW, II, 344 on 7:15, lines 19–20.

[6] Copernicus' *doctiores* does not mean "students" (as in StC VIII, 197, no. 465/6, which overlooks the adjective's comparative degree). These better-informed persons included Cardinal Schönberg, whose letter to Copernicus received a prominent place in the *Revolutions*; Bishop Tiedemann Giese, who is mentioned by name in its Preface as having "repeatedly encouraged" Copernicus to publish the *Revolutions*; and Rheticus,

whose *Narratio prima*, comparing Copernicus with Ptolemy, spoke of "this truly admirable structure of new hypotheses wrought by my teacher" (TCT, p. 109/8 up; 162/12–13).

[7] Copernicus' *quod tamen dictitant me doctiores esse aliquid* is mistranslated by StC VIII, 197, no. 465/6, as "which, however, some students will attribute to Copernicus" (what they will attribute to Copernicus — whether the epigram or the *Revolutions* — is naturally left unclear).

Letter XVI remained in Lidzbark more than two centuries. But after Varmia was absorbed into the kingdom of Prussia by the first partition of Poland in 1772, the outlook for the episcopal archives became uncertain. Under these circumstances the last independent bishop of Varmia, Ignacy Krasicki (1735–1801), transferred Letter XVI to his private collection, which was dispersed after his death. Letter XVI then passed into the Berlin Library (at that time the Royal Library, now the National Library in West Berlin-Dahlem, Darmstaedter Collection, manuscript no. J 1530). Facsimiles of Letter XVI: *Isographie des hommes célèbres*, I (1843); *Journal für reine und angewandte Mathematik*, 1845, *29*, Heft 2; Baranowski (penultimate page); Wołyński (Plate IX); Wasiutyński (facing p. 465); and NCCW IV, Plate XV. The text was printed for the first time by Baranowski, p. 593; then by Hipler, p. 206; PII, 168; and StC XVIII, 238. Previous translations include Baranowski, p. 593 (reprinted by Polkowski, I, 86–87, and in part by Wasiutyński, p. 464) and StC XVIII, 238 (the foregoing into Polish); and Koestler, p. 188 (into English, with Frauenberg instead of Frauenburg).

XVII

Unaware that an epitaph had already been carved on Watzenrode's tombstone, Dantiscus set about composing one for his predecessor, thrice removed, as bishop of Varmia. Lacking the requisite information, Dantiscus sought it from Watzenrode's nephew, Copernicus. In Letter XI, dated 11 January 1539, Copernicus supplied Dantiscus with the information he had requested. Shortly thereafter Dantiscus sent Copernicus the epitaph he had composed and written with his own hand. Copernicus made a copy and kept it. But it is not clear what he did with the original, which could not be used since an epitaph was already in place. After waiting more than two years, during which he had heard nothing about his epitaph, Dantiscus asked Copernicus to return it. On 28 September 1541 Copernicus replied as follows:

My lord, Most Reverend Father
in Christ, most gracious lord:

Your Most Reverend Lordship recently asked me to return to your Most Reverend Lordship the epitaph for my uncle, the late Bishop Lucas, a predecessor of your Most Reverend Lordship.[1] You had once sent me this epitaph, which was composed by yourself.[2] Since I do not have [the original written] by your own hand, Most Reverend Lordship, I am therefore sending [you] a copy.[3] I regret that it did not fulfill the purpose for which it was intended, since a different [epitaph] having little flavor and less taste had already been carved on the tombstone.[4] But what was commissioned was of this nature, and was composed for him in accordance with his wishes while he was still alive.[5]

I hope and pray that your Most Reverend Lordship, to whom I pledge my services, may long [enjoy] good health for the benefit of your church and the solace of your [people].

Gynopolis[6] [Frombork], 28 September,
first year of the 579th Olympiad[7] [1541]

Your Most Reverend Lordship[8]

                                                   most devoted
                                          Nicolas Copernicus

My lord, Most Reverend Father
in Christ, Johannes [Dantiscus],
by the grace of God bishop of Varmia,
my most gracious lord

                                      [received 29 September]

---

[1] Dantiscus' request for the return of his epitaph has not been preserved.

[2] Neither Dantiscus' epitaph has been preserved, nor the accompanying letter, which was written presumably in 1539.

[3] Copernicus lacked not only the "signature" (as in StC VIII, 202, no. 477) but also the entire epitaph as written by Dantiscus' own hand. Copernicus' copy of Dantiscus' epitaph accompanied Letter XVII. Yet the copy has not survived, although Letter XVII, having been acquired by the Czartoryski family, is now preserved in the Czartoryski Library in Cracow (Catalogus Czartoryski, II, 278: MS 1619, pp. 99–100).

[4] The tombstone and its epitaph have not been preserved. Copernicus told Dantiscus that the earlier epitaph had already been carved on the tombstone (*iam antea aliud quoddam sepulcro fuisset insculptum*). Hence there is no historical basis for the statement that Copernicus ordered the carving of the epitaph only after receiving Dantiscus' epitaph (as in StC XVIII, 175). Had it devolved on Copernicus to order the carving of the epitaph, he might well have rejected it, since he regarded it as "having little flavor and less taste." This situation is not similar to the affair of Dantiscus' epigram (as StC XVIII, 156, mistakenly thinks). For in the case of the epitaph, the tombstone had already been carved, while in the matter of the epigram, Copernicus gave it to Rheticus, who printed it.

[5] Copernicus' *quod procuratum sic erat et paratum illi adhuc in humanis volenti* means that Watzenrode commissioned this epitaph and approved it, not that he composed it himself (*verfasste*, as in ZGAE, 6:311).

[6] Copernicus dated Letter X also from Gynopolis.

[7] Copernicus dated the first year of the first Olympiad in the summer of 775 B.C. (NCCW, II, 429/on 278/8). With four years in every Olympiad, 578 Olympiads contain 2312 years. Subtraction of 775 from 2312 leaves the summer of 1537 as the end of the 578th Olympiad and the beginning of the 579th. But Letter XVII was not written on 28 September 1537. For it refers to Dantiscus' epitaph (which was composed after 11 January 1539) as having been sent by Dantiscus to Copernicus long before (*olim*) the date of Letter XVII. Presumably as the result of a slip, Copernicus wrote "first year of the 579th Olympiad" where "last year of the 579th Olympiad" or "first year of the 580th Olympiad" is required as the equivalent of 1541. This date (1541) was correctly introduced by Dominik Szulc, who published Letter XVII for the first time (*Biblioteka Warszawska*, 1857, part IV, page 783). Twelve years later in the same journal (1869, part III, pages 138–139) Alexander Przeździecki printed Letter XVII a second time on the basis of the autograph, which in 1856 he had found in the Polish Library in Paris. It had been taken there, as part of the Czartoryski Library in Puławy, after the Polish insurrection of 1830 had been suppressed by Russia. Strangely unaware of the first printing of Letter XVII, Przeździecki chose an erroneous date (1537) for it on the basis of a miscalculation involving two errors (p. 139, n. 2). He did not, however, opt for 1539, the date misassigned to him by StC VIII, 203, no. 477, and StC XVIII, 207, no. 15. That erroneous date (1539) was given by Hipler, p. 204; Polkowski, I, 87–88; PII, 142/15, 165/10; and MK, p. 400, no. 7. Perhaps baffled by Copernicus' reference to the 579th Olympiad, Catalogus Czartoryski (II, 278:MS 1619, p. 99) dated Letter XVII in 1579, thirty-six years after Copernicus' death, and also in 1529 (in order to avoid the embarrassment of posthumous authorship?). Facsimiles of Letter XVII: Wołyński (Plate VIII, with the erroneous date 1539); NCCW, I, Plate V, and NCCW, IV, Plate XVI. Copernicus' date (775 B.C.) for the first

Olympic games is too late by a year, because he shortened by that amount the first component in his computation; see NCCW, II, 141/14–21; 390/on 141/14, 20.

[8] The first letter in Copernicus' abbreviation ($E\ r\ d\ v$) was mistakenly expanded as *Episcopalis* by Przeździecki (*Biblioteka Warszawska*, 1869, part 3, p. 139), followed by PII, 165. The correct expansion is rather *Eiusdem*.

The Latin text of Letter XVII was published first by Szulc, then by Przeździecki, Hipler, PII, and StC XVIII, 239. Letter XVII was translated into Polish by Szulc (reprinted by Polkowski, I, 87–88), Przeździecki, and StC XVIII, 239.

BIBLIOGRAPHY

Album studiosorum universitatis cracoviensis, II (Cracow, 1892)

Biskup, Marian, "Problem autografu listu Mikołaja Kopernika do Feliksa Reicha z roku 1528," KMW, 1975, pp. 257–261; facsimile of Copernicus' letter to Reich between pp. 260–261

Burmeister, Karl Heinz, *Georg Joachim Rhetikus*, 3 vols. (Wiesbaden, 1967–1968)

Catalogus Czartoryski *Catalogus codicum manu scriptorum Musei principum Czartoryski*, 2 vols. (Cracow, 1887–1913); indexes (Cracow, 1928; Tessin, 1931)

Corpus iuris polonici, IV (Cracow, 1910), ed. Oswald Balzer

Forstreuter, Kurt, "Aktenaustausch zwischen dem Staatsarchiv in Königsberg und den ermländischen Archiven in Frauenburg," *Archivalische Zeitschrift*, 1931, 40:267–269

[Hoffmanowa, Klementyna] in *Rozrywki dla Dzieci*, 1826, 5:81–84; 1828, pp. 81–84

Hubatsch, Walter, *Albrecht von Brandenburg-Anspach* (Heidelberg, 1960; Studien zur Geschichte Preussens, Band 8)

Jamiołkowska, Danuta, "Memoriale Łukasza Watzenrodego — analiza paleograficzna," KMW, 1972, pp. 633–648

Joachim, Erich, *Die Politik des letzten Hochmeisters in Preussen Albrecht von Brandenburg* (Leipzig, 1892–1895; Publicationen aus der k. preussischen Staatsarchiv, Bd. 50, 58, 61)

Kopernikus-Forschungen, eds. Johannes Papritz and Hans Schmauch (Leipzig, 1943; Deutschland und der Osten, Quellen und Forschungen zur Geschichte ihrer Beziehungen, Vol. 22)

Le Branchu, Jean-Yves, *Ecrits notables sur la monnaie*, 2 vols. (Paris, 1934)

Moore, George Albert, *Copernicus, Treatise on Coining Money* (Chevy Chase, Md [1965])

Niemcewicz, Julian Ursyn, *Zbior pamiętnikow historycznych o dawnéy Polszcze* (1st ed. Warsaw, 1822–1833; 2nd ed. Leipzig, 1838–1840; Collection of Historical Diaries concerning Poland in Olden Times)

Oko, Jan, "Paweł Deusterwalt," *Ateneum Wileńskie*, 1930, 7:786–798

Prowe, Leopold, *Mittheilungen aus schwedischen Archiven und Bibliotheken* (Berlin, 1853)

Schmauch, Hans, "Neue Funde zum Lebenslauf des Coppernicus," ZGAE, 1943, 28:53–99

Schultheiss, Tassilo, *Nicolaus Coppernicus aus Thorn Ueber die Umdrehungen der Himmelskörper* (Posen, 1923)

Sikorski, Jerzy, *Prywatne życie Mikołaja Kopernika* (Olsztyn, 1973; The Private Life of Nicholas Copernicus)

Sommerfeld, Erich, *Die Geldlehre des Nicolaus Copernicus* (Vaduz, 1978)

Triller, Anneliese, "Das dy helige gerechtikeit nicht vorhindert worde," ZGAE, 1972, 36:124–133

Wołyński, Arturo, *Autografi di Niccolò Copernico* (Florence, 1879)

INDEX TO THEOPHYLACTUS

Achates 27
Achilles 42, 43
Achilles Tatius 8, 59
Aeacus 43
Aegirus 31, 44
Aegyptus 46, 65, 67
Aelian 8, 51, 53–55, 65, 66, 68, 69
Aeneas 27
Aeschines 47
Aeschylus 8, 60
Aesop 6, 8, 61
Agesilaus 36
Africa 8
Alcibiades 16, 49
Ambrose 24
Ampelinus 50
Ampelius 38, 61
Ampelo 49
Anaxarchus 40
Anthesterion 14
Anthia 45
Anthinus 50
Anthusa 50
Antigonus 44
Antimachus 16, 50
Antipater 30
Antisthenes 35, 37
Aphrodisias 7
Archimedes 16, 48
Aristides 34
Aristo 33
Aristophanes 8, 58, 61, 68
Aristoxenus 48
Arnon 42
Arrian 8, 64
Aspasia 41
Astachyon 36
Atalanta 34
Athenaeus 53
Atlantic Ocean 46
Atlas 27, 51
Attica 11, 15, 36
Augeas 34, 67
Augustus, emperor 51
Aurivillius, P. F. 18
Axiochus 25, 36, 47, 58

Baltic Sea 19
Basil 4
Bauch, G. 20, 22
Boethius 63
Boissonade, J.-F. 7, 25, 26, 65
Brion 66
Bubalio 46, 68

Bucephalus 42, 64
Bucolio 49
Byzantine Empire 15

Calamo 43
Callicomos 49, 70
Callicrates 31, 53
Callimachus 38
Callimachus, the poet 53
Calliope 37, 47, 49, 61
Callistachus 32
Caria 7
Casaubon, I. 26
Caucasian Mountains 28
Cecropis 37
Celtic River 32, 54
Celts 54
Cepias 45
Chance 35, 52, 57
Chloazo 35
Chmiel, A. 19, 20
Chryses 35, 46, 57
Chrysippa 43, 50
Chrysippus 39
Chrysippus River 43
Chrysogone 33, 42, 55
Chrysosthenes 44
Cimedoncius 7, 25, 51
Cimon 40
Cissybius 46
Cleon 43
Colchis 28
Coriannus 45, 47
Corinna 34
Corvinus, A. 22, 27, 28
Corydo 42, 46, 67
Crete 50
Critias 30
Cromylo 49
Cujas, J. 25, 26
Cupid 15, 36, 39, 40, 48–50, 60, 69
Cyclops 41
Cyparisso 32, 33
Cyparissus 47

Damalus 33, 55, 56
Damascius 44
Danae 49, 70
Danaus 46, 67
Daphno 16, 31, 44
Darius, king of Persia 35, 58
Demonicus 41
Demosthenes 7
Dexicrates 32, 37

355

Diodorus 40, 62
Diodota 46
Diogenes 8, 35, 41, 42, 45, 48, 63
Diogenes Laertius 8, 25, 64
Diomedes 41, 63
Dionysius 49, 50
Donauwörth 22
Dorcon 30, 48
Dresden 24

Elapho 48
Erasmius 40, 62
Erato 31
Eratosthenes 47
Erotylus 34
Etna 9, 36
Euripa 32
Euripides 8, 56, 58, 68
Europe 3
Euryades 40
Eurydice 30
Evagoras 30

Fortune 9, 42, 49, 52
Füldener, J. J. v. 24

Galatea 38–40
Gaza, Th. 14
Germanicus, Caesar 29, 51
Gibraltar, Strait of 29
Götze, J. C. 24
Gorgias 34, 46, 47
Greece 7, 8
Grégoire, H. 7
Gregory of Nazianzus 4
Gruter, J. 25

Habrotonon 43
Hall, A. R. 60
Haller, J. 22, 23, 26, 50, 71
Hanke, M. 24
Hanseatic League 19
Heidelberg 25
Helen of Troy 31, 53, 56
Helios 27
Hephaestio 38, 46
Heraclides 37
Hercher, R. 26, 51, 60
Hermagoras 32, 33
Hermus River 28
Herodotus 65
Hindu 33
Hippolytus 47, 58, 68
Homer 8, 13–15, 41, 44, 50, 57, 59, 60, 63, 66, 70, 71
Horus 9, 30, 52
Hypsipyle 34, 43

Isocrates 25, 49

Jadwiga, St. 51
Janocki, J. D. 24
Jason 44, 65
Julian the Apostate 8, 54
Jupiter 15, 35, 39, 49, 70

Kimmendonck, J. 25
Kowalski, J. 18
Krumbacher, K. 8
Kytzler, B. 26, 51

Lachano 37
Laertes 14
Lais 36, 45, 47
Leander 42, 49
Leiden 25
Leningrad 18
Leokorion 9
Leonides 42, 63, 64
Leos 9
Leucippa 36
Leucippus 34, 56
Libanius 4, 60
Linceus 61
Lopho 34, 56
Lubin, E. 25
Lucanians 47
Lyceum 69
Lydus 41
Lysis 4
Lysistratus 40, 45, 46

Macedonia 35, 42
Maia 51
Manuzio, A. 4, 13, 14, 17
Marathon 41
Maurice, Byzantine emperor 6
Meco 35
Medea 8, 36, 38, 44, 65
Melanippe 40, 62
Melanippides 46
Meles 13, 14
Melpomene 33, 55
Menander 8, 60
Mercury 51
Midas 35, 57
Migne 63
Mikhailov, A. A. 18
Milo 36, 58
Mintho 43
Molhuysen, P. C. 25
Morstin, L. H. 26
Moschio 39
Moschon 30

Muses 35
Myron 16, 31, 32
Myronides 33, 39, 49, 62

Neumarkt 19
Nicias 33
Nile 65
Niobe 36, 59
Nissen, Th. 17–19
Nymph 15, 37

Olympus 28
Oława 28, 51
Orelli, J. K. 25
Orestes 37, 60, 68
Orion 45, 66
Ortygo 40
Ovid 29, 51

Pan 15, 30, 37
Parandowski, J. 26
Parke, H. W. 61
Parmenides 44
Parrhasius 31, 53
Pediades 34
Pegano 37, 41
Peitho 47
Peleus 15
Penelope 45, 50, 66, 70
Perdiccas 35
Periander 42
Pericles 8, 35, 41, 49
Perry, B. E. 61, 66
Persians 8, 38
Phaedra 36, 58
Philip 41
Philip II, king of Macedonia 35
Philonides 41, 44, 63
Photius 5–7
Phrygia 35
Phrygius 45
Piraeus 43
Pithou, F. 26
Platanus 31
Plato 7–9, 25, 33, 41, 47, 50, 55, 57
Plotinus 8, 30, 36, 51, 58
Plutarch 8, 58, 60
Poas 38, 61
Podlecki, A. J. 25
Poimnio 42, 64
Polyxene 48
Porphyrio 46
Porphyry 8, 58, 59, 63
Praximille 33
Priam 15, 42, 43
Priamides 42
Proclus 48, 60, 69

Prosna 28
Providence 15
Ptaśnik, J. 22
Pylades 42
Pyrrhias 44
Pythagoras 25
Pythia 15

Rhine see Celtic River
Rhizo 43
Rhodina 37, 61
Rhodoclea 43, 47, 65
Rhodon 47
Rhodope 35

Scaliger, J. J. 26
Scot, J. C. 60
Seruga, J. 22
Servius 51
Seutlion 47
Sicily 9, 36
Simichidas 32, 33
Siren 35, 50, 57, 70
Smit, B. de 25
Smith, L. F. 25
Smith, N. 60
Socrates 9, 25, 41, 43, 44, 49, 65, 67
Solon 31
Sopater 48
Sophroniscus 9
Sosipater 31, 32, 36, 50, 71
Sostrate 45
Sostratus 34, 46, 56, 67
Sotion 48
Sparta 33
Speyer, W. 8
Spiro 36, 43

Tantalus 47
Telesilla 36, 48
Terpander 31, 42
Terpsithea 31
Tettigo 16, 40, 46, 63
Thales 38, 61
Theano 30
Themistocles 25, 39
Theophrastus 69
Theophylactus Simocatta, *History* 28, 29
　　　　　　　　　, *Letters* 27, 29, 30
　　　　　　　　　, *Natural Questions* 7, 57, 60
Theristro 36
Thetis 15, 38, 40, 42
Tolstoi, I. I. 18
Trygias 8
Tycanias 39, 62

Ulysses 14, 40, 41, 44, 49, 50, 63, 65, 66, 70
Urceo, A. (Codro) 4, 5, 17, 18

Vatican Library 25
Venus 50, 63
Vergil 51
Vulcanius see Smit, B. de

Westermann, A. 26
Whitby, L. M. 53

Wormell, D. E. W. 61

Xanthippus 16, 48

Zanetto, G. 15, 18, 51, 60, 61, 65, 67
Zeuxis 53

INDEX OF PERSONS

Achtnicht, M. 266, 279, 290–292, 297, 299
Agrippa 149, 161, 162
Albert of Brudzewo 123, 126
Albert of Prussia, duke 302, 311, 343, 345, 347, 348
Albert the forester 233, 238, 245
Aldejorge 229, 243
Alex 237
Alexander VI, Pope 220
Alexander, king of Poland 270, 271
Alexander the Great 8, 35, 42, 57, 58, 64, 145, 146, 153–155, 161
Alexius 249
Alfonso X, the Wise 98, 99, 145
Alfonso de Corduba 98, 99
Allen, C. 313
Andreas 228
Andreas of Darethen 233
Andres of Micken 233
Andres the overseer 255
Andrhe 229, 243
Anselm, Bishop 261, 279
Anthony, Brethren of St. 266, 279
Antoninus Pius 145, 146, 153, 155
Apian 129
Apollonius 111–113
Aratus 159
Argelata, P. 301
Argyropoulos 155
Aristarchus 42, 94, 159, 161
Aristotle 3, 20, 53–55, 63, 69, 76, 92, 122, 123, 126, 134, 145, 151, 152, 155, 157, 163, 206
Aristyllus 148, 149, 159, 161
Arnold of Villanova 301, 302
Arpád, K. 330
"Arsatilis" 159
Asman 228, 243
Athanasius 333
Augustin of Plauczk 232, 238
Augustin of Spigelberg 228
Augustinus 237, 248
Averroes (Ibn Rushd) 123, 126
Avicenna 303

Bachmann, S. 128, 129, 151
Bahr, E. 219, 222
Bak, J. M. 330
Bańkowski, P. 268, 276
Baranowski, H. 139, 174
Baranowski, J. 26, 62, 209, 310–314, 318, 319, 321, 325–327, 329, 330, 332–335, 338, 339, 343, 349, 352
Barone, F. 207
Bart, C. 304
Bartholomew, St. 290–292, 300

Bartolmis 237, 249
Bartosch 229
Basil III 311
Battani, Al- 83, 98, 99
Baytz, M. 230, 236, 243
Belger, J. 279
Benedict 237
Benedict of Maura 265, 277
Benjamin, F. S. 110
Bentkowski, F. 171, 174, 207, 209, 211
Benzi, U. 303
Berg, A. 304
Bernt 236
Bessarion, Cardinal 151
Bieda, K. 207, 213
Birkenmajer, A. 134
Birkenmajer, L. A. 18, 26, 98, 123, 130, 135, 140, 141, 151, 174, 198, 207, 213, 304, 349
Biskup, M. 215, 242, 268, 274, 319, 353
Blagrave 129
Bodner, H. 228
Bogener, M. 266, 279
Boniface IX, Pope 263, 270
Brahe, T. 76–80, 91, 93–97, 99, 107, 110, 112, 116–122, 124, 134, 136, 138, 144, 158, 307
Broch, B. 237
Brodrick, G. C. 141
Bronzino, 222, 347
Brosche 236, 248
Brożek, J. 130, 138, 164, 308, 325, 338, 339
Brucaeus (Brock), H. 77, 78
Brun, P. 239, 250
Brusien 228
Buczek, K. 131
Bujak, F. 205, 207
Burghley, Lord 141
Buridan, J. 123, 126
Burmeister, K. H. 129–131, 133, 136, 137, 151, 351, 353

Caerellius, Q. 146, 154
Calau, H. 232
Calixtus III, Pope 278
Callippus 81, 91, 122
Campanus of Novara 110
Casaubon, M. 135
Casche, C. 234
Caseler, M. 228
Caselius, J. 77
Casimir IV, king of Poland 208, 265, 270, 274, 276, 293
Censorinus 146, 154
Charles IV, emperor 261, 268
Charles V, emperor 329, 330

359

Charles V, king of France 215, 270
Charles XII, king of Sweden 307
Christopher of Delau (Delen) 266, 278
Christopher of Suchten 246
Chrysostom 333
Clauke, H. 235
Cleban, N. 228, 243
Clement VI, Pope 270, 279
Cleomedes 94
Cletz, A. 288, 294
Collinus, M. 135
Colo, A. 220, 222
Conrad of Jungingen 186, 209
Copernicus, A. 257, 321, 322
Copetz, Cz. 233
Corvinus, L. (Rabe) 19, 20, 22–24, 80
Cosman 233, 238, 245
Craig, J. 77
Cranzel, 235
Cristof 236, 241, 252
Crix, B. 230, 234
Cucuc, H. 235
Curtze, M. 17, 79, 80, 136
Cusche, J. 233
Czacki, T. 275
Czartoryski, family 174, 215, 326, 330, 334, 342, 343, 354
Czartoryski, P. 22, 51, 153, 333
Czartoryski, W. 326
Czehm, A. 286, 287, 289, 294–296, 336
Czepan, G. 226, 230
Czichotzinsky, S. 240

Dam(brow)itz, C. 286, 287, 293, 294, 296
Dantiscus, Bishop Johannes 164, 302, 321–342, 349–353
Daumschen, A. 228
Day, W. 326
Decius, J. L. 200, 201, 204, 205, 207, 214, 215
Dee, J. 135, 141
Delambre, J. B. J. 152, 153, 162
Deusterwald, P. 207, 338
Dioscorides 302, 333
Dmochowski, J. 172, 174, 198, 199, 205, 207, 211–215, 244, 247, 249–251, 254, 281–284, 310–314, 317–319
Doppelmayr, J. G. 139
Drauschwitz, C. 255, 310, 311
Dunajewski, H. 205, 207, 214

Eberhard, Bishop 280
Eberhard of Wesenthau 272
Ebert, P. 238
Eglof, A. 239
Einwald, Bishop Paul 265, 274, 277
Eis, G. 304

Eler, B. 240
Elias of Darethen 279
Em(m)erich, F. 164, 165, 302–304, 335, 336
Erasmus, D. 333
Eriksen, J. 78, 79
Erler, G. 222, 338
Eubel, C. 222, 276
Euclid 129
Eudoxus 81, 91, 122

Faber, B. 231
Faber, K. 173, 207, 319
Fabian of Vusen 273, 294
Faulhaber, J. 292, 300
Feldstedt, R. 312–314
Ferber, Bishop Maurice 199–201, 292, 302, 313, 314, 317, 320, 321, 331, 339, 340
Ferber, E. 321
Ferber, J. 199, 321
Ferdinand, king of Bohemia 219, 327–330
Fischer, F. 174, 196
Fleming, Bishop Henry 270, 272
Flis, S. 304
Forstreuter, K. 311, 353
Francis I, king of France 330
Frederici, G. 256
Frederick 230
Freundt, A. 302
Friedländer, E. 132
Fröhlich see Bogener

Gadel, G. 232, 245
Gaillard, G.-H. 330
Galilei, G. 159, 162, 307
Gansiniec, R. 18, 26, 100, 110, 118
Gardenal, G. 80
Gassendi, P. 138, 139
Gasser, A. P. 132, 133
Gemma, R. 324
George, St. 261, 270
George of Baisen 262, 270, 286, 287, 293
George of Delau 258, 259, 274
George of Kunheim 344–348
George of Schliven 265, 273, 277, 278
Gerard of Cremona 154
Gerber, S. 235, 246
Gesner, C. 131
Giese, Bishop Tiedemann 240, 250–252, 259, 275, 281, 285, 286, 291, 293, 294, 296, 299, 300, 313, 317, 321–323, 325, 329, 331, 338, 351
Gilmeister, F. 239
Glande, P. 228, 232, 244
Gnik(x), A. 238, 249
Gosławski see Kelbas
Grant, E. 135
Grauert, H. 127

INDEX OF PERSONS

Grażyński, M. 215
Greger of Hogenwalt 238
Gregory XI, Pope 261, 269, 273
Greusing, P. 266, 278, 309–311
Greve, C. 288
Grimm, W. 208, 213
Gunter, P. 237, 248
Gunter, U. 237, 248
Gustavus Adolphus 307
Gutco 265

Haggag ibn Yusuf ibn Matar 154
Hajdukiewicz, L. 75
Hájek, S. 134, 135, 139, 142, 144, 163, 164
Hájek, T. 76, 79, 80, 134–136, 139–142, 144
Han, M. 237
Hand, S. B. 304
Haneman, M. 239
Hans of Lesser Cleberg 232, 233, 244
Hans of Schonebrugk 237
Hans the overseer 237
Hartmann, G. 129, 130
Hausberg, J. 238, 249
Haytham, Ibn al- 124, 125
Hegeland, H. 169, 207
Heinrich of Plauen 186, 196, 209, 211
Henkel, G. 288, 294
Henry of Sonnenberg 267, 280
Hensel 228
Herold, M. 128, 129, 151
Hevelius (Hewelke) 78, 140
Hilfstein, E. 130, 136, 138, 143, 308
Hillebrant, U. 234, 235
Hinderliter, R. H. 207
Hinzke, H. 232
Hinzke, L. 232, 245
Hipler, F. 16, 18, 19, 26, 140, 151, 172–174, 211, 224, 226, 227, 242, 243, 247, 249, 251, 252, 254, 256, 275, 280, 283, 297, 310, 311, 313, 314, 317–319, 325–327, 330, 332–335, 338, 339, 341–343, 349, 350, 352–354
Hipparchus 83, 97–99, 147, 152, 158, 161
Hirsch, C. C. 128
Hirschberg, A. 207
Hjörter, O. P. 141
Hoffmanowa, K. 326, 334, 342, 353
Hofmeister, A. 77
Hohenberg, C. 270
Holz, M. 304
Horn, P. 329
Horský, Z. 134
Hosius, S. 329, 339–341
Hoveman, A. 238, 250
Hoveman, S. 239, 250
Hubatsch, W. 311, 353
Hun, M. 239, 250
Hurmuzaki, E. de 330

Innocent IV, Pope 268
Innocent VI, Pope 261, 268, 270
Innocent VII, Pope 279
Innocent VIII, Pope 264, 265, 274, 277
Isidore of Seville 98

Jabir ibn Aflah 83, 98, 99
Jacob of Jomendorf 233, 241, 245, 252
Jacob of Lesser Scaibot 234
Jacob the herdsman 236, 252
Jacob the overseer 256
Jamiołkowska, D. 207, 353
Jan of Greseling 228
Jan of Natternen 236
Jan of Vindica 228, 233, 235, 243, 245, 247
Jerome 226, 228, 231, 232, 238, 243, 244
Jeschke 241
Joachim 228, 242
Joachim, E. 310, 311, 353
Jod(e), M. 220, 222
Johann of Tiefen 208
Johann of Werden 343, 344
John III, king of Portugal 329
John XXII, Pope 270
John Albert, king of Poland 271
Johnson, Ch. 207, 209
Jorge 233
Josten, C. H. 135

Kabe 271
Karliński, F. 139, 140, 143
Kaussler, E. 207, 214
Kelbas, Bishop Vincent 264, 276
Kelley, E. 141
Kepler, J. 95, 102, 112, 124, 307
Keserynne 286, 287, 294
Kętrzyński, W. 326
Kist, J. 127–129, 151
Knobel, G. 228
Knobel, P. 228
Koestler, A. 26, 323, 325, 327–330, 334, 350, 352
Konopacki, G. 339–341, 343
Konopacki, R. 339–343
Konopka, S. 304
Krasicki, I. 352
Krentzheim, L. 127
Kressel, H. 127–129, 151
Kreysig, G. C. 219
Küchmeister, M. 187, 209, 210
Kuhschmalz, Bishop Franz 266, 278
Kunin, A. S. 304
Kunitzsch, P. 154, 158, 159, 162
Kycol, M. 233

Ladislaus II, king of Poland 264, 275, 276
Langermann, T. 124

Le Branchu, J.-I. 171, 207, 212, 213, 319, 353
Legendorf, P. 265, 276–278
Leibniz, G. W. 105
Lemay, R. 162
Lengnich, G. 171, 207
Lewicka-Kamińska, A. 133
Lewicki, A. 268, 275
Leze, M. 232
Libenwald, B. 265, 277
Liddel, D. 77, 78, 80
Lilienthal, J. 317
Linde, S. B. 173, 174
Lindhagen, C. A. 76, 79, 99, 102, 104
Lochner, G. W. K. 129
Longomontanus, C. S. 78, 80
Lorencz 229
Lorenz 234
Lossainen, Bishop Fabian of 262, 264, 268, 270, 273, 276, 302, 310
Lossau, B. 254–256
Luckerath, C. A. 210
Ludica, P. 241
Lupáč, P. 135
Lurenz 234, 249
Lutherans 214, 330, 351

Maciejowski, W. A. 140
Macleod, H. D. 171, 207
Maestlin, M. 102, 119, 120
Maimonides 126
Makowski, A. 139, 144
Malagola, C. 17, 132
Marques, A. 329
Martianus Capella 126
Martin V, Pope 266
Martin the overseer 236
Martzin 232
Martzyn 233, 237, 245, 249
Mattheus 241
Matthew of Miechow 75, 76
Matthias, king of Hungary 264, 276
Matz of Degten 237
Matz of New Cleberg 233
Matz of Plauczk 229
Melczer, A. 266, 278
Menelaus 149, 162
Merriman, R. B. 330
Merten of Godekendorf 235, 247
Merten of New Cleberg 241
Merten of Old Cukendorf 234, 242
Mertin 230
Messing, T. 239
Michel of Licosa 240
Michel of New Cleberg 238
Micher, M. 241
Millás Vallicrosa, J. M. 124
Moldyth, T. 254, 255

Molner, H. 238
Monroe, A. E. 171, 207
Monsterberg, N. 220, 221
Moore, G. A. 207, 319, 353
Moraux, P. 76
Morneweg, K. 127
Müller, A. 117, 118
Muczkowski, J. 19, 347

Nabonassar 145, 146, 153–156, 163
Nebuchadnezzar 153–155, 162
Neudorfer, J. 129
Neugebauer, O. 153
Newton, R. R. 153
Nicholas, Abbot 272
Niclas 238
Nicolaus 229
Niedźwiedzki, L. 140, 143, 144
Niemcewicz, J. U. 327, 330, 353
Nimsgar, J. 230, 244
Nöbel, W. 209, 210
Nomi, F. 171, 173, 207, 209–213
North, J. D. 123, 129
Nosche, G. 232
Nunes, P. 137

Obłąk, J. 258, 259, 268, 270–280
Oko, J. 207, 338, 353
Oporowski, A. 266, 278
Oresme, N. 205–207, 215
Ortroy, F. v. 324
Otto of Russen 266, 279

Padluche, J. 272
Paipo, C. 311
Palm the overseer 255
Panzer, G. W. 24, 128
Papritz, J. 353
Paschalides, T. N. 127
Paul III, Pope 329
Paul of Rusdorf 210, 266, 278
Paul of Vindica 235
Pavel, overseer in Greseling 237
Pawel of Greseling 237
Pawel of Stolpe 233
Perlbach, M. 338
Peter of Caldeborn 236
Peter of Hoensteyn 241
Peter of Lycosa 241
Peter of Old Cukendorf 234
Peter of Plutzk 231
Peter the overseer 236, 247
Petrus the herdsman 233, 246
Petrus the overseer 241
Peucer, C. 137, 138
Peurbach, G. 3, 98, 123

Pippelk, B. 230
Pippelk, N. 230, 243
Pius II, Pope 265, 276, 278
Plastevig, A. 288
Plastewigk, G. 256
Plastwich, J. 272
Plato 76, 134
Pliny the Elder 53, 80, 111, 246
Płotowski, P. 341, 349
Polen, T. 241, 266, 272, 278
Polkowski, I. 26, 134, 139, 140, 143, 151, 174, 312, 314, 318, 319, 326, 327, 330, 332, 334, 339, 342, 343, 349, 352–354
Poppe, J. 235
Praetorius, J. 136–138, 161
Pregers 286, 287
Preus, P. 225, 226, 231
Preusse, A. 243
Proclus 125, 126, 162
Prowe, L. 17–19, 26, 173, 174, 208, 313, 314, 319, 332, 349, 353
Prussia, West, Estates of 169–171, 182, 198–202, 204–206, 208, 209, 211–213, 215, 276, 315, 317, 318, 327
Prussian League 265, 266, 278
Przeździecki, A. 353, 354
Ptolemy 3, 4, 76–79, 81, 84, 92, 93, 98, 99, 110–113, 133, 141, 145–150, 152–154, 156–160, 162, 163, 323, 352
Pythagoreans 82, 92, 94, 104

Quedlitz, A. 262, 271

Rabe, N. 241
Rabener, J. G. 140
Rabenwalt, S. 262, 271
Radau, A. 239
Radymiński, M. 164
Rape, J. 237
Rase, S. 234
Rastawiecki, E. 152
Rautenberg, G. 292
Redinger, L. 220, 221, 223
Regiomontanus, J. 3, 92, 111–113, 137
Reich, F. 173, 174, 196, 199, 201, 204, 205, 209, 214, 275, 276, 281, 282, 293, 300, 302, 315, 317–319, 321, 324, 325, 332–343
Reicke, E. 151, 172
Reinhold, E. 98, 137
Reiss, T. J. 207
Reuss, H. 209, 211
Rex, J. 273, 280
Rheticus, G. J. 76, 79, 80, 129–133, 136, 137, 144, 151, 350, 351, 354
Riccioli, G. B. 92
Richard of Wallingford 123, 126
Rigaud, S. P. 136, 144

Rigoni, E. 219, 222, 223
Römer, K. 214
Roman, J. 230
Rosen, E. 125, 268
Rossmann, F. 104, 105
Ruche, N. 241
Rudolph II, emperor 76
Rupoldus, J. 329
Ryman 239
Rytel, A. 304

Sack, G. 288, 294
Sacrobosco 139
Salah, Ibn as 154
Salmon 233, 246
Salomon 233, 246
Sander of Vusen 265
Santke, M. 235
Savile, H. 135, 141, 144
Savonarola, G. 302
Savonarola, M. 302, 303
Saxonius, P. 137
Scaibot, B. 266, 278
Schäfer, E. 77
Schmauch, H. 169, 170, 173, 207, 214, 215, 219, 222–224, 227, 242–252, 268, 274, 283, 311, 319, 330, 353
Schneeberger, A. 130
Schönberg, N. 310, 351
Schöner, J. 129
Schöttgen, C. 219
Scholze, M. 239
Schonsze(e), G. 255–257
Schottenloher, K. 127, 151
Schreckenfuchs, E. O. 139
Schütz, C. 198, 199, 202, 207, 208, 210
Schultheiss, T. 242, 312, 314, 319, 353
Schwinkowski, W. 207, 208
Sculteti, A. 201, 257, 289, 291–296, 336
Sculteti, J. 274
Scultetus, U. 254, 256
Semrau, A. 207, 214, 215
Sergius 154
Sforza, Bona, queen of Poland 339, 340
Sforza, Ch. 330
Sforza, F. 330
Shatir, Ibn al- 104, 113
Sienkiewicz, K. 326
Sigismund I, king of Poland 28, 51, 152, 164, 171, 199, 207, 212, 215, 262, 264, 270, 271, 274, 293, 310, 311, 317, 318, 333
Sigismund, emperor 265, 277
Sikorski, J. 224, 244–246, 249, 257, 297, 300, 312, 317, 322, 338, 349, 353
Silvatico, M. 301
Simler, J. 131
Simon of Greseling 235, 237

Simon of Vindica 238
Simon the overseer 236
Sixtus IV, Pope 265
Skimina, S. 350, 351
Slander, M. 236, 248
Smith, H. 144, 238
Snellenberg, H. 312–314
Solfa, J. B. 219, 329, 347, 348
Sommerfeld, E. 169, 172, 207, 212, 215, 354
Sorbelli, A. 132
Sorbom, Bishop Henry 265, 276
Spencer, G. 326
Stanislaus 235
Starowolski, S. 75, 136, 138
Steffen 241
Steinpik, V. 173
Stenczel the herdsman 238, 245
Stenczel the overseer 235, 247
Stenczel the taverner 236, 247
Stentzel 233
Stenzel 232
Stenzel the overseer 233, 245
Stockfisch, B. 253–258, 272, 276, 338
Stoke, S. 235, 247
Streifrock, Bishop Johannes 268, 279
Strewbyr, G. 238
Suetonius 351
Suleiman the Magnificent 330
Sylvester, Archbishop 272
Szebulski, A. 226, 229–231, 233, 244, 254
Szelągowski, A. 207, 213
Szulc, D. 164, 341, 342, 353, 354

Tahmasp I 329
Tapiau, Ch. 263, 273
Tapiau, Z. 290–292, 297, 299
Taylor, J. 207
Teige, J. 135
Teschner, P. 139
Teutonic Knights 169–171, 176, 195, 196, 206, 208, 251, 258, 259, 261, 264–266, 268, 269, 275–279, 309–312, 314
Teutonic Knights, Estates of 210
Tewes of Jonikendorf 235
Thabit ibn Qurra 160
Theiner, A. 268–270, 273, 278, 279
Thimm, W. 252
Thunert, F. 268, 276
Tile, U. 238, 249
Timmermann, J. 331
Timocharis 149, 150, 159, 161–163
Tolk(e), F. 263, 273
Tolkesdorf, M. 238
Toomer, G. J. 110
Trenck, A. 292, 316–318, 335
Tressler, J. 327, 329, 330
Treter, J. 239

Triller, A. 310–312, 354
Trokelle, A. 249
Trokelle, B. 231, 238, 244
Tungen, Bishop Nicholas 264, 274, 276, 288, 290–292, 294, 297, 298
Tusi, Nasir al Din 115, 116
Tynappel, K. 266, 279

Ulrich of Jungingen 186, 209
Urban V, Pope 270
Urben, A. 232, 244

Valentin 228, 242
Valentin, H. 304
Valescus of Taranta 303
Valla, G. 4, 80, 94, 110, 162
Varmia Chapter 169, 171–173, 201, 224–227, 240, 242–244, 246, 250, 251, 253–255, 258, 260, 262–264, 266, 268, 270–280, 285–292, 297, 301, 302, 307, 308, 310, 311, 314, 315, 318, 320, 322, 325, 327, 329, 331, 333, 337, 340, 341, 345, 346, 349
Varmia, Estates of 315, 317
Varro, M. 146, 154
Venrade, A. 264
Venturato, S. 219, 223
Vetter, Q. 151
Vicke, N. 254, 255
Vogelsang, Bishop Henry 275, 276
Vogt, J. 288, 289, 294, 295
Voitec 236, 248
Vossberg, B. A. 207, 208
Voteg, J. 230
Voyteg, J. 230, 232
Voyteg, M. 230
Voytek 234, 236

Walgast, J. 237
Wallace, W. A. 159
Wanczke, M. 235, 247
Wapowski, B. 131–134, 136, 138, 139, 144, 145, 152, 158
Waschinski, E. 207, 208, 210–212
Wasiutyński, J. 281, 283, 310, 312, 314, 318, 319, 326, 327, 330, 334, 339, 342, 349, 352
Watzenrode, Bishop Lucas 5, 19, 27, 29, 219, 222, 262, 264, 270–272, 274, 276, 279, 302, 308, 313, 327, 338, 339, 352, 353
Wayner, J. 230, 234
Wayner, Martzyn 236
Wayner, Michel 234
Wayner, Z. 230
Weise, E. 268, 275, 276
Werner, J. 127–145, 151, 152, 154–156, 158–160, 162–165
Winrich of Kniprode 186, 196, 208, 209
Wisłocki, W. 20, 141

INDEX OF PERSONS

Wölky, C. P. 253
Wolf, C. 130, 131
Wolf, J. 237
Wolowski, L. 171, 173, 174, 205, 208–213
Wołyński, A. 330, 332, 334, 343, 349, 352, 353
Woppe, H. 232
Woyteck, J. 232

Zápolya, John 330
Zimmermann, G. 222, 329
Zimmermann, W. 304
Zinner, E. 92, 128, 129, 137, 155, 304
Zupky, S. 236

Żurawicki, S. 208, 214

INDEX OF PLACES

Here, prominent places are indexed according to their contemporary designation. The corresponding Copernican designation will be found in "Polish Places in Copernicus' Life and Work," which also lists less prominent places.

Aberdeen 77, 78
Abestich 235, 247
Alexandria 149, 161
Altdorf 137
Appelaw 288, 294, 298
Athens 14, 126, 150

Baisen 262, 263, 270, 286, 287, 289–291
Balga 272
Basel 263
Bebir 262, 271
Beirut 162
Benatky 78
Berlin 137, 144, 172, 307, 308, 349, 352
Berting Teutonica 229, 234, 235
Birckpusch 262, 272
Bithynia 149, 161
Bohemia 134, 328
Bologna 3–5, 17, 18, 92, 132, 154, 222, 302, 347
Borniten 262, 272
Brandenburg 182, 188, 210, 265, 345, 347, 348
Braniewo 214, 262, 263, 265, 349
Braunswaldt 232, 234, 236, 237, 241, 246, 249
Brussels 324

Caldeborn 236
Caldemflis 263, 272
Caleberg 262
Carsau 262, 270, 271
Castile 99
Chełmno 276, 278, 322, 325, 330
Claukendorf 262, 270, 292
Cleeberg (Greater) 236, 248
Cleeberg (Lesser) 232, 233, 238, 241, 244–246, 249, 252
Cleefelt 240
Codien 262, 270, 286, 287, 291
Colmensche 288, 291, 299
Comain 238, 250
Conradswalt 262, 271
Copenhagen 112
Coseler 237
Courland 264
Cracow 3, 4, 19, 20, 23, 24, 50, 75, 79, 92, 110, 123, 126, 130–132, 134, 136–138, 140, 141, 151, 164, 174, 199, 200, 214, 222, 292, 307, 308, 310, 321, 326, 330, 334–336, 338–340, 342, 343, 347, 353

Crebisdorf 262, 270
Crumse 288, 294, 297
Cukendorf (New) 262
Cukendorf (Old) 234
Czartoryski Library 205, 214, 274, 275, 277, 307, 310, 321, 325–327, 330, 334, 336, 338, 339, 342, 343, 353
Czartoryski Museum 174

Dareten 233, 245, 263
Degten 237, 266, 279
Dewiten 236, 237, 249, 279
Ditterichswalt 237, 248, 254, 256

Edinburgh 325, 326
Egypt 67, 84
Elbląg 170, 199–201, 208, 211, 261, 265, 269–271, 288, 292, 293, 313, 320, 321, 338
Elditen 267, 280
Engelswalt 263, 273
England 141

Flanders 274
Florence 127, 302
Frankfurt am Main 78
Freiburg-im-Breisgau 139
Frombork 5, 22–24, 76, 132, 150, 172, 173, 215, 251, 252, 256–258, 274, 275, 280, 283, 286, 288, 289, 292, 294, 295, 297, 307, 310, 311, 313, 314, 317, 320, 324, 327–329, 331, 333, 334, 336, 337, 341, 342, 345–350, 352, 353
Frombork Cathedral 253, 255, 256, 258, 261, 264, 266–269, 279, 280, 286, 288, 297, 325

Gabelen 262, 271
Gdańsk 19, 169, 171, 198, 199, 201, 202, 205, 208, 214, 215, 251, 262, 270, 303, 312–314
Germany 76, 78, 130, 131, 174, 187
Glanden 262, 271
Glandemansdorf 235, 241
Godekendorf 228, 235, 243, 247, 249, 255
Göttingen 174, 307, 349
Great Britain 307, 325, 326
Greseling 228, 235, 237, 248, 249
Grunwald 186, 275, 277
Gutstadt 265, 271
Gynopolis 333, 334, 353

INDEX OF PLACES

Haselau 262, 287, 288, 291, 294, 297
Hinrichsdorf 267
Hoenberg 279
Hoensteyn 241
Hogenwalt 232, 238
Hungary 130, 276, 327, 328, 337

Ingolstadt 127, 128
Italy 19, 80, 219, 328, 329

Jagiellonian Library 123, 132, 133, 140, 151, 164, 308, 335
Jagiellonian University 200
Jomendorf 241, 245, 252
Jonikendorf 228, 234, 235, 247

Kabe 271
Koenigsberg 173, 174, 208, 269, 277, 289, 293, 295, 307, 310, 311, 319, 320, 344, 345, 349
Košice 328, 330
Kynappel 240

Laisse 239, 250
Leipzig 222, 338
Leon 99
Leynau 231, 242, 244
Libenau 239
Libentail 238, 250
Licosa 233, 240, 241, 245, 252
Lidzbark 5, 19, 20, 164, 172–174, 200, 209, 261, 269, 275, 276, 278, 282, 301, 307, 308, 310, 311, 314, 318, 325–327, 330, 332, 334, 335, 342, 352
Lithuania 264, 277, 340
Livonia 265, 277
Lockerat 292
Louvain 324
Lubawa 322, 323, 325
Lutterfelt 239
Lvov 213, 326

Malbork 197, 200, 201, 272, 286, 287, 311, 322, 327
Marquardshoffen 229
Masovia 340
Maybom 262, 271
Meissen 297
Melsac 224, 227, 238, 239, 243, 244, 250, 251, 267, 275, 280, 309–311
Merton College 141
Mica 230, 233, 236, 244
Milan 328, 330
Millemberg 239, 250
Monsterberg 235
Montiken(dorf) 234, 237, 247, 249
Moravia 328
Münster 319
Munich 128

Naglanden 230, 235, 236, 244, 247
Natternen 236, 247
Neudorf 250
Neuenburg 347, 348
Neuhoff 240, 250
Nuremberg 127–131, 145
   , Germanic Museum 128
   , Stadtbibliothek 92

Oliva 272
Olmütz 269
Olsztyn 224, 243, 244, 248–251, 256, 258, 259, 262, 263, 266, 267, 271–273, 275, 276, 278, 282, 309, 310, 313, 335–337
   , Castle 243, 254, 255, 257, 258, 260, 261, 271, 277, 280, 285, 286, 288, 297
Ossoliński Library 213
Osterode 261
Oxford 135, 144

Padelochen 263, 272
Padua 92, 98, 125, 219, 221, 301, 347
Paris 58, 123, 140, 333, 341, 354
Passau 220, 221
Passenhaim 228, 242
Persia 115, 328, 329
Petrica 230
Peuthuen 263, 272
Pilgrimsdorf 262, 271
Piotrków 337
Pisdecaim 235, 236
Plauczk 229, 231, 238, 249
Plauten 266
Poland 14, 22, 28, 169, 171, 192, 196, 206, 208, 245, 277, 293, 307–309, 311, 313, 326, 340, 349, 352
Pomerania 182, 213, 330, 340–342
Pomesania 264, 328, 330
Portugal 328, 329
Posorten 262, 272
Prague 78, 134–136, 139–141, 219, 329
Prussia 27, 170, 173, 174, 182, 188, 189, 191, 192, 196, 198, 199, 203, 205, 206, 208, 211, 245, 261, 268, 277, 307, 319, 340, 349, 352
   (Ducal, East) 170, 171, 174, 196, 200, 201, 211, 293, 295, 349
   (Royal, West) 169, 171, 174, 196, 200, 206, 208, 211, 285, 322
Puławy 307, 325–327, 330, 353

Quedlitz 271, 273

Rabusen 267
Radecaim 241
Radochońce 132
Reberg 262, 270, 286, 287, 291
Regensburg 76
Reichnaw 271

Resil 266, 278
Riga 264, 272
Rome 127, 132, 135, 149, 209, 224, 257, 279, 310, 322, 330
Rosenau 258
Rosengarten 262
Rostock 25, 77, 78
Rügen 347, 348
Russia 211, 340, 353

Samland 261, 264
Samos 161
Sandecaim 263, 272
Santoppen 261, 267, 269
Scaibot(h) 230, 234, 244, 263, 273
    (Lesser) 234, 248, 271
    (Old) 236, 248
Scharfenberg 262, 270
Schoneberg (New) 232
Schonebrugk 233, 237, 238, 250
Schonebuche 262, 271
Schonemburch 271
Schonewalt 231
Schouffsberg 263, 273
Schweinfurt 137, 144
Seeburg 265, 277, 278
Seefelt 240
Seville 98
Silesia 20, 22, 23, 27, 28, 51, 329
Sonnenwalt 239, 250
Soviet Union 174, 307
Spain 99
Spi(e)gelberg 228, 242, 249
Steemboth 238, 250
Stegemansdorf 239, 250
Steinburg 344
Stockholm 78
Stolpe 233
Strasbourg 139, 140, 142, 163, 333
Stuhm 262, 271, 327
Stynekyn 255, 256
Sweden 5, 51, 101, 141, 173, 307, 311, 314, 326, 327, 330, 332, 334, 342

Świecie 343

Tannenberg 186, 209, 275, 277
Tapiau 344–346, 348
Thebes 150, 163
Thomasdorf 233, 235
Tolkemit 262, 264, 270, 271, 286, 287, 292, 293, 313
Toruń 19, 22, 27, 170, 198–201, 208, 213, 222, 289, 338, 339
Tours 76
Trynckus (Old) 254
Turkey 328
Tus 115

Uppsala 5, 17, 101, 141, 144, 281, 314, 332

Vadang 230, 237
Vangaiten 266, 279
Varmia 19, 27, 29, 169, 201, 224, 245, 252, 261, 262, 264–266, 268–270, 274, 276–279, 285, 307, 309, 310, 312, 313, 322, 329, 343, 345, 352
Venice 3, 4, 80, 98, 125, 301, 303
Vienna 78, 79, 136, 144, 329
Vierzighuben 267
Vindica 233, 235, 238, 243, 245, 247
Vistula 323
Voppen 239, 240, 250
Voytsdorf 228, 232, 234, 242, 244, 246, 263
Vuriten 249
Vusen 238, 263, 266, 273, 288, 294, 298

Warsaw 143, 173
Wendland 347, 348
Westphalia 319
Wissembourg 200
Wittenberg 129, 130, 136, 137, 351
Włocławek 220, 221, 274
Wrocław 19, 20, 22, 27, 28, 51, 213, 219, 220, 222, 258, 328, 329, 351

Zcauwer (Zagern) 254–257, 273
Zürich 130, 131

INDEX OF SUBJECTS

advowson 329
anomaly 84, 86, 111
apogee 109, 110, 120, 137, 138
apse 84, 86–89, 99, 110, 118, 119
archdeaconate 264, 274
Aristotelianism 134
aspect, trine 111, 113
astrology 351
axiom 92, 112

calendar 99, 138
cannon 129
census 224, 242
concentricity 81, 116
conjunction 84, 102

David's Harp 333
declination 83, 85, 88, 105, 106, 118, 164
deferent 84–89, 100, 101, 104, 105, 108, 109, 112, 116, 119, 122
deviation 86, 88, 101, 118, 121, 122

earth 75, 76, 81–85, 88, 89, 93, 94, 104, 105, 120, 121, 164
, revolution of 82, 92, 95, 105, 106
, rotation of 81, 86, 87, 96, 105
eccentricity 107, 112, 113
eclipse 75, 127, 158, 308
ecliptic 83, 85, 86, 89, 105, 114, 137, 148, 159, 160, 164
equant 81, 92, 112, 114
equation see prosthaphaeresis
equinox 83, 96, 97, 99, 106, 111, 146, 155, 162

gravity 81

hermeticism 134
Holy Cross 261, 270
horoscope 127
hypothesis 76–78, 112, 126, 148

Jupiter 82, 85–87, 90, 94, 107

latitude 85, 87–89, 104, 115, 118, 121, 122
libration 88, 115–117
light 109

mansus 224, 242
mark, as coin 180–186, 188, 189, 191, 193, 194, 197–199, 201, 208–211, 286–289, 292, 316, 317
, as weight 180, 181, 187, 193, 194, 198, 300
, Prussian 170, 180, 194, 195, 202, 212, 213, 299
Mars 82, 85–87
Mercury 82, 87, 89, 117, 119–122

meteorology 129
meteoroscope 129, 130
moon 84–86, 99, 101, 102, 104, 108, 116, 117, 127, 146, 155
, full 102
, half 102, 103
, new 102
motion 81, 82, 84, 86, 105, 106, 111, 115, 125, 126, 133, 137, 146–150, 155–162

Neoplatonism 134
neo-Pythagoreanism 134
node 86–88, 105, 117
numeral 138

opposition 84, 102
Orb, Grand 82–87, 89, 94, 95, 100, 105, 107, 114, 117, 119

parallax 85, 92
perigee 137, 138
plague 128
pound 181, 191, 194
precession 97, 99, 105, 106
prosthaphaeresis 137, 138, 160
Pythagoreanism 134

quadrature 84, 85, 103

reflexion 88, 118
Reformation 332
rent 225, 226, 242, 250, 251, 253–257, 280, 288

Saturn 82, 85–87, 90, 94, 107, 110
scholastry 219, 220, 222
shilling 170, 180–189, 193, 194, 199, 203, 208–213, 281, 283, 284, 286–289, 295, 316–317
sign, zodiacal 83, 105, 109, 351
skoter 171, 180, 182, 185, 186, 188, 189, 191, 193, 197, 208–210, 212, 213, 281, 284, 299, 300, 315–317
sphere, solid 106, 108, 123–126
Spike of the Virgin 84, 89, 99, 149, 162, 163
stars 82–84, 93, 111, 133, 145–147, 149, 156, 157, 159, 160, 162–164
sun 81–83, 89, 92, 94, 98, 100, 104, 106, 107, 111, 116, 137, 146, 155

Tables, Alfonsine 82, 91, 94, 95, 98, 99, 109, 110, 114, 145, 153, 154, 160
, *Prussian* 98, 102
, *Toledan* 110
Thirteen Years' War 277, 293
trepidation 148, 150, 160, 164

INDEX OF SUBJECTS

Tusi couple 115, 116
twinkling 146, 157

value, face 186, 189, 192, 204, 211, 316
    , intrinsic 183, 184, 186, 192, 193, 204, 211, 316
    , market 183, 316
Venus 82, 87–90, 94, 95, 116–119, 121, 122
vicariate 279, 280, 290–292, 299

World War II 227, 307

Year, Egyptian 84, 145, 153–155, 157, 159, 162
    , natural 121
    , Roman 143, 146, 154, 155, 163
    , seasonal 121
    , sidereal 99, 121
    , tropical 83, 99, 121

POLISH PLACES IN COPERNICUS' LIFE AND WORK

COPERNICAN	CONTEMPORARY	CONTEMPORARY	COPERNICAN
Abestich	Łupstych	Bartąg	Berting Teutonica
Allenstein	Olsztyn	Bażyny	Baisen
Appelaw	—	Bornity	Borniten
Baisen	Bażyny	Braniewo	Braunsberg
Bebir	—	Brąswałd	Braunswaldt
Berting Teutonica	Bartąg	Cerkiewnik	Monsterberg
Birckpusch	—	Chełmno	Kulm
Borniten	Bornity	Chwalęcin	Stegemansdorf
Braunsberg	Braniewo	Dajtki	Degten
Braunswaldt	Brąswałd	Dągi	Marquardshoffen
Breslau	Wrocław	Dobre Miasto	Gutstadt
Caldeborn	Kaborno	Dorotowo	Dareten
Caldemflis	Żórawno	Dywity	Dewiten
Caleberg	—	Elbląg	Elbing
Carsau	Karszewo	Ełdyty	Elditen
Claukendorf	Wodynia	Frombork	Frauenburg
Cleeberg (Greater)	Klebark Wielki	Gdańsk	Danzig
Cleeberg (Lesser)	Klebark Mały	Gietrzwałd	Ditterichswalt
Cleefelt	Glebisko	Glądy	Glanden
Codien	Kadyny	Glebisko	Cleefelt
Colmensche	Chełmża	Gryźliny	Greseling
Comain	Kumajny	Gutkowo	Godekendorf
Conradswalt	Chojnowo	Jaroty	Jomendorf
Coseler	Kieźliny	Jeziorany	Seeburg
Crebisdorf	—	Jeziorko	Seefelt
Crumse	—	Jonkowo	Jonikendorf
Cukendorf (New)	Nowe Kawkowo	Kaborno	Caldeborn
Cukendorf (Old)	Stare Kawkowo	Kadyny	Codien
Danzig	Gdańsk	Kaliningrad (Królewiec)	Koenigsberg
Dareten	Dorotowo		
Degten	Dajtki	Karszewo	Carsau
Dewiten	Dywity	Kawkowo, Nowe	(New) Cukendorf
Dirsau	Tczew	Kawkowo, Stare	(Old) Cukendorf
Ditterichswalt	Gietrzwałd	Kieźliny	Coseler
Elbing	Elbląg	Klebark Mały	(Lesser) Cleeberg
Elditen	Ełdyty	Klebark Wielki	(Greater) Cleeberg
Engelswalt	Sawity	Kumajny	Comain
Frauenburg	Frombork	Lidzbark Warmiński	Heilsberg
Gabelen	—	Likusy	Licosa
Glandemansdorf	—	Linowo	Leynau
Glanden	Ględy	Lubawa	Loebau
Godekendorf	Gutkowo	Lubianka	Libentail
Greseling	Gryźliny	Lubnowo	Libenau
Gutstadt	Dobre Miasto	Łajsy	Laisse
Haselau	Zajączkowo	Łoźnik	Lutterfelt
Heilsberg	Lidzbark Warmiński	Ługwałd	Hogenwalt
Hinrichsdorf	Jędrychowo	Łupstych	Abestich
Hoenberg	—	Majewo	Maybom
Hoensteyn	Olsztynek	Malbork	Marienburg
Hogenwalt	Ługwałd	Mątki	Montiken(dorf)
Jomendorf	Jaroty	Miłkowo	Millemberg

COPERNICAN	CONTEMPORARY	CONTEMPORARY	COPERNICAN
Jonikendorf	Jonkowo	Myki	Mica
Koenigsberg	Kaliningrad (Królewiec)	Naglady	Naglanden
		Naterki	Natternen
Kulm	Chełmno	Nowe	Neuenburg
Kynappel	—	Nowy Dwór	Neuhoff
Laisse	Łajsy	Olsztyn	Allenstein
Leynau	Linowo	Olsztynek	Hoensteyn
Libenau	Lubnowo	Ostróda	Osterode
Libentail	Lubianka	Pagórki	Reberg
Licosa	Likusy	Pajtuny	Peuthuen
Loebau	Lubawa	Pasym	Passenheim
Lutterfelt	Łoźnik	Patryki	Petrica
Marienburg	Malbork	Pielgrzymowo	Pilgrimsdorf
Marquardshoffen	Dągi	Pieniężno	Melsac
Maybom	Majewo	Pistki	Pisdecaim
Melsac	Pieniężno	Pluski	Plauczk
Mica	Myki	Pluty	Plauten
Millemberg	Miłkowo	Podlechy	Padelochen
Monsterberg	Cerkiewnik	Pozorty	Posorten
Montiken(dorf)	Mątki	Redykajny	Radecaim
Naglanden	Naglady	Reszel	Resil
Natternen	Naterki	Różaniec	Rosengarten
Neuenburg	Nowe	Sądkowo	Sandecaim
Neuhoff	Nowy Dwór	Sątopy	Santoppen
Osterode	Ostróda	Skajboty	Scaibot(h)
Padelochen	Podlechy	Słupy	Stolpe
Passau	—	Spręcowo	Spiegelberg
Passenheim	Pasym	Stękiny	Stynekyn
Petrica	Patryki	Sząbruk	Schonebrugk
Peuthuen	Pajtuny	Szczęsne	Schonewalt
Pilgrimsdorf	Pielgrzymowo	Sztum	Stuhm
Pisdecaim	Pistki	Tczew	Dirsau
Plauczk	Pluski	Tolkmicko	Tolkemit
Plauten	Pluty	Tomaszkowo	Thomasdorf
Posorten	Pozorty	Toruń	Thorn
Quedlitz	—	Wadąg	Vadang
Rabusen	Robusy	Węgajty	Vangaiten
Radecaim	Redykajny	Włóczyska	Vierzighuben
Reberg	Pagórki	Wołowno	Vindica
Resil	Reszel	Wopy	Voppen
Rosengarten	Różaniec	Wójtowo	Voytsdorf
Sandecaim	Sądkowo	Wrocław	Breslau
Santoppen	Sątopy	Zawierz	Zcauwer
Scaibot(h)	Perbady		
Scharfenberg	—		
Schoneberg (New)	Porbady		
Schonebrugk	Sząbruk		
Schonebuche	Połapin		
Schonewalt	Szczęsne		
Schouffsberg	Baranówka		
Seeburg	Jeziorany		
Seefelt	Jeziorko		
Sonnenwalt	Radziejewo		
Spi(e)gelberg	Spręcowo		

POLISH PLACES IN COPERNICUS' LIFE AND WORK

COPERNICAN	CONTEMPORARY
Steemboth	Pełty
Stegemansdorf	Chwalęcin
Steinburg	—
Stolpe	Słupy
Stuhm	Sztum
Stynekyn	Stękiny
Tannenberg	—
Thomasdorf	Tomaszkowo
Thorn	Toruń
Tolkemit	Tolkmicko
Vadang	Wadąg
Vangaiten	Węgajty
Vierzighuben	Włóczyska
Vindica	Wołowno
Voppen	Wopy
Voytsdorf	Wójtowo
Vuriten	—
Vusen	Osetnik
Zcauwer	Zawierz

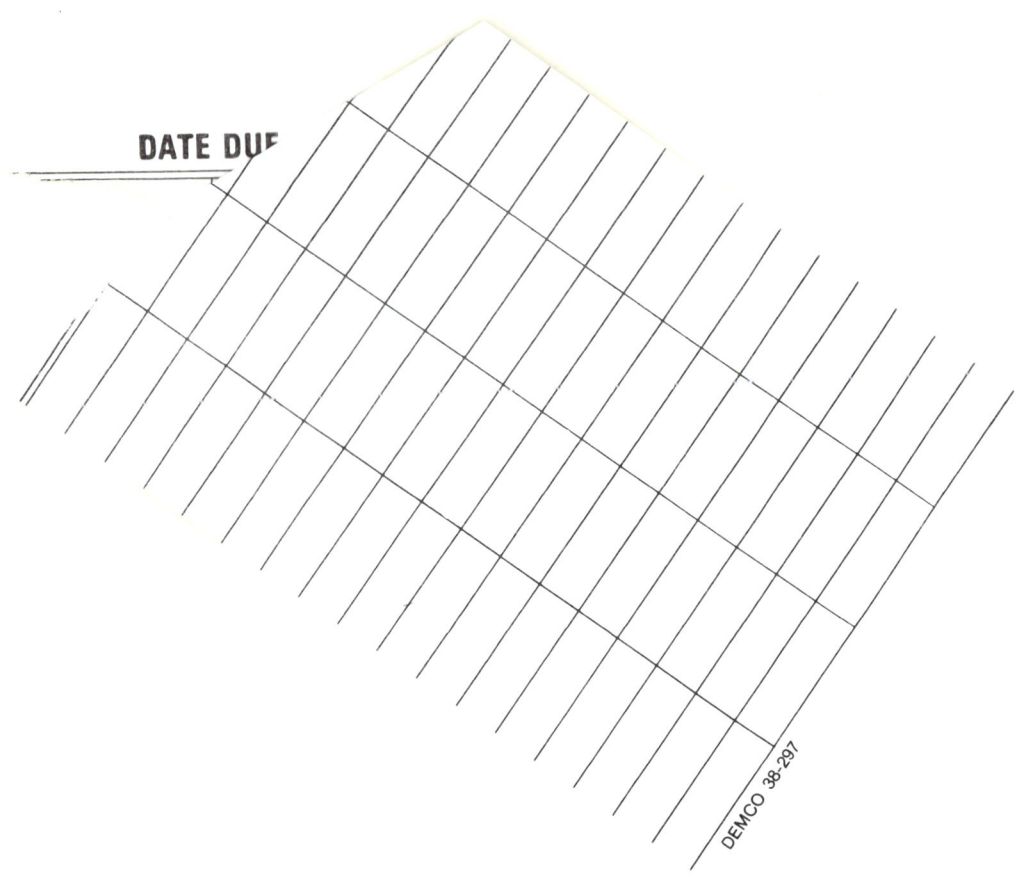